Studies in Big Data

Volume 26

Series Editor

Janusz Kacprzyk, Polish Academy of Sciences, Warsaw, Poland
e-mail: kacprzyk@ibspan.waw.pl

About this Series

The series "Studies in Big Data" (SBD) publishes new developments and advances in the various areas of Big Data – quickly and with a high quality. The intent is to cover the theory, research, development, and applications of Big Data, as embedded in the fields of engineering, computer science, physics, economics and life sciences. The books of the series refer to the analysis and understanding of large, complex, and/or distributed data sets generated from recent digital sources coming from sensors or other physical instruments as well as simulations, crowd sourcing, social networks or other internet transactions, such as emails or video click streams and other. The series contains monographs, lecture notes and edited volumes in Big Data spanning the areas of computational intelligence including neural networks, evolutionary computation, soft computing, fuzzy systems, as well as artificial intelligence, data mining, modern statistics and Operations research, as well as self-organizing systems. Of particular value to both the contributors and the readership are the short publication timeframe and the world-wide distribution, which enable both wide and rapid dissemination of research output.

More information about this series at http://www.springer.com/series/11970

S. Srinivasan
Editor

Guide to Big Data Applications

Springer

Editor
S. Srinivasan
Jesse H. Jones School of Business
Texas Southern University
Houston, TX, USA

ISSN 2197-6503 ISSN 2197-6511 (electronic)
Studies in Big Data
ISBN 978-3-319-53816-7 ISBN 978-3-319-53817-4 (eBook)
DOI 10.1007/978-3-319-53817-4

Library of Congress Control Number: 2017936371

Printed on acid-free paper

This Springer imprint is published by Springer Nature
The registered company is Springer International Publishing AG
The registered company address is: Gewerbestrasse 11, 6330 Cham, Switzerland

To my wife Lakshmi and grandson Sahaas

Foreword

It gives me great pleasure to write this Foreword for this timely publication on the topic of the ever-growing list of Big Data applications. The potential for leveraging existing data from multiple sources has been articulated over and over, in an almost infinite landscape, yet it is important to remember that in doing so, domain knowledge is key to success. Naïve attempts to process data are bound to lead to errors such as accidentally regressing on noncausal variables. As Michael Jordan at Berkeley has pointed out, in Big Data applications the number of combinations of the features grows exponentially with the number of features, and so, for any particular database, you are likely to find some combination of columns that will predict perfectly any outcome, just by chance alone. It is therefore important that we do not process data in a hypothesis-free manner and skip sanity checks on our data.

In this collection titled "Guide to Big Data Applications," the editor has assembled a set of applications in science, medicine, and business where the authors have attempted to do just this—apply Big Data techniques together with a deep understanding of the source data. The applications covered give a flavor of the benefits of Big Data in many disciplines. This book has 19 chapters broadly divided into four parts. In Part I, there are four chapters that cover the basics of Big Data, aspects of privacy, and how one could use Big Data in natural language processing (a particular concern for privacy). Part II covers eight chapters that look at various applications of Big Data in environmental science, oil and gas, and civil infrastructure, covering topics such as deduplication, encrypted search, and the friendship paradox.

Part III covers Big Data applications in medicine, covering topics ranging from "The Impact of Big Data on the Physician," written from a purely clinical perspective, to the often discussed deep dives on electronic medical records. Perhaps most exciting in terms of future landscaping is the application of Big Data application in healthcare from a developing country perspective. This is one of the most promising growth areas in healthcare, due to the current paucity of current services and the explosion of mobile phone usage. The tabula rasa that exists in many countries holds the potential to leapfrog many of the mistakes we have made in the west with stagnant silos of information, arbitrary barriers to entry, and the lack of any standardized schema or nondegenerate ontologies.

In Part IV, the book covers Big Data applications in business, which is perhaps the unifying subject here, given that none of the above application areas are likely to succeed without a good business model. The potential to leverage Big Data approaches in business is enormous, from banking practices to targeted advertising. The need for innovation in this space is as important as the underlying technologies themselves. As Clayton Christensen points out in *The Innovator's Prescription*, three revolutions are needed for a successful disruptive innovation:

1. A technology enabler which "routinizes" previously complicated task
2. A business model innovation which is affordable and convenient
3. A value network whereby companies with disruptive mutually reinforcing economic models sustain each other in a strong ecosystem

We see this happening with Big Data almost every week, and the future is exciting.

In this book, the reader will encounter inspiration in each of the above topic areas and be able to acquire insights into applications that provide the flavor of this fast-growing and dynamic field.

Atlanta, GA, USA
December 10, 2016

Gari Clifford

Preface

Big Data applications are growing very rapidly around the globe. This new approach to decision making takes into account data gathered from multiple sources. Here my goal is to show how these diverse sources of data are useful in arriving at actionable information. In this collection of articles the publisher and I are trying to bring in one place several diverse applications of Big Data. The goal is for users to see how a Big Data application in another field could be replicated in their discipline. With this in mind I have assembled in the "Guide to Big Data Applications" a collection of 19 chapters written by academics and industry practitioners globally. These chapters reflect what Big Data is, how privacy can be protected with Big Data and some of the important applications of Big Data in science, medicine and business. These applications are intended to be representative and not exhaustive. For nearly two years I spoke with major researchers around the world and the publisher. These discussions led to this project. The initial Call for Chapters was sent to several hundred researchers globally via email. Approximately 40 proposals were submitted. Out of these came commitments for completion in a timely manner from 20 people. Most of these chapters are written by researchers while some are written by industry practitioners. One of the submissions was not included as it could not provide evidence of use of Big Data. This collection brings together in one place several important applications of Big Data. All chapters were reviewed using a double-blind process and comments provided to the authors. The chapters included reflect the final versions of these chapters.

I have arranged the chapters in four parts. Part I includes four chapters that deal with basic aspects of Big Data and how privacy is an integral component. In this part I include an introductory chapter that lays the foundation for using Big Data in a variety of applications. This is then followed with a chapter on the importance of including privacy aspects at the design stage itself. This chapter by two leading researchers in the field shows the importance of Big Data in dealing with privacy issues and how they could be better addressed by incorporating privacy aspects at the design stage itself. The team of researchers from a major research university in the USA addresses the importance of federated Big Data. They are looking at the use of distributed data in applications. This part is concluded with a chapter that shows the importance of word embedding and natural language processing using Big Data analysis.

In Part II, there are eight chapters on the applications of Big Data in science. Science is an important area where decision making could be enhanced on the way to approach a problem using data analysis. The applications selected here deal with Environmental Science, High Performance Computing (HPC), friendship paradox in noting which friend's influence will be significant, significance of using encrypted search with Big Data, importance of deduplication in Big Data especially when data is collected from multiple sources, applications in Oil & Gas and how decision making can be enhanced in identifying bridges that need to be replaced as part of meeting safety requirements. All these application areas selected for inclusion in this collection show the diversity of fields in which Big Data is used today. The Environmental Science application shows how the data published by the National Oceanic and Atmospheric Administration (NOAA) is used to study the environment. Since such datasets are very large, specialized tools are needed to benefit from them. In this chapter the authors show how Big Data tools help in this effort. The team of industry practitioners discuss how there is great similarity in the way HPC deals with low-latency, massively parallel systems and distributed systems. These are all typical of how Big Data is used using tools such as MapReduce, Hadoop and Spark. Quora is a leading provider of answers to user queries and in this context one of their data scientists is addressing how the Friendship paradox is playing a significant part in Quora answers. This is a classic illustration of a Big Data application using social media.

Big Data applications in science exist in many branches and it is very heavily used in the Oil and Gas industry. Two chapters that address the Oil and Gas application are written by two sets of people with extensive industry experience. Two specific chapters are devoted to how Big Data is used in deduplication practices involving multimedia data in the cloud and how privacy-aware searches are done over encrypted data. Today, people are very concerned about the security of data stored with an application provider. Encryption is the preferred tool to protect such data and so having an efficient way to search such encrypted data is important. This chapter's contribution in this regard will be of great benefit for many users. We conclude Part II with a chapter that shows how Big Data is used in noting the structural safety of nation's bridges. This practical application shows how Big Data is used in many different ways.

Part III considers applications in medicine. A group of expert doctors from leading medical institutions in the Bay Area discuss how Big Data is used in the practice of medicine. This is one area where many more applications abound and the interested reader is encouraged to look at such applications. Another chapter looks at how data scientists are important in analyzing medical data. This chapter reflects a view from Asia and discusses the roadmap for data science use in medicine. Smoking has been noted as one of the leading causes of human suffering. This part includes a chapter on comorbidity aspects related to smokers based on a Big Data analysis. The details presented in this chapter would help the reader to focus on other possible applications of Big Data in medicine, especially cancer. Finally, a chapter is included that shows how scientific analysis of Big Data helps with epileptic seizure prediction and control.

Part IV of the book deals with applications in Business. This is an area where Big Data use is expected to provide tangible results quickly to businesses. The three applications listed under this part include an application in banking, an application in marketing and an application in Quick Serve Restaurants. The banking application is written by a group of researchers in Europe. Their analysis shows that the importance of identifying financial fraud early is a global problem and how Big Data is used in this effort. The marketing application highlights the various ways in which Big Data could be used in business. Many large business sectors such as the airlines industry are using Big Data to set prices. The application with respect to a Quick Serve Restaurant chain deals with the impact of Yelp ratings and how it influences people's use of Quick Serve Restaurants.

As mentioned at the outset, this collection of chapters on Big Data applications is expected to serve as a sample for other applications in various fields. The readers will find novel ways in which data from multiple sources is combined to derive benefit for the general user. Also, in specific areas such as medicine, the use of Big Data is having profound impact in opening up new areas for exploration based on the availability of large volumes of data. These are all having practical applications that help extend people's lives. I earnestly hope that this collection of applications will spur the interest of the reader to look at novel ways of using Big Data.

This book is a collective effort of many people. The contributors to this book come from North America, Europe and Asia. This diversity shows that Big Data is a truly global way in which people use the data to enhance their decision-making capabilities and to derive practical benefits. The book greatly benefited from the careful review by many reviewers who provided detailed feedback in a timely manner. I have carefully checked all chapters for consistency of information in content and appearance. In spite of careful checking and taking advantage of the tools provided by technology, it is highly likely that some errors might have crept in to the chapter content. In such cases I take responsibility for such errors and request your help in bringing them to my attention so that they can be corrected in future editions.

Houston, TX, USA
December 15, 2016

S. Srinivasan

Acknowledgements

A project of this nature would be possible only with the collective efforts of many people. Initially I proposed the project to Springer, New York, over two years ago. Springer expressed interest in the proposal and one of their editors, Ms. Mary James, contacted me to discuss the details. After extensive discussions with major researchers around the world we finally settled on this approach. A global Call for Chapters was made in January 2016 both by me and Springer, New York, through their channels of communication. Ms. Mary James helped throughout the project by providing answers to questions that arose. In this context, I want to mention the support of Ms. Brinda Megasyamalan from the printing house of Springer. Ms. Megasyamalan has been a constant source of information as the project progressed. Ms. Subhashree Rajan from the publishing arm of Springer has been extremely cooperative and patient in getting all the page proofs and incorporating all the corrections. Ms. Mary James provided all the encouragement and support throughout the project by responding to inquiries in a timely manner.

The reviewers played a very important role in maintaining the quality of this publication by their thorough reviews. We followed a double-blind review process whereby the reviewers were unaware of the identity of the authors and vice versa. This helped in providing quality feedback. All the authors cooperated very well by incorporating the reviewers' suggestions and submitting their final chapters within the time allotted for that purpose. I want to thank individually all the reviewers and all the authors for their dedication and contribution to this collective effort.

I want to express my sincere thanks to Dr. Gari Clifford of Emory University and Georgia Institute of Technology in providing the Foreword to this publication. In spite of his many commitments, Dr. Clifford was able to find the time to go over all the Abstracts and write the Foreword without delaying the project.

Finally, I want to express my sincere appreciation to my wife for accommodating the many special needs when working on a project of this nature.

Contents

List of Reviewers

Maruthi Bhaskar
Jay Brandi
Jorge Brusa
Arnaub Chatterjee
Robert Evans
Aly Farag
Lila Ghemri
Ben Hu
Balaji Janamanchi
Mehmed Kantardzic
Mark Kerzner
Ashok Krishnamurthy
Angabin Matin
Hector Miranda
P. S. Raju
S. Srinivasan
Rakesh Verma
Daniel Vrinceanu
Haibo Wang
Xuqing Wu
Alec Yasinsac

Part I
General

Chapter 1
Strategic Applications of Big Data

Joe Weinman

1.1 Introduction

For many people, big data is somehow virtually synonymous with one application—marketing analytics—in one vertical—retail. For example, by collecting purchase transaction data from shoppers based on loyalty cards or other unique identifiers such as telephone numbers, account numbers, or email addresses, a company can segment those customers better and identify promotions that will boost profitable revenues, either through insights derived from the data, A/B testing, bundling, or the like. Such insights can be extended almost without bound. For example, through sophisticated analytics, Harrah's determined that its most profitable customers weren't "gold cuff-linked, limousine-riding high rollers," but rather teachers, doctors, and even machinists (Loveman 2003). Not only did they come to understand *who* their best customers were, but *how* they behaved and responded to promotions. For example, their target customers were more interested in an offer of $60 worth of chips than a total bundle worth much more than that, including a room and multiple steak dinners in addition to chips.

While marketing such as this is a great application of big data and analytics, the reality is that big data has numerous strategic business applications across every industry vertical. Moreover, there are many sources of big data available from a company's day-to-day business activities as well as through open data initiatives, such as data.gov in the U.S., a source with almost 200,000 datasets at the time of this writing.

To apply big data to critical areas of the firm, there are four major generic approaches that companies can use to deliver unparalleled customer value and

J. Weinman (✉)
Independent Consultant, Flanders, NJ 07836, USA
e-mail: joeweinman@gmail.com

S. Srinivasan (ed.), *Guide to Big Data Applications*, Studies in Big Data 26,
DOI 10.1007/978-3-319-53817-4_1

achieve strategic competitive advantage: better processes, better products and services, better customer relationships, and better innovation.

1.1.1 Better Processes

Big data can be used to optimize processes and asset utilization in real time, to improve them in the long term, and to generate net new revenues by entering new businesses or at least monetizing data generated by those processes. UPS optimizes pickups and deliveries across its 55,000 routes by leveraging data ranging from geospatial and navigation data to customer pickup constraints (Rosenbush and Stevens 2015). Or consider 23andMe, which has sold genetic data it collects from individuals. One such deal with Genentech focused on Parkinson's disease gained net new revenues of fifty million dollars, rivaling the revenues from its "core" business (Lee 2015).

1.1.2 Better Products and Services

Big data can be used to enrich the quality of customer solutions, moving them up the experience economy curve from mere products or services to experiences or transformations. For example, Nike used to sell sneakers, a product. However, by collecting and aggregating activity data from customers, it can help transform them into better athletes. By linking data from Nike products and apps with data from ecosystem solution elements, such as weight scales and body-fat analyzers, Nike can increase customer loyalty and tie activities to outcomes (Withings 2014).

1.1.3 Better Customer Relationships

Rather than merely viewing data as a crowbar with which to open customers' wallets a bit wider through targeted promotions, it can be used to develop deeper insights into each customer, thus providing better service and customer experience in the short term and products and services better tailored to customers as individuals in the long term. Netflix collects data on customer activities, behaviors, contexts, demographics, and intents to better tailor movie recommendations (Amatriain 2013). Better recommendations enhance customer satisfaction and value which in turn makes these customers more likely to stay with Netflix in the long term, reducing churn and customer acquisition costs, as well as enhancing referral (word-of-mouth) marketing. Harrah's determined that customers that were "very happy" with their customer experience increased their spend by 24% annually; those that were unhappy decreased their spend by 10% annually (Loveman 2003).

1.1.4 *Better Innovation*

Data can be used to accelerate the innovation process, and make it of higher quality, all while lowering cost. Data sets can be published or otherwise incorporated as part of an open contest or challenge, enabling ad hoc solvers to identify a best solution meeting requirements. For example, GE Flight Quest incorporated data on scheduled and actual flight departure and arrival times, for a contest intended to devise algorithms to better predict arrival times, and another one intended to improve them (Kaggle n.d.). As the nexus of innovation moves from man to machine, data becomes the fuel on which machine innovation engines run.

These four business strategies are what I call *digital disciplines* (Weinman 2015), and represent an evolution of three customer-focused strategies called *value disciplines*, originally devised by Michael Treacy and Fred Wiersema in their international bestseller *The Discipline of Market Leaders* (Treacy and Wiersema 1995).

1.2 From Value Disciplines to Digital Disciplines

The value disciplines originally identified by Treacy and Wiersema are *operational excellence*, *product leadership*, and *customer intimacy*.

Operational excellence entails processes which generate customer value by being lower cost or more convenient than those of competitors. For example, Michael Dell, operating as a college student out of a dorm room, introduced an assemble-to-order process for PCs by utilizing a direct channel which was originally the phone or physical mail and then became the Internet and eCommerce. He was able to drive the price down, make it easier to order, and provide a PC built to customers' specifications by creating a new assemble-to-order process that bypassed indirect channel middlemen that stocked pre-built machines en masse, who offered no customization but charged a markup nevertheless.

Product leadership involves creating leading-edge products (or services) that deliver superior value to customers. We all know the companies that do this: Rolex in watches, Four Seasons in lodging, Singapore Airlines or Emirates in air travel. Treacy and Wiersema considered innovation as being virtually synonymous with product leadership, under the theory that leading products must be differentiated in some way, typically through some innovation in design, engineering, or technology.

Customer intimacy, according to Treacy and Wiersema, is focused on segmenting markets, better understanding the unique needs of those niches, and tailoring solutions to meet those needs. This applies to both consumer and business markets. For example, a company that delivers packages might understand a major customer's needs intimately, and then tailor a solution involving stocking critical parts at their distribution centers, reducing the time needed to get those products to their customers. In the consumer world, customer intimacy is at work any time a tailor adjusts a garment for a perfect fit, a bartender customizes a drink, or a doctor diagnoses and treats a medical issue.

Traditionally, the thinking was that a company would do well to excel in a given discipline, and that the disciplines were to a large extent mutually exclusive. For example, a fast food restaurant might serve a limited menu to enhance operational excellence. A product leadership strategy of having many different menu items, or a customer intimacy strategy of customizing each and every meal might conflict with the operational excellence strategy. However, now, the economics of information—storage prices are exponentially decreasing and data, once acquired, can be leveraged elsewhere—and the increasing flexibility of automation—such as robotics—mean that companies can potentially pursue multiple strategies simultaneously.

Digital technologies such as big data enable new ways to think about the insights originally derived by Treacy and Wiersema. Another way to think about it is that digital technologies plus value disciplines equal digital disciplines: operational excellence evolves to *information excellence*, product leadership of standalone products and services becomes *solution leadership* of smart, digital products and services connected to the cloud and ecosystems, customer intimacy expands to *collective intimacy*, and traditional innovation becomes *accelerated innovation*. In the digital disciplines framework, innovation becomes a separate discipline, because innovation applies not only to products, but also processes, customer relationships, and even the innovation process itself. Each of these new strategies can be enabled by big data in profound ways.

1.2.1 Information Excellence

Operational excellence can be viewed as evolving to information excellence, where digital information helps optimize physical operations including their processes and resource utilization; where the world of digital information can seamlessly fuse with that of physical operations; and where virtual worlds can replace physical. Moreover, data can be extracted from processes to enable long term process improvement, data collected by processes can be monetized, and new forms of corporate structure based on loosely coupled partners can replace traditional, monolithic, vertically integrated companies. As one example, location data from cell phones can be aggregated and analyzed to determine commuter traffic patterns, thereby helping to plan transportation network improvements.

1.2.2 Solution Leadership

Products and services can become sources of big data, or utilize big data to function more effectively. Because individual products are typically limited in storage capacity, and because there are benefits to data aggregation and cloud processing, normally the data that is collected can be stored and processed in the cloud. A good example might be the GE GEnx jet engine, which collects 5000 data points each second from each of 20 sensors. GE then uses the data to develop better predictive maintenance algorithms, thus reducing unplanned downtime for airlines. (GE Aviation n.d.) Mere product leadership becomes solution leadership, where standalone products become cloud-connected and data-intensive. Services can also become solutions, because services are almost always delivered through physical elements: food services through restaurants and ovens; airline services through planes and baggage conveyors; healthcare services through x-ray machines and pacemakers. The components of such services connect to each other and externally. For example, healthcare services can be better delivered through connected pacemakers, and medical diagnostic data from multiple individual devices can be aggregated to create a patient-centric view to improve health outcomes.

1.2.3 Collective Intimacy

Customer intimacy is no longer about dividing markets into segments, but rather dividing markets into individuals, or even further into multiple personas that an individual might have. Personalization and contextualization offers the ability to not just deliver products and services tailored to a segment, but to an individual. To do this effectively requires current, up-to-date information as well as historical data, collected at the level of the individual and his or her individual activities and characteristics down to the granularity of DNA sequences and mouse moves. Collective intimacy is the notion that algorithms running on collective data from millions of individuals can generate better tailored services for each individual. This represents the evolution of intimacy from face-to-face, human-mediated relationships to virtual, human-mediated relationships over social media, and from there, onward to virtual, algorithmically mediated products and services.

1.2.4 Accelerated Innovation

Finally, innovation is not just associated with product leadership, but can create new processes, as Walmart did with cross-docking or Uber with transportation, or new customer relationships and collective intimacy, as Amazon.com uses data to better upsell/cross-sell, and as Netflix innovated its Cinematch recommendation engine. The latter was famously done through the Netflix Prize, a contest with a million dollar award for whoever could best improve Cinematch by at least 10% (Bennett and Lanning 2007). Such accelerated innovation can be faster, cheaper, and better than traditional means of innovation. Often, such approaches exploit technologies such as the cloud and big data. The cloud is the mechanism for reaching multiple potential solvers on an ad hoc basis, with published big data being the fuel for problem solving. For example, Netflix published anonymized customer ratings of movies, and General Electric published planned and actual flight arrival times.

Today, machine learning and deep learning based on big data sets are a means by which algorithms are innovating themselves. Google DeepMind's AlphaGo Go-playing system beat the human world champion at Go, Lee Sedol, partly based on learning how to play by not only "studying" tens of thousands of human games, but also by playing an increasingly tougher competitor: itself (Moyer 2016).

1.2.5 Value Disciplines to Digital Disciplines

The three classic value disciplines of operational excellence, product leadership and customer intimacy become transformed in a world of big data and complementary digital technologies to become information excellence, solution leadership, collective intimacy, and accelerated innovation. These represent four generic strategies that leverage big data in the service of strategic competitive differentiation; four generic strategies that represent the horizontal applications of big data.

1.3 Information Excellence

Most of human history has been centered on the physical world. Hunting and gathering, fishing, agriculture, mining, and eventually manufacturing and physical operations such as shipping, rail, and eventually air transport. It's not news that the focus of human affairs is increasingly digital, but the many ways in which digital information can complement, supplant, enable, optimize, or monetize physical operations may be surprising. As more of the world becomes digital, the use of information, which after all comes from data, becomes more important in the spheres of business, government, and society (Fig. 1.1).

1.3.1 Real-Time Process and Resource Optimization

There are numerous business functions, such as legal, human resources, finance, engineering, and sales, and a variety of ways in which different companies in a variety of verticals such as automotive, healthcare, logistics, or pharmaceuticals configure these functions into end-to-end processes. Examples of processes might be "claims processing" or "order to delivery" or "hire to fire". These in turn use a variety of resources such as people, trucks, factories, equipment, and information technology.

Data can be used to optimize resource use as well as to optimize processes for goals such as cycle time, cost, or quality.

Some good examples of the use of big data to optimize processes are inventory management/sales forecasting, port operations, and package delivery logistics.

Fig. 1.1 High-level architecture for information excellence

Too much inventory is a bad thing, because there are costs to holding inventory: the capital invested in the inventory, risk of disaster, such as a warehouse fire, insurance, floor space, obsolescence, shrinkage (i.e., theft), and so forth. Too little inventory is also bad, because not only may a sale be lost, but the prospect may go elsewhere to acquire the good, realize that the competitor is a fine place to shop, and never return. Big data can help with sales forecasting and thus setting correct inventory levels. It can also help to develop insights, which may be subtle or counterintuitive. For example, when Hurricane Frances was projected to strike Florida, analytics helped stock stores, not only with "obvious" items such as bottled water and flashlights, but non-obvious products such as strawberry Pop-Tarts (Hayes 2004). This insight was based on mining store transaction data from prior hurricanes.

Consider a modern container port. There are multiple goals, such as minimizing the time ships are in port to maximize their productivity, minimizing the time ships or rail cars are idle, ensuring the right containers get to the correct destinations, maximizing safety, and so on. In addition, there may be many types of structured and unstructured data, such as shipping manifests, video surveillance feeds of roads leading to and within the port, data on bridges, loading cranes, weather forecast data, truck license plates, and so on. All of these data sources can be used to optimize port operations in line with the multiple goals (Xvela 2016).

Or consider a logistics firm such as UPS. UPS has invested hundreds of millions of dollars in ORION (On-Road Integrated Optimization and Navigation). It takes data such as physical mapping data regarding roads, delivery objectives for each package, customer data such as when customers are willing to accept deliveries, and the like. For each of 55,000 routes, ORION determines the optimal sequence of an average of 120 stops per route. The combinatorics here are staggering, since there are roughly 10**200 different possible sequences, making it impossible to calculate a perfectly optimal route, but heuristics can take all this data and try to determine the best way to sequence stops and route delivery trucks to minimize idling time, time waiting to make left turns, fuel consumption and thus carbon footprint, and to maximize driver labor productivity and truck asset utilization, all the while balancing out customer satisfaction and on-time deliveries. Moreover real-time data such as geographic location, traffic congestion, weather, and fuel consumption, can be exploited for further optimization (Rosenbush and Stevens 2015).

Such capabilities could also be used to not just minimize time or maximize throughput, but also to maximize revenue. For example, a theme park could determine the optimal location for a mobile ice cream or face painting stand, based on prior customer purchases and exact location of customers within the park. Customers' locations and identities could be identified through dedicated long range radios, as Disney does with MagicBands; through smartphones, as Singtel's DataSpark unit does (see below); or through their use of related geographically oriented services or apps, such as Uber or Foursquare.

1.3.2 Long-Term Process Improvement

In addition to such real-time or short-term process optimization, big data can also be used to optimize processes and resources over the long term.

For example, DataSpark, a unit of Singtel (a Singaporean telephone company) has been extracting data from cell phone locations to be able to improve the MTR (Singapore's subway system) and customer experience (Dataspark 2016). For example, suppose that GPS data showed that many subway passengers were traveling between two stops but that they had to travel through a third stop—a hub—to get there. By building a direct line to bypass the intermediate stop, travelers could get to their destination sooner, and congestion could be relieved at the intermediate stop as well as on some of the trains leading to it. Moreover, this data could also be used for real-time process optimization, by directing customers to avoid a congested area or line suffering an outage through the use of an alternate route. Obviously a variety of structured and unstructured data could be used to accomplish both short-term and long-term improvements, such as GPS data, passenger mobile accounts and ticket purchases, video feeds of train stations, train location data, and the like.

1.3.3 Digital-Physical Substitution and Fusion

The digital world and the physical world can be brought together in a number of ways. One way is substitution, as when a virtual audio, video, and/or web conference substitutes for physical airline travel, or when an online publication substitutes for a physically printed copy. Another way to bring together the digital and physical worlds is fusion, where both online and offline experiences become seamlessly merged. An example is in omni-channel marketing, where a customer might browse online, order online for pickup in store, and then return an item via the mail. Or, a customer might browse in the store, only to find the correct size out of stock, and order in store for home delivery. Managing data across the customer journey can provide a single view of the customer to maximize sales and share of wallet for that customer. This might include analytics around customer online browsing behavior, such as what they searched for, which styles and colors caught their eye, or what they put into their shopping cart. Within the store, patterns of behavior can also be identified, such as whether people of a certain demographic or gender tend to turn left or right upon entering the store.

1.3.4 Exhaust-Data Monetization

Processes which are instrumented and monitored can generate massive amounts of data. This data can often be monetized or otherwise create benefits in creative ways. For example, Uber's main business is often referred to as "ride sharing," which is really just offering short term ground transportation to passengers desirous of rides by matching them up with drivers who can give them rides. However, in an arrangement with the city of Boston, it will provide ride pickup and drop-off locations, dates, and times. The city will use the data for traffic engineering, zoning, and even determining the right number of parking spots needed (O'Brien 2015).

Such inferences can be surprisingly subtle. Consider the case of a revolving door firm that could predict retail trends and perhaps even recessions. Fewer shoppers visiting retail stores means fewer shoppers entering via the revolving door. This means lower usage of the door, and thus fewer maintenance calls.

Another good example is 23andMe. 23andMe is a firm that was set up to leverage new low cost gene sequence technologies. A 23andMe customer would take a saliva sample and mail it to 23andMe, which would then sequence the DNA and inform the customer about certain genetically based risks they might face, such as markers signaling increased likelihood of breast cancer due to a variant in the BRCA1 gene. They also would provide additional types of information based on this sequence, such as clarifying genetic relationships among siblings or questions of paternity.

After compiling massive amounts of data, they were able to monetize the collected data outside of their core business. In one $50 million deal, they sold data from Parkinson's patients to Genentech, with the objective of developing a cure for Parkinson's through deep analytics (Lee 2015). Note that not only is the deal lucrative, especially since essentially no additional costs were incurred to sell this data, but also highly ethical. Parkinson's patients would like nothing better than for Genentech—or anybody else, for that matter—to develop a cure.

1.3.5 Dynamic, Networked, Virtual Corporations

Processes don't need to be restricted to the four walls of the corporation. For example, supply chain optimization requires data from suppliers, channels, and logistics companies. Many companies have focused on their core business and outsourced or partnered with others to create and continuously improve supply chains. For example, Apple sells products, but focuses on design and marketing, not manufacturing. As many people know, Apple products are built by a partner, Foxconn, with expertise in precision manufacturing electronic products.

One step beyond such partnerships or virtual corporations are dynamic, networked virtual corporations. An example is Li & Fung. Apple sells products such as iPhones and iPads, without owning any manufacturing facilities. Similarly, Li & Fung sells products, namely clothing, without owning any manufacturing facilities. However, unlike Apple, who relies largely on one main manufacturing partner; Li & Fung relies on a network of over 10,000 suppliers. Moreover, the exact configuration of those suppliers can change week by week or even day by day, even for the same garment. A shirt, for example, might be sewed in Indonesia with buttons from Thailand and fabric from S. Korea. That same SKU, a few days later, might be made in China with buttons from Japan and fabric from Vietnam. The constellation of suppliers is continuously optimized, by utilizing data on supplier resource availability and pricing, transportation costs, and so forth (Wind et al. 2009).

1.3.6 Beyond Business

Information excellence also applies to governmental and societal objectives. Earlier we mentioned using big data to improve the Singapore subway operations and customer experience; later we'll mention how it's being used to improve traffic congestion in Rio de Janeiro. As an example of societal objectives, consider the successful delivery of vaccines to remote areas. Vaccines can lose their efficacy or even become unsafe unless they are refrigerated, but delivery to outlying areas can mean a variety of transport mechanisms and intermediaries. For this reason, it is important to ensure that they remain refrigerated across their "cold chain." A low-tech method could potentially warn of unsafe vaccines: for example, put a container of milk in with the vaccines, and if the milk spoils it will smell bad and the vaccines are probably bad as well. However, by collecting data wirelessly from the refrigerators throughout the delivery process, not only can it be determined whether the vaccines are good or bad, but improvements can be made to the delivery process by identifying the root cause of the loss of refrigeration, for example, loss of power at a particular port, and thus steps can be taken to mitigate the problem, such as the deployment of backup power generators (Weinman 2016).

1.4 Solution Leadership

Products (and services) were traditionally standalone and manual, but now have become connected and automated. Products and services now connect to the cloud and from there on to ecosystems. The ecosystems can help collect data, analyze it, provide data to the products or services, or all of the above (Fig. 1.2).

1.4.1 Digital-Physical Mirroring

In product engineering, an emerging approach is to build a data-driven engineering model of a complex product. For example, GE mirrors its jet engines with "digital twins" or "virtual machines" (unrelated to the computing concept of the same name). The idea is that features, engineering design changes, and the like can be made to the model much more easily and cheaply than building an actual working jet engine. A new turbofan blade material with different weight, brittleness, and cross section might be simulated to determine impacts on overall engine performance. To do this requires product and materials data. Moreover, predictive analytics can be run against massive amounts of data collected from operating engines (Warwick 2015).

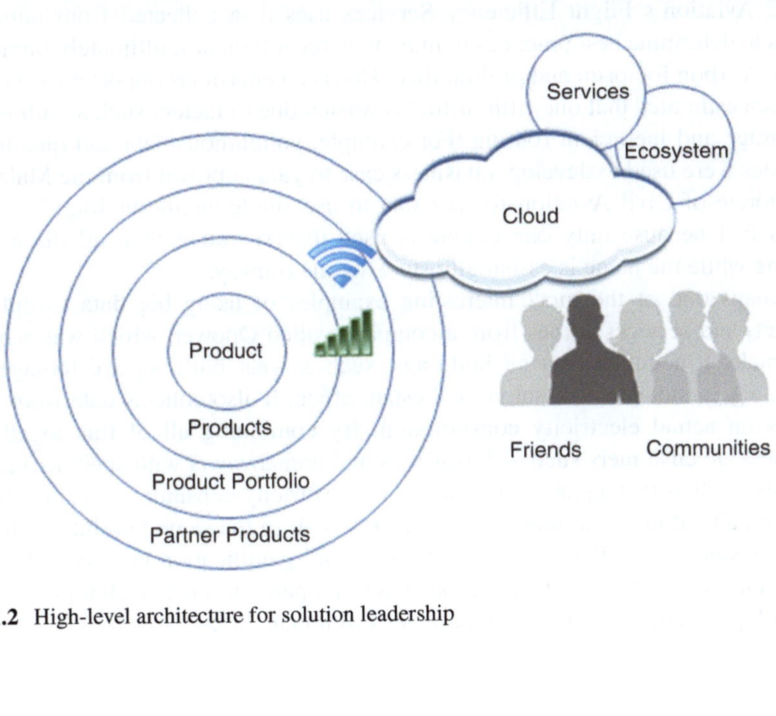

Fig. 1.2 High-level architecture for solution leadership

1.4.2 Real-Time Product/Service Optimization

Recall that solutions are smart, digital, connected products that tie over networks to the cloud and from there onward to unbounded ecosystems. As a result, the actual tangible, physical product component functionality can potentially evolve over time as the virtual, digital components adapt. As two examples, consider a browser that provides "autocomplete" functions in its search bar, i.e., typing shortcuts based on previous searches, thus saving time and effort. Or, consider a Tesla, whose performance is improved by evaluating massive quantities of data from all the Teslas on the road and their performance. As Tesla CEO Elon Musk says, "When one car learns something, the whole fleet learns" (Coren 2016).

1.4.3 Product/Service Usage Optimization

Products or services can be used better by customers by collecting data and providing feedback to customers. The Ford Fusion's EcoGuide SmartGauge provides feedback to drivers on their fuel efficiency. Jackrabbit starts are bad; smooth driving is good. The EcoGuide SmartGauge grows "green" leaves to provide drivers with feedback, and is one of the innovations credited with dramatically boosting sales of the car (Roy 2009).

GE Aviation's Flight Efficiency Services uses data collected from numerous flights to determine best practices to maximize fuel efficiency, ultimately improving airlines carbon footprint and profitability. This is an enormous opportunity, because it's been estimated that one-fifth of fuel is wasted due to factors such as suboptimal fuel usage and inefficient routing. For example, voluminous data and quantitative analytics were used to develop a business case to gain approval from the Malaysian Directorate of Civil Aviation for AirAsia to use single-engine taxiing. This conserves fuel because only one engine is used to taxi, rather than all the engines running while the plane is essentially stuck on the runway.

Perhaps one of the most interesting examples of using big data to optimize products and services comes from a company called Opower, which was acquired by Oracle. It acquires data on buildings, such as year built, square footage, and usage, e.g., residence, hair salon, real estate office. It also collects data from smart meters on actual electricity consumption. By combining all of this together, it can message customers such as businesses and homeowners with specific, targeted insights, such as that a particular hair salon's electricity consumption is higher than 80% of hair salons of similar size in the area built to the same building code (and thus equivalently insulated). Such "social proof" gamification has been shown to be extremely effective in changing behavior compared to other techniques such as rational quantitative financial comparisons (Weinman 2015).

1.4.4 Predictive Analytics and Predictive Maintenance

Collecting data from things and analyzing it can enable predictive analytics and predictive maintenance. For example, the GE GEnx jet engine has 20 or so sensors, each of which collects 5000 data points per second in areas such as oil pressure, fuel flow and rotation speed. This data can then be used to build models and identify anomalies and predict when the engine will fail.

This in turn means that airline maintenance crews can "fix" an engine before it fails. This maximizes what the airlines call "time on wing," in other words, engine availability. Moreover, engines can be proactively repaired at optimal times and optimal locations, where maintenance equipment, crews, and spare parts are kept (Weinman 2015).

1.4.5 Product-Service System Solutions

When formerly standalone products become connected to back-end services and solve customer problems they become product-service system solutions. Data can be the glue that holds the solution together. A good example is Nike and the Nike+ ecosystem.

Nike has a number of mechanisms for collecting activity tracking data, such as the Nike+ FuelBand, mobile apps, and partner products, such as Nike+ Kinect which is a video "game" that coaches you through various workouts. These can collect data on activities, such as running or bicycling or doing jumping jacks. Data can be collected, such as the route taken on a run, and normalized into "NikeFuel" points (Weinman 2015).

Other elements of the ecosystem can measure outcomes. For example, a variety of scales can measure weight, but the Withings Smart Body Analyzer can also measure body fat percentage, and link that data to NikeFuel points (Choquel 2014). By linking devices measuring outcomes to devices monitoring activities—with the linkages being data traversing networks—individuals can better achieve their personal goals to become better athletes, lose a little weight, or get more toned.

1.4.6 Long-Term Product Improvement

Actual data on how products are used can ultimately be used for long-term product improvement. For example, a cable company can collect data on the pattern of button presses on its remote controls. A repeated pattern of clicking around the "Guide" button fruitlessly and then finally ordering an "On Demand" movie might lead to a clearer placement of a dedicated "On Demand" button on the control. Car companies such as Tesla can collect data on actual usage, say, to determine how many batteries to put in each vehicle based on the statistics of distances driven; airlines can determine what types of meals to offer; and so on.

1.4.7 The Experience Economy

In the Experience Economy framework, developed by Joe Pine and Jim Gilmore, there is a five-level hierarchy of increasing customer value and firm profitability. At the lowest level are commodities, which may be farmed, fished, or mined, e.g., coffee beans. At the next level of value are products, e.g., packaged, roasted coffee beans. Still one level higher are services, such as a corner coffee bar. One level above this are experiences, such as a fine French restaurant, which offers coffee on the menu as part of a "total product" that encompasses ambience, romance, and professional chefs and services. But, while experiences may be ephemeral, at the ultimate level of the hierarchy lie transformations, which are permanent, such as a university education, learning a foreign language, or having life-saving surgery (Pine and Gilmore 1999).

1.4.8 Experiences

Experiences can be had without data or technology. For example, consider a hike up a mountain to its summit followed by taking in the scenery and the fresh air. However, data can also contribute to experiences. For example, Disney MagicBands are long-range radios that tie to the cloud. Data on theme park guests can be used to create magical, personalized experiences. For example, guests can sit at a restaurant without expressly checking in, and their custom order will be brought to their table, based on tracking through the MagicBands and data maintained in the cloud regarding the individuals and their orders (Kuang 2015).

1.4.9 Transformations

Data can also be used to enable transformations. For example, the Nike+ family and ecosystem of solutions mentioned earlier can help individuals lose weight or become better athletes. This can be done by capturing data from the individual on steps taken, routes run, and other exercise activities undertaken, as well as results data through connected scales and body fat monitors. As technology gets more sophisticated, no doubt such automated solutions will do what any athletic coach does, e.g., coaching on backswings, grip positions, stride lengths, pronation and the like. This is how data can help enable transformations (Weinman 2015).

1.4.10 *Customer-Centered Product and Service Data Integration*

When multiple products and services each collect data, they can provide a 360° view of the patient. For example, patients are often scanned by radiological equipment such as CT (computed tomography) scanners and X-ray machines. While individual machines should be calibrated to deliver a safe dose, too many scans from too many devices over too short a period can deliver doses over accepted limits, leading potentially to dangers such as cancer. GE Dosewatch provides a single view of the patient, integrating dose information from multiple medical devices from a variety of manufacturers, not just GE (Combs 2014).

Similarly, financial companies are trying to develop a 360° view of their customers' financial health. Rather than the brokerage division being run separately from the mortgage division, which is separate from the retail bank, integrating data from all these divisions can help ensure that the customer is neither over-leveraged or underinvested.

1.4.11 *Beyond Business*

The use of connected refrigerators to help improve the cold chain was described earlier in the context of information excellence for process improvement. Another example of connected products and services is cities, such as Singapore, that help reduce carbon footprint through connected parking garages. The parking lots report how many spaces they have available, so that a driver looking for parking need not drive all around the city: clearly visible digital signs and a mobile app describe how many—if any—spaces are available (Abdullah 2015).

This same general strategy can be used with even greater impact in the developing world. For example, in some areas, children walk an hour or more to a well to fill a bucket with water for their families. However, the well may have gone dry. Connected, "smart" pump handles can report their usage, and inferences can be made as to the state of the well. For example, a few pumps of the handle and then no usage, another few pumps and then no usage, etc., is likely to signify someone visiting the well, attempting to get water, then abandoning the effort due to lack of success (ITU and Cisco 2016).

1.5 Collective Intimacy

At one extreme, a customer "relationship" is a one-time, anonymous transaction. Consider a couple celebrating their 30th wedding anniversary with a once-in-a-lifetime trip to Paris. While exploring the left bank, they buy a baguette and some Brie from a hole-in-the-wall bistro. They will never see the bistro again, nor vice versa.

At the other extreme, there are companies and organizations that see customers repeatedly. Amazon.com sees its customers' patterns of purchases; Netflix sees its customers' patterns of viewing; Uber sees its customers' patterns of pickups and drop-offs. As other verticals become increasingly digital, they too will gain more insight into customers as individuals, rather than anonymous masses. For example, automobile insurers are increasingly pursuing "pay-as-you-drive," or "usage-based" insurance. Rather than customers' premiums being merely based on aggregate, coarse-grained information such as age, gender, and prior tickets, insurers can charge premiums based on individual, real-time data such as driving over the speed limit, weaving in between lanes, how congested the road is, and so forth.

Somewhere in between, there are firms that may not have any transaction history with a given customer, but can use predictive analytics based on statistical insights derived from large numbers of existing customers. Capital One, for example, famously disrupted the existing market for credit cards by building models to create "intimate" offers tailored to each prospect rather than a one-size fits all model (Pham 2015).

Big data can also be used to analyze and model churn. Actions can be taken to intercede before a customer has defected, thus retaining that customer and his or her profits.

In short, big data can be used to determine target prospects, determine what to offer them, maximize revenue and profitability, keep them, decide to let them defect to a competitor, or win them back (Fig. 1.3).

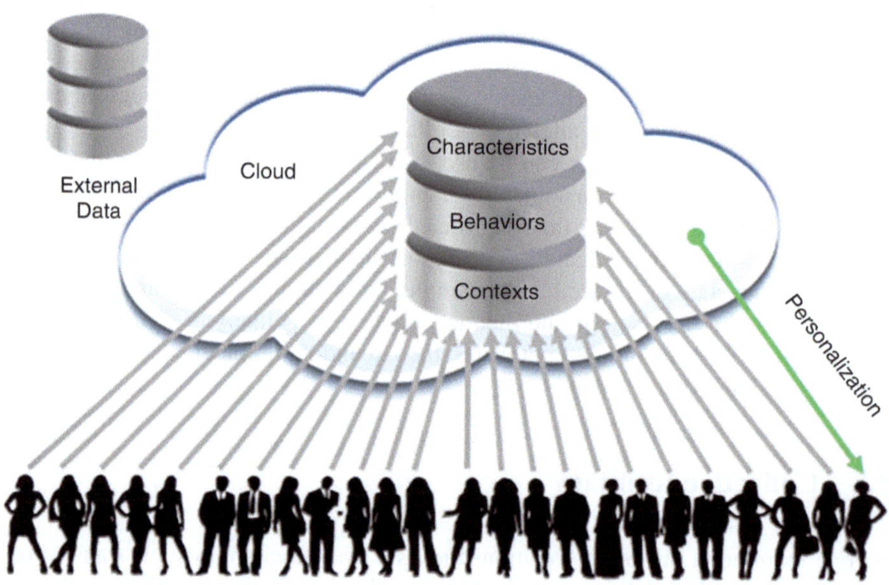

Fig. 1.3 High-level architecture for collective intimacy

1.5.1 Target Segments, Features and Bundles

A traditional arena for big data and analytics has been better marketing to customers and market basket analysis. For example, one type of analysis entails identifying prospects, clustering customers, non-buyers, and prospects as three groups: loyal customers, who will buy your product no matter what; those who won't buy no matter what; and those that can be swayed. Marketing funds for advertising and promotions are best spent with the last category, which will generate sales uplift.

A related type of analysis is market basket analysis, identifying those products that might be bought together. Offering bundles of such products can increase profits (Skiera and Olderog 2000). Even without bundles, better merchandising can goose sales. For example, if new parents who buy diapers also buy beer, it makes sense to put them in the aisle together. This may be extended to product features, where the "bundle" isn't a market basket but a basket of features built in to a product, say a sport suspension package and a V8 engine in a sporty sedan.

1.5.2 Upsell/Cross-Sell

Amazon.com uses a variety of approaches to maximize revenue per customer. Some, such as Amazon Prime, which offers free two-day shipping, are low tech and based on behavioral economics principles such as the "flat-rate bias" and how humans frame expenses such as sunk costs (Lambrecht and Skiera 2006). But they are perhaps best known for their sophisticated algorithms, which do everything from automating pricing decisions, making millions of price changes every day, (Falk 2013) and also in recommending additional products to buy, through a variety of algorithmically generated capabilities such as "People Also Bought These Items". Some are reasonably obvious, such as, say, paper and ink suggestions if a copier is bought or a mounting plate if a large flat screen TV is purchased. But many are subtle, and based on deep analytics at scale of the billions of purchases that have been made.

1.5.3 Recommendations

If Amazon.com is the poster child for upsell/cross-sell, Netflix is the one for a pure recommendation engine. Because Netflix charges a flat rate for a household, there is limited opportunity for upsell without changing the pricing model. Instead, the primary opportunity is for customer retention, and perhaps secondarily, referral marketing, i.e., recommendations from existing customers to their friends. The key to that is maximizing the quality of the total customer experience. This has multiple dimensions, such as whether DVDs arrive in a reasonable time or a streaming video plays cleanly at high resolution, as opposed to pausing to rebuffer frequently. But one very important dimension is the quality of the entertainment recommendations, because 70% of what Netflix viewers watch comes about through recommendations. If viewers like the recommendations, they will like Netflix, and if they don't, they will cancel service. So, reduced churn and maximal lifetime customer value are highly dependent on this (Amatriain 2013).

Netflix uses extremely sophisticated algorithms against trillions of data points, which attempt to solve as best as possible the recommendation problem. For example, they must balance out popularity with personalization. Most people like popular movies; this is why they are popular. But every viewer is an individual, hence will like different things. Netflix continuously evolves their recommendation engine(s), which determine which options are presented when a user searches, what is recommended based on what's trending now, what is recommended based on prior movies the user has watched, and so forth. This evolution spans a broad set of mathematical and statistical methods and machine learning algorithms, such as matrix factorization, restricted Boltzmann machines, latent Dirichlet allocation, gradient boosted decision trees, and affinity propagation (Amatriain 2013). In addition, a variety of metrics—such as member retention and engagement time—and experimentation techniques—such as offline experimentation and A/B testing—are tuned for statistical validity and used to measure the success of the ensemble of algorithms (Gomez-Uribe and Hunt 2015).

1.5.4 Sentiment Analysis

A particularly active current area in big data is the use of sophisticated algorithms to determine an individual's sentiment (Yegulalp 2015). For example, textual analysis of tweets or posts can determine how a customer feels about a particular product. Emerging techniques include emotional analysis of spoken utterances and even sentiment analysis based on facial imaging.

Some enterprising companies are using sophisticated algorithms to conduct such sentiment analysis at scale, in near real time, to buy or sell stocks based on how sentiment is turning as well as additional analytics (Lin 2016).

1.5.5 Beyond Business

Such an approach is relevant beyond the realm of corporate affairs. For example, a government could utilize a collective intimacy strategy in interacting with its citizens, in recommending the best combination of public transportation based on personal destination objectives, or the best combination of social security benefits, based on a personal financial destination. Dubai, for example, has released a mobile app called Dubai Now that will act as a single portal to thousands of government services, including, for example, personalized, contextualized GPS-based real-time traffic routing (Al Serkal 2015).

1.6 Accelerated Innovation

Innovation has evolved through multiple stages, from the solitary inventor, such as the early human who invented the flint hand knife, through shop invention, a combination of research lab and experimental manufacturing facility, to the corporate research labs (Weinman 2015). However, even the best research labs can only hire so many people, but ideas can come from anywhere.

The theory of open innovation proposes loosening the firm boundaries to partners who may have ideas or technologies that can be brought into the firm, and to distribution partners who may be able to make and sell ideas developed from within the firm. Open innovation suggests creating relationships that help both of these approaches succeed (Chesbrough 2003).

However, even preselecting relationships can be overly constricting. A still more recent approach to innovation lets these relationships be ad hoc and dynamic. I call it accelerated innovation, but it can be not only faster, but also better and cheaper. One way to do this is by holding contests or posting challenges, which theoretically anyone in the world could solve. Related approaches include innovation networks and idea markets. Increasingly, machines will be responsible for innovation, and we are already seeing this in systems such as IBM's Chef Watson, which ingested a huge database of recipes and now can create its own innovative dishes, and Google DeepMind's AlphaGo, which is innovating game play in one of the oldest games in the world, Go (Fig. 1.4).

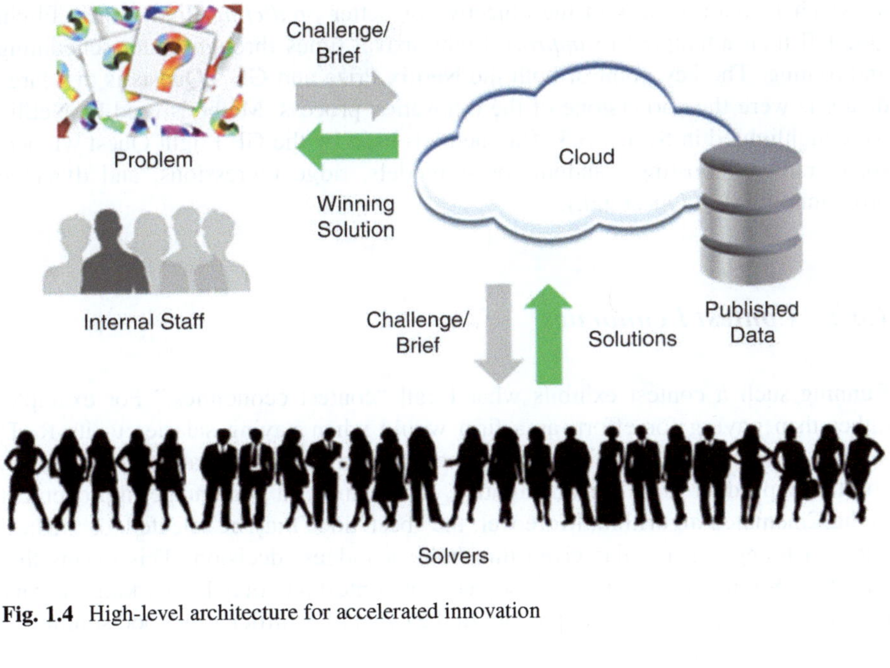

Fig. 1.4 High-level architecture for accelerated innovation

1.6.1 Contests and Challenges

Netflix depends heavily on the quality of its recommendations to maximize customer satisfaction, thus customer retention, and thereby total customer lifetime value and profitability. The original Netflix Cinematch recommendation system let Netflix customers rate movies on a scale of one star to five stars. If Netflix recommended a movie that the customer then rated a one, there is an enormous discrepancy between Netflix's recommendations and the user's delight. With a perfect algorithm, customers would always rate a Netflix-recommended movie a five.

Netflix launched the Netflix Prize in 2006. It was open to anyone who wished to compete, and multiple teams did so, from around the world. Netflix made 100 million anonymized movie ratings available. These came from almost half a million subscribers, across almost 20,000 movie titles. It also withheld 3 million ratings to evaluate submitted contestant algorithms (Bennett and Lanning 2007). Eventually, the prize was awarded to a team that did in fact meet the prize objective: a 10% improvement in the Cinematch algorithm. Since that time, Netflix has continued to evolve its recommendation algorithms, adapting them to the now prevalent streaming environment which provides billions of additional data points. While user-submitted DVD ratings may be reasonably accurate; actual viewing behaviors and contexts are substantially more accurate and thus better predictors. As one example, while many viewers say they appreciate foreign documentaries; actual viewing behavior shows that crude comedies are much more likely to be watched.

GE Flight Quest is another example of a big data challenge. For Flight Quest 1, GE published data on planned and actual flight departures and arrivals, as well as weather conditions, with the objective of better *predicting* flight times. Flight Quest II then attempted to *improve* flight arrival times through better scheduling and routing. The key point of both the Netflix Prize and GE's Quests is that large data sets were the cornerstone of the innovation process. Methods used by Netflix were highlighted in Sect. 1.5.3. The methods used by the GE Flight Quest winners span gradient boosting, random forest models, ridge regressions, and dynamic programming (GE Quest n.d.).

1.6.2 Contest Economics

Running such a contest exhibits what I call "contest economics." For example, rather than paying for effort, as a firm would when paying salaries to its R&D team, it can now pay only for results. The results may be qualitative, for example, "best new product idea," or quantitative, for example, a percentage improvement in the Cinematch algorithm. Moreover, the "best" idea may be selected, or the best one surpassing a particular given threshold or judges' decision. This means that hundreds or tens of thousands of "solvers" or contestants may be working on your problem, but you only need to pay out in the event of a sufficiently good solution.

Moreover, because the world's best experts in a particular discipline may be working on your problem, the quality of the solution may be higher than if conducted internally, and the time to reach a solution may be faster than internal R&D could do it, especially in the fortuitous situation where just the right expert is matched with just the right problem.

1.6.3 Machine Innovation

Technology is evolving to be not just an enabler of innovation, but the source of innovation itself. For example, a program called AlphaGo developed by DeepMind, which has been acquired by Google, bested the European champion, Fan Hui, and then the world champion, Lee Sedol. Rather than mere brute force examination of many moves in the game tree together with a board position evaluation metric, it used a deep learning approach coupled with some game knowledge encoded by its developers (Moyer 2016).

Perhaps the most interesting development, however, was in Game 2 of the tournament between AlphaGo and Sedol. Move 37 was so unusual that the human commentators thought it was a mistake—a bug in the program. Sedol stood up and left the game table for 15 min to regain his composure. It was several moves later that the rationale and impact of Move 37 became clear, and AlphaGo ended up winning that game, and the tournament. Move 37 was "beautiful," (Metz 2016) in retrospect, the way that the heliocentric theory of the solar system or the Theory of Relativity or the concept of quasicrystals now are. To put it another way, a machine innovated beyond what thousands of years and millions of players had been unable to do.

1.6.4 Beyond Business

Of course, such innovation is not restricted to board games. Melvin is a program that designs experiments in quantum physics, which are notoriously counterintuitive or non-intuitive to design. It takes standard components such as lasers and beam splitters, and determines new ways to combine them to test various quantum mechanics hypotheses. It has already been successful in creating such experiments.

In another example of the use of big data for innovation, automated hypothesis generation software was used to scan almost two hundred thousand scientific paper abstracts in biochemistry to determine the most promising "kinases,"—a type of protein—that activate another specific protein, "p53", which slows cancer growth. All but two of the top prospects identified by the software proved to have the desired effect (The Economist 2014).

1.7 Integrated Disciplines

A traditional precept of business strategy is the idea of focus. As firms select a focused product area, market segment, or geography, say, they also make a conscious decision on what to avoid or say "no" to. A famous story concerns Southwest, an airline known for its no frills, low-cost service. Its CEO, Herb Kelleher, in explaining its strategy, explained that every strategic decision could be viewed in the light of whether it helped achieve that focus. For example, the idea of serving a tasty chicken Caesar salad on its flights could be instantly nixed, because it wouldn't be aligned with low cost (Heath and Heath 2007).

McDonald's famously ran into trouble by attempting to pursue operational excellence, product leadership, and customer intimacy at the same time, and these were in conflict. After all, having the tastiest burgers—product leadership—would mean foregoing mass pre-processing in factories that created frozen patties—operational excellence. Having numerous products, combinations and customizations such as double patty, extra mayo, no onions—customer intimacy—would take extra time and conflict with a speedy drive through line—operational excellence (Weinman 2015).

However, the economics of information and information technology mean that a company can well orient itself to more than one discipline. The robots that run Amazon.com's logistics centers, for example, can use routing and warehouse optimization programs—operational excellence—that are designed once, and don't necessarily conflict with the algorithms that make product recommendations based on prior purchases and big data analytics across millions or billions of transactions.

The efficient delivery of unicast viewing streams to Netflix streaming subscribers—operational excellence—doesn't conflict with the entertainment suggestions derived by the Netflix recommender—collective intimacy—nor does it conflict with the creation of original Netflix content—product leadership—nor does it impact Netflix's ability to run open contests and challenges such as the Netflix Prize or the Netflix Cloud OSS (Open Source Software) Prize—accelerated innovation.

In fact, not only do the disciplines not conflict, but, in such cases, data captured or derived in one discipline can be used to support the needs of another in the same company. For example, Netflix famously used data on customer behaviors, such as rewind or re-watch, contexts, such as mobile device or family TV, and demographics, such as age and gender, that were part of its collective intimacy strategy, to inform decisions made about investing in and producing *House of Cards*, a highly popular, Emmy-Award-winning show, that supports product leadership.

The data need not even be restricted to a single company. Uber, the "ride-sharing" company, entered into an agreement with Starwood, the hotel company (Hirson 2015). A given Uber customer might be dropped off at a competitor's hotel, offering Starwood the tantalizing possibility of emailing that customer a coupon for 20% off their next stay at a Sheraton or Westin, say, possibly converting a lifelong competitor customer into a lifelong Starwood customer. The promotion could be extremely targeted, along the lines of, say, "Mr. Smith, you've now stayed at our competitor's hotel at least three times. But did you know that the Westin Times Square is rated 1 star higher than the competitor? Moreover, it's only half as far away from your favorite restaurant as the competitor, and has a health club included in the nightly fee, which has been rated higher than the health club you go to at the competitor hotel."

Such uses are not restricted to analytics for marketing promotions. For example, Waze and the City of Rio de Janeiro have announced a collaboration. Waze is a mobile application that provides drivers information such as driving instructions, based not only on maps but also real-time congestion data derived from all other Waze users, a great example of crowdsourcing with customer intimacy. In a bidirectional arrangement with Rio de Janeiro, Waze will improve its real time routing by utilizing not only the data produced by Waze users, but additional data feeds offered by the City. Data will flow in the other direction, as well, as Rio uses data collected by Waze to plan new roads or to better time traffic signals (Ungerleider 2015).

1.8 Conclusion

A combination of technologies such as the cloud, big data and analytics, machine learning, social, mobile, and the Internet of Things is transforming the world around us. Information technologies, of course, have information at their nexus, and consequently data, and the capabilities to extract information and insight and make decisions and take action on that insight, are key to the strategic application of information technology to increase the competitiveness of our firms, enhance value created for our customers, and to excel beyond these domains also into the areas of government and society.

Four generic strategies—information excellence, solution leadership, collective intimacy, and accelerated innovation—can be used independently or in combination to utilize big data and related technologies to differentiate and create customer value—for better processes and resources, better products and services, better customer relationships, and better innovation, respectively.

References

Abdullah, Z. (2015). *New app to help motorists find available parking*. http://www.straitstimes.com/singapore/new-app-to-help-motorists-find-available-parking. Accessed 15 September 2016.

Al Serkal, M. (2015). *Dubai to launch smart app for 2000 government services*. http://gulfnews.com/news/uae/government/dubai-to-launch-smart-app-for-2-000-government-services-1.1625556. Accessed 15 September 2016.

Amatriain, X. (2013). Big and personal: data and models behind Netflix recommendations. In *Proceedings of the 2nd International Workshop on Big Data, Streams and Heterogeneous Source Mining: Algorithms, Systems, Programming Models and Applications* (pp. 1–6). ACM.

Bennett, J., & Lanning, S. (2007). The Netflix prize. In *Proceedings of KDD Cup and Workshop* (Vol. 2007, p. 35).

Chesbrough, H. W. (2003). *Open innovation: The new imperative for creating and profiting from technology*. Boston: Harvard Business School Press.

Choquel, J. (2014). *NikeFuel total on your Withings scale*. http://blog.withings.com/2014/07/22/new-way-to-fuel-your-motivation-see-your-nikefuel-total-on-your-withings-scale. Accessed 15 September 2016.

Combs, V. (2014). *An infographic that works: I want dose watch from GE healthcare. Med City News*. http://medcitynews.com/2014/05/infographic-works-want-ges-dosewatch. Accessed 15 September 2016.

Coren, M. (2016). *Tesla has 780 million miles of driving data, and adds another million every 10 hours*. http://qz.com/694520/tesla-has-780-million-miles-of-driving-data-and-adds-another-million-every-10-hours/. Accessed 15 September 2016.

Dataspark (2016) *Can data science help build better public transport?* https://datasparkanalytics.com/insight/can-data-science-help-build-better-public-transport. Accessed 15 September 2016.

Falk, T. (2013). *Amazon changes prices millions of times every day. ZDnet.com*. http://www.zdnet.com/article/amazon-changes-prices-millions-of-times-every-day. Accessed 15 September 2016.

GE Aviation (n.d.). http://www.geaviation.com/commercial/engines/genx/. Accessed 15 September 2016.

GE Quest (n.d.). http://www.gequest.com/c/flight. Accessed 15 November 2016.

Gomez-Uribe, C., & Hunt, N. (2015). The netflix recommender system: algorithms, business value, and innovation. *ACM Transactions on Management Information Systems, 6*(4), 1–19.

Hayes, C. (2004). What Wal-Mart knows about customers' habits. *The New York Times*. http://www.nytimes.com/2004/11/14/business/yourmoney/what-walmart-knows-about-customers-habits.html. Accessed 15 September 2016.

Heath, C., & Heath, D. (2007). *Made to Stick: Why some ideas survive and others die*. New York, USA: Random House.

Hirson, R. (2015). *Uber: The big data company*. http://www.forbes.com/sites/ronhirson/2015/03/23/uber-the-big-data-company/#2987ae0225f4. Accessed 15 September 2016.

ITU and Cisco (2016). *Harnessing the Internet of Things for Global Development*. http://www.itu.int/en/action/broadband/Documents/Harnessing-IoT-Global-Development.pdf. Accessed 15 September 2016.

Kaggle (n.d.) *GE Tackles the industrial internet*. https://www.kaggle.com/content/kaggle/img/casestudies/Kaggle%20Case%20Study-GE.pdf. Accessed 15 September 2016.

Kuang, C. (2015). Disney's $1 Billion bet on a magical wristband. *Wired*. http://www.wired.com/2015/03/disney-magicband. Accessed 15 September 2016.

Lambrecht, A., & Skiera, B. (2006). Paying too much and being happy about it: Existence, causes and consequences of tariff-choice biases. *Journal of Marketing Research, XLIII*, 212–223.

Lee, S. (2015). *23 and Me and Genentech in deal to research Parkinson's treatments*. SFgate, January 6, 2015. http://www.sfgate.com/health/article/23andMe-and-Genentech-in-deal-to-research-5997703.php. Accessed 15 September 2016.

Lin, D. (2016). *Seeking and finding alpha—Will cloud disrupt the investment management industry?* https://thinkacloud.wordpress.com/2016/03/07/seeking-and-finding-alpha-will-cloud-disrupt-the-investment-management-industry/. Accessed 15 September 2016.

Loveman, G. W. (2003). Diamonds in the data mine. *Harvard Business Review, 81*(5), 109–113.

Metz, C. (2016). In two moves, AlphaGo and lee sedol redefined the future. *Wired*. http://www.wired.com/2016/03/two-moves-alphago-lee-sedol-redefined-future/. Accessed 15 September 2016.

Moyer, C. (2016). *How Google's AlphaGo beat a Go world champion*. http://www.theatlantic.com/technology/archive/2016/03/the-invisible-opponent/475611/. Accessed 15 September 2016.

O'Brien, S. A. (2015). *Uber partners with Boston on traffic data*. http://money.cnn.com/2015/01/13/technology/uber-boston-traffic-data/. Accessed 15 September 2016.

Pham, P. (2015). *The Impacts of big data that you may not have heard of.* forbes.com. http://www.forbes.com/sites/peterpham/2015/08/28/the-impacts-of-big-data-that-you-may-not-have-heard-of/#3b1ccc1c957d. Accessed 15 September 2016.
Pine, J., & Gilmore, J. (1999). *The experience economy: Work is theatre and every business a stage*. Boston: Harvard Business School Press.
Rosenbush, S., & Stevens, L. (2015). At UPS, the algorithm is the driver. *The Wall Street Journal*. http://www.wsj.com/articles/at-ups-the-algorithm-is-the-driver-1424136536. Accessed 15 September 2016.
Roy, R., (2009). *Ford's green goddess grows leaves*. http://www.autoblog.com/2009/10/29/ford-smart-gauge-engineer/. Accessed 15 September 2016.
Skiera, B., & Olderog, T. (2000). The benefits of bundling strategies. *Schmalenbach Business Review, 52*, 137–159.
The Economist (2014). *Computer says try this*. http://www.economist.com/news/science-and-technology/21621704-new-type-software-helps-researchers-decide-what-they-should-be-looking. Accessed 15 September 2016.
Treacy, M., & Wiersema, F. (1995). *The discipline of market leaders*. Reading, USA: Addison-Wesley.
Ungerleider, N. (2015). *Waze is driving into city hall*. http://www.fastcompany.com/3045080/waze-is-driving-into-city-hall. Accessed 15 September 2016.
Warwick, G. (2015). *GE advances analytical maintenance with digital twins*. http://aviationweek.com/optimizing-engines-through-lifecycle/ge-advances-analytical-maintenance-digital-twins. Accessed 15 September 2016.
Weinman, J. (2015). *Digital disciplines*. New Jersey: John Wiley & Sons.
Weinman, J. (2016). The internet of things for developing economies. *CIO*. http://www.cio.com/article/3027989/internet-of-things/the-internet-of-things-for-developing-economies.html. Accessed 15 September 2016.
Wind, J., Fung, V., & Fung, W. (2009). Network orchestration: Creating and managing global supply chains without owning them. In P. R. Kleindorfer, Y. (Jerry) R. Wind, & R. E. Gunther (Eds.), *The Network Challenge: Strategy, Profit, and Risk in an Interlinked World* (pp. 299–314). Upper Saddle River, USA: Wharton School Publishing.
Withings (2014). *NikeFuel total on your Withings scale*. http://blog.withings.com/2014/07/22/new-way-to-fuel-your-motivation-see-your-nikefuel-total-on-your-withings-scale/. Accessed 15 September 2016.
Xvela (2016). https://xvela.com/solutions.html. Accessed 15 September 2016.
Yegulalp, S. 2015. *IBM's Watson mines Twitter for sentiments*. 17 Mar 2015. Infoworld.com. http://www.infoworld.com/article/2897602/big-data/ibms-watson-now-mines-twitter-for-sentiments-good-bad-and-ugly.html. Accessed 15 September 2016.

Chapter 2
Start with Privacy by Design in All Big Data Applications

Ann Cavoukian and Michelle Chibba

2.1 Introduction

The evolution of networked information and communication technologies has, in one generation, radically changed the value of and ways to manage data. These trends carry profound implications for privacy. The creation and dissemination of data has accelerated around the world, and is being copied and stored indefinitely, resulting in the emergence of Big Data. The old information destruction paradigm created in an era of paper records is no longer relevant, because digital bits and bytes have now attained near immortality in cyberspace, thwarting efforts to successfully remove them from "public" domains. The practical obscurity of personal information—the data protection of yesteryear—is disappearing as data becomes digitized, connected to the grid, and exploited in countless new ways. We've all but given up trying to inventory and classify information, and now rely more on advanced search techniques and automated tools to manage and "mine" data. The combined effect is that while information has become cheap to distribute, copy, and recombine; personal information has also become far more available and consequential. The challenges to control and protect personal information are significant. Implementing and following good privacy practices should not be a hindrance to innovation, to reaping societal benefits or to finding the means to reinforce the public good from Big Data analytics—in fact, by doing so, innovation is fostered with doubly-enabling, win–win outcomes. The privacy solution requires a combination of data minimization techniques, credible safeguards, meaningful individual participation in data processing life cycles, and robust accountability measures in place by organizations informed by an enhanced and enforceable set of

A. Cavoukian (✉) • M. Chibba
Faculty of Science, Privacy and Big Data Institute, Ryerson University, 350 Victoria Street, Toronto, ON M5B 2K3, Canada
e-mail: ann.cavoukian@ryerson.ca; michelle.chibba@ryerson.ca

S. Srinivasan (ed.), *Guide to Big Data Applications*, Studies in Big Data 26,
DOI 10.1007/978-3-319-53817-4_2

universal privacy principles better suited to modern realities. This is where Privacy by Design becomes an essential approach for Big Data applications. This chapter begins by defining information privacy, then it will provide an overview of the privacy risks associated with Big Data applications. Finally, the authors will discuss Privacy by Design as an international framework for privacy, then provide guidance on using the Privacy by Design Framework and the 7 Foundational Principles, to achieve both innovation and privacy—not one at the expense of the other.

2.2 Information Privacy Defined

Information privacy refers to the right or ability of individuals to exercise control over the collection, use and disclosure by others of their personal information (Clarke 2000). The ability to determine the fate of one's personal information is so important that the authors wish to bring to the attention of the readers, the term "informational self-determination" which underpins the approach taken to privacy in this chapter. This term was established in 1983 in Germany when the Constitutional Court ruled that individuals, not governments, determine the fate of their personal information. Since this time, in December 2013, the United Nations General Assembly adopted resolution 68/167 (UN 2016), which expressed deep concern at the negative impact that surveillance and interception of communications may have on human rights. The General Assembly affirmed that the rights held by people offline must also be protected online, and it called upon all States to respect and protect the right to privacy in digital communication.

Information privacy makes each of us 'masters' of the data that identifies each of us – as individual, citizen, worker, consumer, patient, student, tourist, investor, parent, son, or daughter. For this, the notions of empowerment, control, choice and self-determination are the very essence of what we refer to as information privacy. As 'custodians' of our information, we expect governments and business can be trusted with its safekeeping and proper use.

There have also been references to statements such as "If you have nothing to hide, you have nothing to fear." (Solove 2007) Privacy is not about secrecy. It is about the freedom to exercise one's right to decide who to choose to share the personal details of one's life with. Democracy does not begin with intrusions into one's personal sphere—it begins with human rights, civil liberties and privacy—all fundamental to individual freedom.

Sometimes, safekeeping or information security is taken to mean that privacy has been addressed. To be clear, information security does not equal privacy. While data security certainly plays a vital role in enhancing privacy, there is an important distinction to be made—security is about protecting data assets. It is about achieving the goals of confidentiality, integrity and availability. Privacy related goals developed in Europe that complement this security triad are: unlinkability, transparency and intervenability. In other words, information privacy incorporates a much broader set of protections than security alone. We look to the work on 'contextual integrity' (Dwork 2014) that extends the meaning of privacy to a much broader class of transmission principles that cannot be presumed unless warranted by other context-specific parameters influenced by other actors and information types. Privacy relates not only to the way that information is protected and accessed, but also to the way in which it is collected and used. For example, user access controls protect personal information from internal threats by preventing even the possibility of accidental or intentional disclosure or misuse. This protection is especially needed in the world of Big Data.

2.2.1 Is It Personally Identifiable Information?

Not all data gives rise to privacy concerns. An important first step for any Big Data application is to determine whether the information involved falls under the definition of personally identifiable information (PII). Privacy laws around the world include a definition of personal information and it is this definition which is integral to whether or not the rules apply. Although there are privacy laws around the world, each with a definition of personal information, we will use the NIST definition, where personal information (also known as personally identifiable information) may be defined as any information, recorded or otherwise, relating to an identifiable individual (NIST 2010). It is important to note that almost any information (e.g. biographical, biological, genealogical, historical, transactional, locational, relational, computational, vocational, or reputational), may become personal in nature. Privacy laws and associated rules will apply to information if there is a reasonable possibility of identifying a specific individual—whether directly, indirectly, or through manipulation or data linkage.

Understanding the different forms of non-personal data helps to better understand what constitutes personal information. One example is de-identified or anonymous information, which will be dealt with in more detail later in this chapter. NIST defines de-identified information as records that have had enough personal information removed or obscured in some manner such that the remaining information does not identify an individual, and there is no reasonable basis to believe that the information can be used to identify an individual (NIST 2015). As an illustration, under a U.S. law known as the Health Insurance Portability and Accountability Act (HIPAA), a set of standards exist to determine when health-care information is no longer 'individually identifiable' or de-identified (HHS 2012). If this standard is achieved, then the health-care information would not be subject to this law governing the privacy of health care information. Another example is the EU General Data Protection Regulation (GDPR) that similarly, excludes anonymous information (EU Commission 2015). Of interest, however, is that this European law introduces the concept of "pseudonymization" defined as the processing of personal data in such a way as to prevent attribution to an identified or identifiable person without additional information that may be held separately.[1] For research and statistical purposes, certain requirements under the GDPR are relaxed if the personal data is pseudonymized, which is considered an appropriate safeguard alongside encryption (Official Journal of the European Union 2016).

Another form is when personal information is aggregated. Aggregation refers to summary data that have been generated by performing a calculation across all individual units as a whole. For example, medical researchers may use aggregated patient data to assess new treatment strategies; governments may use aggregated population data for statistical analysis on certain publicly funded programs for reporting purposes; companies may use aggregated sales data to assist in determining future product lines. Work has also been done on privacy-preserving data aggregation in wireless sensor networks, especially relevant in the context of the Internet of Things (Zhang et al. 2016). By using aggregated data, there is a reduced risk of connecting this information to a specific person or identify an individual.

Lastly, while personal information may be classified as confidential, not all confidential information should be governed under privacy rules. Confidential information includes information that should not be publicly available and often holds tremendous value and importance for organizations, such as strategic business plans, interim revenue forecasts, proprietary research, or other intellectual property. The distinction is that while the theft or loss of such confidential information is of grave concern for an organization it would not constitute a privacy breach because it does not involve personal information—rather, it is business information.

The growth in Big Data applications and other information communication technologies have added to the challenges of definition of personal information. There are times when information architectures, developed by engineers to ensure the smooth functioning of computer networks and connectivity, lead to unforeseen uses that have an impact on identity and privacy. These changes present challenges to what constitutes personal information, extending it from obvious tombstone data (name, address, telephone number, date of birth, gender) to the innocuous computational or metadata once the purview of engineering requirements for communicating between devices (Cameron 2013; Mayer et al. 2016).

Metadata, for example, is information generated by our communications devices and our communications service providers as we use landline or mobile phones, computers, tablets, or other computing devices. Metadata is essentially information about other information—in this case, relating to our communications (Mayer et al. 2016). Using metadata in Big Data analysis requires understanding of context.

---

[1]NIST (2015) defines 'pseudonymization' as a specific kind of transformation in which the names and other information that directly identifies an individual are replaced with pseudonyms. Pseudonymization allows linking information belonging to an individual across multiple data records or information systems, provided that all direct identifiers are systematically pseudonymized. Pseudonymization can be readily reversed if the entity that performed the pseudonymization retains a table linking the original identities to the pseudonyms, or if the substitution is performed using an algorithm for which the parameters are known or can be discovered.

Metadata reveals detailed pattern of associations that can be far more invasive of privacy than merely accessing the content of one's communications (Cavoukian 2013a, b). Addresses, such as the Media Access Control (MAC) number that are designed to be persistent and unique for the purpose of running software applications and utilizing Wi-Fi positioning systems to communicate to a local area network can now reveal much more about an individual through advances in geo-location services and uses of smart mobile devices (Cavoukian and Cameron 2011). Another good example in the mobile environment would be a unique device identifier such as an International Mobile Equipment Identity (IMEI) number: even though this does not name the individual, if it is used to treat individuals differently it will fit the definition of personal data (Information Commissioner's Office ICO 2013).

No doubt, the mobile ecosystem is extremely complex and architectures that were first developed to ensure the functioning of wireless network components now act as geo-location points, thereby transforming the original intent or what might be an unintended consequence for privacy. As noted by the International Working Group on Data Protection in Telecommunications (IWGDPT 2004) "The enhanced precision of location information and its availability to parties other than the operators of mobile telecommunications networks create unprecedented threats to the privacy of the users of mobile devices linked to telecommunications networks." When a unique identifier may be linked to an individual, it often falls under the definition of "personal information" and carries with it a set of regulatory responsibilities.

2.3 Big Data: Understanding the Challenges to Privacy

Before moving into understanding the challenges and risks to privacy that arise from Big Data applications and the associated data ecosystem, it is important to emphasize that these should not be deterrents to extracting value from Big Data. The authors believe that by understanding these privacy risks early on, Big Data application developers, researchers, policymakers, and other stakeholders will be sensitized to the privacy issues and therefore, be able to raise early flags on potential unintended consequences as part of a privacy/security threat risk analysis.

We know that with advances in Big Data applications, organizations are developing a more complete understanding of the individuals with whom they interact because of the growth and development of data analytical tools, and systems available to them. Public health authorities, for example, have a need for more detailed information in order to better inform policy decisions related to managing their increasingly limited resources. Local governments are able to gain insights never before available into traffic patterns that lead to greater road and pedestrian safety. These examples and many more demonstrate the ability to extract insights from Big Data that will, without a doubt, be of enormous socio-economic significance. These challenges and insights are further examined in the narrative on the impact of Big Data on privacy (Lane et al. 2014).

With this shift to knowledge creation and service delivery, the value of information and the need to manage it responsibly have grown dramatically. At the same time, rapid innovation, global competition and increasing system complexity present profound challenges for informational privacy. The notion of informational self-determination seems to be collapsing under the weight, diversity,. speed and volume of Big Data processing in the modern digital era. When a Big Data set is comprised of identifiable information, then a host of customary privacy risks apply. As technological advances improve our ability to exploit Big Data, potential privacy concerns could stir a regulatory backlash that would dampen the data economy and stifle innovation (Tene and Polonetsky 2013). These concerns are reflected in, for example, the debate around the new European legislation that includes a 'right to be forgotten' that is aimed at helping individuals better manage data protection risks online by requiring organizations to delete their data if there are no legitimate grounds for retaining it (EU Commission 2012). The genesis of the incorporation of this right comes from a citizen complaint to a data protection regulator against a newspaper and a major search engine concerning outdated information about the citizen that continued to appear in online search results of the citizen's name. Under certain conditions now, individuals have the right to ask search engines to remove links with personal information about them that is "inaccurate, inadequate, irrelevant or excessive." (EU Commission 2012)

Big Data challenges the tenets of information security, which may also be of consequence for the protection of privacy. Security challenges arise because Big Data involves several infrastructure layers for data processing, new types of infrastructure to handle the enormous flow of data, as well as requiring non-scalable encryption of large data sets. Further, a data breach may have more severe consequences when enormous datasets are stored. Consider, for example, the value of a large dataset of identifiable information or confidential information for that matter, that could make it a target of theft or for ransom—the larger the dataset, the more likely it may be targeted for misuse. Once unauthorized disclosure takes place, the impact on privacy will be far greater, because the information is centralized and contains more data elements. In extreme cases, unauthorized disclosure of personal information could put public safety at risk.

Outsourcing Big Data analytics and managing data accountability are other issues that arise when handling identifiable datasets. This is especially true in a Big Data context, since organizations with large amounts of data may lack the ability to perform analytics themselves and will outsource this analysis and reporting (Fogarty and Bell 2014). There is also a growing presence of data brokers involved in collecting information, including personal information, from a wide variety of sources other than the individual, for the purpose of reselling such information to their customers for various purposes, including verifying an individual's identity, differentiating records, marketing products, and preventing financial fraud (FTC 2012). Data governance becomes a sine qua non for the enterprise and the stakeholders within the Big Data ecosystem.

2.3.1 Big Data: The Antithesis of Data Minimization

To begin, the basis of Big Data is the antithesis of a fundamental privacy principle which is data minimization. The principle of data minimization or the limitation principle (Gürses et al. 2011) is intended to ensure that no more personal information is collected and stored than what is necessary to fulfil clearly defined purposes. This approach follows through the fully data lifecycle where personal data must be deleted when it is no longer necessary for the original purpose. The challenge to this is that Big Data entails a new way of looking at data, where data is assigned value in itself. In other words, the value of the data is linked to its *future and potential* uses.

In moving from data minimization to what may be termed data maximization or Big Data, the challenge to privacy is the risk of creating automatic data linkages between seemingly non-identifiable data which, on its own, may not be sensitive, but when compiled, may generate a sensitive result. These linkages can result in a broad portrait of an individual including revelations of a sensitive nature—a portrait once inconceivable since the identifiers were separated in various databases. Through the use of Big Data tools, we also know that it is possible to identify patterns which may predict people's dispositions, for example related to health, political viewpoints or sexual orientation (Cavoukian and Jonas 2012).

By connecting key pieces of data that link people to things, the capability of data analytics can render ordinary data into information about an identifiable individual and reveal details about a person's lifestyle and habits. A telephone number or postal code, for example, can be combined with other data to identify the location of a person's home and work; an IP or email address can be used to identify consumer habits and social networks.

An important trend and contribution to Big Data is the movement by government institutions to open up their data holdings in an effort to enhance citizen participation in government and at the same time spark innovation and new insights through access to invaluable government data (Cavoukian 2009).[2]

With this potential for Big Data to create data linkages being so powerful, the term "super" data or "super" content has been introduced (Cameron 2013). "Super" data is more powerful than other data in a Big Data context, because the use of one piece of "super" data, which on its own would not normally reveal much, can spark new data linkages that grow exponentially until the individual is identified. Each new transaction in a Big Data system would compound this effect and spread identifiability like a contagion.

Indeed, to illustrate the significant implications of data maximization on privacy we need only look at the shock of the Snowden revelations and the eventual repercussions. A top EU court decision in 2015 declared the longstanding Safe Harbor

[2]There are many government Open Data initiatives such as U.S. Government's Open Data at www.data.gov; Canadian Government's Open Data at http://open.canada.ca/en/open-data; UN Data at http://data.un.org/; EU Open Data Portal at https://data.europa.eu/euodp/en/data/. This is just a sample of the many Open Data sources around the world.

data transfer agreement between Europe and the U.S. invalid (Lomas 2015). The issues had everything to do with concerns about not just government surveillance but the relationship with U.S. business and their privacy practices. Eventually, a new agreement was introduced known as the EU-U.S. Privacy Shield (US DOC 2016) (EU Commission 2016). This new mechanism introduces greater transparency requirements for the commercial sector on their privacy practices among a number of other elements including U.S. authorities affirming that collection of information for intelligence is focussed and targeted.

The authors strongly believe that an important lesson learned for Big Data success is that when the individual participant is more directly involved in information collection, the accuracy of the information's context grows and invariably increases the quality of the data under analysis. Another observation, that may seem to be contradictory, is that even in Big Data scenarios where algorithms are tasked with finding connections within vast datasets, data minimization is not only essential for safeguarding personally identifiable information—it could help with finding the needle without the haystack by reducing extraneous irrelevant data.

2.3.2 Predictive Analysis: Correlation Versus Causation

Use of correlation analysis may yield completely incorrect results for individuals. Correlation is often mistaken for causality (Ritter 2014). If the analyses show that individuals who like X have an eighty per cent probability rating of being exposed to Y, it is impossible to conclude that this will occur in 100 per cent of the cases. Thus, discrimination on the basis of statistical analysis may become a privacy issue (Sweeney 2013). A development where more and more decisions in society are based on use of algorithms may result in a "Dictatorship of Data", (Cukier and Mayer-Schonberger 2013) where we are no longer judged on the basis of our actual actions, but on the basis of what the data indicate will be our probable actions. In a survey undertaken by the Annenberg Public Policy Center, the researchers found that most Americans overwhelmingly consider forms of price discrimination and behavioral targeting ethically wrong (Turow et al. 2015). Not only are these approaches based on profiling individuals but using personal information about an individual for purposes the individual is unaware of. The openness of data sources and the power of not just data mining but now predictive analysis and other complex algorithms also present a challenge to the process of de-identification. The risks of re-identification are more apparent, requiring more sophisticated de-identification techniques (El Emam et al. 2011). In addition, while the concept of "nudging" is gaining popularity, using identifiable data for profiling individuals to analyse, predict, and influence human behaviour may be perceived as invasive and unjustified surveillance.

Data determinism and discrimination are also concerns that arise from a Dictatorship of Data. Extensive use of automated decisions and prediction analyses may actually result in adverse consequences for individuals. Algorithms are not neutral, but reflect choices, among others, about data, connections, inferences, interpretations, and thresholds for inclusion that advances a specific purpose. The concern is that Big Data may consolidate existing prejudices and stereotyping, as well as reinforce social exclusion and stratification (Tene and Polonetsky 2013; IWGDPT 2014; FTC 2016). This is said to have implications for the quality of Big Data analysis because of "echo chambers"[3] in the collection phase (Singer 2011; Quattrociocchi et al. 2016).

2.3.3 Lack of Transparency/Accountability

As an individual's personal information spreads throughout the Big Data ecosystem amongst numerous players, it is easy to see that the individual will have less control over what may be happening to the data. This secondary use of data raises privacy concerns. A primary purpose is identified at the time of collection of personal information. Secondary uses are generally permitted with that person's consent, unless otherwise permitted by law. Using personal information in Big Data analytics may not be permitted under the terms of the original consent as it may constitute a secondary use—unless consent to the secondary use is obtained from the individual. This characteristic is often linked with a lack of transparency. Whether deliberate or inadvertent, lack of openness and transparency on how data is compiled and used, is contrary to a fundamental privacy principle.

It is clear that organizations participating in the Big Data ecosystem need to have a strong privacy program in place (responsible information management). If individuals don't have confidence that their personal information is being managed properly in Big Data applications, then their trust will be eroded and they may withdraw or find alternative mechanisms to protect their identity and privacy. The consequences of a privacy breach can include reputational harm, legal action, damage to a company's brand or regulatory sanctions and disruption to internal operations. In more severe cases, it could cause the demise of an organization (Solove 2014). According to TRUSTe's Consumer Privacy Confidence Index 2016, 92 per cent of individuals worry about their privacy online, 44 per cent do not trust companies with their personal information, and 89 per cent avoid doing business with companies that they believe do not protect their privacy (TRUSTe/NCSA 2016).

Despite the fact that privacy and security risks may exist, organizations should not fear pursuing innovation through data analytics. Through the application of privacy controls and use of appropriate privacy tools privacy risks may be mitigated, thereby enabling organizations to capitalize on the transformative potential of Big Data—while adequately safeguarding personal information. This is the central

[3] In news media an echo chamber is a metaphorical description of a situation in which information, ideas, or beliefs are amplified or reinforced by transmission and repetition inside an "enclosed" system, where different or competing views are censored, disallowed, or otherwise underrepresented. The term is by analogy with an acoustic echo chamber, where sounds reverberate.

motivation for Privacy by Design, which is aimed at preventing privacy violations from arising in the first place. Given the necessity of establishing user trust in order to gain public acceptance of its technologies, any organization seeking to take advantage of Big Data must apply the Privacy by Design framework as new products and applications are developed, marketed, and deployed.

2.4 Privacy by Design and the 7 Foundational Principles

The premise of Privacy by Design has at its roots, the Fair Information Practices or FIPs. Indeed, most privacy laws around the world are based on these practices. By way of history, the Code of Fair Information Practices (FIPs) was developed in the 1970s and based on essentially five principles (EPIC n.d.):

1. There must be no personal data record-keeping systems whose very existence is secret.
2. There must be a way for a person to find out what information about the person is in a record and how it is used.
3. There must be a way for a person to prevent information about the person that was obtained for one purpose from being used or made available for other purposes without the person's consent.
4. There must be a way for a person to correct or amend a record of identifiable information about the person.
5. Any organization creating, maintaining, using, or disseminating records of identifiable personal data must assure the reliability of the data for their intended use and must take precautions to prevent misuses of the data.

FIPs represented an important development in the evolution of data privacy since they provided an essential starting point for responsible information management practices. However, many organizations began to view enabling privacy via FIPs and associated laws as regulatory burdens that inhibited innovation. This zero-sum mindset viewed the task of protecting personal information as a "balancing act" of competing business and privacy requirements. This balancing approach tended to overemphasize the significance of notice and choice as the primary method for addressing personal information data management. As technologies developed, the possibility for individuals to meaningfully exert control over their personal information became more and more difficult. It became increasingly clear that FIPs were a necessary but not a sufficient condition for protecting privacy. Accordingly, the attention of privacy protection had begun to shift from reactive compliance with FIPs to proactive system design.

With advances in technologies, it became increasingly apparent that systems needed to be complemented by a set of norms that reflect broader privacy dimensions (Damiani 2013). The current challenges to privacy related to the dynamic relationship associated with the forces of innovation, competition and the global adoption of information communications technologies. These challenges have been mirrored in security by design. Just as users rely on security engineers to ensure the adequacy of encryption key lengths, for example, data subjects will rely on privacy engineers to appropriately embed risk-based controls within systems and processes. Given the complex and rapid nature of these developments, it becomes apparent that privacy has to become the default mode of design and operation.

Privacy by Design (PbD), is a globally recognized proactive approach to privacy. It is a framework developed in the late 1990s by co-author Dr. Ann Cavoukian (Cavoukian 2011). Privacy by Design is a response to compliance-based approaches to privacy protection that tend to focus on addressing privacy breaches after-the-fact. Our view is that this reactive approach does not adequately meet the demands of the Big Data era. Instead, we recommend that organizations consciously and proactively incorporate privacy strategies into their operations, by building privacy protections into their technology, business strategies, and operational processes.

By taking a proactive approach to privacy and making privacy the default setting, PbD can have a wide-ranging impact across an organization. The approach can result in changes to governance structures, operational and strategic objectives, roles and accountabilities, policies, information systems and data flows, decision-making processes, relationships with stakeholders, and even the organization's culture.

PbD has been endorsed by many public- and private-sector authorities in the United States, the European Union, and elsewhere (Harris 2015). In 2010, PbD was unanimously passed as a framework for privacy protection by the International Assembly of Privacy Commissioners and Data Protection Authorities (CNW 2010). This approach transforms consumer privacy issues from a pure policy or compliance issue into a business imperative. Since getting privacy right has become a critical success factor to any organization that deals with personal information, taking an approach that is principled and technology-neutral is now more relevant than ever. Privacy is best interwoven proactively and to achieve this, privacy principles should be introduced early on—during architecture planning, system design, and the development of operational procedures. Privacy by Design, where possible, should be rooted into actual code, with defaults aligning both privacy and business imperatives.

The business case for privacy focuses on gaining and maintaining customer trust, breeding loyalty, and generating repeat business. The value proposition typically reflects the following:

1. Consumer trust drives successful customer relationship management (CRM) and lifetime value—in other words, business revenues;
2. Broken trust will result in a loss of market share and revenue, translating into less return business and lower stock value; and
3. Consumer trust hinges critically on the strength and credibility of an organization's data privacy policies and practices.

In a marketplace where organizations are banding together to offer suites of goods and services, trust is clearly essential. Of course, trust is not simply an end-user issue. Companies that have done the work to gain the trust of their customers cannot risk losing it as a result of another organization's poor business practices.

2.4.1 The 7 Foundational Principles

Privacy by Design Foundational Principles build upon universal FIPPs in a way that updates and adapts them to modern information management needs and requirements. By emphasizing proactive leadership and goal-setting, systematic and verifiable implementation methods, and demonstrable positive-sum results, the principles are designed to reconcile the need for robust data protection and an organization's desire to unlock the potential of data-driven innovation. Implementing PbD means focusing on, and living up to, the following 7 Foundational Principles, which form the essence of PbD (Cavoukian 2011).

Principle 1: Use proactive rather than reactive measures, anticipate and prevent privacy invasive events *before* they happen (*Proactive* not Reactive; *Preventative* not Remedial).

Principle 2: Personal data must be automatically protected in any given IT system or business practice. If an individual does nothing, their privacy still remains intact (Privacy as the *Default*). Data minimization is also a default position for privacy, i.e. the concept of always starting with the minimum personal data possible and then justifying additional collection, disclosure, retention, and use on an exceptional and specific data-by-data basis.

Principle 3: Privacy must be embedded into the design and architecture of IT systems and business practices. It is not bolted on as an add-on, after the fact. Privacy is integral to the system, without diminishing functionality (Privacy *Embedded* into Design).

Principle 4: All legitimate interests and objectives are accommodated in a positive-sum manner (Full Functionality—*Positive-Sum* [win/win], not Zero-Sum [win/lose]).

Principle 5: Security is applied throughout the entire lifecycle of the data involved—data is securely retained, and then securely destroyed at the end of the process, in a timely fashion (End-to-End Security—*Full Lifecycle Protection*).

Principle 6: All stakeholders are assured that whatever the business practice or technology involved, it is in fact, operating according to the stated promises and objectives, subject to independent verification; transparency is key (*Visibility* and *Transparency*—Keep it *Open*).

Principle 7: Architects and operators must keep the interests of the individual uppermost by offering such measures as strong privacy defaults, appropriate notice, and empowering user-friendly options (*Respect* for User Privacy—Keep it *User-Centric*).

2.5 Big Data Applications: Guidance on Applying the PbD Framework and Principles

While the 7 Foundational Principles of PbD should be applied in a holistic manner as a broad framework, there are specific principles worthy of pointing out because they are what defines and distinguishes this approach to privacy. These are principles 1 (Proactive and Preventative), 2 (By Default/Data Minimization), 3 (Embedded in Design) and 4 (Positive-sum). Although the two examples provided below are specific to mobile apps, they are illustrative of the Privacy by Design approach to being proactive, focussing on data minimization and embedding privacy by default.

2.5.1 Being Proactive About Privacy Through Prevention

Privacy by Design aspires to the highest global standards of practical privacy and data protection possible and to go beyond compliance and achieve visible evidence and recognition of leadership, regardless of jurisdiction. Good privacy doesn't just happen by itself—it requires proactive and continuous goal-setting at the earliest stages. Global leadership in data protection begins with explicit recognition of the benefits and value of adopting strong privacy practices, early and consistently (e.g., preventing data breaches or harms to individuals from occurring in the first place).

> Your app's main purpose is to display maps. These maps are downloaded by a mobile device from your central server. They are then later used on the device, when there may be no network connection available. You realise that analytics would be useful to see which maps are being downloaded by which users. This in turn would allow you to make targeted suggestions to individual users about which other maps they might want to download. You consider using the following to identify individuals who download the maps: i) the device's IMEI number; ii) the MAC address of the device's wireless network interface; and iii) the mobile phone number used by the device. You realise that any of those identifiers may constitute personal data, so for simplicity you decide not to take on the responsibility of dealing with them yourself. Instead, you decide to gain users' consent for the map suggestions feature. When a user consents, they are assigned a randomly generated unique identifier, solely for use by your app. (Excerpted from Information Commissioner's Office ICO 2013)

2.5.2 *Data Minimization as the Default Through De-identification*

Personal information that is not collected, retained, or disclosed is data that does not need to be protected, managed, or accounted for. If the personal information does not exist, then it cannot be accessed, altered, copied, enriched, shared, lost, hacked, or otherwise used for secondary and unauthorized purposes. Privacy by Design is premised on the idea that the starting point for designing information technologies and systems should always be maximally privacy-enhancing. The default configuration or settings of technologies, tools, platforms, or services offered to individuals should be as restrictive as possible regarding use of personally identifiable data.

When Big Data analytics involves the use of personally identifiable information, data minimization has the biggest impact on managing data privacy risks, by effectively eliminating risk at the earliest stage of the information life cycle. Designing Big Data analytical systems at the front end with *no* collection of personally identifiable information—unless and until a specific and compelling purpose is defined, is the ideal. For example, use(s) of personal information should be limited to the intended, primary purpose(s) of collection and only extended to other, non-consistent uses with the explicit consent of the individual (Article 29 Data Protection Working Party 2013). In other cases, organizations may find that summary or aggregate data may be more than sufficient for their needs.

> Your app uses GPS location services to recommend interesting activities near to where the user is. The database of suggested activities is kept on a central server under your control. One of your design goals is to keep the amount of data your app downloads from the central server to a minimum. You therefore design your app so that each time you use it, it sends location data to the central server so that only the nearest activities are downloaded. However, you are also keen to use less privacy-intrusive data where possible. You design your app so that, by default, the device itself works out where the nearest town is and uses this location instead, avoiding the need to send exact GPS co-ordinates of the user"s location back to the central server. Users who want results based on their accurate location can change the default behaviour. (Excerpted from Information Commissioner's Office ICO 2013)

De-identification strategies are considered data minimization. De-identification provides for a set of tools or techniques to strip a dataset of all information that could be used to identify an individual, either directly or indirectly, through linkages to other datasets. The techniques involve deleting or masking "direct identifiers," such as names or social insurance numbers, and suppressing or generalizing indirect identifiers, such as postal codes or birthdates. Indirect identifiers may not be personally identifying in and of themselves, but when linked to other datasets that contain direct identifiers, may personally identify individuals. If done properly, de-identified data can be used for research purposes and data analysis—thus contributing new insights and achieving innovative goals—while minimizing the risk of disclosure of the identities of the individuals behind the data (Cavoukian and El Emam 2014).

This is not to suggest, of course, that data should be collected exclusively in instances where it may become useful or that data collected for one purpose may be repurposed at will. Rather, in a big data world, the principle of data minimization should be interpreted differently, requiring organizations to de-identify data when possible, implement reasonable security measures, and limit uses of data to those that are acceptable from not only an individual but also a societal perspective (Tene and Polonetsky 2013).

2.5.3 *Embedding Privacy at the Design Stage*

When privacy commitments and data protection controls are embedded into technologies, operations, and information architectures in a holistic, integrative manner, innovation and creativity are often by-products (Cavoukian et al. 2014a, b). By holistic, we mean that broader contexts should always be considered for a proper assessment of privacy risks and remedies. An integrative approach takes into consideration all stakeholder interests as part of the development dialogue. Sometimes, having to re-look at alternatives because existing solutions are unacceptable from a privacy perspective spurs innovative and creative thinking. Embedding privacy and data protection requires taking a systematic, principled approach—one that not only relies on accepted standards and process frameworks, but that can stand up to external reviews and audits. All of the 7 Foundational Principles should be applied with equal rigour, at every step in design and operation. By doing so, the privacy impacts of the resulting technology, process, or information architecture, and their uses, should be demonstrably minimized, and not easily degraded through use, misconfiguration, or error. To minimize concerns of untoward data usage, organizations should disclose the logic underlying their decision-making processes to the extent possible without compromising their trade secrets or intellectual property rights.

The concept of "user-centricity" may evoke contradictory meanings in networked or online environments. Through a privacy lens, it contemplates a right of control by an individual over his or her personal information when online, usually with the help of technology. For most system designers, it describes a system built with individual users in mind that may perhaps incorporate users' privacy interests, risks and needs. The first may be considered libertarian (informational self-determination), the other, paternalistic. Privacy by Design embraces both. It acknowledges that technologies, processes and infrastructures must be designed not just for individual users, but also structured by them. Users are rarely, if ever, involved in every design decision or transaction involving their personal information, but they are nonetheless in an unprecedented position today to exercise a measure of meaningful control over those designs and transactions, as well as the disposition and use of their personal information by others.

User interface designers know that human-computer interface can often make or break an application. Function (substance) is important, but the way in which that function is delivered is equally as important. This type of design embeds an effective user privacy experience. As a quid pro quo for looser data collection and minimization restrictions, organizations should be prepared to share the wealth created by individuals' data with those individuals. This means providing individuals with access to their data in a "usable" format and allowing them to take advantage of third party applications to analyze their own data and draw useful conclusions (e.g., consume less protein, go on a skiing vacation, invest in bonds) (Tene and Polonetsky 2013).

2.5.4 *Aspire for Positive-Sum Without Diminishing Functionality*

In Big Data scenarios, networks are more complex and sophisticated thereby undermining the dominant "client-server" transaction model because individuals are often far removed from the client side of the data processing equation. How could privacy be assured when the collection, disclosure, and use of personal information might not even involve the individual at all? Inevitably, a zero-sum paradigm prevails where more of one good (e.g., public security, fraud detection, operational control) cancels out another good (individual privacy, freedom). The authors challenge the premise that privacy and data protection necessarily have to be ceded in order to gain public, personal, or information security benefits from Big Data. The opposite of zero-sum is positive-sum, where multiple goals may be achieved concurrently.

Many security technologies and information systems could be designed (or redesigned) to be effective while minimizing or even eliminating their privacy-invasive features. This is the positive-sum paradigm. We need only look to the work of researchers in the area of privacy preserving data mining (Lindell and Pinkas 2002). In some cases, however, this requires broadening the scope of application from only information communication technologies (ICTs) to include the "soft" legal, policy, procedural, and other organizational controls and operating contexts in which privacy might be embedded.

De-identification tools and techniques are gaining popularity and there are several commercially available products. Nonetheless, furthering research into de-identification continues (El Emam 2013a, b). Some emerging research-level technologies hold much promise for enabling privacy and utility of Big Data analysis to co-exist. Two of these technologies are differential privacy and synthetic data.

Differential privacy is an approach that injects random noise into the results of dataset queries to provide a mathematical guarantee that the presence of any one individual in the dataset will be masked—thus protecting the privacy of each individual in the dataset. Typical implementations of differential privacy work by creating a query interface or "curator" that stands between the dataset's personal information and those wanting access to it. An algorithm evaluates the privacy risks of the queries. The software determines the level of "noise" to introduce into the analysis results before releasing it. The distortion that is introduced is usually small enough that it does not affect the quality of the answers in any meaningful way—yet it is sufficient to protect the identities of the individuals in the dataset (Dwork 2014).

At an administrative level, researchers are not given access to the dataset to analyze themselves when applying differential privacy. Not surprisingly, this limits the kinds of questions researchers can ask. Given this limitation, some researchers are exploring the potential of creating "synthetic" datasets for researchers' use. As long as the number of individuals in the dataset is sufficiently large in comparison to the number of fields or dimensions, it is possible to generate a synthetic dataset comprised entirely of "fictional" individuals or altered identities that retain the statistical properties of the original dataset—while delivering differential privacy's mathematical "noise" guarantee (Blum et al. 2008). While it is possible to generate such synthetic datasets, the computational effort required to do so is usually extremely high. However, there have been important developments into making the generation of differentially private synthetic datasets more efficient and research continues to show progress (Thaler et al. 2010).

2.6 Conclusion

There are privacy and security risks and challenges that organizations will face in the pursuit of Big Data nirvana. While a significant portion of this vast digital universe is not of a personal nature, there are inherent privacy and security risks that cannot be overlooked. Make no mistake, organizations must seriously consider not just the use of Big Data but also the implications of a failure to fully realize the potential of Big Data. Big data and big data analysis, promise new insights and benefits such as medical/scientific discoveries, new and innovative economic drivers, predictive solutions to otherwise unknown, complex societal problems. Misuses and abuses of personal data diminish informational self-determination, cause harms, and erode the confidence and trust needed for innovative economic growth and prosperity. By examining success stories and approaches such as Privacy by Design, the takeaway should be practical strategies to address the question of 'How do we achieve the value of Big Data and still respect consumer privacy?' Above all, Privacy by Design requires architects and operators to keep the interests of the individual uppermost by offering such measures as strong privacy defaults, appropriate notice, and empowering user-friendly options. Keep it user-centric!

References

Article 29 Data protection working party (2013). *Opinion 03/2013 on purpose limitation*. http://ec.europa.eu/justice/data-protection/index_en.htm. Accessed 2 August 2016.

Blum, A., Ligett, K., Roth, A. (2008). A learning theory approach to non-interactive database privacy. In *Proceedings of the 40th ACM SIGACT Symposium on Theory of Computing* (pp. 609–618).

Cameron, K. (2013). Afterword. In M. Hildebrandt et al. (Eds.), *Digital Enlightenment Yearbook 2013*. Amsterdam: IOS Press.

Cavoukian, A. (2009). *Privacy and government 2.0: the implications of an open world*. http://www.ontla.on.ca/library/repository/mon/23006/293152.pdf. Accessed 22 November 2016.

Cavoukian, A. (2011). *Privacy by Design: The 7 Foundational Principles*. Ontario: IPC.

Cavoukian, A. (2013a). *A Primer on Metadata: Separating Fact from Fiction*. Ontario: IPC. http://www.ipc.on.ca/images/Resources/metadata.pdf.

Cavoukian, A. (2013b). Privacy by design: leadership, methods, and results. In S. Gutwirth, R. Leenes, P. de Hert, & Y. Poullet (Eds.), *Chapter in European Data Protection: Coming of Age* (pp. 175–202). Dordrecht: Springer Science & Business Media Dordrecht.

Cavoukian, A., & Cameron, K. (2011). *Wi-Fi Positioning Systems: Beware of Unintended Cosnequences: Issues Involving Unforeseen Uses of Pre-Existing Architecture*. Ontario: IPC.

Cavoukian, A., & El Emam. (2014). *De-identification Protocols: Essential for Protecting Privacy*, Ontario: IPC.

Cavoukian, A., & Jonas, J. (2012). *Privacy by Design in the Age of Big Data*. Ontario: IPC.

Cavoukian, A., Bansal, N., & Koudas, N. (2014a). *Building Privacy into Mobile Location Analytics (MLA) through Privacy by Design*. Ontario: IPC.

Cavoukian, A., Dix, A., & El Emam, K. (2014b). *The Unintended Consequences of Privacy Paternalism*. Ontario: IPC.

Clarke, R. (2000). *Beyond OECD guidelines; privacy protection for the 21st century*. Xamax Consultancy Pty Ltd. http://www.rogerclarke.com/DV/PP21C.html. Accessed 22 November 2016.

CNW (2010). Landmark resolution passed to preserve the future of privacy. Press Release. Toronto, ON, Canada. http://www.newswire.ca/news-releases/landmark-resolution-passed-to-preserve-the-future-of-privacy-546018632.html. Accessed 22 November 2016.

Cukier, K., & Mayer-Schonberger, V. (2013). The dictatorship of data. *MIT Technology Review*. https://www.technologyreview.com/s/514591/the-dictatorship-of-data/. Accessed 22 November 2016.

Damiani, M. L. (2013). Privacy enhancing techniques for the protection of mobility patterns in LBS: research issues and trends. In S. Gutwirth, R. Leenes, P. de Hert, & Y. Poullet (Eds.), *Chapter in european data protection: coming of age* (pp. 223–238). Dordrecht: Springer Science & Business Media Dordrecht.

Department of Commerce (US DOC) (2016). *EU-U.S. privacy shield fact sheet. Office of public affairs, US department of commerce*. https://www.commerce.gov/news/fact-sheets/2016/02/eu-us-privacy-shield. Accessed 22 November 2016.

Dwork, C. (2014). Differential privacy: a cryptographic approach to private data analysis. In J. Lane, V. Stodden, S. Bender, & H. Nissenbaum (Eds.), *Privacy, big data, and the public good: Frameworks for engagement*. New York: Cambridge University Press.

El Emam, K. (2013a). Benefiting from big data while protecting privacy. In K. El Emam (Ed.), *Chapter in risky business: sharing health data while protecting privacy*. Bloomington, IN: Trafford Publishing.

El Emam, K. (2013b). In K. El Emam (Ed.), *Who's afraid of big data? chapter in risky business: Sharing health data while protecting privacy*. Bloomington, IN, USA: Trafford Publishing.

El Emam, K., Buckeridge, D., Tamblyn, R., Neisa, A., Jonker, E., & Verma, A. (2011). The re-identification risk of Canadians from longitudinal demographics. *BMC Medical Informatics and Decision Making, 11*:46. http://bmcmedinformdecismak.biomedcentral.com/articles/10.1186/1472-6947-11-46. Accessed 22 November 2016.

EPIC (n.d.). Website: https://epic.org/privacy/consumer/code_fair_info.html. Accessed 22 November 2016.

EU Commission (2012). *Fact sheet on the right to be forgotten*. http://ec.europa.eu/justice/data-protection/files/factsheets/factsheet_data_protection_en.pdf. Accessed 22 November 2016.

EU Commission (2015). *Fact sheet—questions and answers—data protection reform*. Brussels. http://europa.eu/rapid/press-release_MEMO-15-6385_en.htm. Accessed 4 November 2016.

EU Commission (2016). *The EU data protection reform and big data factsheet*. http://ec.europa.eu/justice/data-protection/files/data-protection-big-data_factsheet_web_en.pdf. Accessed 22 November 2016.

Fogarty, D., & Bell, P. C. (2014). Should you outsource analytics? *MIT Sloan Management Review, 55*(2), Winter.

FTC (2012). *Protecting consumer privacy in an era of rapid change: Recommendations for businesses and policymakers*. https://www.ftc.gov/sites/default/files/documents/reports/federal-trade-commission-report-protecting-consumer-privacy-era-rapid-change-recommendations/120326privacyreport.pdf Accessed August 2016.

FTC (2016). *Big data: A tool for inclusion or exclusion? Understanding the Issues*. https://www.ftc.gov/system/files/documents/reports/big-data-tool-inclusion-or-exclusion-understanding-issues/160106big-data-rpt.pdf. Accessed 23 November 2016.

Gürses, S.F. Troncoso, C., & Diaz, C. (2011). *Engineering privacy by design, Computers, Privacy & Data Protection*. http://www.cosic.esat.kuleuven.be/publications/article-1542.pdf. Accessed 19 November 2016.

Harris, M. (2015). *Recap of covington's privacy by design workshop. inside privacy: updates on developments in data privacy and cybsersecurity*. Covington & Burlington LLP, U.S. https://www.insideprivacy.com/united-states/recap-of-covingtons-privacy-by-design-workshop/. Accessed 19 November 2016.

HHS (2012). *Guidance regarding methods for de-identification of protected health information in accordance with the health insurance portability and accountability act (HIPPA) privacy rule*. http://www.hhs.gov/hipaa/for-professionals/privacy/special-topics/de-identification/index.html. Accessed 2 August 2016.

Information Commissioner's Office (ICO) (2013). *Privacy in Mobile Apps: Guide for app developers*. https://ico.org.uk/media/for-organisations/documents/1596/privacy-in-mobile-apps-dp-guidance.pdf Accessed 22 November 2016.

International Working Group on Data Protection in Telecommunications (IWGDPT) (2004) *Common position on privacy and location information in mobile communications services*. https://datenschutz-berlin.de/content/europa-international/international-working-group-on-data-protection-in-telecommunications-iwgdpt/working-papers-and-common-positions-adopted-by-the-working-group. Accessed 22 November 2016.

International Working Group on Data Protection in Telecommunications (IWGDPT) (2014). Working Paper on Big Data and Privacy: Privacy principles under pressure in the age of Big Data analytics. *55th Meeting*. https://datenschutz-berlin.de/content/europa-international/international-working-group-on-data-protection-in-telecommunications-iwgdpt/working-papers-and-common-positions-adopted-by-the-working-group. Accessed 22 November 2016.

Lane, J., et al. (2014). *Privacy, big data and the public good: frameworks for engagement.* Cambridge: Cambridge University Press.

Lindell, Y., & Pinkas, B. (2002). Privacy preserving data mining. *Journal of Cryptology, 15*, 177–206. International Association for Cryptologic Research.

Lomas, N. (2015). *Europe's top court strikes down safe Harbor data-transfer agreement with U.S. Techcrunch.* https://techcrunch.com/2015/10/06/europes-top-court-strikes-down-safe-harbor-data-transfer-agreement-with-u-s/. Accessed 22 November 2016.

Mayer, J., Mutchler, P., & Mitchell, J. C. (2016). Evaluating the privacy properties of telephone metadata. *Proceedings of the National Academies of Science, U S A, 113*(20), 5536–5541.

NIST. (2010). *Guide to protecting the confidentiality of personally identifiable information (PII). NIST special publication 800–122.* Gaithersburg, MD: Computer Science Division.

NIST (2015). *De-identification of Personal Information. NISTR 8053.* This publication is available free of charge from: http://dx.doi.org/10.6028/NIST.IR.8053. Accessed 19 November 2016.

Official Journal of the European Union (2016). *Regulation (EU) 2016/679 Of The European Parliament and of the Council.* http://ec.europa.eu/justice/data-protection/reform/files/regulation_oj_en.pdf. Accessed 19 November 2016.

Quattrociocchi, W. Scala, A., & Sunstein, C.R. (2016) *Echo Chambers on Facebook. Preliminary draft, not yet published.* Available at: http://ssrn.com/abstract=2795110. Accessed 19 November 2016.

Ritter, D. (2014). *When to Act on a correlation, and when Not To. Harvard Business Review.* https://hbr.org/2014/03/when-to-act-on-a-correlation-and-when-not-to. Accessed 19 November 2016.

Singer, N. (2011). The trouble with the echo chamber online. *New York Times online.* http://www.nytimes.com/2011/05/29/technology/29stream.html?_r=0. Accessed 19 November 2016.

Solove, D. J. (2007). I've got nothing to hide' and other misunderstandings of privacy. *San Diego Law Review, 44*, 745.

Solove, D. (2014). Why did in bloom die? A hard lesson about education privacy. Privacy + Security Blog. TeachPrivacy. Accessed 4 Aug 2016. https://www.teachprivacy.com/inbloom-die-hard-lesson-education-privacy/

Sweeney, L. (2013) *Discrimination in online ad delivery.* http://dataprivacylab.org/projects/onlineads/1071-1.pdf. Accessed 22 November 2016.

Tene, O., & Polonetsky, J. (2013). Big data for all: Privacy and user control in the age of analytics. *New Journal of Technology and Intellectual Property, 11*(5), 239–272.

Thaler, J., Ullman, J., & Vadhan, S. (2010). PCPs and the hardness of generating synthetic data. *Electronic Colloquium on Computational Complexity, Technical Report*, TR10–TR07.

TRUSTe/NCSA (2016). *Consumer privacy infographic—US Edition.* https://www.truste.com/resources/privacy-research/ncsa-consumer-privacy-index-us/. Accessed 4 November 2016.

Turow, J., Feldman, L, & Meltzer, K. (2015). *Open to exploitation: american shoppers online and offline. A report from the Annenberg Public Policy Center of the University of Pennsylvania.* http://www.annenbergpublicpolicycenter.org/open-to-exploitation-american-shoppers-online-and-offline/. Accessed 22 November 2016.

United Nations General Assembly (2016). *Resolution adopted by the General Assembly. The right to privacy in the digital age (68/167).* http://www.un.org/ga/search/view_doc.asp?symbol=A/RES/68/167. Accessed 4 November 2016.

Zhang, Y., Chen, Q., & Zhong, S. (2016). Privacy-preserving data aggregation in mobile phone sensing. *Information Forensics and Security IEEE Transactions on, 11*, 980–992.

Chapter 3
Privacy Preserving Federated Big Data Analysis

Wenrui Dai, Shuang Wang, Hongkai Xiong, and Xiaoqian Jiang

3.1 Introduction

With the introduction of electronic health records (EHRs), massive patient data have been involved in biomedical researches to study the impact of various factors on disease and mortality. Large clinical data networks have been developed to facilitate analysis and improve treatment of diseases by collecting healthcare data from a variety of organizations, including healthcare providers, government agencies, research institutions and insurance companies. The National Patient-Centered Clinical Research Network, PCORnet (n.d.), facilitates clinical effectiveness research to provide decision support for prevention, diagnosis and treatment with the data gathered nationwide. PopMedNet (n.d.) enables the distributed analyses of EHR held by different organizations without requiring a central repository to collect data. HMORNnet (Brown et al. 2012) combines PopMedNet platform to provide a shared infrastructure for distributed querying to allow data sharing between multiple HMO Research Network projects. Integrating PopMedNet, ESPnet achieves disease

W. Dai (✉)
Department of Biomedical Informatics, University of California San Diego, La Jolla, CA 92093, USA

Department of Electronic Engineering, Shanghai Jiao Tong University, Shanghai 200240, China
e-mail: wed004@ucsd.edu

S. Wang • X. Jiang
Department of Biomedical Informatics, University of California San Diego, La Jolla, CA 92093, USA
e-mail: shw070@ucsd.edu; x1jiang@ucsd.edu

H. Xiong
Department of Electronic Engineering, Shanghai Jiao Tong University, Shanghai 200240, China
e-mail: xionghongkai@sjtu.edu.cn

S. Srinivasan (ed.), *Guide to Big Data Applications*, Studies in Big Data 26,
DOI 10.1007/978-3-319-53817-4_3

surveillance by collecting and analyzing EHRs owned by different organizations in a distributed fashion.

Although data sharing can benefit both biomedical discovery and public health, it would also pose risks for disclosure of sensitive information and consequent breach of individual privacy. Leakage of demographic, diagnostic, phenotypic and genotypic information would lead to unexpected implications like discrimination by employers and health insurance companies. To protect the individually identifiable health information, Health Insurance Portability and Accountability Act (HIPAA) (n.d.) was enacted in the United States, in which the security and privacy of protected health information (PHI) are guaranteed under the standards and regulations specified by HIPAA Privacy Rule. It defines two methods, Expert Determination and Safe Harbor (Lafky 2010), to meet with the de-identification standard. In practice, the Safe Harbor method is widely adopted, where specific information should be removed and suppressed according to a predefined checklist. However, these de-identification standards defined by HIPAA Privacy Rule do not provide adequate privacy protection for healthcare data, as argued by McGraw (2008). Taking advantage of publicly available background information about an individual, it is possible to infer sensitive information like predisposition to disease and surnames from de-identified data. Homer et al. (2008) utilized aggregated allele frequencies in genome-wide association studies (GWAS) to re-identify individual patients in a case group based on the reference population from the International HapMap Project. Wang et al. (2009) extended Homer's attack with two models to identify patients from a smaller subset of published statistics or under limited precision and availability of statistics. Sweeney et al. (2013) showed that most (84–97%) patients could be exactly identified by linking their profiles in the Personal Genome Project (PGP) with publicly available records like voter lists. Gymrek et al. (2013) inferred the surnames from personal genome data sets by profiling Y-chromosome haplotypes based on recreational genetic genealogy databases, which are public and online accessible. For Healthcare Cost and Utilization Project (HCUP), Vaidya et al. (2013) demonstrated the vulnerability of its querying system by making query inference attacks to infer patient-level information based on multiple correlated queries.

Privacy concerns have presented a challenge to efficient collaborative prediction and analysis for biomedical research that needs data sharing. Patients would be unwilling to provide their data to research projects or participate in treatments under the insufficient privacy protection. For data custodian, data utility might be degraded to lower the potential privacy risk, as they take responsibility for the security and confidentiality of their data. Due to institutional policies and legislation, it is not viable to explicitly transfer patient-level data to a centralized repository or share them among various institutions in many scenarios. For example, without specific institutional approval, the U.S. Department of Veterans Affairs requires all patient data to remain in its server. Naive procedure for exchanging patient-level data would also be restricted in international cross-institutional collaboration. The Data Protection Act (1998) in the UK, in line with the EU Data Protection Directive, prohibits transferring clinical data outside the European Economic Area, unless the protection for data security is sufficiently guaranteed. Therefore, it is of necessity to develop models and algorithms for cross-institutional collaboration with sufficient protection of patient privacy and full compliance with institutional policies.

Federated data analysis has been developed as an alternative for cross-institutional collaboration, which proposes to exchange aggregated statistics instead of patient-level data. It enables a variety of privacy-preserving distributed algorithms that facilitate computation and analysis with a guarantee of prediction accuracy and protection of privacy and security of patient-level data. These algorithms are commonly designed to perform regression, classification and evaluation over data with different types of partition, i.e. horizontally partitioned data (Kantarcioglu 2008) and vertically partitioned data (Vaidya 2008), respectively. Horizontally partitioned data are composed of data of different patients with the same clinical variables. Thus, multiple institutions share the sample types of attributes for all the patients in the federated database. Horizontally partitioned data would be suitable for collaborative analysis and computation over patients from organizations in different geographical areas, especially for studies of diseases that require a large number of examples (patient-level data). On the other hand, for vertically partitioned data, each institution owns a portion of clinical variables for the same patients. Attributes from all the institutions are collected and aligned as a federated database for computation and analysis. In fact, vertically partitioned data would facilitate collaboration among organizations owning different types of patient-level data. For example, the PCORnet clinical data research network (CDRN), pSCANNER (Ohno-Machado et al. 2014), allows distributed analysis over data of 31 million patients that are vertically partitioned across Centers for Medicare and Medicaid Services, Department of Veteran Affairs, insurance companies, and health systems. The studies of diseases can be jointly performed based on the diagnostic information from medical centers, demographic and financial data from insurance companies and genomic data from laboratories. Figure 3.1 provides an illustrative example for horizontally and vertically partitioned data, respectively.

In this chapter, we review the privacy-preserving federated data analysis algorithms for large-scale distributed data, especially biomedical data. To collaborate on distributed data analysis in a privacy-preserving fashion, institutions might not be able to explicitly share their patient-level data. Privacy-preserving federated data analysis algorithms aim to establish global models for analysis and prediction based on non-sensitive local statistics, e.g., intermediary results for Hessian matrix and kernel matrix, instead of explicitly transferring sensitive patient-level data to a central repository. Over the past few decades, a series of federated analysis algorithms have been developed for regression, classification and evaluation of distributed data with a protection of data privacy and security. Enlightened by the principle of sharing model without sharing data, federated data modeling techniques have been proposed to securely derive global model parameters and perform statistical tests over horizontally and vertically partitioned data. In comparison to centralized realizations, federated modeling techniques achieved equivalent accuracy in model parameter estimation and statistical tests with no exchange of patient-level data. Server/client and decentralized architectures have been developed to realize federated data analysis in a distributed and privacy-preserving fashion.

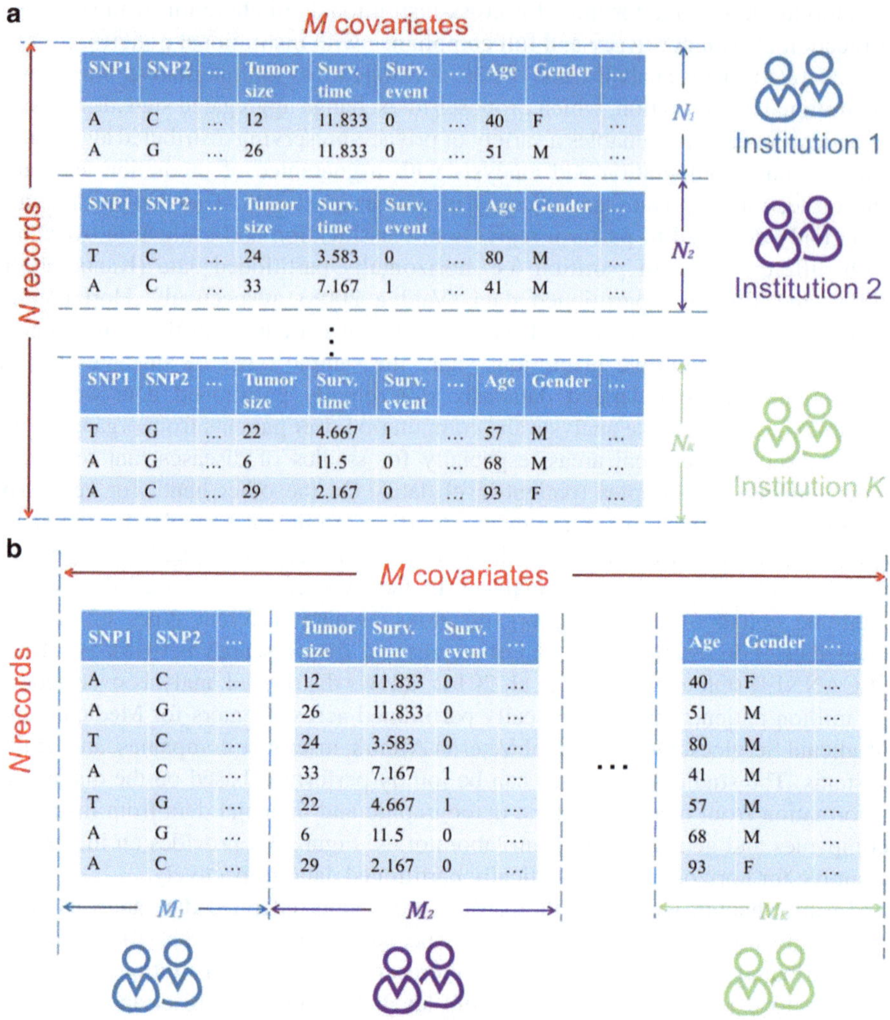

Fig. 3.1 Illustrative examples for horizontally and vertically partitioned data in federated data analysis. N records with M covariates are distributed across K institutions. (**a**) Horizontally partitioned data. The K institutions have different patients sharing the same type of covariates; (**b**) Vertically partitioned data. Covariates from the same patients are distributed across all the K institutions

For regression tasks, distributed optimization was efficiently achieved using the Newton-Ralphson method. To further improve distributed optimization in federated models, alternating direction method of multipliers (ADMM) was integrated to formulate decomposable minimization problems with additional auxiliary variables. Inheriting the convergence properties of methods of Lagrangian multipliers, it is robust for a variety of distributed analyses under horizontal and vertical data partitioning. Recognizing that communication between the server and clients was not protected, secure multiparty computation (SMC) protocols were adopted for stronger data security and privacy. Secure protocols were widely considered for distributed analysis like regression, classification and evaluation. The intermediary results from multiple institutions were aggregated with secure summation, product and reordering to support the server/client or decentralized architecture. To handle real-world network conditions, we discussed the asynchronous optimization for distributed optimization under server/client or decentralized architecture. To support privacy-preserving data sharing and analysis, this paper summarizes the relevant literature to present the state-of-the-art algorithms and applications and prospect the promising improvements and extensions along current trend.

The rest of the paper is organized as follows. Section 3.2 overviews the architecture and optimization for federated modeling analysis over horizontally and vertically partitioned data. In Sect. 3.3, we review the applications in regression and classification based on the Newton-Raphson method and ADMM framework. Section 3.3 integrates secure multiparty computation protocols with federated data analysis for protection of intermediary results in distributed analysis and computation. In Sect. 3.5, we present the asynchronous optimization for general fixed-point problem and specific coordinate gradient descent and ADMM-based methods. Finally, Sect. 3.6 makes discussion and conclusion.

3.2 Federated Data Analysis: Architecture and Optimization

In this section, we overview the architectures and optimization methods for federated data analysis. Server/client and decentralized architectures are well established in privacy-preserving analysis. Under these architectures, the Newton-Raphson method and alternating direction method of multipliers (ADMM) framework are leveraged for distributed computation.

3.2.1 Architecture

3.2.1.1 Server/Client Architecture

In the federated models, the server/client architecture has been established to estimate global model parameters and perform statistical test over horizontally and vertically partitioned data, as shown in Fig. 3.2. Under this architecture, federated data modeling shares models rather than patient-level information. The server iteratively optimizes the global model parameters based on aggregated intermediary results that are decomposable over the clients. Each client can utilize its local data to separately calculate corresponding intermediary results. Subsequently, instead of sharing the sensitive patient-level data, these intermediary results are exchanged for secure computation and analysis. Taking maximum likelihood estimation (MLE) of binary logistic regression for example, each institution calculates and exchanges

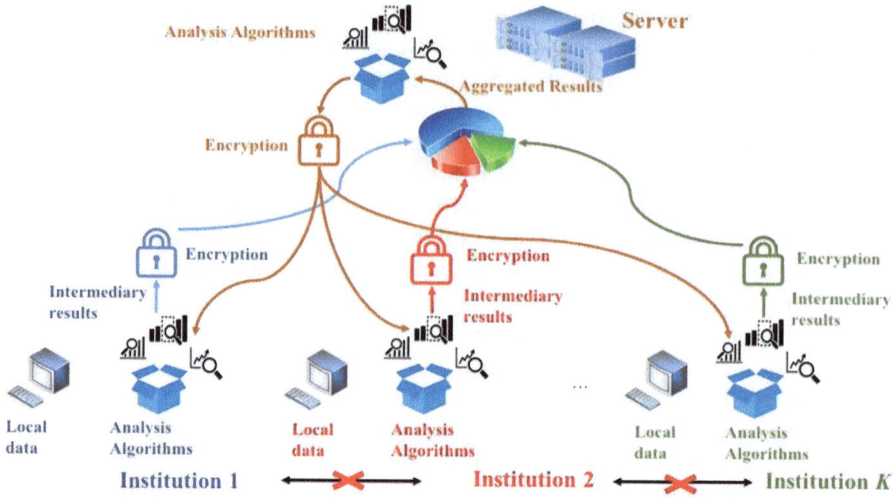

Fig. 3.2 Server/client architecture for federated data analysis. Each institution only exchanges intermediary results for the estimation of global model parameters and statistical tests. Each institution does not communicate with the others to prevent unexpected information leakage

the partial Hessian matrix derived from its local records for horizontally partitioned data and partial kernel matrix of its local attributes for vertically partitioned data. Thus, federated models only exchanged aggregated intermediary results rather than collected raw data in a central repository to make parameter estimation. Moreover, clients will not collude to infer the raw data, as each client separately performs the computation. These facts imply that distributed optimization can be performed in a privacy-preserving fashion, as the raw data tend not to be recovered from the aggregated intermediary results. It should be noted that the accuracy of parameter estimation could be guaranteed in federated models, as aggregated intermediary results do not lose any information in comparison to the centralized methods. Furthermore, the security of federated models can be further improved by integrating secure protocols and encryption methods to protect the intermediary results exchanged between the server and clients.

For logistic regression and multinomial regression models, federated models can also support distributed statistical tests over horizontally partitioned data, including goodness-of-fit test and AUC score estimation (Chambless and Diao 2006). Besides model parameters, variance-covariance matrix could be similarly obtained by aggregating the decomposable intermediary results. Using global model parameters and variance-covariance matrix, federated models were able to estimate the statistics of logistic and multinomial regression, including confidence intervals (CIs), standard error, Z-test statistics and p-values. Furthermore, goodness-of-fit test and AUC score estimation can be achieved in a distributed manner. The Hosmer and Lemeshow (H-L) test (Hosmer et al. 2013) is considered to check model fitness, where each institution only shares its number of records with positive patient outcomes per

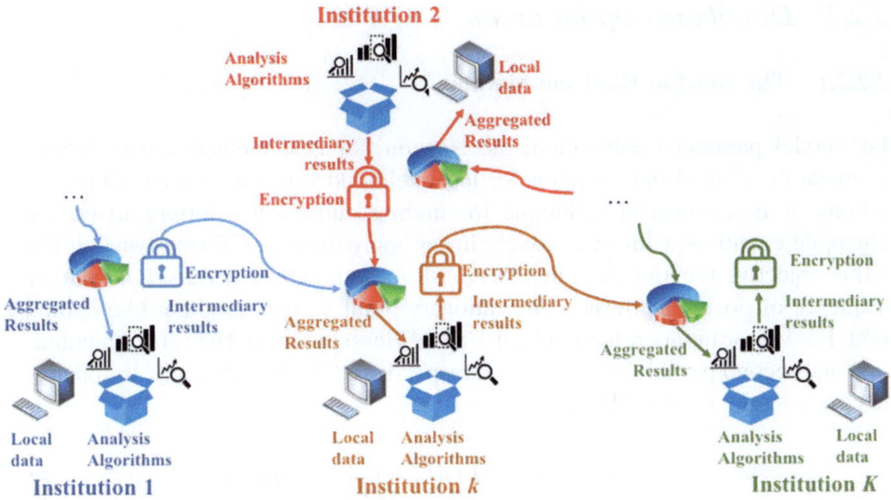

Fig. 3.3 Decentralized architecture for federated data analysis. To estimate global model parameters and perform statistical tests, each institution exchanges intermediary results with its neighboring institutions and updates its local model with received aggregated results

decile. Thus, patient-level information and estimate outcomes will not be exchanged for privacy-preserving consideration. For computation of AUC score, raw data and estimated outcomes of patients can be protected with peer-to-peer communication.

3.2.1.2 Decentralized Architecture

Figure 3.3 illustrates the decentralized architecture for federated analysis over horizontally and vertically partitioned data. Contrary to server/client architectures, decentralized architectures do not require a central node (server) to collect aggregated intermediary results from all the institutions and make global model parameter estimation and statistical tests. Each institution only communicates with its neighbors to exchange messages, e.g. institutions with linked health records. To prevent leakage of patient-level information, institutions would exchange intermediary results rather than raw data. At each iteration, each institution derives its intermediary results from local data and the aggregated results from its neighboring institutions. Taking global consensus optimization under ADMM framework (Boyd et al. 2011) for example, local model parameters are exchanged among neighboring institutions for global consensus in applications like sparse linear regression (Mateos et al. 2010), principal component analysis (PCA) (Schizas and Aduroja 2015) and support vector machine (SVM) (Forero and Giannakis 2010). Under such architecture, distributed optimization can be performed in a privacy-preserving manner, as patient-level information would never be exchanged. It is worth mentioning that communication cost would be reduced in the decentralized architecture, as messages are only exchanged among neighboring institutions.

3.2.2 Distributed Optimization

3.2.2.1 The Newton-Raphson Method

For model parameter estimation, the Newton-Raphson method can be extended to make distributed optimization for log-likelihood functions over multiple institutions. It is a powerful technique for finding numerical solutions to nonlinear algebraic equations with successively linear approximations. Given twice differentiable objective function $f(x)$, the Newton-Raphson method iteratively constructs a sequence of positions towards the stationary point x^* with gradient-like optimization. For vector inputs x, the gradient ∇f and Hessian matrix $\mathbf{H}(f)$ are computed for first and second partial derivatives of f, respectively. At the t-th step, x is updated by maximizing the log-likelihood function f.

$$\boldsymbol{x}^{(t+1)} = \boldsymbol{x}^{(t)} - \left[\boldsymbol{H}\left(f\left(\boldsymbol{x}^{(t)}\right)\right)\right]^{-1} \nabla f\left(\boldsymbol{x}^{(t)}\right) \tag{3.1}$$

In federated data modeling, to enable distributed optimization, the first and second partial derivatives are required to be decomposable over multiple institutions. Thus, the gradient and Hessian matrix can be derived from the aggregated intermediary results separately obtained from all the institutions. The intermediary results would vary for different tasks and data with different partition. For example, each institution holds a portion of records A_k for horizontally partitioned data. Consequently, the intermediary results exchanged in binary logistic regression are $A_k^T \Lambda_k A_k$ with $\Lambda_k = \mathrm{diag}(\pi(A_k, x_k)(1 - \pi(A_k, x_k)))$ related with logit function, while they tend to depend on the set of records at risk for Cox regression model. For vertically distributed logistic regression, it requires Legendre transform for distributed dual optimization, where the kernel matrix $A_k A_k^T$ is separately calculated based on the portion of attributes A_k held by the k-th institution.

The Newton-Raphson method can achieve faster convergence towards a local optimum in comparison to gradient descent, when f is a valid objective function. At the meantime, distributed optimization with the Newton-Raphson method can achieve equivalent accuracy in model parameter estimation, when compared to its centralized realization. However, it requires separable first and second partial derivatives over multiple institutions, which make the Newton-Raphson method restrictive in some distributed optimization scenarios, e.g. distributed Cox regression. In Sect 3.2.1.2, we introduce alternating direction method of multipliers (ADMM) for a ubiquitous distributed optimization framework.

3.2.2.2 Alternating Direction Method of Multipliers

Alternating direction method of multipliers (ADMM) (Boyd et al. 2011) is a variant of the augmented Lagrangian scheme that supports decomposable dual ascent solution for the method of Lagrangian multipliers. It develops a decomposition-coordination procedure to decompose the large-scale optimization problem into a set of small local subproblems. Inheriting the convergence properties of the method of Lagrangian multipliers, ADMM is able to obtain the globally optimal solution. In comparison to general equality-constrained minimization, ADMM introduces auxiliary variables to split the objective function for decomposability. To be concrete, ADMM solves the minimization problem of the summation of two convex functions $f(x)$ and $g(z)$ under the constraint of $Ax + Bz = C$. Thus, the augmented Lagrangian function for optimal solution is formulated with dual variable γ.

$$\min_{x,z,y} L(x,z,y) = f(x) + g(z) + \gamma^T(Ax + Bz - C) + \frac{\rho}{2}|Ax + Bz - C|_2^2 \quad (3.2)$$

where ρ is the positive Lagrangian parameter. The augmented Lagrangian function can be iteratively solved with three steps for partially updating the primal variables x auxiliary variables z and dual variables γ. In the t-th iteration, these three steps are conducted in a sequential manner.

1. x-minimization: partially update x by minimizing the augmented Lagrangian function $L(x, z, \gamma)$ with fixed z and dual variable γ, or $x^{(t+1)} = \underset{x}{\text{argmin}}\, L\left(x, z^{(t)}, \gamma^{(t)}\right)$,

2. z-minimization: partially update z by minimizing the augmented Lagrangian function $L(x, z, \gamma)$ with updated x and fixed γ, or $z^{(t+1)} = \underset{x}{\text{argmin}}\, L\left(x, z^{(t+1)}, \gamma^{(t)}\right)$,

and

3. Update dual variables γ with updated x and z, or $\gamma^{(t+1)} = \gamma^{(t)} + \rho\left(Ax^{(t+1)} + Bz^{(t+1)} - C\right)$

Here, dual variables γ are updated with the step size ρ for each iteration. According to Steps 1–3, x and z are alternately updated in each iteration based on $f(x)$ and $g(z)$. This fact implies that the minimization problem over x and z can be separately solved for distributed data, when $f(x)$ or $g(z)$ are decomposable. Since x and z can be derived from each other based on the dual variables γ, they can be exchanged among multiple institutions for distributed optimization. Therefore, ADMM is desirable for privacy-preserving federated data analysis, where each institution can solve the decomposable optimization problem based on local data and submits intermediary results for a global solution.

Under the ADMM framework, the two generic optimization problems, consensus and sharing, are formulated for distributed analysis over horizontally and vertically partitioned data, respectively. For horizontally partitioned data, the consensus problem splits primal variables x and separately optimizes the decomposable cost function $f(x)$ for all the institutions under the global consensus constraints. Considering that the submatrix $A_k \in R^{N_k \times M}$ of $A \in R^{N \times M}$ corresponds to the local data held by the k-th institution, the primal variables $x_k \in R^{M \times 1}$ for the K institutions are solved by

$$\min_{x,z} \sum_{k=1}^{K} f_k(A_k x_k) + g(z) \qquad \text{s.t. } x_k - z = 0, k = 1, \cdots, K. \quad (3.3)$$

Here, $f_k(A_k x_k)$ is the cost function for the k-th institution and $g(z)$ commonly represents the regularization term for the optimization problem. According to the derived augmented Lagrangian function, x_k is independently solved based on local data A_k and corresponding dual variables y_k, while z is updated by averaging x_k and y_k. Thus, the global model can be established by optimizing over the K institutions under the global consensus constraints.

The sharing problem is considered for vertically partitioned data, where A and x are vertically split into $A_k \in R^{N\times M_k}$ and $x_k \in R^{M_k\times 1}$ for the K institutions. Auxiliary variables $z_k \in R^{N\times 1}$ are introduced for the k-th institution based on A_k and x_k. In such case, the sharing problem is formulated based on the decomposable cost function $f_k(x_k)$.

$$\min_{x,z} \sum_{k=1}^{K} f_k(x_k) + g\left(\sum_{k=1}^{K} z_k\right) \qquad \text{s.t. } A_k x_k - z_k = 0, k = 1, \cdots, K \tag{3.4}$$

Under the ADMM framework, x_k and its dual variables y_k can be separately solved, while z_k is derived from the aggregated results of $A_k x_k$ and y_k. This fact implies that each institution can locally optimize the decomposable cost function using its own portion of attributes and adjust the model parameters according to auxiliary variables derived from global optimization problem. It is worth mentioning that the global consensus constraints implicitly exist for the dual variables in sharing problem.

3.3 Federated Data Analysis Applications

In this section, we review federated data analysis models for regression and classification based on the Newton-Raphson method and ADMM framework.

3.3.1 Applications Based on the Newton-Raphson Method

The Newton-Ralphson method is widely adopted in federated data analysis for generalized linear models, e.g. logistic regression, multinomial regression, and Cox proportional hazard model, which are widely used in biomedicine. Table 3.1 summarizes the existing federated modeling techniques for distributed data with their application scenarios, data partitioning, mechanisms for model parameter estimation, statistical tests and communication protection.

An early federated data analysis paper in biomedical informatics introduced the Grid binary LOgistic REgression (GLORE) framework (Wu et al. 2012). In this work, a binary logistic regression model was developed for model parameter estimation and statistical tests over data horizontally distributed across multiple institutions in a privacy-preserving manner. For security and confidentiality, the proposed model shared models rather than patient-level information. Besides model parameter estimation, distributed algorithms were developed for H-L test and AUC

Table 3.1 Federated modeling techniques for distributed data based on the Newton-Raphson method

	Application scenario	Data partitioning	Parameter estimation	Statistical test	Communication protection
GLORE (Wu et al. 2012)	Logistic regression	Horizontal	MLE	H-L test AUC score estimation	N/A
IPDLR (Wu et al. 2012)					Secure summation protocol
EXPLORER (Wang et al. 2013)			MAP	N/A	SINE protocol
SMAC-GLORE (Shi et al. 2016)			MLE		Garbled circuits
VERTIGO (Li et al. 2016)		Vertical			N/A
Multi-category GLORE (Wu et al. 2015)	Multinomial regression	Horizontal	MLE	H-L test AUC score estimation	N/A
HPPCox (Yu et al. 2008)	Cox regression	Horizontal	MLE	N/A	N/A
WebDISCO (Lu et al. 2014)					

score estimation in GLORE. It was shown to achieve equivalent model parameter estimation and statistical tests over simulated and clinical datasets in comparison to centralized methods. WebGLORE (Jiang et al. 2013) provided a free web service to implement the privacy-preserving architecture for GLORE, where AJAX, JAVA Applet/Servlet and PHP technologies were seamlessly integrated for secure and easy-to-use web service. Consequently, it would benefit biomedical researchers to deploy the practical collaborative software framework in real-world clinical applications.

Inspired by GLORE, a series of extensions and improvements have been made for various regression tasks with different privacy concerns. Despite shedding light on federated data analysis, GLORE still suffered from two main limitations: privacy protection of intermediary results and synchronization for iterative distributed optimization. Wu et al. (2012) considered the institutional privacy for GLORE. During iterative optimization, sensitive information of an institution would be leaked, as its contribution to each matrix of coefficients is known to the server. Therefore, institutional privacy-preserving distributed binary logistic regression (IPDLR) was developed to enhance the institutional privacy in GLORE by masking the ownership of the intermediary results exchanged between the server and institutions. To make all the institutions remain anonymous, a secure summation procedure was developed to integrate all the intermediary results without identifying their ownership. At each iteration, client-to-client communication was conducted to merge the intermediary results on a client basis based on the random matrix assigned by the server. Thus, the server would obtain the aggregated results without knowing the contribution of each institution. The secure summation procedure was also employed in ROC curve plotting to securely integrate local contingency tables derived in the institutions. Wang et al. (2013) proposed a Bayesian extension for GLORE, namely EXpectation Propagation LOgistic REgRession (EXPLORER) model, to achieve distributed privacy-preserving online learning. In comparison to frequentist logistic regression model, EXPLORER made maximum a posteriori (MAP) estimation using expectation propagation along the derived factor graph. The model parameters were iteratively updated based on partial posterior function w.r.t. the records held by each institution (intra-site update) and the messages passing among the server and institutions (inter-site update). As a result, EXPLORER improved the security and flexibility for distributed model learning with similar discrimination and model fit performance. To reduce the information leakage from unprotected intermediary results, EXPLORER exchanged the encrypted posterior distribution of coefficients rather than the intermediary results for model parameter estimation. The sensitive information about individual patient would not be disclosed, as only statistics like mean vector and covariance matrix were shared to represent the aggregated information of the raw data. Moreover, secured intermediate information exchange (SINE) protocol was adopted to further protect aggregation information. To guarantee flexibility, EXPLORER leveraged online learning to update the model based on the newly added records. It also supported asynchronous communication to avoid coordinating multiple institutions, so that it would be robust under the emergence of offline institution and interrupted communication. Shi et al. (2016) developed a grid logistic regression framework based on secure multiparty computation. In addition to raw data, the proposed SMAC-GLORE protected the decomposable intermediary results based on garbled circuits during iterative model learning. Secure matrix multiplication and summation protocols were presented for maximum likelihood estimation using fixed-Hessian methods. For MLE, Hessian matrix inversion problem was securely transferred to a recursive procedure of matrix multiplications and summations using Strassen algorithm, while exp006Fnential function was approximated with Taylor series expansion.

Wu et al. (2015) extended GLORE to address multi-centric modeling of multi-category response, where grid multi-category response models were developed for ordinal and multinomial logistic regression over horizontally partitioned data. Grid Newton method was proposed to make maximum likelihood estimation of model parameters in a privacy-preserving fashion. At each iteration, each institution separately calculated partial gradients and Hessian matrix based on its own data and the server could integrate these intermediary results to derive the global model parameters. Thus, the proposed models could reduce disclosure risk, as patient-level data would not be moved outside the institutions. Furthermore, privacy-preserving distributed algorithms were presented for grid model fit assessment and AUC score computation in ordinary and multinomial logistic regression model by extending the corresponding algorithms for binary response models. The proposed models were demonstrated to achieve the same accuracy with a guarantee of data privacy in comparison to the corresponding centralized models.

Recently, Li et al. (2016) proposed a novel method that leveraged dual optimization to solve binary logistic regression over vertically partitioned data. The proposed vertical grid logistic regression (VERTIGO) derived the global solution with aggregated intermediate results rather than the sensitive patient-level data. In the server/client architecture, the server iterative solved the dual problem of binary logistic regression using the Newton-Raphson method. To compute the Hessian matrix, each institution transmitted the kernel matrix of its local statistics to the server for merging. Dot product kernel matrix was adopted to guarantee that the global gram matrix was decomposable over multiple institutions. For iterative optimization, it was only required to exchange the dot products of patient records and dual parameters between the server and institutions. This fact implies that the patient-level information would not be revealed, when the distribution of covariates is not highly unbalanced. Employed on both synthetic and real datasets, VERTIGO was shown to achieve equivalent accuracy for binary logistic regression in comparison to its centralized counterpart.

Distributed survival analysis is one of the prevailing topics in biomedical research, which studies the development of a symptom, disease, or mortality with distributed time-to-event data. Cox proportional hazard model (Cox 1972) is widely concerned in survival analysis, which evaluates the significance of time-varying covariates with a hazard function. Yu et al. (2008) proposed a privacy-preserving Cox model for horizontally partitioned data, where affine projections of patient data in a lower dimensional space were shared to learn survival model. The proposed HPPCox model utilized a rank-deficient projection matrix to hide sensitive information in raw data, as the lower dimensional projections were commonly irreversible. Project matrix was optimized to minimize loss of information led by these lower dimensional projections, projection matrix was optimized to simultaneously maintain the major structures (properties) and reduce the dimensionality of input data. Thus, feature selection could be enabled to prevent overfitting for scenarios requiring limited training data. This model was shown to achieve nearly optimal predictive performance for multi-centric survival analysis. O'Keefe et al. (2012) presented explicit confidentialisation measures for survival models to avoid exchanging patient-level data in a remote analysis system, but did not consider distributed learning model. The work considered and compared confidentialised outputs for non-parametric survival model with Kaplan-Meier estimates (Kaplan and Meier 1958), semiparametric Cox proportional hazard model and parametric survival model with Weibull distribution (Weibull 1951). The confidentialised outputs would benefit model fit assessment with similar model statistics and significance in comparison to traditional methods. However, the work was focused on securely generating survival outputs, did not perform distributed model learning. Lu et al. (2014) proposed a web service WebDISCO for distributed Cox proportional hazard model over horizontally partitioned data. The global Cox model was established under the server/client architecture, where each institution separately calculated its non-sensitive intermediary statistics for model parameter estimation in server. WebDISCO investigated the technical feasibility of employing federated data analysis on survival data. The distributed Cox model was shown to be identical in mathematical formulation and achieve an equivalent precision for model learning in comparison to its centralized realization.

3.3.2 *Applications Based on ADMM*

In this subsection, we review the ADMM-based distributed algorithms for regression and classification with a brief overview on the convergence analysis for ADMM-based methods. Table 3.2 summarizes the ADMM-based algorithms for horizontally and vertically partitioned data and most of them have not been applied in the context of biomedical informatics.

Table 3.2 Federated modeling techniques for distributed data based on ADMM

	Application scenario	Data partitioning
Boyd et al. (2011)	Logistic regression	Horizontal/Vertical
	Lasso/Group Lasso	Vertical
	Support vector machine	Vertical
Mateos et al. (2010)	Linear regression	Horizontal
Mateos and Schizas (2009)	Recursive least-squares	Horizontal
Mateos and Giannakis (2012)	Recursive least-squares	Horizontal
Forero and Giannakis (2010)	Support vector machine	Horizontal
Schizas and Aduroja (2015)	Principal component analysis	Horizontal
Scardapane et al. (2016)	Recurrent neural networks	Horizontal

3.3.2.1 Regression

Boyd et al. (2011) summarized the ADMM-based distributed ℓ_1-penalized logistic regression model for horizontally and vertically partitioned data, respectively. For horizontally partitioned data, model parameters were separately solved for each institution by minimizing an ℓ_2-regularized log-likelihood function over local data. Subsequently, auxiliary variable for global consensus was found to minimize the combination of its ℓ_1-norm and the squared difference between the auxiliary variable and averaged primal and dual variables. It should be noted that the auxiliary variable could be computed for each attribute in parallel for improved efficiency. Each institution would not leak its sensitive information, as only its local model parameters were exchanged for global consensus. When data are vertically distributed across multiple institutions, a Lasso problem based on the local attributes was formulated for each institution under the ADMM framework. Aggregating the intermediary results from all the institutions, the auxiliary variables were derived from the ℓ_2-regularized logistic loss function. The dual variables were updated based on the aggregated intermediary results and averaged auxiliary variables and remained same for all the institutions. The distributed logistic regression could be performed in a privacy-preserving manner, as local data owned by each institution would not be inferred from the aggregated intermediary results. In both cases, the ℓ_2-regularized minimization based on log-sum-exp. functions can be iteratively solved using the Newton-Raphson method or L-BFGS algorithm.

Mateos et al. (2010) leveraged alternating direction method of multipliers (ADMM) to make model parameter estimation in sparse linear regression. The centralized Lasso model was transformed into a decomposable consensus-based minimization problem for horizontally partitioned data. The derived minimization problem can be iteratively solved with ADMM in a decentralized manner, where each institution only communicates with its neighboring institutions to protect data privacy. To balance computational complexity and convergence rate, three iterative algorithms, DQP-Lasso, DCD-Lasso and D-Lasso, were developed based on ADMM. DQP-Lasso introduced strictly convex quadratic terms to constrain the model parameters and auxiliary variables for each institution and its neighbors in augmented Lagrangian function. Thus, quadratic programming (QP) is adopted for each institution to obtain its model parameters by minimizing the decomposable Lagrangian function in a cyclic manner. To improve efficiency of iterative optimization, DCD-Lasso introduced coordinate descent to simplify the QP process for each institution. At each iteration, DCD-Lasso updated the model parameters for each institution with coordinate descent rather than iteratively obtained the exact solution to the corresponding QP problem. Furthermore, D-Lasso enabled parallelized computation of model parameters corresponding to multiple institutions to enhance convergence speed. D-Lasso can achieve equivalent model estimation in comparison to DCD-Lasso without additional relaxation terms. It is demonstrated that the model parameters locally derived by these algorithms are convergent to the global solution to Lasso.

Similar to sparse linear regression, Mateos and Schizas (2009) adopted ADMM to develop a distributed recursive least-squares (D-RLS) algorithm for time series data horizontally distributed across multiple institutions. The proposed algorithm reformulated the exponentially weighted least-squares estimation to a consensus-based optimization problem by introducing auxiliary variables for corresponding institutions. The reformulated minimization problem was decomposed into a series of quadratic optimization problems that were recursively solved for each institution. Furthermore, Mateos and Giannakis (2012) improved the efficiency of D-RLS by avoiding explicit matrix inversion in recursively solving quadratic problems for each institution. The proposed algorithms are demonstrated to be stable for time series data with sufficient temporal samples under the metric of means and mean squared errors (MSE). These ADMM-based linear regression and least-squares estimation were validated over the wireless sensor networks.

The sharing formulations have been also studied for Lasso and group Lasso problems over vertically partitioned data. Similar to distributed logistic regression model, sparse least-squares estimation was formulated for each institution to independently obtain the model parameters corresponding to its own attributes. In group Lasso, institutions would adopt various regularization parameters for ℓ_1-regularized problem. Since the auxiliary variables were updated analytically with a linear combination of the aggregated intermediary results and dual variables, the computational complexity mainly depended on the decomposable ℓ_1-regularized problem for multiple institutions. To improve the efficiency, the x-minimization for certain institution could be skipped, when its attributes were not considered to be involved in distributed optimization based on a threshold w.r.t. the regularization parameters and Lagrangian multiplier.

3.3.2.2 Classification

Forero and Giannakis (2010) leveraged ADMM to develop a distributed support vector machine (DSVM) classifier for training data horizontally distributed across multiple institutions. Introducing auxiliary variables for the local model parameters at each node, the linear SVM classifier was decomposed into a series of convex sub-problems over these auxiliary variables under the consensus constraints. For each node, its local model parameters were independently derived from the corresponding sub-problem. The decomposable optimization was iteratively performed to obtain the unique optimal model parameters in a decentralized manner. Under the ADMM framework, the global SVM classifier was trained in a privacy-preserving fashion, as each node exchanges its locally estimated model parameters rather than the training data it owns. To handle sequential and asynchronous learning tasks, the linear DSVM classifier support online update for time-varying training data. The classifier could be partially updated to adapt with the cases that training samples were added to and removed from the training set. The ADMM-based distributed optimization was also generalized to nonlinear SVM classifiers, where consensus constraints on discriminant functions of local model parameters were shrunk to a rank-deficient subspace of the reproducing kernel Hilbert space. In analogy to generalized linear regression, ADMM was also adopted for SVM classifier over vertically partitioned data. The centralized SVM classifier was decomposed into a series of quadratic programming problems to separately derive the model parameters corresponding to the attributes owned by each institution. These model parameters were iteratively adjusted with the auxiliary variables obtained by soft thresholding over the aggregated intermediary results and averaged dual variables. Therefore, each institution only needs to share the aggregated intermediary results of its attributes for each patient's record.

Schizas and Aduroja (2015) proposed a distributed principal component analysis (PCA) framework based on ADMM under the metric of mean-square error (MSE). An equivalent constrained optimization problem was formulated for the classical PCA framework, where each node used local covariance matrix to separately estimate its principal eigenspace in a recursive manner. Coordinate descents were adopted in the ADMM framework to iteratively minimize the augmented Lagrangian function under the consensus constraints. The proposed framework enabled distributed dimensionality reduction over horizontally partitioned data in a privacy-preserving fashion, as each node only exchanged aggregated intermediary results with its neighboring nodes. Under sufficient iterative optimization, the estimated principal eigenspace is demonstrated to asymptotically converge to the subspace spanned by the actual principal eigenvectors. For validation, the proposed distributed PCA framework was employed in data denoising over wireless sensor networks, where synthetic and practical evaluations showed it could achieve enhanced convergence rate and improved performance for noise resilience.

Scardapane et al. (2016) adopted ADMM in the decentralized training of recurrent neural networks (RNNs) that optimized the global loss function over the data horizontally distributed across all the nodes. The proposed distributed algorithm was designed specifically for Echo State Networks. The decomposable optimization under consensus constraints was formulated for each node to separately calculate the model parameter based on the ℓ_2-regularized least-squares estimation. For each node, its auxiliary variable for global consensus was obtained with a weighted average of model parameters from its neighboring nodes. Thus, it was not required to share the training data or the hidden matrix representing the current states of neural networks among multiple nodes. Since the auxiliary variables were not calculated based on all the nodes in the neural network, it did not necessarily depend on a server to perform global optimization. Evaluations over large scale synthetic data showed that the proposed distributed algorithm could achieve comparable classification performance in comparison to its centralized counterpart.

3.3.2.3 Convergence and Robustness for Decentralized Data Analysis

Ling et al. (2015) proposed a decentralized linearized ADMM (DLM) method to reduce computational complexity and enhance convergence speed for standard ADMM. The proposed DLM method simplified the decomposable optimization problems with linearized approximation. Thus, the computational cost for implementation based on gradient descent could be significantly reduced for applications like distributed logistic regression and least-squares estimation. It is demonstrated that the DLM method can converge to the optimal solution with a linear rate, when strongly convex and the Lipschitz continuity conditions are satisfied for the decomposable cost functions and its gradients, respectively. Furthermore, Mokhtari et al. (2016) considered quadratic approximation of ADMM-based formulation for logistic regression model to improve the accuracy of model parameter estimation. The proposed DQM method was shown to obtain more accurate estimates for model parameters with a linear convergence rate in comparison to DLM.

Investigations have also been made to improve the robustness of ADMM for various application scenarios. Wang and Banerjee (2012) introduced online algorithm into ADMM to yield an enhance convergence rate. Goldfarb et al. (2012) developed a fast first-order linearized augmented Lagrangian minimization to accelerate the convergence of ADMM algorithm. Considering the insufficient accessibility of true data values, Ouyang et al. (2013) presented a stochastic ADMM algorithm aiming to minimize non-smooth composite objective functions.

3.4 Secure Multiparty Computation

The previous sections introduced federated technologies reducing the privacy risks through model decomposition so that we build global models only on locally aggregated statistics (without accessing the patient-level data). To further mitigate the privacy risk, we need to ensure the confidentiality of transmitted summary statistics in each institution as well. This can be achieved using Secure Multiparty Computation (SMC), which ensures computation and communication security via advance cryptographic protocols. Many state-of-the-art SMC schemes are based upon the idea of translating an algorithm to a binary circuit. Scalable SMC protocols like Yao's garbled circuit (Yao 1982) could represent arbitrary functions with a Boolean circuit to enable masking of inputs and outputs of each gate for sensitive information. In federated data analysis, two architectures are widely considered to integrate secure multiparty computation protocols, namely spoke-hub and peer-to-peer architectures, as shown in Fig. 3.4. The spoke-hub architecture requires one or more non-colluding institution to perform securely computation and analysis based on the data collected from the other institutions, while the peer-to-peer architecture allows secure exchange of encrypted data in a decentralized manner. In this section, we will introduce some secure multiparty communication protocols for regression, classification and evaluation tasks.

3.4.1 Regression

Fienberg et al. (2006) proposed privacy-preserving binary logistic regression for horizontally partitioned data with categorical covariates. Secure summation and data integration protocols were adopted to securely compute the maximum likelihood

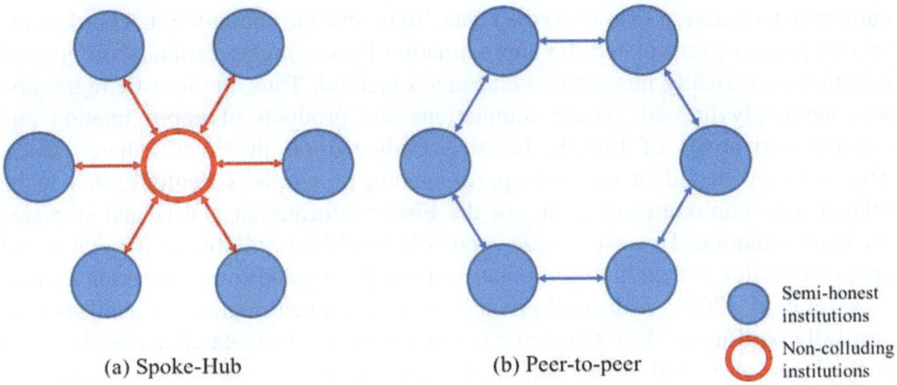

Fig. 3.4 Illustration of architectures for federated data analysis using secure multiparty computation (SMC) protocols. (**a**) Spoke-Hub architecture. The non-colluding institution is required to perform global computation and operation based the encrypted data collected from the other institutions; (**b**) Peer-to-peer architecture. All the institutions exchanged and aggregated encrypted data with each other in a successive manner to achieve federated data analysis

and perform contingency table analysis for log-linear models. Under the categorical covariates, the logistic regression model could be constructed from the log-linear models in a distributed and privacy-preserving fashion. Slavkovic and Nardi (2007) presented a secure logistic regression approach on horizontally and vertically partitioned data without actually combining them. Secure multiparty computation protocols were developed for the Newton-Raphson method to solve binary logistic regression with quantitative covariates. At each iteration, multiparty secure matrix summation and product protocols were employed to compute the gradients and Hessian matrix. Here, the inverse of Hessian was derived by recursively performing secure matrix product over its sub-block matrices. The proposed protocol would protect the privacy of each institution under the secure multiparty protocols, when the matrices of raw data are all with a dimensionality greater than one. However, Fienberg et al. (2009) showed that it would lead to serious privacy breach when intermediary results are shared among multiple institutions in numerous iterations, even though they can be protected by mechanism like encryption with random shares. Nardi et al. (2012) developed a secure protocol to fit logistic regression over vertically partitioned data based on maximum likelihood estimation (MLE). In comparison to previous works, it enhanced the security by that only providing the final results for private computation, as there would be a chance to compromise the confidentiality of raw data held by each institution from the shared intermediate values. Two protocols were proposed for approximating logistic function using existing cryptographic primitives including secret sharing, secure summation and product with random shares, secure interval membership evaluation and secure matrix inversion. The first protocol approximated the logistic function with the summation of step functions. However, this protocol would be computationally prohibitive for high-dimensional large-scale problems due to the secret sharing and comparison operation over encrypted data. To relieve the computational burden, the second protocol formulated the approximation by solving an ordinary differential equation numerically integrated with Euler's method. Thus, the logistic regression was iteratively fit with secure summations and products of approximation and average derivatives of logistic function evaluated on the intermediate values. The accuracy bound of the two approximation protocols is demonstrated to be related with minimum eigenvalue of the Fisher information matrix and step size for approximation. However, these protocols would be prohibitive for real-world applications due to their high computational complexity and communication rounds.

Sanil et al. (2004) addressed privacy-preserving linear regression analysis over vertically partitioned data. Quadratic optimization was formulated derive the exact coefficients of global regression model over multiple institutions. Distributed computation of regression coefficients was achieved by implementing Powell's algorithm under the secure multiparty computation framework. At each iteration, each institution updated its local regression coefficients and its own subset of search directions. Subsequently, a common vector was generated by aggregating the products of local attributes and search directions using secure summation protocol for computation in next iteration. Finally, the proposed algorithm obtained the global coefficients and the vector of residuals. Thus, global linear regression could be made based on the entire dataset collected from all the institutions without actually exchanging the raw data owned by each institution. Moreover, basic goodness-of-fit diagnostics could be performed with the coefficient of determination that measured the strength of the linear relationship. However, this algorithm would be unrealistic due to the assumption that the institution holding the response attribute should share it with the other institutions. Karr et al. (2009) proposed a protocol for secure linear regression over data vertically distributed across multiple institutions. The proposed protocol was able to estimate the regression coefficients and related statistics like standard error in a privacy-preserving manner. In distributed optimization, multiple institutions collaborated to calculate the off-diagonal blocks of the global covariance matrix using secure matrix products, while the diagonal blocks were obtained based on the corresponding local attributes. The global covariance matrix was shared among multiple institutions for secure linear regression and statistical analyses. Moreover, model diagnostic measures based on residuals can be similarly derived from the global covariance matrix using secure summation and matrix products protocols. Remarkably, the proposed protocol could be generalized to a variety of regression models, e.g. weighted least squares regression, stepwise regression and ridge regression, under the constraint that sample means and co-variances are sufficient statistics.

3.4.2 Classification

Naïve Bayes classifier is an effective Bayes learning method with consistent and reasonable performance, which is commonly adopted as benchmark for classification methods to be evaluated. Vaidya and Clifton (2003b) developed a privacy-preserving

Naive Bayes classifier for vertically partitioned data, where multiple institutions collaborated to achieve classification with random shares of the global model. In the proposed Bayes classifier, each institution is only required to share the class of each instance, rather than the distribution of classes or attribute values. Secure protocols were developed for training and classification under the secure multiparty computation framework. In training, random shares of the conditionally independent probabilities for nominal and numeric attributes were computed for model parameter estimation. To be concrete, the probability estimates for classes of instances given the attributes were computed for shares of nominal attributes. For numeric attributes, mean and variance for the probability density function of Normal distribution are required. For each class and attribute value, the shares can be obtained from the institutions owning them. For evaluation, each new instance was classified by maximizing its posterior probability using Bayes theorem under the assumption of conditionally independent attribute values. Here, the probabilities of class conditioned on attributes are derived based on the model parameters. Furthermore, it is demonstrated that these protocols for training and evaluation are able to securely compute shares of nominal and numeric attributes and classify the classes, respectively. Furthermore, Vaidya et al. (2008) introduced secure logarithm primitives from secure multiparty computation for naïve Bayes classification over horizontally partitioned data. The model parameters for normal and numeric attributes were directly computed based on the local counts using secure sum protocols, as each institution held all the attributes required for classifying an instance. Therefore, classification could be made locally for each institution.

K-means clustering is popular for cluster analysis, which partitions observations into the clusters with the nearest means. For data vertically distributed across multiple institutions, privacy-preserving K-means clustering (Vaidya and Clifton 2003a) was studied to perform clustering without sharing raw data. Using the proposed K-means clustering algorithm, each institution can obtain its projection of the cluster means and learn the cluster assignment of each record without revealing its exact attributes. Since high dimensional problem cannot be simply decomposed into a combination of lower dimensional problems for each institution, cooperation between multiple institutions is required to learn the cluster that each record belongs to. To achieve the privacy-preserving clustering algorithm, secure multiparty computation framework using homomorphic encryption is introduced for multiple institutions. For common distance metrics like Euclidean and Manhattan, the distances between each record and the means of K clusters can be split over these institutions. For security, these distances were disguised with random values from a uniform random distribution and non-colluding institutions were adopted to compare the randomized distances and permute the comparison results. The secure permutation algorithm is performed in an asymmetric two-party manner according to the permutation owned by the non-colluding institution. It should be noted that the initial values of the K means were assigned to their local shares for the institutions to obtain a feasible solution. As a result, non-colluding institutions would only know the selected cluster in the permutation, while the exact attributes owned by each institution would not be disclosed to the others.

Liang et al. (2013) developed a distributed PCA algorithm to estimate the global covariance matrix for principal components. Each institution leveraged standard PCS algorithm to determine its principal components over the local data. These local principal components were exchanged and collected to obtain global covariance matrix. The proposed algorithm integrated the distributed coreset-based clustering to guarantee that the number of vectors for communication was independent of size and dimension of the federated data. It is demonstrated that the divergence between approximations on projected and original data for k-means clustering can be upper bounded. Guo et al. (2013) developed a covariance-free iterative distributed principal component analysis (CIDPCA) algorithm for vertically partitioned high-dimensional data. Instead of approximating global PCA with sampled covariance matrix, the proposed CIDPCA algorithm is designed to directly determine principal component by estimating their eigenvalues and eigenvectors. The first principal component corresponding to the maximum eigenvalues of the covariance matrix was derived by maximizing the Rayleigh quotient using gradient ascent method. The iterative method is demonstrated to converge in an exponential rate under arbitrary initial values of principal components. Subsequently, the remaining principal components could be iteratively calculated in the orthogonal complement of the subspace spanned by the previously derived principal components. In comparison to previous distributed PCA methods, it is shown to achieve higher accuracy in estimating principal components and better classification performance with a significant reduction on communication cost. This conclusion is also validated with a variety of studies over real-world datasets.

Du et al. (2004) studied multivariate statistical analysis for vertically partitioned data in the secure two-party computation framework, including secure two-party multivariate linear regression (S2-MLR) and classification (S2-MC). These problems were addressed in a privacy-preserving manner with secure two-party matrix computation based on a set of basic protocols. A new security model was proposed to lower security requirements for higher efficiency, where each institution was allowed to reveal a part of information about its raw data under the guarantee that the raw data would not be inferred from the disclosed information. To securely perform matrix computation, building blocks for matrix product, matrix inverse and matrix determinant were presented. Thus, S2-MLR and S2-MC could be securely solved with these building blocks. It is demonstrated that, in the proposed two-party multivariate statistical analysis, it would be impossible to infer the raw data owned by each institution, when less than half of its disguised matrix is revealed.

Yu et al. (2006) proposed an efficient privacy-preserving support vector machine (SVM) classification method, namely PP-SVMV, for vertically partitioned data. In the proposed method, the global SVM model was constructed from local SVM models rather than directly exchanging the local data. Thus, both local SVM models and their corresponding local data for each institution were not disclosed. For linear kernels, the global kernel matrix is computed by directly merging gram matrices from multiple institutions to solve the dual problem. This result can also be extended to ordinary non-linear kernels that can be represented by dot products of covariates, i.e. polynomial and radial basis function (RBF) kernels.

To guarantee data and model privacy, merge of local models is performed with secure summation of scalar integers and matrices. Experimental results demonstrated the accuracy and scalability of PP-SVMV in comparison to the centralized SVM over the original data. Similarly, Yu et al. (2006) presented a privacy-preserving solution to support non-linear SVM classification over horizontally partitioned data. It required that the nonlinear kernel matrices could be directly calculated based on the gram matrix. Thus, widely-used nonlinear kernel matrices like polynomial and RBF kernels can be derived from the dot products of all data pairs using the proposed solution. Secure set intersection cardinality was adopted as an equivalency to these dot products based on the data horizontally distributed across the institutions. Thus, commutative one-way hash functions were utilized to securely obtain the set intersection cardinality. The proposed method was shown to achieve equivalent classification performance in comparison to the centralized SVM classifiers. Mangasarian et al. (2008) constructed a reduced kernel matrix with the original data and a random matrix to perform classification under the protection of local data. The random kernel based SVM classifier could support both horizontally and vertically partitioned data. Yunhong et al. (2009) proposed a privacy-preserving SVM classifier without using secure multiparty computation. The proposed SVM classifier built its kernel matrix by combining local gram matrices derived from the original data owned by the corresponding institutions. It shows that local gram matrices would not reveal the original data using matrix factorization theory, as it is not unique for each institution to infer its covariates from local gram matrix. The proposed classification algorithm is developed for SVM classifier with linear and nonlinear kernels, where the accuracy of distributed classification is comparable to the ordinary global SVM classifier. Que et al. (2012) presented a distributed privacy-preserving SVM (DPP-SVM), where server/client collaborative learning framework is developed to securely estimate parameters of covariates based on the aggregated local kernel matrices from multiple institutions. For security, all the model operations are performed on the trusted server, including service layer for server/client communication, task manager for data validation and computation engine for parameter estimation.

Vaidya et al. (2008) presented a generalized privacy-preserving algorithm for building ID3 decision tree over data vertically distributed across multiple institutions. For efficient and secure classification using the ID3 decision tree, the proposed algorithm only revealed the basic structure of the tree and the specific institution responsible for decision making at each node, rather than the exact values of attributes. For each node in the tree, the basic structure includes its number of branches and the depths of its subtrees, which represented the number of distinct values for corresponding attributes. Thus, it is not necessary for each institution to introduce complex cryptographic protocol at each possible level to securely classify an instance. It should be noted that the proposed algorithm only needs to assign class attribute needs to one institution, but each interior node could learn the count of classes. Consequently, the institution owning class attribute estimated the distributions throughout the decision tree based on the derived transaction counts, which would not disclose much new information. The distribution and majority class was determined based on the cardinality of set intersection protocol. Given multiple institutions, classification of an instance was conducted by exchanging the control information based on the decision made for each node, but not the attribute values. To further enhance the efficiency of privacy-preserving ID3 algorithm, Vaidya et al. (2014) developed random decision tree (RDT) framework to fit parallel and distributed architecture. Random decision tree is desirable for privacy-preserving distributed data mining, as it can achieve equivalent effect as perturbation without diminishing the utility of information from data mining. For horizontally partitioned data, the structure of RDTs was known to all the institutions that held the same types of attributes. The RDTs were constructed by considering the accessibility of the global class distribution vector for leaf nodes. Each institution could derive its local distribution vector from its own data and submitted the encrypted versions for aggregation. When the class distribution vectors were known to all the institutions or the institution owning the RDTs, the aggregation could be directly made based on homomorphically encrypted data. If the class distribution vectors were forced to remain unrevealed, a secure electronic voting protocol was presented to make decision based on the collected encrypted local vectors. For vertically partitioned data, fully distributed trees with a specified total number were considered, so that the sensitive attribute information for each institution was not revealed. Each random tree was split among multiple institutions and constructed recursively using the *BuildTree* procedure in a distributed fashion. It is worth mentioning that this procedure does not require the transaction set. Subsequently, the statistics of each node was securely updated based on the training set from multiple institutions using additively homomorphic encryption. Similarly, instance classification was achieved by averaging the estimated probabilities from multiple RDTs in a distributed manner. The proposed RDT algorithm is secure, as neither the attribute values nor the RDT structure is shared during RDT construction and instance classification.

3.4.3 Evaluation

Sorting algorithm is essential for privacy-preserving distributed data analysis, including ranked elements query, group-level aggregation and statistical tests. In the secure multiparty computation framework, oblivious sorting algorithm can be implement by hiding the propagation of values in the sorting network or directly using sorting algorithm as a basis. Bogdanov et al. (2014) investigated four different oblivious sorting algorithms for vertically partitioned data, where two algorithms improved the existing sorting network and quicksort algorithms and the other two ones were developed to achieve low round count for short vectors and low communication cost for large inputs, respectively. For short vectors, a naive sorting protocol NaiveCompSort was presented based on oblivious shuffling of input data and vectorized comparison of shuffled data. Given large inputs, oblivious radix sorting protocol was developed as an efficient alternative. It leveraged binary count sorting algorithm to rearrange the input integer vectors based on the sorted digits in the same positions. Thus, the oblivious radix sorting protocol is efficient, as it does not require oblivious comparisons. Furthermore, optimization methods were proposed to improve the efficiency of the oblivious sorting algorithms. For example, bitwise shared representation and vectorization would allow data parallelization to reduce the communication cost and complexity for SMC. For sorting network structure, shuffling the inputs and re-using the generated network could optimize its implementation, while uniqueness transformation for comparison-based sorting protocols could avoid information leakage in the sortable vector. It should be noted that these sorting algorithms could also be generalized to support matrix sorting. The complexity and performance analysis for all the four sorting algorithms, including detailed running-time, network and memory usage, was also presented.

Makri et al. (2014) proposed a privacy-preserving statistical verification for clinical research based on the aggregated results from statistical computation. It leveraged secure multiparty computation primitives to perform evaluations for Student's and Welch's t-test, ANOVA (F-test), chi-squared test, Fisher's exact test and McNemar's test over horizontally partitioned data. The proposed statistical verification could be outsourced to a semi-host third party, namely verifiers, as no private data were exchanged during the verification process. Secure protocols based on secret sharing were utilized to compute the means and variance. Consequently, Student's t-test, Welch's test and F-test could be evaluated based on the derived means and variance. At the meantime, chi-squared test, Fisher's exact test and McNemar's test could be performed based on the frequencies in the contingency table, which would not reveal the individual records in the group of data held by certain institutions. The proposed mechanism is proven to protect the data security and privacy in the semi-honest model using secure multiparty computation protocols from Shamir's secret sharing.

3.5 Asynchronous Optimization

In Sect. 3.2, we discussed the Newton-Raphson method and ADMM framework for distributed optimization in federated data analysis. In these methods, all institutions are commonly synchronized for computation at each iteration. However, these synchronous methods would fail due to unexpected communication delay and interruption under practical network conditions. Asynchronous optimization algorithms have been developed to make distributed and parallel optimization based on local updates from institutions with various delays, e.g. institutions with different computation and processing speeds and access frequencies. Thus, server and institutions can proceed without waiting for prerequisite information as in synchronous methods.

3.5.1 Asynchronous Optimization Based on Fixed-Point Algorithms

Chazan and Miranker (1969) firstly proposed asynchronous parallel method to solve linear systems of equations with chaotic relaxation. The proposed method developed totally (infinite delay) and partially (bounded delay) asynchronous parallel algorithms to improve the efficiency of iterative schemes based on limited number of updates with a guarantee of convergence. Baudet (1978) extended totally asynchronous chaotic relaxation to solve fixed-point problems with contracting operators. The proposed methods were guaranteed with the sufficient condition of contracting operators for convergence to fixed point.

Bertsekas and Tsitsiklis (1989) introduced gradient-like optimization based on totally and partially asynchronous parallel algorithms in unconstrained and constrained optimization problems. Totally asynchronous gradient algorithm was studied for optimization problems with contraction mapping under the metric of weighted maximum norm. Its convergence is guaranteed when the diagonal dominance condition is satisfied for its Hessian matrix. Provided finite asynchrony for communication and computation, the partially asynchronous parallel optimization is shown to converge when the step size for iterative optimization is sufficiently small in comparison to the asynchrony measure. Tseng (1991) analyzed the rate of convergence for partially asynchronous gradient projection algorithm. Given an objective function with Lipschitz continuous gradient, a linear convergence rate can be achieved, when its isocost surfaces can be discriminated and upper Lipschitz property holds for at least one of its multivalued functions. Tai and Tseng (2002) studied the rate of convergence for strongly convex optimization problems based on asynchronous domain decomposition methods.

Recently, Peng et al. (2016) developed an algorithmic framework AROCK for asynchronous optimization related to non-expansive operator T with fixed point. Under the assumption of atomic update, AROCK randomly selected a component of primal variables x and updated it without memory locking by using sub-operator $S = I - T$ based on the newly updated variables $\widehat{x}$ within a bounded delay.

$$x^{(t+1)} = x^{(t)} - \eta S_{i_t}\left(\widehat{x}^{(t)}\right) \tag{3.5}$$

Here i_t is the random variable indicating the variable for atomic writing at time t. AROCK is demonstrated to achieve convergence under finite-dimensional operator. It is also guaranteed to converge to the fixed point with a linear rate, when the difference between the identity matrix and non-expansive operator is quasi-strongly monotone. Furthermore, Peng et al. (2016) studied coordinate update methods for specific optimization problems with coordinate friendly operators. Remarkably, coordinate update methods could derive and leverage a variety of coordinate friendly operators to make asynchronous realization for coordinate descent, proximal gradient, ADMM and primal-dual methods.

3.5.2 Asynchronous Coordinate Gradient Descent

A series of partially asynchronous parallel methods have been developed for coordinate gradient descent. Niu et al. (2011) presented an asynchronous strategy HOGWILD! to perform stochastic gradient descent in parallel without memory locking. The presented strategy enabled multiple processors to make gradient updates in the shared memory. For sparse learning problem, HOGWILD! achieves a sublinear convergence rate, as the error led by gradient update could be trivial. Liu et al. (2015) proposed an asynchronous stochastic coordinate descent algorithm AsySCD to improve parallel minimization of smooth convex functions. When essential strong convexity condition is satisfied, the minimization can be solved in a linear convergence rate. Later, Liu and Wright (2015) developed an asynchronous parallel algorithm AsySPCD to minimize the composite objective function composed of smooth convex functions using stochastic proximal coordinate descent. AsySPCD considered the scenario that data were simultaneously accessed and updated by multiple institutions. Under the relaxed optimal strong convexity condition, the proposed algorithm can achieve a linear convergence rate. Remarkably, both AsySCD and AsySPCD could be accelerated in a nearly linear rate under enough number of processors (institutions). Hsieh et al. (2015) developed a family of parallel algorithms PASSCoDe for stochastic dual coordinate descent based on the LIBLINEAR software to efficiently solve -regularized empirical risk minimization problems. At each iteration, PASSCoDe allowed each institution to randomly select a dual variable to update the primal variables with coordinate descent. Under asynchronous settings, three parallel algorithms were developed to handle the possible violation of primal-dual relationship for -regularized minimization. PASSCoDe-Lock and PASSCoDe-Atomic were proposed to maintain the primal-dual relationship with memory locking and atomic writing. PASSCoDe-Wild achieved nearly optimal performance with no locking and atomic operations by performing backward error analysis for memory conflicts.

3.5.3 Asynchronous Alternating Direction Method of Multipliers

Recently, asynchronous ADMM framework with consensus constraints has been studied for distributed optimization under practical network conditions. Similar to SMC, two architectures have been developed to realize asynchronous optimization under the ADMM framework. In the peer-to-peer architecture, institutions are interconnected with their delay or inconsistency indicated by asynchronous clock, so that optimization is made in a decentralized manner. Iutzeler et al. (2013) adopted randomized Gauss-Seidel iterations of Douglas-Rachford operator to develop a generalized asynchronous ADMM framework over multiple institutions. The proposed asynchronous optimization methods allowed partially activation of isolated decision variables (without involved in the active constraints) to make distributed minimization without coordinating all the institutions. Based on the monotone operator theory, a randomized ADMM algorithm was formulated for various institutions with different frequencies of activation. It is demonstrated to converge to the optimal minimum under the connectivity assumption for the graph derived from institutions. Wei and Ozdaglar (2013) proposed an asynchronous algorithm based on ADMM for separable linearly constrained optimization over multiple institutions. It established a general ADMM-based formulation under asynchronous settings to iteratively determine a pair of random variables representing active constraints and coupled decision variables and utilize them to update primal and dual variables. Relating the iterative results under asynchronous setting to those based on all the institutions, the proposed asynchronous algorithm is demonstrated to converge with a rate inversely proportional to the number of iteration.

Spoke-hub architecture maintains a master node (server) to make consensus over the primal variables distributed across multiple institutions. Zhang and Kwok (2014) proposed an asynchronous ADMM algorithm that introduced partial barrier and bounded delay to control asynchrony and guarantee convergence. The proposed async-ADMM algorithm updated the local primal and dual variables for each institution with most recent consensus variable. At the meantime, a master node required only a partial subset of updated primal and dual variables with bounded delay to update the consensus variable in an asynchronous manner. Under the bounded delay, institutions with different computation and processing speed can be properly counted for global consensus. The convergence property can be held for a composition of non-smooth convex local objective functions. Hong (2014) presented a generalized proximal ADMM framework to solve non-convex optimization problems with nonsmooth objective functions in an asynchronous and distributed manner. An incremental algorithm was developed for iterative asynchronous optimization with varying active constraints. At each iteration, the server updated all the primal and dual variables with the most recent gradients of local objective functions from institutions. The consensus variables were derived with proximal operation to compute the gradient in each institution. It should be noted that the server only required a subset of institutions with newly-computed gradients to update the primal and dual variables. Thus, the proposed Async-PADMM algorithm is robust to handle the asynchrony led by communication delay and interruption.

3.6 Discussion and Conclusion

In this chapter, we review the privacy-preserving federated data analysis algorithms in the context of biomedical research. Federated data analysis algorithms aim to facilitate biomedical research and clinical treatments based on large-scale healthcare data collected from various organizations with full compliance with the institutional policies and legislation. Federated models are developed to bring computation to horizontally or vertically partitioned data rather than explicitly transfer them to a central repository. Server/client and decentralized architectures are established to estimate global model parameters and perform statistical test based on the aggregated intermediary results from multiple institutions. As a result, the patient-level data would not be inferred, when they are not distributed in an extremely unbalanced way. Distributed optimization for federated data analysis are achieved and generalized by integrating the Newton-Raphson method and ADMM framework. For generalized linear regression, the Newton-Raphson method is demonstrated to achieve equivalent accuracy in model parameter estimation and statistical tests in comparison to its centralized counterparts. ADMM framework splits the objective function for distributed optimization with auxiliary variables to formulate decomposable cost function for multiple institutions. Since it is not necessarily required to derive the global consensus from all the local data, ADMM is possible to support decentralized analysis without data server. The separate estimates from the institutions are guaranteed to converge under the consensus constraints based on auxiliary variables. To further protect the communication among servers and clients, secure multiparty communication protocols were adopted for the applications like regression, classification and evaluation over distributed data. It protects the process of computation and communication for improved data security and privacy. Finally, asynchronous parallel optimization can be incorporated to handle real-world network conditions and improve the efficiency for parallel computation.

Privacy-preserving federated data analysis can be further studied for improved security and efficiency and generic formulation for practice. One major challenge for the federated data analysis is distributed model parameter estimation and statistical tests for vertically partitioned data. Although federated models are shown to be effective for logistic regression models and SVM classifiers over vertically partitioned data, an analogical formulation based on dual optimization would fail for the analysis tasks with non-separable objective functions like the Cox proportional hazard model. Since each institution only holds a portion of covariates, the kernel matrix cannot be explicitly decomposed across the institutions. Under such cases, ADMM tends to be a promising alternative. However, ADMM commonly suffers from low convergence rate and consequent high communication cost to solve the iterative optimization with decomposable subproblems. Two-loop recursion would be required to solve the optimization problem based on augmented Lagrangian functions and regularized optimization problem in the server or client side. Thus, it is imperative to develop privacy-preserving algorithms to consider both accuracy and efficiency for distributed analysis over vertically partitioned data.

The second challenge is the communication cost between institutions and server. A server is required to collect the data information or computation results from each institution and output the final results, while all the institutions collaborate in a round robin manner without a server in the second architecture. El Emam et al. (2013) showed that privacy risk might exist even for exchanging intermediary results that are aggregated from a local patient cohort. Taking horizontally partitioned data for example, there is a chance to identify sensitive patient-level data from gram matrix for linear models or indicating variables, covariance matrix, intermediary results from multiple iterations, or models. On the other hand, encryption methods can prevent information leakage in exchanging data but will noticeably affect computational and storage efficiency. For example, homomorphic encryption would significantly increase computation and storage costs, when securely outsourcing data and computations. To protect all raw data and intermediary results, SMC-based distributed analysis methods require a high communications cost to deploy oblivious transfer protocols. Moreover, many existing encryption-based methods cannot be scaled up to handle large biomedical data. As a result, proper protocol or mechanism is crucial to balance security and efficiency in institution/server communication.

Acknowledgement This research was supported by the Patient-Centered Outcomes Research Institute (PCORI) under contract ME-1310-07058, the National Institute of Health (NIH) under award number R01GM118574, R01GM118609, R00HG008175, R21LM012060, and U01EB023685.

References

Act, D. P. (1998). *Data protection act*. London: The Stationery Office.

Baudet, G. M. (1978). Asynchronous iterative methods for multiprocessors. *Journal of the ACM, 25*(2), 226–244.

Bertsekas, D. P., & Tsitsiklis, J. N. (1989). *Parallel and distributed computation: numerical methods* (Vol. 23). Englewood Cliffs, NJ: Prentice hall.

Bogdanov, D., Laur, S., & Talviste, R. (2014). A practical analysis of oblivious sorting algorithms for secure multi-party computation. In K. Bernsmed & S. Fischer-Hübner (Eds.), *Secure IT systems* (pp. 59–74). Springer International Publishing.

Boyd, S., Parikh, N., Chu, E., Peleato, B., & Eckstein, J. (2011). Distributed optimization and statistical learning via the alternating direction method of multipliers. *Foundations and Trends in Machine Learning, 3*(1), 1–122.

Brown, J., Balaconis, E., Mazza, M., Syat, B., Rosen, R., Kelly, S., et al. (2012). PS1-46: HMORNnet: Shared infrastructure for distributed querying by HMORN collaboratives. *Clinical Medicine & Research, 10*(3), 163.

Chambless, L. E., & Diao, G. (2006). Estimation of time-dependent area under the ROC curve for long-term risk prediction. *Statistics in Medicine, 25*(20), 3474–3486.

Chazan, D., & Miranker, W. (1969). Chaotic relaxation. *Linear Algebra and Its Applications, 2*(2), 199–222.

Cox, D. R. (1972). Regression models and life-tables. *Journal of the Royal Statistical Society, Series B, Statistical Methodology, 34*(2), 187–220.

Du, W., Han, Y. S., & Chen, S. (2004). Privacy-preserving multivariate statistical analysis: Linear regression and classification. In *Proceedings of the 2004 SIAM international conference on data mining* (pp. 222–233).

El Emam, K., Samet, S., Arbuckle, L., Tamblyn, R., Earle, C., & Kantarcioglu, M. (2013). A secure distributed logistic regression protocol for the detection of rare adverse drug events. *Journal of the American Medical Informatics Association: JAMIA, 20*(3), 453–461.

Fienberg, S. E., Fulp, W. J., Slavkovic, A. B., & Wrobel, T. A. (2006). "Secure" log-linear and logistic regression analysis of distributed databases. In J. Domingo-Ferrer & L. Franconi (Eds.), *Privacy in statistical databases* (pp. 277–290). Berlin: Springer.

Fienberg, S. E., Nardi, Y., & Slavković, A. B. (2009). Valid statistical analysis for logistic regression with multiple sources. In C. S. Gal, P. B. Kantor, & M. E. Lesk (Eds.), *Protecting persons while protecting the people* (pp. 82–94). Berlin: Springer.

Forero, P. a., & Giannakis, G. B. (2010). Consensus-based distributed support vector machines. *Journal of Machine Learning Research: JMLR, 11*, 1663–1707.

Goldfarb, D., Ma, S., & Scheinberg, K. (2012). Fast alternating linearization methods for minimizing the sum of two convex functions. *Mathematical Programming A Publication of the Mathematical Programming Society, 141*(1–2), 349–382.

Guo, Y.-F., Lin, X., Teng, Z., Xue, X., & Fan, J. (2013). A covariance-free iterative algorithm for distributed principal component analysis on vertically partitioned data. *Pattern Recognition, 45*(3), 1211–1219.

Gymrek, M., McGuire, A. L., Golan, D., Halperin, E., & Erlich, Y. (2013). Identifying personal genomes by surname inference. *Science, 339*(6117), 321–324.

Health Insurance Portability and Accountability Act (HIPAA). (n.d.). Retrieved from http://www.hhs.gov/ocr/hipaa.

Homer, N., Szelinger, S., Redman, M., Duggan, D., Tembe, W., Muehling, J., et al. (2008). Resolving individuals contributing trace amounts of DNA to highly complex mixtures using high-density SNP genotyping microarrays. *PLoS Genetics, 4*(8), e1000167.

Hong, M. (2014). *A distributed, asynchronous and incremental algorithm for nonconvex optimization: An ADMM based approach. arXiv [cs.IT]*. Retrieved from http://arxiv.org/abs/1412.6058.

Hosmer, D. W., Lemeshow, S., & Sturdivant, R. X. (2013). *Applied logistic regression*. New York, NY: Wiley.

Hsieh C-J., Yu H-F., & Dhillon I. S. (2015). Passcode: Parallel asynchronous stochastic dual coordinate descent. In *Proceedings of the 32nd international conference on machine learning (ICML-15)* (pp. 2370–2379).

Iutzeler, F., Bianchi, P., Ciblat, P., & Hachem, W. (2013). Asynchronous distributed optimization using a randomized alternating direction method of multipliers. In *52nd IEEE conference on decision and control* (pp. 3671–3676). ieeexplore.ieee.org.

Jiang, W., Li, P., Wang, S., Wu, Y., Xue, M., Ohno-Machado, L., & Jiang, X. (2013). WebGLORE: A web service for Grid LOgistic REgression. *Bioinformatics, 29*(24), 3238–3240.

Kantarcioglu, M. (2008). A survey of privacy-preserving methods across horizontally partitioned data. In C. C. Aggarwal & P. S. Yu (Eds.), *Privacy-preserving data mining* (pp. 313–335). New York, NY: Springer.

Kaplan, E. L., & Meier, P. (1958). Nonparametric estimation from incomplete observations. *Journal of the American Statistical Association, 53*(282), 457–481.

Karr, A. F., Lin, X., Sanil, A. P., & Reiter, J. P. (2009). Privacy-preserving analysis of vertically partitioned data using secure matrix products. *Journal of Official Statistics, 25*(1), 125–138.

Lafky, D. (2010). The Safe Harbor method of de-identification: An empirical test. In *Fourth national HIPAA summit west*.

Liang, Y., Balcan, M. F., & Kanchanapally, V. (2013). Distributed PCA and k-means clustering. In *The big learning workshop at NIPS 2013* (pp. 1–8).

Ling, Q., Shi, W., Wu, G., & Ribeiro, A. (2015). DLM: Decentralized linearized alternating direction method of multipliers. *IEEE Transactions on Signal Processing: A Publication of the IEEE Signal Processing Society, 63*(15), 4051–4064.

Liu, J., & Wright, S. J. (2015). Asynchronous stochastic coordinate descent: Parallelism and convergence properties. *SIAM Journal on Optimization, 25*(1), 351–376.

Liu, J., Wright, S. J., Ré, C., Bittorf, V., & Sridhar, S. (2015). An asynchronous parallel stochastic coordinate descent algorithm. *Journal of Machine Learning Research JMLR, 16*, 285–322. Retrieved from http://www.jmlr.org/papers/volume16/liu15a/liu15a.pdf.

Li, Y., Jiang, X., Wang, S., Xiong, H., & Ohno-Machado, L. (2016). VERTIcal Grid lOgistic regression (VERTIGO). *Journal of the American Medical Informatics Association, 23*(3), 570–579.

Lu, C.-L., Wang, S., Ji, Z., Wu, Y., Xiong, L., & Jiang, X. (2014). WebDISCO: A web service for distributed cox model learning without patient-level data sharing. *Journal of the American Medical Informatics Association, 22*(6), 1212–1219.

Makri, E., Everts, M. H., de Hoogh, S, Peter, A., H. op den Akker., Hartel, P., & Jonker, W. (2014). Privacy-preserving verification of clinical research (pp. 481–500). Presented at the Sicherheit 2014 - Sicherheit, Schutz und Zuverlässigkeit, Beiträge der 7. Jahrestagung des Fachbereichs Sicherheit der Gesellschaft für Informatik e.V. (GI), Bonn, Germany: Gesellschaft für Informatik.

Mangasarian, O. L., Wild, E. W., & Fung, G. M. (2008). Privacy-preserving classification of vertically partitioned data via random kernels. *ACM Transactions on Knowledge Discovery from Data, 2*(3), 12:1–12:16.

Mateos, G., Bazerque, J. A., & Giannakis, G. B. (2010). Distributed sparse linear regression. *IEEE Transactions on Signal Processing: A Publication of the IEEE Signal Processing Society, 58*(10), 5262–5276.

Mateos, G., & Giannakis, G. B. (2012). Distributed recursive least-squares: Stability and performance analysis. *IEEE Transactions on Signal Processing: A Publication of the IEEE Signal Processing Society, 60*(612), 3740–3754.

Mateos, G., & Schizas, I. D. (2009). Distributed recursive least-squares for consensus-based in-network adaptive estimation. *IEEE Transactions on Signal Processing, 57*(11), 4583–4588. Retrieved from http://ieeexplore.ieee.org/xpls/abs_all.jsp?arnumber=5061644.

McGraw, D. (2008). Why the HIPAA privacy rules would not adequately protect personal health records: Center for Democracy and Technology (CDT) brief. http, 1–3.

Mokhtari, A., Shi, W., Ling, Q., & Ribeiro, A. (2016). DQM: Decentralized quadratically approximated alternating direction method of multipliers. *IEEE Transactions on Signal Processing: A Publication of the IEEE Signal Processing Society, 64*(19), 5158–5173.

Nardi, Y., Fienberg, S. E., & Hall, R. J. (2012). Achieving both valid and secure logistic regression analysis on aggregated data from different private sources. *Journal of Privacy and Confidentiality, 4*(1), 9.

Niu, F., Recht, B., Re, C., & Wright, S. (2011). Hogwild: A lock-free approach to parallelizing stochastic gradient descent. In J. Shawe-Taylor, R. S. Zemel, P. L. Bartlett, F. Pereira, & K. Q. Weinberger (Eds.), *Advances in neural information processing systems 24* (pp. 693–701). Red Hook, NY: Curran Associates, Inc.

Ohno-Machado, L., Agha, Z., Bell, D. S., Dahm, L., Day, M. E., & Doctor, J. N. (2014). pSCANNER: Patient-centered scalable national network for effectiveness research. *Journal of the American Medical Informatics Association: JAMIA, 21*(4), 621–626.

O'Keefe, C. M., Sparks, R. S., McAullay, D., & Loong, B. (2012). Confidentialising survival analysis output in a remote data access system. *Journal of Privacy and Confidentiality, 4*(1), 6.

Ouyang, H., He, N., Tran, L., & Gray, A. (2013). Stochastic alternating direction method of multipliers. In *Proceedings of the 30th international conference on machine learning* (pp. 80–88).

PCORnet: The National Patient-Centered Clinical Research Network. (n.d.). Retrieved from http://www.pcornet.org/.

Peng, Z., Wu, T., Xu, Y., Yan, M., & Yin, W. (2016a). Coordinate-friendly structures, algorithms and applications. *Annals of Mathematical Sciences and Applications, 1*(1), 57–119.

Peng, Z., Xu, Y., Yan, M., & Yin, W. (2016b). ARock: An algorithmic framework for asynchronous parallel coordinate updates. *SIAM Journal of Scientific Computing, 38*(5), A2851–A2879.

PopMedNet. (n.d.). Retrieved from http://www.popmednet.org/.

Que, J., Jiang, X., & Ohno-Machado, L. (2012). A collaborative framework for distributed privacy-preserving support vector machine learning. In *AMIA annual symposium* (pp. 1350–1359). Chicago, IL: AMIA.

Sanil, A. P., Karr, A. F., Lin, X., & Reiter, J. P. (2004). Privacy Preserving regression modelling via distributed computation. In *Proceedings of the tenth ACM SIGKDD international conference on knowledge discovery and data mining* (pp. 677–682). New York, NY, USA: ACM.

Scardapane, S., Wang, D., & Panella, M. (2016). A decentralized training algorithm for Echo State Networks in distributed big data applications. *Neural Networks: The Official Journal of the International Neural Network Society, 78*, 65–74.

Schizas, I. D., & Aduroja, A. (2015). A distributed framework for dimensionality reduction and denoising. *IEEE Transactions on Signal Processing: A Publication of the IEEE Signal Processing Society, 63*(23), 6379–6394.

Shi, H., Jiang, C., Dai, W., Jiang, X., Tang, Y., Ohno-Machado, L., & Wang, S. (2016). Secure multi-pArty computation grid LOgistic REgression (SMAC-GLORE). *BMC Medical Informatics and Decision Making, 16*(Suppl 3), 89.

Slavkovic, A. B., & Nardi, Y. (2007). "Secure" logistic regression of horizontally and vertically partitioned distributed databases. *Seventh IEEE International*. Retrieved from http://ieeexplore.ieee.org/xpls/abs_all.jsp?arnumber=4476748.

Sweeney, L., Abu, A., & Winn, J. (2013). Identifying participants in the personal genome project by name. Available at SSRN 2257732. https://doi.org/10.2139/ssrn.2257732.

Tai, X.-C., & Tseng, P. (2002). Convergence rate analysis of an asynchronous space decomposition method for convex minimization. *Mathematics of Computation, 71*(239), 1105–1135.

Tseng, P. (1991). On the rate of convergence of a partially asynchronous gradient projection algorithm. *SIAM Journal on Optimization, 1*(4), 603–619.

Vaidya, J. (2008). A survey of privacy-preserving methods across vertically partitioned data. In C. C. Aggarwal & P. S. Yu (Eds.), *Privacy-preserving data mining* (pp. 337–358). New York, NY: Springer.

Vaidya, J., & Clifton, C. (2003a). Privacy-preserving K-means clustering over vertically partitioned data. In *Proceedings of the ninth ACM SIGKDD international conference on knowledge discovery and data mining* (pp. 206–215). New York, NY: ACM.

Vaidya, J., & Clifton, C. (2003b). Privacy preserving naive bayes classifier for vertically partitioned data. In *Proceedings of the ninth acm SIGKDD international conference on knowledge discovery and data mining.*

Vaidya, J., Clifton, C., Kantarcioglu, M., & Patterson, A. S. (2008a). Privacy-preserving decision trees over vertically partitioned data. *ACM Transactions on Knowledge Discovery from Data, 2*(3), 14:1–14:27.

Vaidya, J., Kantarcıoğlu, M., & Clifton, C. (2008b). Privacy-preserving naive bayes classification. *Journal on Very Large Data Bases, 17*(4), 879–898.

Vaidya, J., Shafiq, B., Fan, W., Mehmood, D., & Lorenzi, D. (2014). A random decision tree framework for privacy-preserving data mining. *IEEE Transactions on Dependable and Secure Computing, 11*(5), 399–411.

Vaidya, J., Shafiq, B., Jiang, X., & Ohno-Machado, L. (2013). Identifying inference attacks against healthcare data repositories. In *AMIA joint summits on translational science proceedings. AMIA summit on translational science, 2013*, (pp. 262–266).

Wang, H., & Banerjee, A. (2012). Online alternating direction method. In J. Langford & J. Pineau (Eds.), *Proceedings of the 29th international conference on machine learning (ICML-12)* (pp. 1119–1126). New York, NY: Omnipress.

Wang, R., Li, Y. F., Wang, X., Tang, H., & Zhou, X. (2009). Learning your identity and disease from research papers: information leaks in genome wide association study. In *Proceedings of the 16th ACM conference on computer and communications security* (pp. 534–544). New York, NY, USA: ACM.

Wang, S., Jiang, X., Wu, Y., Cui, L., Cheng, S., & Ohno-Machado, L. (2013). EXpectation propagation LOgistic REgRession (EXPLORER): Distributed privacy-preserving online model learning. *Journal of Biomedical Informatics, 46*(3), 1–50.

Weibull, W. (1951). A statistical distribution function of wide applicability. *Journal of Applied Mathematics, 18*(3), 293–297.

Wei, E., & Ozdaglar, A. (2013). On the O(1 = k) convergence of asynchronous distributed alternating direction method of multipliers. In *Global conference on signal and information processing (GlobalSIP), 2013 IEEE* (pp. 551–554). ieeexplore.ieee.org.

Wu, Y., Jiang, X., Kim, J., & Ohno-Machado, L. (2012a). Grid binary LOgistic REgression (GLORE): Building shared models without sharing data. *Journal of the American Medical Informatics Association, 2012*(5), 758–764.

Wu, Y., Jiang, X., & Ohno-Machado, L. (2012b). Preserving institutional privacy in distributed binary logistic regression. In *AMIA annual symposium* (pp. 1450–1458). Chicago, IL: AMIA.

Wu, Y., Jiang, X., Wang, S., Jiang, W., Li, P., & Ohno-Machado, L. (2015). Grid multi-category response logistic models. *BMC Medical Informatics and Decision Making, 15*(1), 1–10.

Yao, A. C. (1982). Protocols for secure computations. In *Foundations of computer science, 1982. SFCS '08. 23rd annual symposium on* (pp. 160–164). ieeexplore.ieee.org.

Yu, H., Jiang, X., & Vaidya, J. (2006a). Privacy-preserving SVM using nonlinear kernels on horizontally partitioned data. In *Proceedings of the 2006 ACM symposium on applied computing* (pp. 603–610). New York, NY: ACM.
Yu, H., Vaidya, J., & Jiang, X. (2006b). Privacy-preserving SVM classification on vertically partitioned data. *Lecture Notes in Computer Science, 3918 LNAI*, 647–656.
Yunhong, H., Liang, F., & Guoping, H. (2009). Privacy-Preserving svm classification on vertically partitioned data without secure multi-party computation. In *2009 fifth international conference on natural computation* (Vol. 1, pp. 543–546). ieeexplore.ieee.org.
Yu, S., Fung, G., Rosales, R., Krishnan, S., Rao, R. B., Dehing-Oberije, C., & Lambin, P. (2008). Privacy-preserving cox regression for survival analysis. In *Proceedings of the 14th ACM SIGKDD international conference on Knowledge discovery and data mining* (pp. 1034–1042). New York, NY: ACM.
Zhang, R., & Kwok, J. (2014). Asynchronous distributed ADMM for consensus optimization. In *Proceedings of the 31st international conference on machine learning* (pp. 1701–1709).

Chapter 4
Word Embedding for Understanding Natural Language: A Survey

Yang Li and Tao Yang

4.1 Introduction

Natural language understanding from text data is an important field in Artificial Intelligence. As images and acoustic waves can be mathematically modeled by analog or digital signals, we also need a way to represent text data in order to process it automatically. For example, the sentence "The cat sat on the mat." can not be processed or understood directly by the computer system. The easiest way is to represent is through a sparse discrete vector $\{(i_{cat}, 1), (i_{mat}, 1), (i_{on}, 1), (i_{sit}, 1), (i_{the}, 2)\}$, where i_w denotes the index of word w in the vocabulary. This is called one-hot embedding. However, there are several disadvantages for this simple model. First, it generates high dimensional vectors whose length depends on the volume of the vocabulary, which is usually very large. Meanwhile, the semantic relationship between words (e.g., "sit on") cannot be reflected by these separate counts. Due to the subjectivity of languages, the meaning of word (phrase) varies in different contexts. This makes it a more challenging task to automatically process text data.

A goal of language modeling is to learn the joint probability that a given word sequence occurs in texts. A larger probability value indicates that the sequence is more commonly used. One candidate solution to involve the relationship between words is called "n-gram" model, where n is the hyperparameter manually chosen. The "n-gram" model takes into consideration the phrases of n consecutive words. In the example above, "the cat", "cat sat", "sat on", "on the", "the mat" will be counted, if we set $n = 2$. The disadvantage of "n-gram" model is that it is constrained by the parameter n. Moreover, the intrinsic difficulty of language modeling is that: a word sequence that is used to test the model is likely to be different from all the sequences in the training set (Bengio et al. 2003). To make it more concrete, suppose

Y. Li • T. Yang (✉)
School of Automation, NorthWestern Polytechnical University, Xi'an, Shanxi 710072, P.R. China
e-mail: liyangnpu@mail.nwpu.edu.cn; yangtao107@nwpu.edu.cn

S. Srinivasan (ed.), *Guide to Big Data Applications*, Studies in Big Data 26,
DOI 10.1007/978-3-319-53817-4_4

we want to model the joint distribution of 5-word sequences in a vocabulary of 100,000 words, the possible number of combinations is $100000^5 - 1 = 10^{25} - 1$, which is also the number of free parameters to learn. It is prohibitively large for further processing. This phenomenon is called "the curse of dimensionality". The root of this curse is the "generalization" problem. We need an effective mechanism to extrapolate the knowledge obtained during training to the new cases. For discrete spaces mentioned above, the structure of generalization is not obvious, i.e. any change on the discrete variables, as well as their combinations, will have a drastic influence on the value of joint distribution to be estimated.

To solve the dilemma, we can model the variables in a continuous space, where we can think of how training points trigger probability mass and distribute it smoothly to the neighborhood around them. Then it goes back to the problem of word embedding, whose concept was first introduced in Hinton (1986). Word embedding, sometimes named as word representation, is a collective name for a set of language models and feature selection methods. Its main goal is to map textual words or phrases into a low-dimensional continuous space. Using the example above, "cat" can be denoted as $[0.1, 0.3, \ldots, 0.2]^M$ and "mat" can be expressed as $[0.1, 0.2, \ldots, 0.4]^M$, where M is the hyperparameter. After that, advanced NLP tasks can be processed implemented based on these real-valued vectors. Word embedding encodes the semantic and syntactic information of words, where semantic information mainly correlates with the meaning of words, while syntactic information refers to their structural roles. Is a basic procedure in natural language processing. From the high level, most of the models try to optimize a loss function trying to minimize the discrepancy between prediction values and target values. A basic assumption is that words in similar context should have similar meaning (Harris 1954). This hypothesis emphasizes the bound of $(w, \tilde{w})$ for common word-context pairs (word w and its contextual word $\tilde{w}$ usually appear together) and weaken the correlation of rare ones.

The existing word embedding approaches are diverse. There are several ways to group them into different categories. According to Sun et al. (2015) and Lai et al. (2015), the models can be classified as either paradigmatic models or syntagmatic models, based on the word distribution information. The text region where words co-occur is the core of the syntagmatic model, but for the paradigmatic model it is the similar context that matters. Take "The tiger is a fierce animal." and "The wolf is a fierce animal." as examples, "tiger-fierce" and "wolf-fierce" are the syntagmatic words, while "wolf-tiger" are the paradigmatic words. Shazeer et al. (2016) divides the models into two classes, matrix factorization and slide-window sampling method, according to how word embedding is generated. The former is based on the word co-occurrence matrix, where word embedding is obtained from matrix decomposition. For the latter one, data sampled from sliding windows is used to predict the context word.

In this chapter, word embedding approaches are introduced according to the method of mapping words to latent spaces. Here we mainly focus on the Neural Network Language Model (NNLM) (Xu and Rudnicky 2000; Mikolov et al. 2013) and the Sparse Coding Approach (SPA) (Yogatama et al. 2014a). Vector Space

Model aims at feature expression. A word-document matrix is first constructed, where each entry counts the occurrence frequency of a word in documents. Then embedding vectors containing semantic information of words are obtained through probability generation or matrix decomposition. In the Neural Network Language Model, fed with training data, word embedding is encoded as the weights of a certain layer in the neural network. There are several types of network architectures, such as Restricted Boltzmann Machine, Recurrent Neural Network, Recursive Neural Network, Convolutional Neural Network and Hierarchical Neural Network. They are able to capture both the semantic and syntactic information. Sparse coding model is another state-of-the-art method to get word embedding. Its goal is to discover a set of bases that can represent the words efficiently.

The main structure of this paper is as follows: the models mentioned above and the evaluation methods are introduced in Sect. 4.2; the applications of word embedding are included in Sect. 4.3; conclusion and future work are in Sect. 4.4.

4.2 Word Embedding Approaches and Evaluations

Generally speaking, the goal of word embedding is mapping the words in unlabeled text data to a continuously-valued low dimensional space, in order to capture the internal semantic and syntactic information. In this section, we first introduce the background of text representation. Then, according to the specific methods for generating the mapping, we mainly focus on the Neural Network Language Model and Sparse Coding Approach. The former is further introduced in three parts, depending on the network structures applied in the model. In the end, we provide evaluation approaches for measuring the performance of word embedding models.

4.2.1 Background

One of the widely used approach for expressing the text documents is the Vector Space Model (VSM), where documents are represented as vectors. VSM was originally developed for the SMART information retrieval system (Salton et al. 1997). Some classical VSMs can be found in Deerweste et al. (1990) and Hofmann (2001).

There are various ways of building VSMs. In scenarios of information retrieval where people care more about the textual features that facilitate text categorization, various features selection methods such as document frequency (DF), information gain (IG), mutual information (MI), χ^2-test (CHI-test) and term strength (TS) have different effects on text classification (Yang and Pedersen 1997). These approaches help reduce the dimension of the text data, which could be helpful for the subsequent processing. DF refers to the number of documents that a word appears. IG measures the information loss between presence and absence term in the document. MI is the ratio between term-document joint probability and the product of their marginal probability. CHI-test applies the sums of squared errors and tries to find out the significant difference between observed word frequency and expected word frequency. TS estimates how likely a term will appear in "closely-related" documents. In information retrieval systems, these methods are useful for transforming the raw text to the vector space, and tremendous improvement has been achieved for information classification. However, it does not work well alone in applications that require both semantic and syntax information of words, since part of the original text information is already lost in feature selection procedures. However, various word embedding models, such as sparse coding, are built based on the VSM.

In Deerweste et al. (1990), Latent Semantic Analysis (LSA) is proposed to index and retrieve the information from text data automatically. It applies SVD on the term-document association data to gain the embedding output. Work in Landauer et al. (1998) and Landauer and Dumais (1997) gives an explanation to the word similarity that derives from the Test of English as a Foreign Language (TOEFL). After that, Landauer (2002) tries to detect the synonym through LSA. Furthermore, Yih et al. (2012) uses LSA to distinguish between synonyms and antonyms in the documents and its extension Multiview LSA (MVLSA) (Rastogi et al. 2015) supports the fusion of arbitrary views of data. However, all of them have to confront the limitation that depends on a single matrix of term-document co-occurrences. Usually the word embedding from those models is not flexible enough, because of the strong reliance on the observed matrix.

Latent Dirichlet Allocation (LDA) (Bengio et al. 2003) is another useful model for feature expression. It assumes that some latent topics are contained in documents. The latent topic T is derived from the conditional distribution $p(w|T)$, i.e. the probability that word w appears in T. Word embedding from traditional LDA models can only capture the topic information but not the syntactic one, which does not fully achieve its goal. Moreover, traditional LDA-based model is only used for topic discovery, but not word representation. But we can get the k-dimensional word embedding by training a k-topic model, the word embedding is from the rows of word-topic matrix and the matrix is filled by $p(w|T)$ values.

Although both of LSA and LDA utilize the statistical information directly for word embedding generation, there are some differences between them. Specifically, LSA is based on matrix factorization and subject to the non-negativity constraint. LDA relies on the word distribution, and it is expressed by the Dirichlet priori distribution which is the conjugate of multinomial distribution.

4.2.2 Neural Network Language Model

The concept of word embedding was first introduced with the Neural Networks Language Model (NNLM) in Xu and Rudnicky (2000) and Bengio et al. (2003). The motivation behind is that words are more likely to share similar meaning if they are in the same context (Harris 1954). The probability that the word sequence W occurs can be formulated according to the Bayes rule:

$$P(W) = \prod_{t=1}^{N} P(w_t|w_1, \ldots, w_{t-1}) = \prod_{t=1}^{N} P(w_t|h_t) \tag{4.1}$$

where *P(W)* is the joint distribution of sequence W, and h_t denotes the context words around word w_t. The goal is to evaluate the probability that the word w_t appears given its context information. Because there are a huge number of possible combinations of a word and its context, it is impractical to specify all $P(w_t|h_t)$. In this case, we can use a function Φ to map the context into some equivalent classes, so that we have

$$P(w_t|h_t) = P(w_t|\Phi(h_t)) \tag{4.2}$$

where all of the words are statistically independent.

If we use the n-gram model to construct a table of conditional probability for each word, then the last n-1 words are combined together as the context information:

$$P(w_t|h_t) \approx P(w_t|\tilde{w}_{t-n+1}^{t-1}) \tag{4.3}$$

where n is the number of the words considered, while 1 and 2 are the most commonly used value. Only the successive words occurring before the current word w_t are taken into account.

Most NNLM-based approaches belong to the class of unsupervised models. Networks with various architectures such as Restrict Boltzmann Machine (RBM), Convolutional Neural Network (CNN), Recurrent Neural Network (RNN) and Long-Short Term Memory (LSTM) can be used to build word embedding. Log-Bilinear (LBL) model, SENNA, and Word2Vector are some of the most representative examples of NNLM.

Usually the goal of the NNLM is to maximize or minimize the function of Log-Likelihood, sometimes with additional constraints. Suppose we have a sequence $w_1, w_2, \ldots, w_n$ in the corpus, and we want to maximize the log-likelihood of $P(w_t|\tilde{w}_{t-n+1}^{t-1})$ in the Feed Forward Neural Network (Bengio et al. 2003). Let x be the embedding vector of proceeding words, so that

$$x = [e(w_{t-n+1}), \ldots, e(w_{t-2}), e(w_{t-1})] \tag{4.4}$$

where $e(w)$ represents the embedding of word w. In Feed Forward Neural Network (without the direct edge in the structure) with one hidden layer, the layer function can be expressed as:

$$y = b + U(tanh(d + Wx)) \tag{4.5}$$

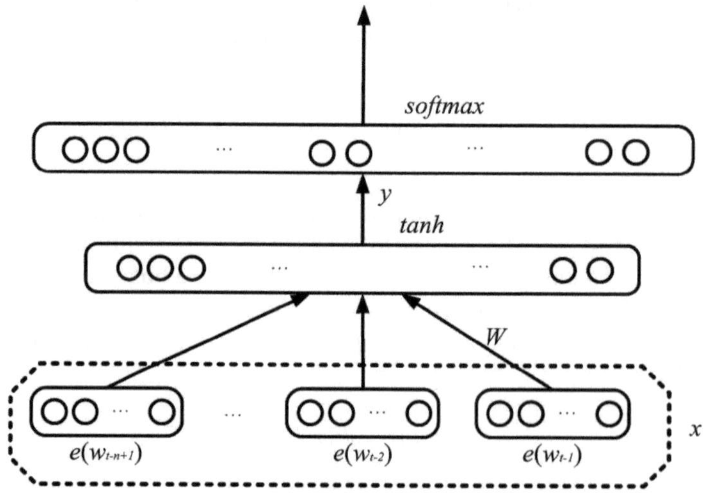

Fig. 4.1 The basic structure of the Feed Forward Neural Network

where U is the transformation matrix, W is the weight matrix of hidden layer, b and d are bias vectors. Finally, y is fed into a softmax layer to obtain the probability of the target word, and the basic structure of this model is shown in Fig. 4.1. The parameters θ in this model are (b, U, d, W). However, since we need to explicitly normalize (e.g. use softmax) all of the values in the vocabulary when computing the probability for the next word, it will be very costly for training and testing (Morin and Bengio 2005). After estimating the conditional probability in Eq. (4.3), the total probability $P(w_{t-n+1}, .., w_{t-1}, w_t)$ can be obtained directly by locating the mark of the phrase that is constructed by the words that appears in the same window (Collobert et al. 2011). The resultant probability value quantifies the normality if such a sentence appears in the natural language (more details can be found in Sect. 4.2.2.3).

Instead of using matrix multiplication on the final layer (Bengio et al. 2003; Mnih and Hinton 2007) applies the hierarchical structure (see Sect. 4.2.2.4) to reduce the computational complexity. It constructs the Log-BiLinear (LBL) model by adding bilinear interaction between word vectors and hidden variables on the basis of the energy function (see Sect. 4.2.2.1).

Recently, Word2Vector proposed by Mikolov et al. (2013) which contains Skip-Gram and (Continuous Bag-of-Words) CBOW these two frameworks is very popular in NLP tasks. It can be seen as a two-layer Neural Network Language Model. Furthermore, applying Noise Contrastive Estimation (NCE) increases its efficiency. However, we can also add the priori information into this model, Liu et al. (2015) extends Skip-Gram by treating topical information as important priori knowledge for training word embedding and proposes the topical word embedding (TWE) model, where text words and their affiliated topic (context) derived from LDA are combined to obtain the embedding.

In the subsections below, we will introduce some basic architectures of NNLM for word embedding.

4.2.2.1 Restricted Boltzmann Machine (RBM)

The ability of capturing the latent features among words usually depends on the model structure. Boltzmann Machine (BM) originates from the log-linear Markov Random Field (MRF) and its energy function is linearly related to free parameters. According to the statistical dynamics, energy function is useful in word embedding generation (Mnih and Hinton 2007). Restricted Boltzmann Machine (RBM), which prunes the visible-visible and hidden-hidden connections, is a simplified version of BM. A diagram of RBM is given in Fig. 4.2.

The main components of the energy function in RBM include binary variables (hidden unites h and visible unites v), weights $W = (w_{i,j})$ that establish connections between h_j and v_i, biases a_i for the visible units and b_j for the hidden ones. Specifically, the energy function is formulated as follows,

$$E(v, h) = -v^T W h - a^T v - b^T h \tag{4.6}$$

During the generation process word embedding, the energy function is used as the probability distribution as shown in Eq. (4.7).

$$P(v, h) = e^{-E(v,h)}/Z \tag{4.7}$$

where Z is a partition function defined as the sum of $e^{-E(h,v)}$.

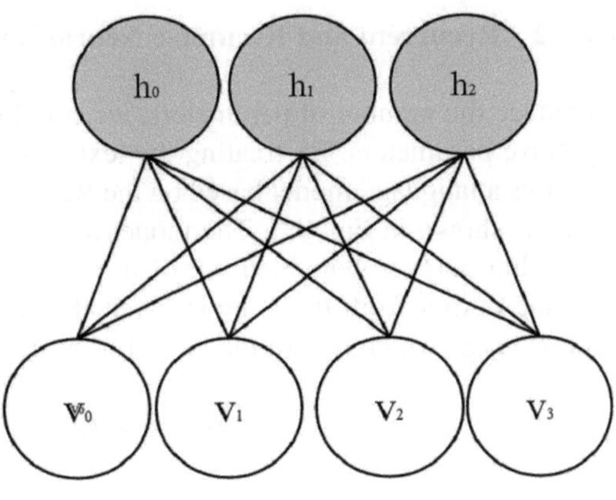

Fig. 4.2 The basic structure of the RBM

Mnih and Hinton (2007) proposes three models on the basis of RBM by adding connections to the previous status. Starting from the undirected graphical model, they try to estimate the latent conditional distribution of words. To compute the probability of the next word in a sequence, the energy function is defined as follows:

$$E(w_t, h; w_{t-n+1:t-1}) = -(\sum_{i=t-n+1}^{t-1} v_i^T RW_i)h - b_h^T h - b_t^T R^T v_n - b_v^T v_n \tag{4.8}$$

where $v_i^T R$ denotes the vector that represents word w_i from the word dictionary R, W_i in hidden units denotes the similarity between two words, and b_h, b_v, b_r are the biases for the hidden units, words and word features respectively.

The disadvantage of this class of models is that it is costly to estimate massive parameters during the training process. To reduce the number of parameters, Mnih and Hinton (2007) extends the factored RBM language model by adding connections C among words. It also removes the hidden variables and directly applies the stochastic binary variables. The modified energy function is as below:

$$E(w_t; w_{t-n+1:t-1}) = -(\sum_{i=t-n+1}^{t-1} v_i^T RC_i)R^T v_n - b_r^T R^T v_n - b_v^T v_n \tag{4.9}$$

Here C_i denotes correlation between word vectors w_i and w_t, b_r and b_v denotes the word biases. This function quantifies the bilinear interaction between words, which is also the reason why models of this kind are called Log-BiLinear Language (LBL) models. In Eq. (4.9), entries of vector $h = \sum_{i=t-n+1}^{t-1} v_i^T RC_i$ correspond to the nodes in the hidden layer, and $y_i = hR^T v_n$ represents the set of nodes in the output layer. We can see that the syntax information is contained in the hidden layer, for C_i here can be seen as the contribution of word i to current word n, which is like the cosine similarity between two words.

4.2.2.2 Recurrent and Recursive Neural Network

To reduce the number of parameters, we could unify the layer functions with some repetitive parameters. By treating the text as sequential data, Mikolov et al. (2010) proposes a language model based on the Recurrent Neural Network, the structure of which is shown in Fig. 4.3. The model has a circular architecture. At time t, word embedding $w(t)$ is generated out of the first layer. Then, it is transferred together with the output from the context layer at time $t-1$ as the new input $x(t)$ to the context layer, which is formulated as follows:

$$x(t) = w(t) + s(t-1) \tag{4.10}$$

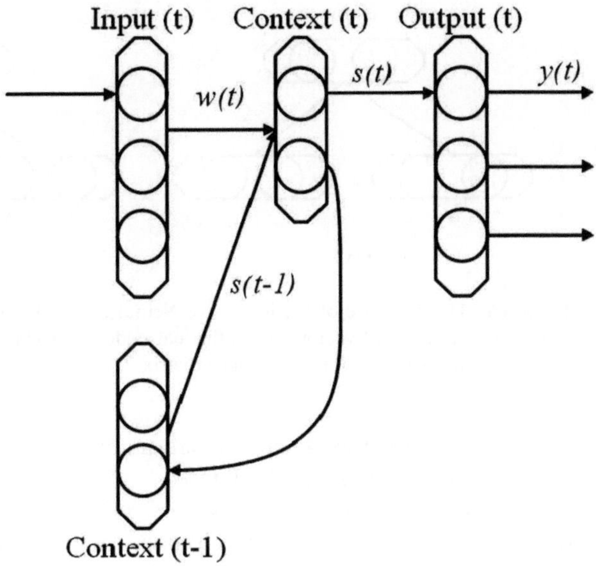

Fig. 4.3 The structure of the recurrent neural network

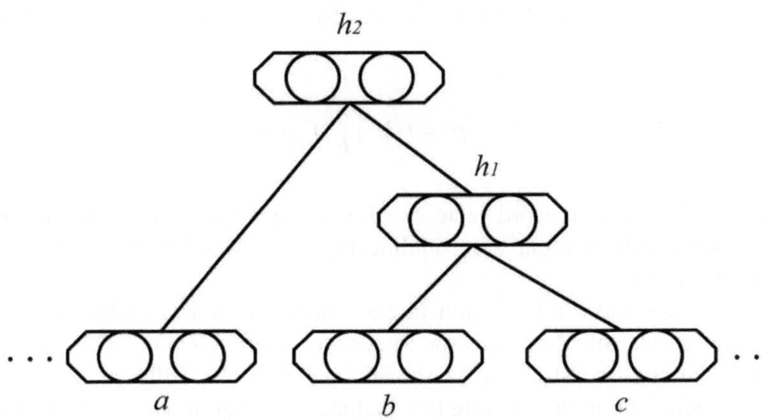

Fig. 4.4 The structure of the Recursive Neural Network

In each cycle, the outcomes of context layer and output layer are denoted as $s(t)$ and $y(t)$ respectively. Inspired by Bengio et al. (2003), the output layer also applies the softmax function.

Unlike Recurrent Neural Network, Recursive Neural Network (Goller and Kuchler 1996) is a linear chain in space as shown in Fig. 4.4. To fully analyze the sentence structure, the process iterates from two word vectors in leaf nodes b, c to merge into a hidden output h_1, which will continue to concatenate another input word vector a as the next input. The computing order is determined either by the sentence structure or the graph structure. Not only could hidden layers capture the context information, they could also involve the syntax information. Therefore,

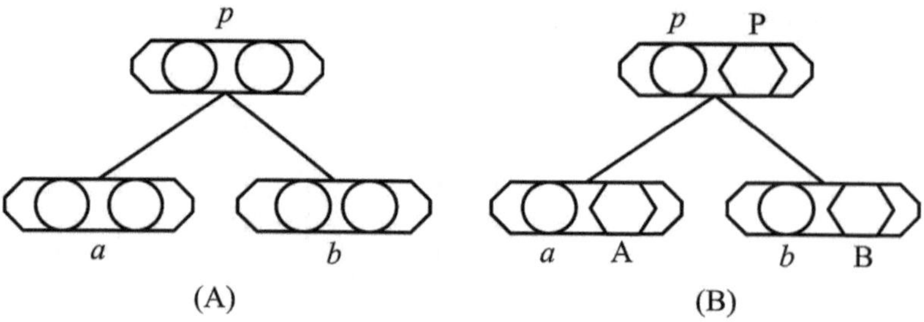

Fig. 4.5 (**A**) The structure of the Recursive Neural Network model where each node represents a vector and all the word vectors are in the leaf nodes. (**B**) The structure of the MV-RNN model where each node consists of a vector and a matrix

Recursive Neural Network can capture more information than the context-based model, which is desirable for NLP tasks.

For the sequential text data in Recursive Neural Network, Socher et al. (2011, 2013, 2012) parses the sentences into tree structures through the tree bank models (Klein and Manning 2003; Antony et al. 2010). The model proposed by Socher et al. (2011) employs the Recursive Neural Network, and its structure is shown in Fig. 4.5A,

$$p = f(W \begin{bmatrix} a \\ b \end{bmatrix}) \tag{4.11}$$

where $a, b \in R^{d\times 1}$ are the word vectors, parent node $p \in R^{d\times 1}$ is the hidden output, and $f = tanh$ adds element-wise nonlinearity to the model. $W \in R^{d\times 2d}$ is the parameter to learn.

To incorporate more information in each node, work in (Socher et al. 2012) proposes the model of Matrix-Vector Recursive Neural Network (MV-RNN), for which the parsing-tree structure is shown in Fig. 4.5B. Different from that in Fig. 4.5A, each node in this parsing tree includes a vector and a matrix. The vector represents the word vector in leaf nodes and phrase in non-leaf ones. The matrix is applied to neighboring vectors, and it is initialized by the identity matrix I plus a Gaussian noise. The vectors a, b, p capture the inherent meaning transferred in the parsing tree, and the matrixes A, B, P are to capture the changes of the meaning for the neighboring words or phrases. Apparently, the drawback of this model is the large number of parameters to be estimated. So in the work of Socher et al. (2013), the model of Recursive Neural Tensor Network (RNTN) uses the tensor-based composition function to avoid this problem.

In comparison, the main differences between recurrent and recursive models are the computing order and the network structure. The structure of the former has a closed loop dealing with the time sequence, while the latter has an open loop tackling the spatial sequence.

4.2.2.3 Convolutional Neural Network (CNN)

Convolutional Neural Network (CNN) consists of several feature extraction layers, which is inspired by biological process (Matsugu et al. 2003). It has already been successfully applied in many image recognition problems. The main advantage of CNN is that it does not depend on the prior knowledge or human efforts when doing feature selection.

CNN can also be applied to extract latent features from text data (Collobert and Weston 2008; Collobert et al. 2011). As shown in Fig. 4.6, three different convolutional kernels, in three different colors, select different features from word embedding separately. The feature vector is the combination of the max values from the rows of the selected features.

In the models above, supervised learning is applied to train the whole neural network, and the convolutional layer here is to model the latent feature of the initialized word embedding. Since supervised training method is applied here, it is able to generate word embedding suitable for the special task. Though the word embedding is usually a by-product in this case, it still has a sound performance in capturing semantic and syntactic information.

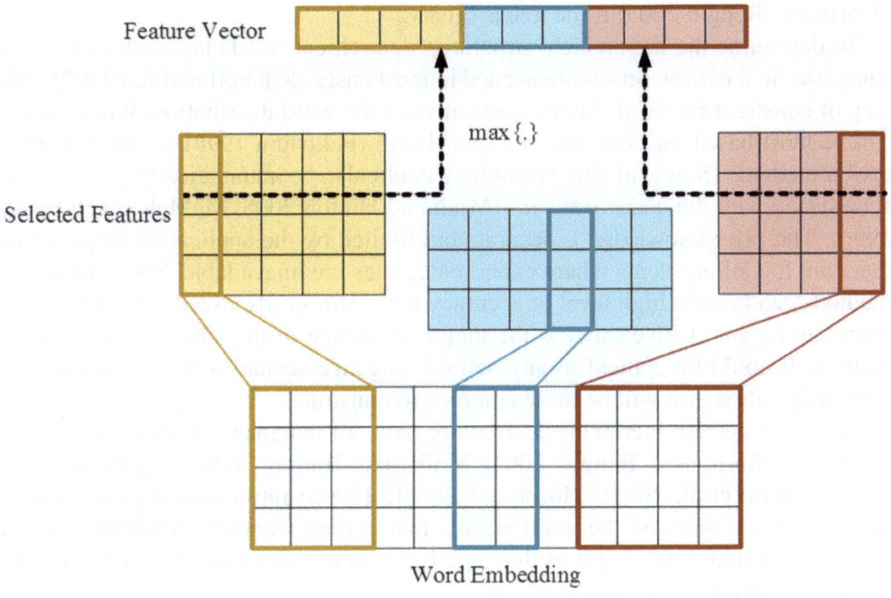

Fig. 4.6 The structure of CNN for feature extraction by using convolutional kernels

4.2.2.4 Hierarchical Neural Language Model (HNLM)

There are many tricks to speed up the process of training NNLM, such as short list, hash table for the word and stochastic gradient descent (Bengio et al. 2003). But when facing with datasets of large scale, further improvement in model efficiency is still needed.

Transferring the knowledge from a small portion of observed data examples to other cases can save a lot of computational efforts, thus speeding up the learning process. However, it is a difficult task for neural language models which involve computing the joint probability of certain word combinations. This is because the possible combinations of words from a vocabulary is immensely larger than the ones from all potential texts. So it is necessary to look for some priori information before searching all possible combinations. Based on human knowledge or statistical information, we can cluster all the words into several classes where some similar statistical properties are shared by the words in each class. Some hierarchical models are constructed based on this idea (Mnih and Hinton 2008; Morin and Bengio 2005).

Hierarchical structures has a big influence on the algorithm complexity. For example, Mnih and Hinton (2008) and Morin and Bengio (2005) reduces the complexity of language models by clustering words into a binary tree. Specifically, given a dictionary V containing $|V|$ words, the speed up will be $O(|V|/log|V|)$ (Morin and Bengio 2005) if the tree is binary.

To determine the hierarchical structure, hierarchical word classes (lexical word categories in a narrow sense) are needed in most cases (Rijkhoff and Jan 2007). The way of constructing word classes is sensitive to the word distribution. Word classes can be built based on prior (expert) knowledge (Fellbaum 1998) or through data-driven methods (Sun et al. 2012), but they could also be automatically generated in accordance with the usage statistics (Mnih and Hinton 2008; McMahon and Smith 1996). The prior knowledge is accurate but limited by the application range, since there are lots of problems where expert categories are unavailable. The data-driven methods can boost a high level of accuracy but it still needs a seed corpus which is manually tagged. Universality is the major advantage of the automatic generation method. It could be applied to any natural language scenarios without the expert knowledge, though it will be more complex to construct.

By utilizing the hierarchical structure, lots of language models have been proposed (Morin and Bengio 2005; Mnih and Hinton 2008; Yogatama et al. 2014a; Djuric et al. 2015). Morin and Bengio (2005) introduces the hierarchical decomposition bases on the word classes that extract from the WordNet. Similar with Bengio et al. (2003), it uses the layer function to extract the latent features, the process is shown in Eq. (4.12).

$$P(d_i = 1|q_i, w_{t-n+1:t-1}) = sigmoid(U(tanh(d + Wx)_{q_i}) + b_i) \tag{4.12}$$

Here x is the concatenation of the context word [same as Eq. (4.4)], b_i is the biases vector, U, W, b_i, x plays the same roles as in Eq. (4.5). The disadvantage is the procedure of tree construction which has to combine the manual and data-driving processing (Morin and Bengio 2005) together. Hierarchical Log BiLinear (HLBL) model which is proposed by Mnih and Hinton (2008) overcomes the disadvantage by using a boosting method to generate the tree automatically. The binary tree with words as leaves consists of two components: the words in the leaves which can be represented by a sequential binary code uniquely from top to down, as well as the probability for decision making at each node. Each non-leaf node in the tree also has an associated vector q which is used for discrimination. The probability of the predicted word w in HLBL is formulated as follows

$$P(w_t = w|w_{t-n+1:t-1}) = \prod_i P(d_i|q_i, w_{t-n+1:t-1}) \tag{4.13}$$

where d_i is t^{th} digit in the binary code sequence for word w, and q_i is the feature vector for the ith node in the path to word w from the root. In each non-leaf node, the probability of the decision is given by

$$P(d_i = 1|q_i, w_{t-n+1:t-1}) = \tau((\sum_{j=t-n+1}^{t-1} C_j \acute{w}_j)\dot{q}_i + b_i) \tag{4.14}$$

where $\tau(x)$ is the logistic function for decision making, C_j is the weight matrix corresponding to the context word w_j, and b_i is the bias to capture the context-independent tendency to one of its children when leaving this node.

There are both advantages and disadvantages of the hierarchical structure. Many models benefit from the reduction of computation complexity, while the consuming time on the structure building is the main drawback.

4.2.3 Sparse Coding Approach

Sparse Coding is an unsupervised model that learns a set of over-complete bases to represent data efficiently. It generates base vectors $\{\phi_i\}$ such that the input vector $x \in R^n$ can be represented by a linear combination of them $x = \sum_{i=1}^{k} \alpha_i \phi_i$, where k is the number of base vectors and $k > n$. Although there are many techniques such as Principal Component Analysis (PCA) that helps learn a complete set of basis vectors efficiently, Sparse Coding aims to use the least bases to represent the input x. The over-complete base vectors are able to capture the structures and patterns inherent in the data, so the nature characteristic of the input can be seized through Sparse Coding.

To be specific, sparse Coding constructs an over-complete dictionary $D \in R^{L \times K}$ of basis vectors, together with a code matrix $A \in R^{K \times V}$, to represent V words in C contents $X \in R^{L \times V}$ by minimizing such function as follows:

$$\arg\min_{D,A} \|X - DA\|_2^2 + \lambda \Omega(A) \tag{4.15}$$

where λ is a regularization hyperparameter and Ω is the regularizer. It applies the squared loss for the reconstructed error, but loss functions like L_1-regularized quadratic function can also be an efficient alternative for this problem (Schökopf et al. 2007).

Plenty of algorithms have been proposed to extend the sparse coding framework starting from the general loss function above. If we apply a non-negativity constraint on A (i.e. $A_{i,j} \geq 0$), the problem is turned into the Non-Negative Sparse Embedding (NNSE). Besides adding constraints on A, if we also requires that all entries in D are bigger than 0, then the problem can be transformed into the Non-Negative Sparse Coding (NNSC). NNSC is a matrix factorization technique previously studied in machine learning communities (Hoyer 2002).

For $\Omega(A)$, Yogatama et al. (2014b) designs a forest-structured regularizer that enables the mixture use of the dimension. The structure of this model is in Fig. 4.7, there are seven variables in the tree which describes the order that variables 'enter the model'. In this work, it sets the rule that a node may take a nonzero value only if its ancestors do. For example, nodes 3 and 4 may only be nonzero if nodes 1 and 2 are also nonzero. So the regularizer for A shows in Eq. (4.16).

$$\Omega(a_v) = \sum_{i=1}^{V} \left\| \langle a_{v,i}, a_{v,Descendants(i)} \rangle \right\|_2 \tag{4.16}$$

where a_v is the vth column of A.

Faruqui et al. (2015) proposes two methods by adding restriction on the dictionary D:

$$\underset{D,A}{\arg\min} \, \|X - DA\|_2^2 + \lambda\Omega(A) + \tau \, \|D\|_2^2 \tag{4.17}$$

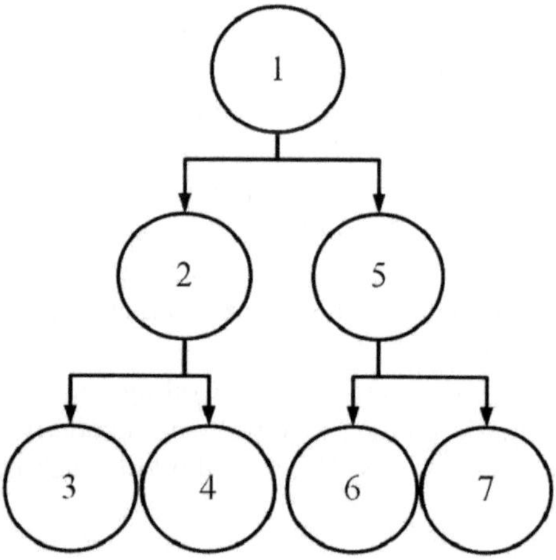

Fig. 4.7 An example of a regularization forest that governs the order in which variables enter the model

where τ is the regularization hyperparameter. The first method applies l_1 penalty on A, so that the Function 4.17 can be broken into the loss for each word vector, which makes it possible for parallel optimization.

$$\arg\min_{D,A} \sum_{i=1}^{V} \|x_i - Da_i\|_2^2 + \lambda \|a_i\|_1 + \tau \|D\|_2^2 \tag{4.18}$$

Here x_i and a_i are the ith column vector of matrix X and A respectively. The second method is to add non-negativity constraints on variables so that:

$$\arg\min_{D\in R_{\geq 0}^{L\times K}, A\in R_{\geq 0}^{K\times V}} \sum_{i=1}^{V} \|x_i - Da_i\|_2^2 + \lambda \|a_i\|_1 + \tau \|D\|_2^2 \tag{4.19}$$

Apart from the work based on Sparse Coding, Sun et al. (2016) adds the l_1 regularizer into Word2Vector. The challenge in this work lies in the optimization method, for stochastic gradient descent (SGD) cannot produce sparse solution directly with l_1 regularizer in online training. The method of Regularized Dual Averaging (RDA) proposed by Lin (2009) keeps track of online average subgradients at each update, and optimizes l_1 regularized loss function based on Continuous Bag-of-Words (CBOW) Model (Mikolov et al. 2013) in online learning.

4.2.4 Evaluations of Word Embedding

Measurements for word embedding usually depend on the specific applications. Some representative examples include perplexity, analogy precision and sentiment classification precision.

Perplexity is an evaluation method which originates from information theory. It measures how well a distribution or a model can predict the sample. Bengio et al. (2003) uses the perplexity of the corpus as the target, where the lower of the perplexity value is, the higher quality the word embedding conveys since the information is more specific. Word or phrase analogy analysis is another important assessment way. Work in (Mikolov et al. 2013) designs the precision of semantic and syntactic prediction as the standard measurement to evaluate the quality of word embedding. It is applied in phrase and word level independently. All the assessment methods mentioned above are from the view of linguistic phenomenon that the distance between two words in vector space reflects the correlation between them.

Besides the measurements of linguistic phenomenon, a lot of work evaluates word embedding by using the task where they are applied. For example, if the word embedding is used in sentiment classification, then the precision of the classification can be used for evaluation. If it is applied in machine translation, then the precision of the translation is the one that matters. This rule also holds for other tasks like POS tagging, named entity recognize, textual entailment, and so on.

There is no best measurement method for all scenarios. The most useful way is to combine it with the target of the task to achieve a more reasonable evaluation result.

4.3 Word Embedding Applications

There are many applications for word embedding, especially in NLP tasks where word embedding is fed as input data or the features of the text data directly. In this section, we will discuss how word embedding can be applied in the scenarios such as semantic analysis, syntax analysis, idiomaticity analysis, Part Of the Speech (POS) tagging, sentiment analysis, named entity recognition, textual entailment as well as machine translation.

The goal of syntax analysis is to extract the syntactic structure from sentences. Recent works (Socher et al. 2011; Huang et al. 2012; Collobert and Weston 2008; Mikolov et al. 2013) have taken the syntax information into consideration to obtain word embedding, and the result in Andreas and Dan (2014) shows that word embedding can entail the syntactic information directly. A fundamental problem to syntax analysis is part of the speech tagging, which is about labeling the words to different categories (e.g., plural, noun, adverb). It requires the knowledge of definitions and contextual information. Part of the speech tagging is a word-level NLP task. Collobert and Weston (2008) and Collobert et al. (2011) apply word embedding for POS tagging and achieve state-of-the-art results. In some scenarios, in order to figure out the syntax of text, we need to first recognize the named entities in it. Named entity recognition aims to find out the names of persons, organizations, time expressions, monetary values, locations and so on. Works in Collobert and Weston (2008), Collobert et al. (2011), Zou et al. (2013), Luo et al. (2014), Pennington et al. (2014) show the expressiveness of word embedding in these applications.

As a subsequent task, semantic analysis (Goddard 2011) relates the syntactic structures of words to their language-independent meanings.[1] In other words, it reveals words that are correlated with each other. For example, the vector ("Madrid" − "Spain" + "France") is close to ("Paris"). Previous approaches (Scott et al. 1999; Yih et al. 2012) mainly use the statistical information of the text. The works in Mikolov et al. (2013), Socher et al. (2011), Huang et al. (2012), Collobert and Weston (2008) apply the analogy precision to measure the quality of word embedding (see in Sect. 4.2.4). It is shown that word embedding can manifest the semantic information of words in the text. The semantics of each word can help us understand the combinatory meaning of several words. The phenomenon of multi-words expression idiomaticity is common in nature language, and it is difficult to be

[1] https://en.wikipedia.org/wiki/Semantic_analysis_(linguistics).

inferred. For example, the meaning of "ivory tower" could not be inferred from the separate words "ivory" and "tower" directly. Different from semantic analysis and syntax analysis, idiomaticity analysis (Salehi et al. 2015) requires the prediction on the compositionality of the multi-words expression. Once we figure out the syntax and semantics of words, we can do sentiment analysis whose goal is to extract the opinion, sentiment as well as subjectivity from the text. Word embedding can act as the text features in the sentiment classifier (Socher et al. 2012; Dickinson and Hu 2015).

In some NLP problems, we need to deal with more than one type of language corpus. Textual entailment (Saurf and Pustejovsky 2007) is such a task that involves various knowledge sources. It attempts to deduce the meaning of the new text referred from other known text sources. Works in Bjerva et al. (2014) and Zhao et al. (2015) apply word embedding into the score computing in accordance with the clustering procedure. Machine translation is another important field in NLP, which tries to substitute the texts in one language for the corresponding ones in another language. Statistical information is widely used in some previous works (Ueffing et al. 2007). Neural network models are also a class of important approaches. Bahdanau et al. (2014) builds the end-to-end model that is based on the encoder-decoder framework, while it is still common for machine translation to use word embedding as the word expression (Mikolov et al. 2013; Hill et al. 2014). Furthermore, Zou et al. (2013) handles the translation task using bilingual embedding which is shown to be a more efficient method.

According to the type of roles that it acts in the applications, word embedding could be used as the selected feature or in the raw digital format. If word embedding is used as the selected feature, it is usually used for the higher level NLP tasks, like sentiment classification, topic discovery, word clustering. In linguistics tasks, word embedding is usually used in its raw digital format, like semantic analysis, syntax analysis and idiomaticity analysis. Based on the type of machine learning problems involved, applications of word embedding could be divided into regression task, clustering task and classification task. Works in Zhao et al. (2015) prove that word embedding could make improvements for regression and classification problems which attempt to find out the pattern of the input data and make prediction after the fitting. Tasks like semantic analysis, idiomaticity analysis and machine translation belong to this category. Luo et al. (2014) tries to find out the similarity between short texts by applying a regression step in word embedding, where the resultant model could be used for searching, query suggestion, image finding, and so on. Word embedding is also commonly used in classification tasks like sentiment analysis and textual entailment (Amir et al. 2015). POS tagging and named entity recognition discussed above belong to clustering problems.

Besides word embedding, phrase embedding and document embedding are some other choices for expressing the words in the text. Phrase embedding vectorizes the phrases for higher level tasks, such as web document management (Sharma and Raman 2003), paraphrase identification (Yin and Schütze 2016) and machine translation (Zou et al. 2013). Document embedding treats documents as basic units. It can be learned from documents directly (Huang et al. 2013) or aggregated

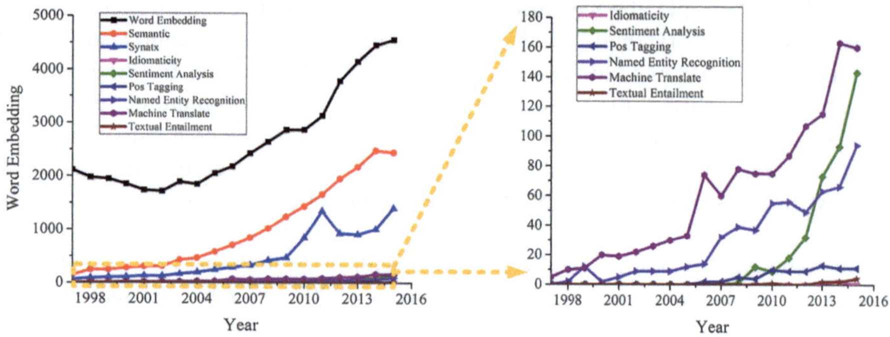

Fig. 4.8 The citation numbers of the topics in each year

by word embedding (Lin and He 2009; Zhou et al. 2015). Similar to phrase embedding, document embedding can also be applied in sentiment analysis and machine translation.

Figure 4.8 summarizes the application distribution of word embedding in different NLP tasks.[2] We can see that word embedding stably gains its popularity starting from 2004, as reflected though the rising curve. One of the most common applications is semantic analysis, in which nearly half of the works of word embedding are involved. Then it comes to the syntax analysis whose popularity dramatically increases between 2009 and 2011. Compared with syntax and semantic analysis, although other applications account for a much less proportion of work, domains like machine translation, names entity recognition and sentiment analysis receive dramatically increasing attention since 2010.

4.4 Conclusion and Future Work

In conclusion, word embedding is a powerful tool for many NLP tasks, especially the ones that require original input as the text features. There are various types of models for building word embedding, and each of them has its own advantages and disadvantages. Word embedding can be regarded as textual features, so that it can be counted as a preprocessing step in more advanced NLP tasks. Not only can it be fed into classifiers, but it can be also used for clustering and regression problems. Regarding the level that embedding represents, word embedding is a fine-grit representation compared with phrase embedding and document embedding.

Word embedding is an attractive research topic worth of further exploration. First, to enrich the information contained in word embedding, we can try to involve various prior knowledge such as synonymy relations between words, domain

[2]The citation numbers are from http://www.webofscience.com.

specific information, sentiment information and topical information. The resultant word embedding generated towards this direction will be more expressive. Then, besides using words to generate embedding, we may want to explore how character-level terms can affect the output. This is due to the reason that words themselves are made up of character-level elements. Such morphological analysis matches the way of how people perceive and create words, thus can help deal with the occurrences of new words. Moreover, as data volume is fast accumulating nowadays, it is necessary to develop techniques capable of efficiently process huge amount of text data. Since text data may arrive in streams, word embedding models that incorporate the idea of online learning are more desirable in this scenario. When the new chunk of data is available, we do not need to learn a new model using the entire data corpus. Instead, we only need to update the original model to fit the new data.

References

Amir, S., Astudillo, R., Ling, W., Martins, B., Silva, M. J., & Trancoso, I. (2015). INESC-ID: A regression model for large scale twitter sentiment lexicon induction. In *International Workshop on Semantic Evaluation.*

Andreas, J., & Dan, K. (2014). How much do word embeddings encode about syntax? In *Meeting of the Association for Computational Linguistics* (pp. 822–827).

Antony, P. J., Warrier, N. J., & Soman, K. P. (2010). Penn treebank. *International Journal of Computer Applications, 7*(8), 14–21.

Bahdanau, D., Cho, K., & Bengio, Y. (2014). Neural machine translation by jointly learning to align and translate. Eprint arxiv.

Bengio, Y., Schwenk, H., Senécal, J. S., Morin, F., & Gauvain, J. L. (2003). A neural probabilistic language model. *Journal of Machine Learning Research, 3*(6), 1137–1155.

Bjerva, J., Bos, J., van der Goot, R., & Nissim, M. (2014). The meaning factory: Formal semantics for recognizing textual entailment and determining semantic similarity. In *SemEval-2014 Workshop*.

Collobert, R., & Weston, J. (2008). A unified architecture for natural language processing: deep neural networks with multitask learning. In *International Conference, Helsinki, Finland, June* (pp. 160–167).

Collobert, R., Weston, J., Bottou, L., Karlen, M., Kavukcuoglu, K., & Kuksa, P. (2011). Natural language processing (almost) from scratch. *Journal of Machine Learning Research, 12*(1), 2493–2537.

Deerweste, S., Dumais, S. T., Furnas, G. W., Landauer, T. K., & Richard (1990). Indexing by latent semantic analysis. *Journal of the American Society for Information Science, 41*, 391–407.

Dickinson, B., & Hu, W. (2015). Sentiment analysis of investor opinions on twitter. *Social Networking, 04*(3), 62–71.

Djuric, N., Wu, H., Radosavljevic, V., Grbovic, M., & Bhamidipati, N. (2015). Hierarchical neural language models for joint representation of streaming documents and their content. In *WWW*.

Faruqui, M., Tsvetkov, Y., Yogatama, D., Dyer, C., & Smith, N. (2015). Sparse overcomplete word vector representations. Preprint, arXiv:1506.02004.

Fellbaum, C. (1998). *WordNet*. Wiley Online Library.

Goddard, C. (2011). *Semantic analysis: A practical introduction*. Oxford: Oxford University Press.

Goller, C., & Kuchler, A. (1996). Learning task-dependent distributed representations by backpropagation through structure. In *IEEE International Conference on Neural Networks* (Vol. 1, pp. 347–352).

Harris, Z. S. (1954). Distributional structure. *Synthese Language Library, 10*(2–3), 146–162.
Hill, F., Cho, K., Jean, S., Devin, C., & Bengio, Y. (2014). Embedding word similarity with neural machine translation. Eprint arXiv.
Hinton, G. E. (1986). Learning distributed representations of concepts. In *Proceedings of CogSci.*
Hofmann, T. (2001). Unsupervised learning by probabilistic latent semantic analysis. *Machine Learning, 42*(1–2), 177–196.
Hoyer, P. O. (2002). Non-negative sparse coding. In *IEEE Workshop on Neural Networks for Signal Processing* (pp. 557–565).
Huang, E. H., Socher, R., Manning, C. D., & Ng, A. Y. (2012). Improving word representations via global context and multiple word prototypes. In *Meeting of the Association for Computational Linguistics: Long Papers* (pp. 873–882).
Huang, P.-S., He, X., Gao, J., Deng, L., Acero, A., & Heck, L. (2013). Learning deep structured semantic models for web search using clickthrough data. In *Proceedings of the 22Nd ACM International Conference on Information & Knowledge Management*, CIKM '13 (pp. 2333–2338). New York, NY: ACM.
Klein, D., & Manning, C. D. (2003). Accurate unlexicalized parsing. In *Meeting on Association for Computational Linguistics* (pp. 423–430).
Lai, S., Liu, K., Xu, L., & Zhao, J. (2015). How to generate a good word embedding? *Credit Union Times*, III(2).
Landauer, T. K. (2002). On the computational basis of learning and cognition: Arguments from lsa. *Psychology of Learning & Motivation, 41*(41), 43–84.
Landauer, T. K., & Dumais, S. T. (1997). A solution to plato's problem: The latent semantic analysis theory of acquisition, induction, and representation of knowledge. *Psychological Review, 104*(2), 211–240.
Landauer, T. K., Foltz, P. W., & Laham, D. (1998). An introduction to latent semantic analysis. *Discourse Processes, 25*(2), 259–284.
Lin, C., & He, Y. (2009). Joint sentiment/topic model for sentiment analysis. In *ACM Conference on Information & Knowledge Management* (pp. 375–384).
Lin, X. (2009). Dual averaging methods for regularized stochastic learning and online optimization. In *Conference on Neural Information Processing Systems 2009* (pp. 2543–2596).
Liu, Y., Liu, Z., Chua, T. S., & Sun, M. (2015). Topical word embeddings. In *Twenty-Ninth AAAI Conference on Artificial Intelligence.*
Luo, Y., Tang, J., Yan, J., Xu, C., & Chen, Z. (2014). Pre-trained multi-view word embedding using two-side neural network. In *Twenty-Eighth AAAI Conference on Artificial Intelligence.*
Matsugu, M., Mori, K., Mitari, Y., & Kaneda, Y. (2003). Subject independent facial expression recognition with robust face detection using a convolutional neural network. *Neural Networks, 16*(5–6), 555–559.
McMahon, J. G., & Smith, F. J. (1996). Improving statistical language model performance with automatically generated word hierarchies. *Computational Linguistics, 22*(2), 217–247.
Mikolov, T., Chen, K., Corrado, G., & Dean, J. (2013). Efficient estimation of word representations in vector space. *CoRR*, abs/1301.3781.
Mikolov, T., Karafiát, M., Burget, L., Cernocký, J., & Khudanpur, S. (2010). Recurrent neural network based language model. In *INTERSPEECH 2010, Conference of the International Speech Communication Association*, Makuhari, Chiba, Japan, September (pp. 1045–1048).
Mnih, A., & Hinton, G. (2007). Three new graphical models for statistical language modelling. In *International Conference on Machine Learning* (pp. 641–648).
Mnih, A., & Hinton, G. E. (2008). A scalable hierarchical distributed language model. In *Advances in Neural Information Processing Systems 21, Proceedings of the Twenty-Second Annual Conference on Neural Information Processing Systems*, Vancouver, British Columbia, Canada, December 8–11, 2008 (pp. 1081–1088).
Morin, F., & Bengio, Y. (2005). Hierarchical probabilistic neural network language model. *Aistats* (Vol. 5, pp. 246–252). Citeseer.
Pennington, J., Socher, R., & Manning, C. (2014). Glove: Global vectors for word representation. In *Proceedings of the 2014 Conference on Empirical Methods in Natural Language Processing (EMNLP).*

Rastogi, P., Van Durme, B., & Arora, R. (2015). Multiview LSA: Representation learning via generalized CCA. In *Conference of the North American chapter of the association for computational linguistics: Human language technologies, NAACL-HLT'15* (pp. 556–566).
Rijkhoff, & Jan (2007). Word classes. *Language & Linguistics Compass, 1*(6), 709–726.
Salehi, B., Cook, P., & Baldwin, T. (2015). A word embedding approach to predicting the compositionality of multiword expressions. In *Conference of the North American Chapter of the Association for Computational Linguistics: Human Language Technologies.*
Salton, G., Wong, A., & Yang, C. S. (1997). *A vector space model for automatic indexing.* San Francisco: Morgan Kaufmann Publishers Inc.
Saurf, R., & Pustejovsky, J. (2007). Determining modality and factuality for text entailment. In *International Conference on Semantic Computing* (pp. 509–516).
Schökopf, B., Platt, J., & Hofmann, T. (2007). Efficient sparse coding algorithms. In *NIPS* (pp. 801–808).
Scott, D., Dumais, S. T., Furnas, G. W., Lauer, T. K., & Richard, H. (1999). Indexing by latent semantic analysis. In *Proceedings of the Fifteenth Conference on Uncertainty in Artificial Intelligence* (pp. 391–407).
Sharma, R., & Raman, S. (2003). Phrase-based text representation for managing the web documents. In *International Conference on Information Technology: Coding and Computing* (pp. 165–169).
Shazeer, N., Doherty, R., Evans, C., & Waterson, C. (2016). Swivel: Improving embeddings by noticing what's missing. Preprint, arXiv:1602.02215.
Socher, R., Huval, B., Manning, C. D., & Ng, A. Y. (2012). Semantic compositionality through recursive matrix-vector spaces. In *Joint Conference on Empirical Methods in Natural Language Processing and Computational Natural Language Learning* (pp. 1201–1211).
Socher, R., Pennington, J., Huang, E. H., Ng, A. Y., & Manning, C. D. (2011). Semi-supervised recursive autoencoders for predicting sentiment distributions. In *Conference on Empirical Methods in Natural Language Processing, EMNLP 2011, 27–31 July 2011, John Mcintyre Conference Centre, Edinburgh, A Meeting of SIGDAT, A Special Interest Group of the ACL* (pp. 151–161).
Socher, R., Perelygin, A., Wu, J. Y., Chuang, J., Manning, C. D., Ng, A. Y., & Potts, C. (2013). Recursive deep models for semantic compositionality over a sentiment treebank. In *Conference on Empirical Methods on Natural Language Processing.*
Sun, F., Guo, J., Lan, Y., Xu, J., & Cheng, X. (2015). Learning word representations by jointly modeling syntagmatic and paradigmatic relations. In *AAAI.*
Sun, F., Guo, J., Lan, Y., Xu, J., & Cheng, X. (2016). Sparse word embeddings using l1 regularized online learning. In *International Joint Conference on Artificial Intelligence.*
Sun, S., Liu, H., Lin, H., & Abraham, A. (2012). Twitter part-of-speech tagging using pre-classification hidden Markov model. In *IEEE International Conference on Systems, Man, and Cybernetics* (pp. 1118–1123).
Ueffing, N., Haffari, G., & Sarkar, A. (2007). Transductive learning for statistical machine translation. In *ACL 2007, Proceedings of the Meeting of the Association for Computational Linguistics*, June 23–30, 2007, Prague (pp. 25–32).
Xu, W., & Rudnicky, A. (2000). Can artificial neural networks learn language models? In *International Conference on Statistical Language Processing* (pp. 202–205).
Yang, Y., & Pedersen, J. O. (1997). A comparative study on feature selection in text categorization. In *Fourteenth International Conference on Machine Learning* (pp. 412–420).
Yih, W.-T., Zweig, G., & Platt, J. C. (2012). Polarity inducing latent semantic analysis. In *Proceedings of the 2012 Joint Conference on Empirical Methods in Natural Language Processing and Computational Natural Language Learning*, EMNLP-CoNLL '12 (pp. 1212–1222). Stroudsburg, PA: Association for Computational Linguistics.
Yin, W., & Schütze, H. (2016). Discriminative phrase embedding for paraphrase identification. Preprint, arXiv:1604.00503.
Yogatama, D., Faruqui, M., Dyer, C., & Smith, N. A. (2014a). Learning word representations with hierarchical sparse coding. Eprint arXiv.

Yogatama, D., Faruqui, M., Dyer, C., & Smith, N. A. (2014b). Learning word representations with hierarchical sparse coding. Eprint arXiv.

Zhao, J., Lan, M., Niu, Z. Y., & Lu, Y. (2015). Integrating word embeddings and traditional NLP features to measure textual entailment and semantic relatedness of sentence pairs. In *International Joint Conference on Neural Networks* (pp. 32–35).

Zhou, C., Sun, C., Liu, Z., & Lau, F. (2015). Category enhanced word embedding. Preprint, arXiv:1511.08629.

Zou, W. Y., Socher, R., Cer, D. M., & Manning, C. D. (2013). Bilingual word embeddings for phrase-based machine translation. In *EMNLP* (pp. 1393–1398).

Part II
Applications in Science

Chapter 5
Big Data Solutions to Interpreting Complex Systems in the Environment

Hongmei Chi, Sharmini Pitter, Nan Li, and Haiyan Tian

5.1 Introduction

The role of big data analysis in various fields has only recently been explored. The amount of data produced in the digital age is profound. Fortunately, the technical capabilities of the twenty-first century is beginning to meet the needs of processing such immense amounts of information. Data collected through social media, marketing transactions, and internet search engines have opened up the path to in depth, real-time quantitative research to the fields of sociology, anthropology, and economics (Boyd and Crawford 2012).

Other fields, e.g., the medical and business fields, have been quick to recognize the utility of rapidly collecting, storing, and analyzing vast amounts of data such as patient records and customer purchasing patterns. Even individuals now have the ability to track their health statistics through ever increasing access to personal tracking devices leading to the Quantified Self Movement (Kelly 2007; Swan 2013).

In the realms of environmental science and ecology the capabilities of big data analysis remain largely unexplored. We have merely scratched the surface of possibilities (Hampton et al. 2013). And yet it is in these areas that we may have the most to gain. The complexity of environmental systems, particularly in relation to human behavior and impact on human life, require the capabilities of modern data analysis.

H. Chi (✉) • S. Pitter
Florida A&M University, Tallahassee, FL 32307, USA
e-mail: hongmei.chi@famu.edu; sharmini.pitter@famu.edu

N. Li
Guangxi Teachers Education University, Nanning, 530001, China
e-mail: nli@yic.ac.cn

H. Tian
University of Southern Mississippi, Hattiesburg, MS 39406, USA
e-mail: haiyan.tian@usm.edu

S. Srinivasan (ed.), *Guide to Big Data Applications*, Studies in Big Data 26,
DOI 10.1007/978-3-319-53817-4_5

By looking at large data sets through platforms that closely resemble neural networks we are no longer restricted to simple 1:1 relationships between environmental variables and effects on economic, recreational, and agricultural variables. It is now possible to look at environmental impacts through the lens of the system. Take for instance the combination of factors that contribute to higher intensity of hurricanes in Florida and the impact on property distribution (coastal vs. inland). Instead of simply asking how increased storm intensity on the Saffir-Simpson hurricane wind scale is affecting property purchasing, it is now possible to understand what factors may be leading to increased storm intensity in order to effectively produce mediation strategies to reduce the impact high intensity storms have on the Florida economy. These types of analysis will become increasingly necessary in order to provide communities adequate information to prepare for and adapt to the regional effects of climate change (Bjarnadottir et al. 2011).

In the case of agriculture a good example of the new way forward is to pool together information regarding crop production, local/regional soil quality, and temperature/climate variation from several farms to strategize ways to adapt to increasing temperatures or shifting growing seasons. Companies such as the Farmers Business Network provide access to data at a much more precise level, allowing farmers to make informed decisions based on data sets collected region by region, farm by farm.

In order to apply information garnered through Big Data analytics (Shiffrin 2016) to real world issues a certain amount of interpretation is necessary. The variables that have been considered or ignored must be taken into account in order to discern what can and cannot be interpreted from any given dataset. For instance in the example of GMO adoption any number of factors could have a substantial effect on the adoption process. Just as with any implementation of a new agricultural methodology, social interaction, economic standing, family structure, community structure, and a plethora of other factors may have a significant effect on any one farmer's likelihood of adopting the method. As Big Data analysis develops these broader impacts on decision-making will likely become clearer. However, as we explore the interdependence of information it is important to avoid drawing direct connections where none exist (Boyd and Crawford 2012).

In this chapter we will explore several possible applications of data analytics in the environmental sciences as well as the data analysis tools RapidMiner, Hadoop, and the statistical package R. The strengths of each analysis tool will be highlighted through two case studies. We will use the examples of hurricane frequency in Florida. The chapter will also include an environmental genomic example. Our hope is that this information will help set the stage for others to delve deeper into the possibilities that big data analytics can provide. Collaborations between data scientists and environmental scientists will lead to increased analyzing capabilities and perhaps a more accurate dissection of the complexity of the systems that environmental scientists seek to understand. Beyond that, the simple examples provided in this chapter should encourage environmental scientists to further their own discovery of the analytical tools available to them.

5.2 Applications: Various Datasets

Data analysis is the process of examining data to uncover hidden patterns, unknown correlations, and other useful information that can be used to make better decisions. Data analytics is playing an ever-increasing role in the process of scientific discovery. EPA and NOAA related datasets are on demand to be analyzed by using data analysis tools. Those data are difficult to handle using traditional database systems.

A wireless sensor network (WSN) is defined as a human-engineered, complex dynamic system of interacting sensor nodes that must combine its understanding of the physical world with its computational and control functions, and operate with constrained resources. These miniature sensor nodes must collectively comprehend the time evolution of physical and operational phenomena and predict their effects on mission execution and then actuate control actions that execute common high-level mission goals. Rapid modern advancements in micro-electromechanical systems (MEMS) and distributed computing have propelled the use of WSNs in diverse applications including education, geological monitoring, ecological habitat monitoring, and healthcare monitoring. Generally, sensor nodes are equipped with modules for sensing, computing, powering, and communicating to monitor specific phenomena via self-organizing protocols, since node positions are not predetermined. Figure 5.1 represents a general architecture for a sensor node, where the microcontroller or computing module processes the data observed by the sensing module, which is then transmitted to a required destination via a wireless link with a communication module.

Some environmental applications of sensor networks include tracking the movements of birds, small animals, and insects; monitoring environmental conditions that affect crops and livestock (Greenwood et al. 2014); monitoring irrigation; the use of macro-instruments for large-scale Earth monitoring and planetary exploration; chemical/biological detection; precision agriculture; biological and environmental monitoring in marine, soil, and atmospheric contexts; forest fire detection; meteorological or geophysical research; flood detection; bio-complexity mapping of the environment; and pollution study. In Sect. 5.2.1 a few of these examples have been further explained.

5.2.1 Sample Applications

Forest fire detection: Since sensor nodes may be strategically, randomly, and densely deployed in a forest, sensor nodes can relay the exact origin of the fire to the end users before the fire is spread uncontrollably. Millions of sensor nodes can be deployed and integrated using radio frequencies/optical systems. The nodes may be equipped with effective power scavenging methods, such as solar cells, because the sensors may be left unattended for months and even years. The sensor nodes will collaborate with each other to perform distributed sensing and overcome obstacles, such as trees and rocks that block wired sensors' line of sight.

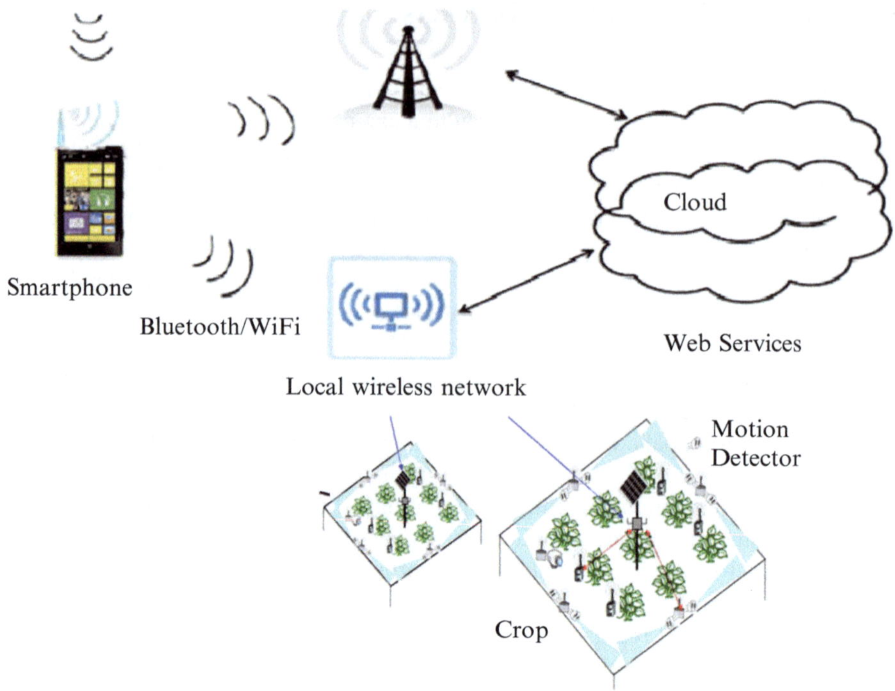

Fig. 5.1 Wireless sensor networks (WSN) used in precision agriculture. These networks allow remote monitoring of field conditions for crops and livestock

Biocomplexity mapping of the environment: This strategy requires sophisticated approaches to integrate information across temporal and spatial scales. Advances of technology in the remote sensing and automated data collection have enabled higher spatial, spectral, and temporal resolution at a geometrically declining cost per unit area. Along with these advances, the sensor nodes also have the ability to connect with the Internet, which allows remote users to control, monitor and observe the biocomplexity of the environment (Khedo et al. 2010a; Khedo et al. 2010b). Although satellite and airborne sensors are useful in observing large biodiversity, e.g., spatial complexity of dominant plant species, they do not have enough resolution to observe small size biodiversity, which makes up most of the biodiversity in an ecosystem. As a result, there is a need for ground level deployment of wireless sensor nodes to observe this level of biocomplexity. One example of biocomplexity mapping of the environment is done at the James Reserve in Southern California (Ravi and Subramaniam 2014). Three monitoring grids, each having 25–100 sensor nodes, will be implemented for fixed view multimedia and environmental sensor data loggers.

Flood detection: An example of flood detection is the ALERT system deployed in the US (Basha et al. 2008). Several types of sensors deployed in the ALERT system are rainfall, water level and weather sensors. These sensors supply information to the centralized database system in a pre-defined way. Research projects, such as the COUGAR Device Database Project at Cornell University and the Data Space project at Rutgers, are investigating distributed approaches in interacting with sensor nodes in the sensor field to provide snapshot and long-running queries.

Precision agriculture: Some of the benefits include the ability to monitor the pesticide levels in soil and groundwater, the level of soil erosion, and the level of air pollution in real-time (Lehmann et al. 2012) (Fig. 5.1).

Every day a large number of Earth Observation spaceborne and airborne sensors from many different countries provide a massive amount of remotely sensed data (Ma et al. 2015). A vast amount of remote sensing data is now freely available from the NASA.

Open Government Initiative (http://www.nasa.gov/open/). The most challenging issues are managing, processing, and efficiently exploiting big data for remote sensing problems.

5.3 Big Data Tools

In this section we will explore the data analysis tools, RapidMiner, Apache Spark and the statistical package R. Examples of the possible uses of RapidMiner and R will be highlighted through two corresponding case studies. There are many open sources for analyzing environmental big datasets.

5.3.1 RapidMiner

RapidMiner is a software platform developed by the company of the same name that provides an integrated environment for machine learning, data mining, text mining, predictive analytics, and business analytics. It is used for business and industrial applications as well as for research, education, training, rapid prototyping, and application development and supports all steps of the data mining process including results visualization, validation, and optimization. In addition to data mining, RapidMiner also provides functionality like data preprocessing and visualization, predictive analytics and statistical modeling, evaluation, and deployment. RapidMiner also provides the ability to run real-time data analysis on a set schedule, which is helpful in analyzing the high velocity and high volume data that is characteristic of big data. Written in the Java Programming language, this tool offers advanced analytics through template-based frameworks. A bonus: Users hardly have to write any code. Offered as a service, rather than a piece of local software, this tool holds top position on the list of data mining tools.

This chapter focuses on the usage of Rapidminer that can be used in analyzing Florida hurricane datasets.

5.3.2 *Apache Spark*

Spark is one of the most active and fastest-growing Apache projects, and with heavy-weights like IBM throwing their weight behind the project and major corporations bringing applications into large-scale production, the momentum shows no signs of letting up. Apache Spark is an open source cluster-computing framework. Originally developed at the University of California, Berkeley's AMPLab.

5.3.3 *R*

R is a language and environment for statistical computing and graphics, developed at Bell Laboratories. One of R's strengths is the ease with which well-designed publication-quality plots can be produced, including mathematical symbols and formulae where needed. R provides an environment within which statistical techniques are implemented. R can be extended easily via packages. Programming with Big Data in R (pbdR) (http://r-pbd.org/) is a series of R packages and an environment for statistical computing with Big Data by using high-performance statistical computation. The significance of pbdR is that it mainly focuses on distributed memory systems. The package pbdR can deal with big data in a timely manner.

5.3.4 *Other Tools*

There are many other open sources for processing big data, Weka, Apache projects: such as MapReduce Spark. A few of tools are built on top of MapReduce, such as GraphLab and Pegasus. All of those open sources are excellent in handling environmental datasets.

5.4 Case Study I: Florida Hurricane Datasets (1950–2013)

5.4.1 *Background*

Due to its large coastal area and the warm Gulf Stream waters that surround it, Florida is particularly vulnerable to hurricanes (Blake et al. 2007; Frazier et al. 2010). Within the past century over $450 billion of damage have occurred in Florida as a result of hurricanes (Malmstadt et al. 2009). Hurricanes pose the greatest meteorological threat to the state. They not only threaten property damage but can also be costly in terms of revenue, employment, and loss of life (Belasen and Polachek 2009; Blake et al. 2007).

The frequency of hurricanes in the state of Florida from the years 1950–2015 was explored. This case study seeks to highlight the utility of public database use. Combining location data with information regarding the landfall location, intensity, and categorization of storm events from 1950–2015 allows us to demonstrate which areas may be the most vulnerable, and thus should invest in proper education and infrastructure for the communities that need them most.

At this time areas of South Florida are already experiencing the effects of climate change in the form of sea level rise and increased high tide flooding events (SFRC Compact 2012; Wdowinski et al. 2016). It is likely that the hurricanes experienced will be of increased intensity and frequency.

With the possibility of extreme sea level rise on the horizon for the state of Florida over the course of the next 100 years this threat is even more severe as studies have shown increased water level in this region will likely result in higher storm surges, causing more damage with each major storm event (Frazier et al. 2010).

5.4.2 *Dataset*

Data for this case study was selected from the Atlantic HURDAT2 Database (NHC Data Archive 2016).

A record of tropical storms, tropical depressions, and hurricanes was explored to demonstrate the potential of analyzing large data sets in understanding the impact of possible environmental disasters. The data utilized was sourced from the Atlantic Hurricane Database (HURDAT2) of the National Hurricane Center (NHC). The National Weather Service originally collected the data. This database is widely used for risk assessment (Landsea and Franklin 2013).

Modern methods of storm data collection include observations measured from Air Force and NOAA aircraft, ships, buoys, coastal stations, and other means (Powell et al. 1998). Data collection in the 2000s has also been enhanced by the use of satellite-based scatterometers, Advanced Microwave Sounding Units, the Advanced Dvorak Technique, and aircraft-based Stepped Frequency Microwave Radiometers (Landsea and Franklin 2013; Powell et al. 2009).

The wind speed categories used are denoted in Table 5.1, which shows categories as they are defined by the NHC. This information is available through the NHC website (http://www.nhc.noaa.gov/aboutsshws.php). Unlike the original Saffir-Simpson scale, this modified version does not include information such as central pressure or storm surge and only determines category based on peak maximum sustained wind speed (Saffir 1973; Simpson and Saffir 1974; Blake et al. 2007). The scale used has been modified over the years (Landsea et al. 2004; Schott et al. 2012). For the purpose of this study the wind speed records reported in the HURDAT2 database were converted into modern hurricane categories. The classifications of Tropical Depression (TD) and Tropical Storm (TS) were assigned to storm events with sustained wind speeds of less than 39 mph and 39-73mph respectively.

Table 5.1 Modified Saffir-Simpson wind scale

Category	Sustained winds	Types of damage due to hurricane winds
TD	<39 mph	
TS	39–73 mph	
1	74–95 mph 64–82 kt 119–153 km/h	*Very dangerous winds will produce some damage*: Well-constructed frame homes could have damage to roof, shingles, vinyl siding and gutters. Large branches of trees will snap and shallowly rooted trees may be toppled. Extensive damage to power lines and poles likely will result in power outages that could last a few to several days
2	96–110 mph 83–95 kt 154–177 km/h	*Extremely dangerous winds will cause extensive damage*: Well-constructed frame homes could sustain major roof and siding damage. Many shallowly rooted trees will be snapped or uprooted and block numerous roads. Near-total power loss is expected with outages that could last from several days to weeks
3 (major)	111–129 mph 96–112 kt 178–208 km/h	*Devastating damage will occur*: Well-built framed homes may incur major damage or removal of roof decking and gable ends. Many trees will be snapped or uprooted, blocking numerous roads. Electricity and water will be unavailable for several days to weeks after the storm passes
4 (major)	130–156 mph 113–136 kt 209–251 km/h	*Catastrophic damage will occur*: Well-built framed homes can sustain severe damage with loss of most of the roof structure and/or some exterior walls. Most trees will be snapped or uprooted and power poles downed. Fallen trees and power poles will isolate residential areas. Power outages will last weeks to possibly months. Most of the area will be uninhabitable for weeks or months
5 (major)	157 mph or higher 137 kt or higher 252 km/h or higher	*Catastrophic damage will occur*: A high percentage of framed homes will be destroyed, with total roof failure and wall collapse. Fallen trees and power poles will isolate residential areas. Power outages will last for weeks to possibly months. Most of the area will be uninhabitable for weeks or months

Source: http://www.nhc.noaa.gov/aboutssh

5.4.3 Data Analysis

Data was selected from the HURDAT2 database based on location to limit storm events that occurred in the state of Florida during the years 1950–2013. As expected storms of lower sustained wind speeds are more likely to occur with category 5 hurricanes comprising less than 1% of the total storms experienced in Florida from the year 1950–2013 (Fig. 5.2, Table 5.2).

In Fig. 5.3 wind speed is plotted against the number of storms reported for a given sustained maximum wind speed. The average maximum sustained wind speed was found to be approximately 45 mph.

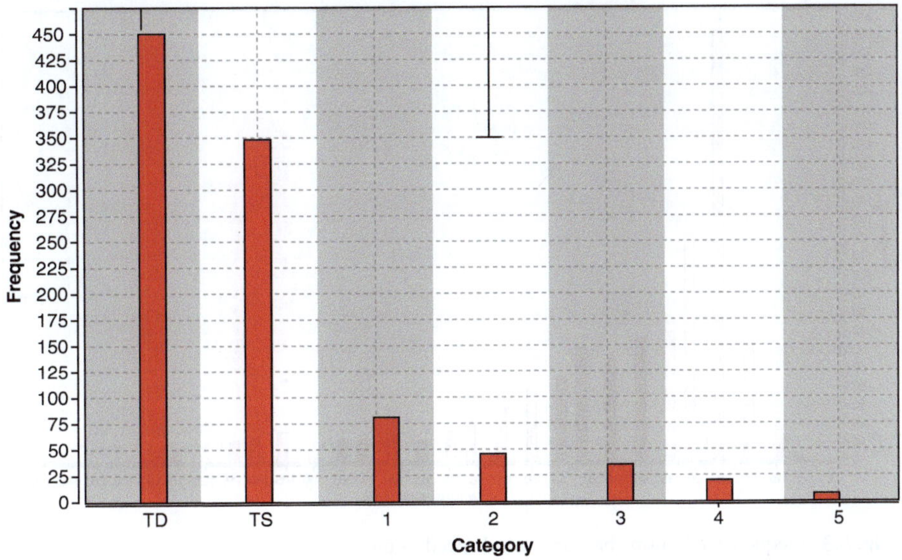

Fig. 5.2 Hurricane frequency values by category from 1950–2015

Table 5.2 Number of storm events in each category

Nominal value	Absolute count
TD	451
TS	348
1	82
2	43
3	35
4	20
5	5

It is also possible to determine the time of year that most storms occur by breaking down storm frequency per month. In Fig. 5.4 this has been further broken down by year to show the distribution of high-powered storms over time.

5.4.4 Summary

This simple example demonstrates how easily large databases can be utilized to discover information that may be useful to the general public. In this instance the high frequency of hurricanes occurring during the months of August and September could affect travel and recreational tourism planning. This information becomes even more valuable when compared to other large datasets. The low average wind speed of the storms (Fig. 5.2) may seem surprising given the reputation of Florida

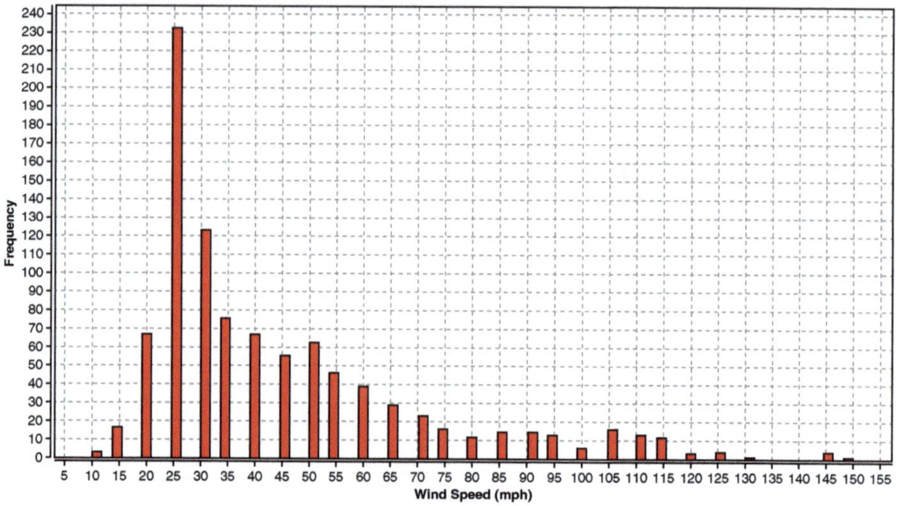

Fig. 5.3 Frequency of storms based on wind speed (mph)

Table 5.3 Key databases for environmental microbiologists

Database name	URL	Brief description
IMG	http://img.jgi.doe.gov/	Comprehensive platform for annotation and analysis of microbial genomes and metagenomes
SEED	http://www.theseed.org/	Portal for curated genomic data and automated annotation of microbial genomes
NCBI	http://www.ncbi.nlm.nih.gov/	A series of databases relevant to biotechnology and biomedicine and an important resource for bioinformatics tools and services. Major databases include GenBank for DNA sequences and PubMed, a bibliographic database for the biomedical literature
Pfam	http://pfam.xfam.org/	Database of protein families
STRING	http://string-db.org/	Database of protein association networks
RDP	http://rdp.cme.msu.edu/	16S rRNA gene database
SILVA	http://www.arb-silva.de/	rRNA gene database

as being highly vulnerable to hurricanes. However, when compared to a study conducted of the Georgia coastline, which experienced only 14 hurricane landfalls during the same time period, we can see why Florida is viewed as highly vulnerable to high-powered storms (Bossak et al. 2014).

Public sharing of data is something that must be addressed (Fig. 5.5). Data that is available for research collaboration is also vital to taking advantage of the true potential of big data analytics to solve complex environmental challenges or even the economic impact of various solution strategies (Bjarnadottir et al. 2011).

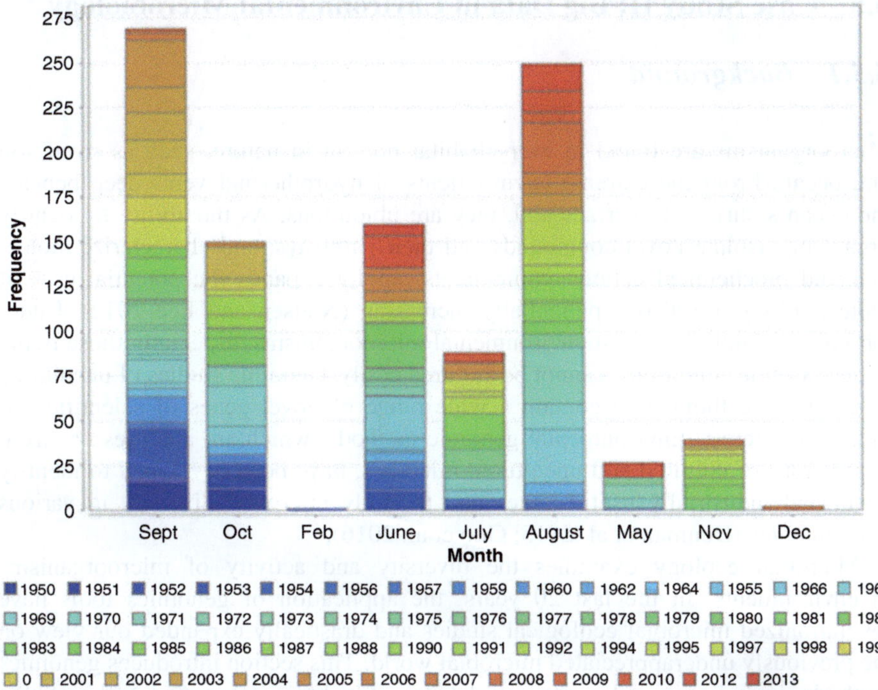

Fig. 5.4 Distribution of storms for a given month from 1950 to 2013

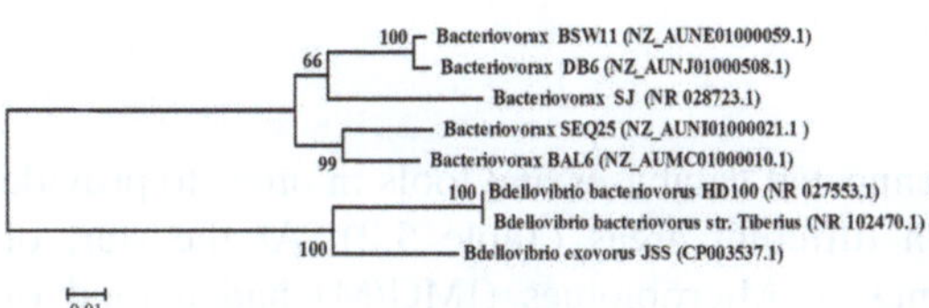

Habites	Cluster*	Size (Mb)	GC%	Gene	ABCs	ABCs/ORFs	BioProject
Marine	IV	3.65	36.9	3491	26	0.007	PRJNA220291, PRJNA210325
Marine	IX	3	37.6	3218	21	0.007	PRJNA220293, PRJNA210327
Marine	III	3.44	36.7	3292	28	0.008	PRJNA82341, PRJNA50431
Marine	V	3.23	36.5	3332	25	0.007	PRJNA220292, PRJNA210326
Marine	X	3.23	36.5	3101	17	0.005	PRJNA220289, PRJNA210328
Fresh water	----	3.78	50.6	3628	47	0.013	PRJNA61595, PRJNA9637
Fresh water	----	3.99	49.9	3779	52	0.013	PRJNA182482, PRJNA70801
Sewage	----	2.66	41.9	2618	53	0.020	PRJNA194119, PRJNA163339

Fig. 5.5 Distribution of ABC systems across the phylogenetic tree of BALOs. The phylogenetic tree was constructed based on their 16S rRNA sequences using the Neighbor-Joining method. The reliability of the tree was evaluated with 1,000 replicates of bootstrapping test and only high bootstrap value scores (>50%) were indicated on the branches. In addition, each strain is followed by its isolation habitat, total number of ORFs, as well as absolute and relative number of ABC systems and other information. *Clusters were identified by previous study[28]. 16s rRNA sequences of strainsBSW 11, DB6, SEQ25 and BAL6 were extracted from their genomic sequences according to the annotation.

5.5 Case Study II: Big Data in Environmental Microbiology

5.5.1 Background

Microorganisms are found in every habitat present in nature, such as river, soil and ocean. From the extreme environments of hydrothermal vents deep beneath the ocean's surface to surface soil, they are ubiquitous. As the ability to identify organisms, isolate novel compounds and their pathways, and characterize molecular and biochemical cellular components rapidly expands the potential uses of biotechnology are also exponentially increasing (Nielsen and Lee 2012). Under laboratory conditions, most environmental microorganisms, especially those living under extreme conditions, cannot be cultured easily. Genomic studies of uncultured organisms are thought to contain a wide range of novel genes of scientific and industrial interest. Environmental genomic methods, which are analyses of mixed populations of cultured and uncultured microbes, have been developed to identify novel and industrially useful genes and to study microbial diversity in various environments (Denman et al. 2015; Guo et al. 2016).

Microbial ecology examines the diversity and activity of microorganisms in environments. In the last 20 years, the application of genomics tools have revolutionized microbial ecological studies and drastically expanded our view on the previously underappreciated microbial world. This section introduces genomics methods, including popular genome database and basic computing technics that have been used to examine microbial communities and evolutionary history.

5.5.2 Genome Dataset

Global resources have many interconnected databases and tools in order to provide convenient services for users from different areas (Table 5.2). At the start of 2016, Integrated Microbial Genomes & Microbiomes (IMG/M) had a total of 38,395 genome datasets from all domains of life and 8077 microbiome datasets, out of which 33,116 genome datasets and 4615 microbiome datasets are publicly available. The National Center for Biotechnology Information (NCBI) at the National Institutes of Health in the United States and the European Molecular Biology Laboratory/European Bioinformatics Institute (EMBL-EBI) are undisputed leaders that offer the most comprehensive suites of genomic and molecular biology data collections in the world. All genomes of bacteria, archaea, eukaryotic microorganisms, and viruses have been deposited to GenBank, EMBL Bank or DNA Data Bank of Japan (DDBJ). Take NCBI for example, environmental microbiologists used NCBI literature resources—PubMed and PubMed Central, to access the full text of peer-reviewed journal articles, as well as NCBI Bookshelf, which has rights to the full text of books and reports. The central features of the NCBI collection are nonredundant (NR) databases of nucleotide and protein sequences and their curated subset, known as Reference Sequences or RefSeq (Pruitt et al. 2007). The NCBI Genome database maintains genome sequencing projects, including all sequenced microbial genomes, and provides links to corresponding records in NR databases and BioProject, which is a central access point to the primary data from sequencing projects. NCBI also maintains the Sequence Read Archive (SRA), which is a public repository for next-generation sequence data (Kodama et al. 2012) and GEO (Gene Expression Omnibus), the archive for functional genomics data sets, which provides an R-based web application to help users analyze its data (Barrett et al. 2009).

The NCBI's Basic Local Alignment Search Tool (BLAST) (Cock et al. 2015) is the most popular sequence database search tool, and it now offers an option to search for sequence similarity against any taxonomic group from its NCBI web page or do it using your local computer. For example, a user may search for similarity only in cyanobacteria, or within a single organism, such as *Escherichia coli*. Alternatively, any taxonomic group or an organism can be excluded from the search. NCBI BLAST also allows users to search sequence data from environmental samples, which providing a way to explore metagenomics data. NCBI Taxonomy database (Federhen 2012) is another useful resource for environmental microbiologists, because it contains information for each taxonomic node, from domain kingdoms to subspecies.

5.5.3 Analysis of Big Data

Because of an unprecedented increase in data information available in public databases, bioinformatics has become an important part of many areas of biology, including environmental microbiology. In the field of genetics and genomics, it helps in digging out sequencing information and annotating genomes. Bioinformatics tools, which developed on R language, help in the comparison of genetic and genomic data and more generally in the understanding of evolutionary aspects of molecular biology. A study case that demonstrates the basic process of analyzing the environmental microbial genomes using various bioinformatics tools is provided herein.

In this study (Li et al. 2015) the FASTA DNA sequencing software package was utilized. Text-based FASTA files representing eight *Bdellovibrio* and like organisms (BALOs) genomes (Fig. 5.4): Bx BSW11 (NZ_AUNE01000059.1), Bx DB6 (NZ_AUNJ01000508.1), Bx SJ (NR 028723.1), Bx SEQ25 (NZ_AUNI01000021.1), Bx BAL6 (NZ_AUMC01000010.1), BD HD100 (NR 027553.1), BD Tiberius (NR 102470.1), BD JSS (CP003537.1), were downloaded from NCBI (ftp://ftp.ncbi.nih.gov/genomes/) on March 23, 2014. ABC systems in these genomes were identified using the bacteria-specific ATP-binding cassette (ABC) systems systems profile HMM and HMMER 3.0 hmmsearch at default settings. Sequences with a domain independent E-value ≤ 0.01 and a score/bias ratio ≥ 10 were accepted. The ABCdb database (https://wwwabcdb.biotoul.fr/) which provides comprehensive information on ABC systems, such as ABC transporter classification and predicted function (Fichant et al. 2006), was used to check predicted ABC systems.

Finally, a total of 269 putative ABC proteins were identified in BALOs. The genes encoding these ABC systems occupy nearly 1.3% of the gene content in freshwater *Bdellovibrio* strains and about 0.7% in their saltwater counterparts (Fig. 5.6). The proteins found belong to 25 ABC systems families based on their structural characteristics and functions. Among these, 16 families function as importers, 6 as exporters and 3 are involved in various cellular processes. Eight of these 25 ABC system families were deduced to be the core set of ABC systems conserved in all BALOs. All *Bacteriovorax* strains have 28 or less ABC systems. To the contrary, the freshwater *Bdellovibrio* strains have more ABC systems, typically around 51. In the genome of *Bdellovibrio exovorus* JSS (CP003537.1), 53 putative ABC systems were detected, representing the highest number among all the BALOs genomes examined in this study. Unexpected high numbers of ABC systems involved in cellular processes were found in all BALOs. Phylogenetic analysis (Fig. 5.6) suggests that the majority of ABC proteins can be assigned into many separate families with high bootstrap supports (>50%). In this study, a general framework of sequence-structure-function connections for the ABC systems in BALOs was revealed providing novel insights for future investigations.

5.5.4 Summary

Genomic (and other "omic") information builds the foundation for a comprehensive analysis of environmental microbes. It becomes very important for environmental microbiologists to know how to utilize genomic resources—databases and computational tools—to enhance their research and to gain and share known knowledge in a useful way. One valuable option is to deposit the results of experimental research, especially high-throughput data, to public repositories. Submission of genomic and metagenomic sequencing and other similar data to public databases has become mandatory. Similarly, when publishing their research papers, it is necessary for authors to use standard database accession numbers that link with genes, proteins, and other data sets described in the paper. Since many electronic journals now provide hyperlinks to genomic databases, one can access relevant data with one click. It is clear that with the development of biological databases, research applications involving large datasets will play an increasingly important role in environmental microbiological discovery.

5.6 Discussion and Future Work

In this chapter we have provided a basic introduction to a few available open source tools and several applications to environmental related research for both

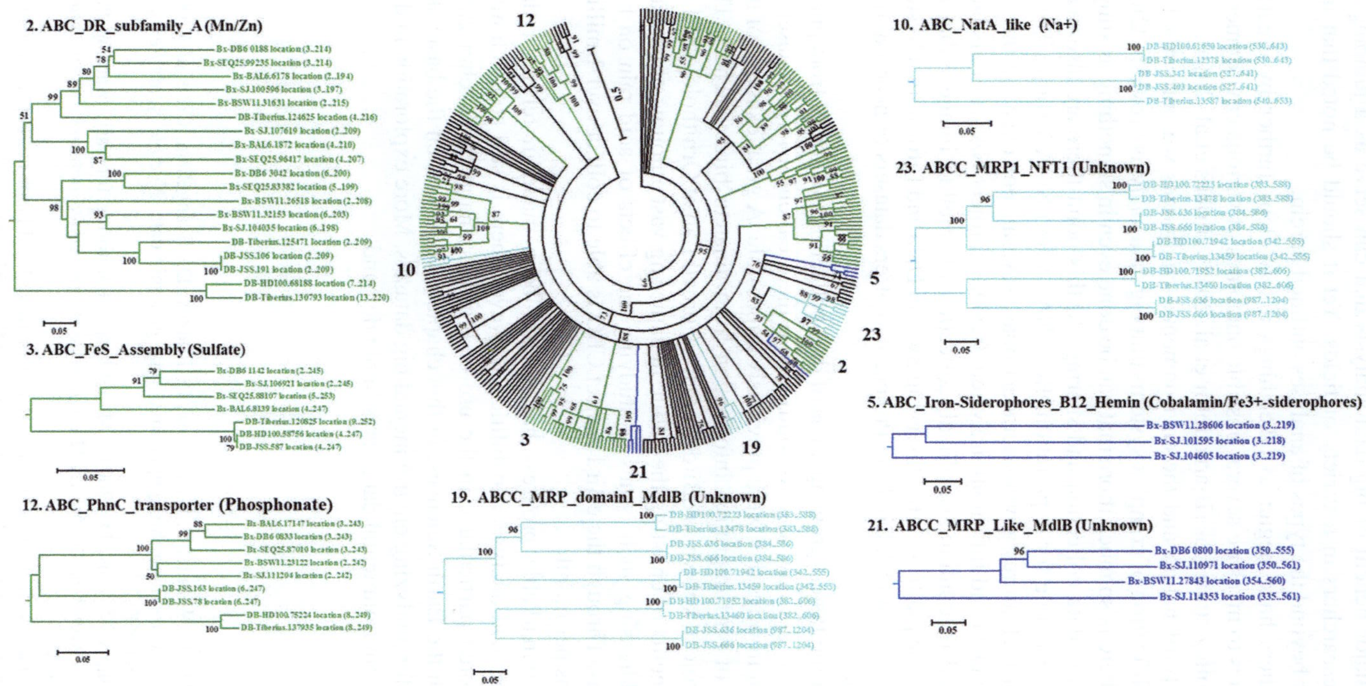

Fig. 5.6 Phylogenetic tree of all of the ABC systems in BALOs. The phylogenetic tree is constructed based on the ABC system domains of ABC systems. Strain names are shortened for brevity on the phylogenetic tree using the Neighbor-Joining method. The branches of 9 common ABC systems families are marked in *deep green*; the branches of expanded freshwater specific groups and salt water specific groups are separately marked in *deep blue* and *light blue*. Representative families were labeled with family name followed by putative substrate in bracket. *BD* Bdellovibrio, *Bx* Bacteriovorax

academic and policy related pursuits. Re-packaging complex information into useful summaries is a major benefit of big data analysis that can serve as a jumping off point for most researchers in a variety of fields. Yet it should be noted that it is possible to go far beyond the types of analyses outlined herein.

Some researchers have begun to take things a few steps farther by creating predictive analytics to measure several possible outcomes of various environmental adaptation and policy strategies (Bjarnadottir et al. 2011; Ulrichs et al. 2015).

The complexity of nature and the environmental challenges we currently face demand the simplification of complex systems that big data analysis provides. On an individual level there is so much more that can be done even in something as simple as increasing public data collection and sharing. At this point there are also many public databases available for exploration. Both NASA and NOAA (https://azure.microsoft.com/en-us/features/gov/noaacrada/) boast several. For example through the Earth Science Data and Information System (ESDIS) Project NASA offers large datasets comprised of continuous, global, collection of satellite imagery that are freely available to the public (Earth Observing System Data and Information System (EOSDIS) 2009). The NOAA Big Data Project also offers datasets of near real time environmental data.

Finally, one of the fastest growing areas of data collection is directly from public sources. Individuals are now able to contribute to the creation of large datasets to increase precision in ways that were previously impossible. A great example of this is the collection of bird migration information from amateur bird watchers. This citizen science project, known as eBird, from the Cornell Lab of Ornithology (CLO) highlights the benefits of enlisting the public in gathering vast amounts of data and has helped in addressing such issues as studying the effects of acid rain on bird migration patterns. Through the eBird project CLO is able to collect over 1 million avian observations per month.

At this time remarkable progress has already been made in terms of data collection, storage, and analysis capabilities. However, there is still so much more that can be explored particularly in the use of big data analytics (Kitchin 2014; Jin et al. 2015) in the Earth Sciences. In this chapter, we just touch the basic skill and applications for analyzing environmental big datasets. More exploration in data analysis tools for environmental datasets is in big demand.

References

Barrett, T., Troup, D. B., Wilhite, S. E., Ledoux, P., Rudnev, D., Evangelista, C., et al. (2009). NCBI GEO: Archive for high-throughput functional genomic data. *Nucleic Acids Research, 37*(suppl 1), D885–D890.

Basha, E. A., Ravela, S., & Rus, D. (2008). Model-based monitoring for early warning flood detection. In *Proceedings of the 6th ACM conference on embedded network sensor systems* (pp. 295–308). ACM.

Belasen, A. R., & Polachek, S. W. (2009). How disasters affect local labor markets the effects of hurricanes in Florida. *Journal of Human Resources, 44*(1), 251–276.

Bjarnadottir, S., Li, Y., & Stewart, M. G. (2011). A probabilistic-based framework for impact and adaptation assessment of climate change on hurricane damage risks and costs. *Structural Safety, 33*(3), 173–185.

Blake, E. S., Rappaport, E. N., & Landsea, C. W. (2007). The deadliest, costliest, and most intense United States tropical cyclones from 1851 to 2006 (and other frequently requested hurricane facts) (p. 43). NOAA/National Weather Service, National Centers for Environmental Prediction, National Hurricane Center.

Bossak, B. H., et al. (2014). Coastal Georgia is not immune: Hurricane history, 1851–2012. *Southeastern Geographer, 54*(3), 323–333.

Boyd, D., & Crawford, K. (2012). Critical questions for big data. *Information, Communication and Society, 15*(5), 662–679.

Cock, P. J., et al. (2015). NCBI BLAST+ integrated into galaxy. *Gigascience, 4*, 39.

Denman, S. E., Martinez Fernandez, G., Shinkai, T., Mitsumori, M., & McSweeney, C. S. (2015). Metagenomic analysis of the rumen microbial community following inhibition of methane formation by a halogenated methane analog. *Frontiers in Microbiology, 6*, 1087.

Earth Observing System Data and Information System (EOSDIS) (2009). Earth Observing System ClearingHOuse (ECHO) /Reverb, Version 10.X [online application]. Greenbelt, MD: EOSDIS, Goddard Space Flight Center (GSFC) National Aeronautics and Space Administration (NASA). URL: http://reverb.earthdata.nasa.gov.

Federhen, S. (2012). The NCBI Taxonomy database. *Nucleic Acids Research, 40*, D136–D143.

Fichant, G., Basse, M. J., & Quentin, Y. (2006). ABCdb: An online resource for ABC transporter repertories from sequenced archaeal and bacterial genomes. *FEMS Microbiology Letters, 256*, 333–339.

Frazier, T. G., Wood, N., Yarnal, B., & Bauer, D. H. (2010). Influence of potential sea level rise on societal vulnerability to hurricane storm-surge hazards, Sarasota County, Florida. *Applied Geography, 30*(4), 490–505.

Greenwood, P. L., Valencia, P., Overs, L., Paull, D. R., & Purvis, I. W. (2014). New ways of measuring intake, efficiency and behaviour of grazing livestock. *Animal Production Science, 54*(10), 1796–1804.

Guo, J., Peng, Y., Fan, L., Zhang, L., Ni, B. J., Kartal, B., et al. (2016). Metagenomic analysis of anammox communities in three different microbial aggregates. *Environmental Microbiology, 18*(9), 2979–2993.

Hampton, S. E., Strasser, C. A., Tewksbury, J. J., Gram, W. K., Budden, A. E., Batcheller, A. L., et al. (2013). Big data and the future of ecology. *Frontiers in Ecology and the Environment, 11*(3), 156–162.

Jin, X., Wah, B. W., Cheng, X., & Wang, Y. (2015). Significance and challenges of Big Data research. *Big Data Research, 2*(2), 59–64.

Kelly, K. (2007). What is the quantified self. *The Quantified Self, 5*, 2007.

Khedo, K. K., Perseedoss, R., & Mungur, A. (2010a). A wireless sensor network air pollution monitoring system. Preprint arXiv:1005.1737.

Khedo, K. K., Perseedoss, R., Mungur, A., & Mauritius. (2010b). A wireless sensor network air pollution monitoring system. *International Journal of Wireless and Mobile Networks, 2*(2), 31–45.

Kitchin, R. (2014). Big Data, new epistemologies and paradigm shifts. *Big Data & Society, 1*(1), 2053951714528481.

Kodama, Y., Shumway, M., & Leinonen, R. (2012). The International Nucleotide Sequence Database Collaboration. The sequence read archive: explosive growth of sequencing data. *Nucleic Acids Research, 40*, D54–D56.

Landsea, C. W., & Franklin, J. L. (2013). Atlantic hurricane database uncertainty and presentation of a new database format. *Monthly Weather Review, 141*(10), 3576–3592.

Landsea, C. W., et al. (2004). The Atlantic hurricane database re-analysis project: Documentation for the 1851–1910 alterations and additions to the HURDAT database. In *Hurricanes and typhoons: Past, present and future* (pp. 177–221).

Lehmann, R. J., Reiche, R., & Schiefer, G. (2012). Future internet and the agri-food sector: State-of-the-art in literature and research. *Computers and Electronics in Agriculture, 89*, 158–174.
Li, N., Chen, H., & Williams, H. N. (2015). Genome-wide comparative analysis of ABC systems in the Bdellovibrio-and-like organisms. *Gene, 562*, 132–137.
Ma, Y., Wu, H., Wang, L., Huang, B., Ranjan, R., Zomaya, A., & Jie, W. (2015). Remote sensing big data computing: challenges and opportunities. *Future Generation Computer Systems, 51*, 47–60.
Malmstadt, J., Scheitlin, K., & Elsner, J. (2009). Florida hurricanes and damage costs. *Southeastern Geographer, 49*(2), 108–131.
NHC Data Archive. Retrieved from <http://www.nhc.noaa.gov/data/hurdat/hurdat2-1851-2015-070616.txt>, June 7, 2016.
Nielsen, J., & Lee, S. Y. (2012). Systems biology: The 'new biotechnology'. *Current Opinion in Biotechnology, 23*, 583–584.
Powell, M. D., Houston, S. H., Amat, L. R., & Morisseau-Leroy, N. (1998). The HRD real-time hurricane wind analysis system. *Journal of Wind Engineering and Industrial Aerodynamics, 77*, 53–64.
Powell, M. D., Uhlhorn, E. W., & Kepert, J. D. (2009). Estimating maximum surface winds from hurricane reconnaissance measurements. *Weather and Forecasting, 24*(3), 868–883.
Pruitt, K. D., Tatusova, T., & Maglott, D. R. (2007). NCBI reference sequences (RefSeq): A curated non-redundant sequence database of genomes, transcripts and proteins. *Nucleic Acids Research, 35*, D61–D65.
Ravi, M., & Subramaniam, P. (2014). Wireless sensor network and its security—A survey. *International Journal of Science and Research (IJSR), 3*, 12.
Saffir, H. S. (1973). Hurricane wind and storm surge, and the hurricane impact scale (p. (423). The Military Engineer: Alexandria, VA.
Schott, T., Landsea, C., Hafele, G., Lorens, J., Taylor, A., Thurm, H., et al. (2012). The Saffir-Simpson hurricane wind scale. National Hurricane Center. National Weather Service. Coordinación General de Protección Civil de Tamaulipas. National Oceanic and Atmospheric Administration (NOAA) factsheet. URL: http://www.nhc.noaa.gov/pdf/sshws.pdf.
Shiffrin, R. M. (2016). Drawing causal inference from Big Data. *Proceedings of the National Academy of Sciences, 113*(27), 7308–7309.
Simpson, R. H., & Saffir, H. (1974). The hurricane disaster potential scale. *Weatherwise, 27*(8), 169.
South Florida Regional Climate Compact (SFRCCC) 2012. Analysis of the vulnerability of Southeast Florida to sea-level rise. Available online: http://www.southeastfloridaclimatecompact.org/wp-content/uploads/2014/09/regional-climate-action-plan-final-ada-compliant.pdf. Accessed 14 August 2016.
Swan, M. (2013). The quantified self: Fundamental disruption in big data science and biological discovery. *Big Data, 1*(2), 85–99.
Ulrichs, M., Cannon, T., Newsham, A., Naess, L. O., & Marshall, M. (2015). Climate change and food security vulnerability assessment. Toolkit for assessing community-level potential for adaptation to climate change. Available online: https://cgspace.cgiar.org/rest/bitstreams/55087/retrieve. Accessed 15 August 2016.
Wdowinski, S., Bray, R., Kirtman, B. P., & Wu, Z., et al. (2016). Increasing flooding hazard in coastal communities due to rising sea level: Case study of Miami Beach, Florida. *Ocean and Coastal Management, 126*, 1–8.

Chapter 6
High Performance Computing and Big Data

Rishi Divate, Sankalp Sah, and Manish Singh

6.1 Introduction

Big Data systems are characterized by variable and changing datasets from multiple sources across language, culture and geo-location. The data could be in various formats such as text, video or audio files. The power of Big Data analytics compared to traditional relational database management systems (RDBMS) or data warehouses is the fact that multiple disparate sources of information can be quickly analyzed to come up with meaningful insight that a customer or an internal user can take advantage of. Companies can build products or services based on the insights provided by Big Data analytics platforms.

Datasets themselves are growing rapidly and organizations are looking at a minimum of 20% increase in data volumes year over year, as a result of which Big Data systems need to be able to match up to the demand to be able to ingest, store and process them. In a survey by VansonBourne (2015), the top two drivers for new Big Data projects are improved customer experience and the need to get new customers.

Any perceived slowness wherein the insight is stale; loses the appeal and the value proposition of Big Data systems in which case customers will move to other providers and internal users will just not trust the data. For example, once you buy a product online from Amazon.com, you instantly get recommendations (Linden et al. 2003) about what else to buy given your recent and past purchases, never mind the kind of data-crunching that goes behind the scenes wherein large datasets are analyzed by highly optimized clusters of machines.

R. Divate • S. Sah • M. Singh (✉)
MityLytics Inc., Alameda, CA 94502, USA
e-mail: rishi@mitylytics.com; sankalp@mitylytics.com; mksingh@mitylytics.com

S. Srinivasan (ed.), *Guide to Big Data Applications*, Studies in Big Data 26,
DOI 10.1007/978-3-319-53817-4_6

In addition, with the growing advent of Internet of Things (IoT) Xia et al. (2012) and mobile phones sending and receiving data in real-time (Gartner n.d.), performance becomes even more critical. Imagine what would happen if you fire-up your favorite ride sharing application on your smartphone and you see information that is a few minutes old. This is inspite of the fact that the application backend needs to first collect location information from all cars in your city, sift out only the ones that are available, and then present the car and model that works best for you. All this has to happen within a few seconds for you to have enough faith to continue using it.

Therefore, when you are starting a Big Data project, the first step is to understand the kind of response time you are looking for; whether it is in hours, minutes, seconds or sub-seconds.

6.2 High Performance in Action

Now that we have established the need for high performance, let us look at how to make this happen.

6.2.1 Defining a Data Pipeline

To start with, let us define a data pipeline which is essentially how data flows through various components in a Big Data deployment system as shown in the figure below. Each component in the pipeline operates by splitting and replicating data up across multiple machines or servers, analyzing each piece of data and combining the results to either store categorized data or to come up with insight (Fig. 6.1).

6.2.1.1 Events

Events are pieces of information that's coming in from various sources often in real-time. This could be from other databases, mobile phones, IoT devices, or user-generated data such as instant messages.

6.2.1.2 Ingestion

Once events are received it is important that they are processed and categorized appropriately. Given that data is potentially coming at a very high rate, it is critical that ingestion works fast and does not miss any data. Example: LinkedIn developed Apache Kafka (Kafka Ecosystem n.d.), a messaging system specifically designed for handling real-time data feeds and that currently processes over a trillion messages a day.

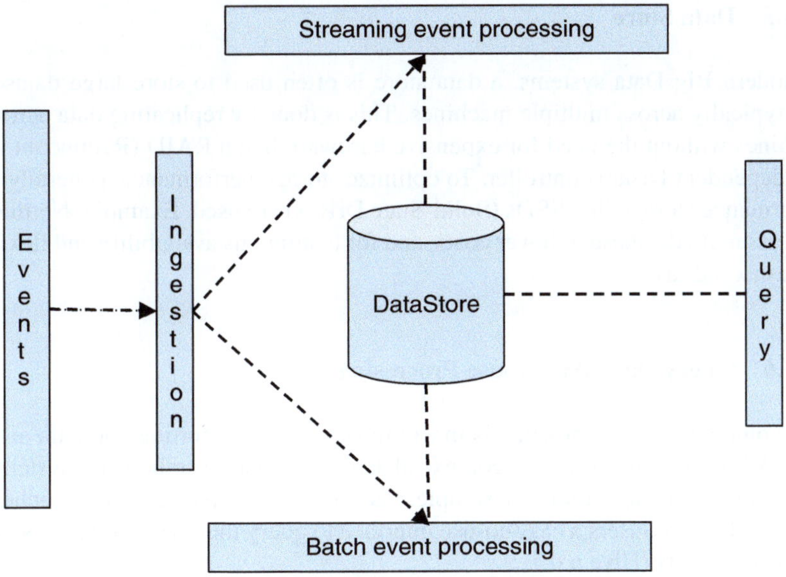

Fig. 6.1 Data pipeline

6.2.1.3 Streaming Event Processing

This component of the pipeline is specifically designed to analyze and store categorized data from the ingestion component in real-time. As data is being received, depending on how critical it is, it may also need to be made available to customers or end-users in a live view before it is stored in a datastore. Stream processing is typically done on smaller sets of data. Response time is sub-second or seconds at the most from when data is received. For example, Pinterest uses Apache Spark (Apache Spark™–Lightning-Fast Cluster Computing n.d.), a streaming processing framework to measure user engagement in real-time (MemSQL n.d.).

6.2.1.4 Batch Event Processing

Batch event processing is used for analyzing and storing high volumes of data in batches in a scheduled manner. In batch processing, it is assumed that the results of processing are not needed for real-time analysis unlike stream event processing. Batch processing can take minutes or hours depending on the size of the dataset. For example, Netflix runs a daily job that looks through the customer base to determine the customers to be billed that day and the amount to be billed by looking at their subscription plans and discounts (Netflix n.d.-b).

6.2.1.5 Data Store

In modern Big Data systems, a data store is often used to store large datasets or files typically across multiple machines. This is done by replicating data across the machines without the need for expensive hardware like a RAID (Redundant Array of Independent Disks) controller. To optimize storage performance, generally high-performance storage like SSDs (Solid-State Drives) are used. Example: Netflix uses the Cassandra database to lower costs, and for continuous availability and flexibility (Datastax n.d.-a).

6.2.1.6 Query/Data Warehouse Processing

This component of the pipeline is meant to be used for reporting and data analysis (Data Warehouse n.d.). In the context of Big Data, data warehousing systems are built upon distributed storage. Example: Facebook built Apache Hive (Apache Hive n.d.) to allow end-users a easy-to-use interface to query their dataset which spanned petabytes of data (Hive n.d.).

6.2.2 Deploying for High Performance

Given the difference in performance across various data processing, querying and storage frameworks, it is first critical to understand the performance requirements (seconds, minutes, hours) of your Big Data deployment, before actually deploying technology.

In subsequent sections, we will discuss performance related tradeoffs based on where you deploy your Big Data stack and the kind of applications that you will run.

6.3 High-Performance and Big Data Deployment Types

There are many different ways an enterprise can choose to deploy their Big Data pipeline. They have an option to choose from the various cloud based configuration-ready vendors, on-premise vendors, cloud based hardware vendors or Big Data as a service providers.

When an enterprise has to choose among the above mentioned options they typically weigh in several different deployment considerations that arise depending on the type of deployment.

Broadly, these considerations can be broken down into the following:

1. Provisioning
2. Deployability
3. Configurability

4. Manageability
5. Costs
6. Supportability

The components in the Big Data stack can be deployed in either one of the following broad categories:

1. Platform as a Service (PaaS) (Platform as a Service n.d.): Cloud based config-ready deployment
2. Cloud-based hardware
3. On-premise
4. Big Data as a Service

We will now start outlining each of these deployments focusing on what they do or do not offer in detail.

6.3.1 Platform as a Service (PaaS): Cloud Based Config-Ready Deployment

For someone looking to get started without spending a lot on hardware, IT teams to monitor and troubleshoot the clusters, cloud based config-ready providers are the best bet. The major issue seen here is the inflexibility. If problems arise on the cluster, it becomes very hard to discover the root cause for the failure given that various software components are pre-configured by the vendor. Another major issue is inconsistent performance seen at various times of the day, due to the multi-tenant nature. Since this type of solution comes with the Big Data software already installed, the operator doesn't have to install and configure most of the software pieces in the data pipeline. If however, you need to have control over what version or type of software needs to be installed, then this solution is not for you. In the short-term, costs are lower as compared to on premise vendors and time to deploy is quick. Example vendors include Amazon Web Services (AWS) (Amazon Web Services n.d.) Elastic MapReduce (EMR) (AWS EMR n.d.) offering and Microsoft Azure's HDInsight (Microsoft Azure n.d.-a) offering.

6.3.2 Cloud-Based Hardware Deployment

This is similar to cloud based config-ready solutions except that you can pick and choose the Big Data software stack. The vendor provides either dedicated bare metal or virtual machines on shared hardware with varied configurations (CPU, memory, network throughput, disk space etc). The operator needs to install and configure the Big Data software components. That is as far as the flexibility goes as compared to config-ready cloud-based solutions. The option to have a single tenant on dedicated bare metal machines gives more predictability and consistency. Example vendors include Packet.net (Premium Bare Metal Servers and Container Hosting–Packet n.d.), Internap (Internap n.d.), IBM Softlayer (SoftLayer n.d.) and AWS EC2 (Amazon Elastic Compute Cloud n.d.) and Microsoft Azure (Microsoft Azure n.d.-b).

6.3.3 *On-Premise Deployment*

With this deployment type, an in house IT team procures and manages all hardware and software aspects of Big Data clusters. Costs and security are generally the top drivers for these kinds of deployments. If data or processing needs change, in-house staff will have to add or remove machines and re-configure the cluster. In the long run, on-premise deployments can be more cost effective, flexible, secure and performant, but time to deploy is much slower compared with cloud based deployments.

6.3.4 *Big Data as a Service (BDaaS)*

Big Data as a Service is an up and coming offering, wherein enterprises outsource the setup and management of the entire Big Data hardware and software stacks to a vendor. Enterprises use the analytical capabilities of the vendor's platform either by directly accessing the vendor's user interface or by developing applications on top of the vendor's analytics platform once the data is ingested.

Internally, BDaaS systems can comprise of:

1. Platform as a Service (PaaS) and Infrastructure as a Service (IaaS) bundled together
2. PaaS and Software as a Service (SaaS) bundled together
3. IaaS, PaaS and SaaS bundled together

The implementation of the platform itself is transparent to the enterprise and is usually highly optimized by the vendor for costs. Time-to-deploy will be much faster than all of the above options but costs will generally be much higher. Increase in performance will also result in an increase in the price of these systems. Example vendors include Qubole and Altiscale.

6.3.5 *Summary*

If Big Data software was given scores based on the six considerations (Provisioning, Deployability, Configurability, Manageability, Costs, Supportability), mentioned earlier, then generally, the cost is higher for higher scores on other factors but performance per dollar is still pretty low.

To elaborate, the performance of the PaaS systems is generally lower per dollar than other systems because today the systems for general availability are still at a stage where applications are being transitioned from older systems and the performance gain is considerable for custom built clusters. For example in banking sectors newer BI applications realize order of magnitude gains with customised cluster deployments. For next-generation systems, which perform real-time analytics on high volumes of data coming in from all the devices in the world performance considerations become paramount, which we see as the biggest challenge for Big Data platforms going forward.

Now, that we have identified deployment types and their tradeoffs, let us look at specific hardware and software considerations. We do not consider pure BDaaS (Big Data as a Service) systems in our analysis, because one generally do not have control over the kind of hardware and software combinations provided by those types of vendors.

6.4 Software and Hardware Considerations for Building Highly Performant Data Platforms

6.4.1 Software Considerations

The Fig. 6.2 below shows sample Big Data ingestion, processing and querying technologies. All of the technologies shown (Hive (Apache Hive n.d.), Flume (Apache Flume n.d.), Pig (Apache Pig! n.d.), Spark (Apache Spark™–Lightning-Fast Cluster Computing n.d.), Hadoop-MR (MapReduce n.d.), Storm (Apache Storm n.d.), Phoenix (Apache Phoenix n.d.) are open-sourced via the Apache software foundation.

Let us look at what drives the requirements for software stacks.

6.4.1.1 Data Ingestion Stacks

Rate of Data Ingestion

Millions of events/second corresponding to tens and sometimes hundreds of writes per day as reported by companies such as Linkedin and Netflix (Netflix n.d.-a). This will likely increase given the expected proliferation of sensors driven by an IoT world. An organization hoping to handle large streams of data should plan on systems that can handle 10s of millions of events/messages per-second.

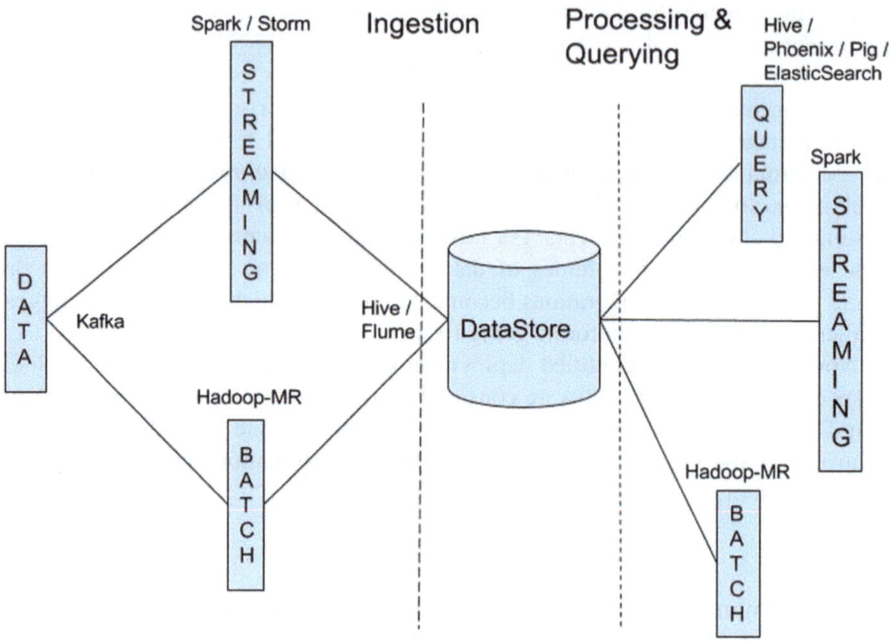

Fig. 6.2 Big Data ingestion technology

Replication for High Availability

Replication is a feature of most Big Data systems, wherein multiple copies of the same data exist across various storage devices. Distributed software stacks such as Kafka are very good at replication and having this feature in the software makes it possible for data pipelines to work at a higher level of abstraction without the need for specialized hardware such as RAID. In order to replicate efficiently one has to establish the amount of replication levels, depending on the access patterns of the data from different clients. Clients may include stream processing engines such as Spark (Apache Spark™–Lightning-Fast Cluster Computing n.d.) and/or batch processing frameworks such as Hadoop (MapReduce n.d.). A normal replication factor is three and if needed it should be increased but this would typically mean that more RAM and storage potentially could be needed.

Replication for Fault-Tolerance

There maybe scenarios in which one requires replication so that if data pipelines fail then data events or messages can be retained to be accessed at a later stage. Typically storage in the order of terabytes should be set aside for this type of fault-tolerance given that typically a day's worth of data would need to be accessible by a batch processing engine.

Low Latency Processing Hooks

The message or event ingestion system should be able to provide hooks for data processing (e.g. Storm spouts) (Apache Storm n.d.), where processing systems can hook into. What is important here is low latency and some amount of buffering. The amount of buffering would typically depend on the latency tolerance of the system. So for example data processing systems that can tolerate hundreds of milliseconds of delay should have that kind of buffering built into the event engine or in the data processing engine. Doing some simple math, 100 milliseconds of delay with 10 million messages coming in per second translates to a million messages worth of data being buffered, which would be amount to ~200 MBytes of buffer space assuming average message size of 200 bytes. Buffering also happens between the storage system and the data ingestion system but that would be typically be hardware buffering.

6.4.1.2 Data Processing Stacks

Low latency (10s of milliseconds of delay) is the maximum delay that should be planned since real-time analytics means no further delays can be tolerated. When considering interfaces for writing to data stores, fast interfaces like Storm (Apache Storm n.d.) bolts are key.

6.4.1.3 Data Stores

A few important considerations for Data stores that are universal are:

1. Read performance—Latency and throughput for reads from the storage system.
2. Write performance—Latency and throughput for writes to the storage system.
3. Query performance—Throughput in queries/second and individual query performance.

6.4.1.4 Indexing and Querying Engine

Indexing and querying engines are increasing being used as front-ends and through well published APIs they are being integrated into applications. They are available on-premise as as well as a service in the cloud. Some of the examples are ELK (ELK Stack n.d.) stack or Elastic, keen.io (Keen n.d.) and Solr their key performance indicators are

- *Latency*—100 milliseconds and speed of indexing
- *Throughput*—A measure of how many concurrent queries can be run
- *Real-Time*—The responses for the system should be in real-time or near real-time

- *Performance with scale*—How the system scales up with more queries and how indexing performs
- *Replication*—An important question to ask is, are replicas real-time too

6.4.2 Getting the Best Hardware Fit for Your Software Stack

The building blocks for each component cluster of the Data pipeline software stack are discussed in detail here. Given that Big Data systems use a distributed model, where data storage and analytics is spread out across multiple machines connected via a high-speed network, the discussion is framed in terms of the following components:

1. Compute nodes
2. Storage
3. Networking

We will be using hardware from AWS (Amazon Web Services n.d.) as an example and will use information that is publicly available from the AWS website (Amazon Web Services n.d.).

6.4.2.1 Data Ingestion Cluster

For a Data ingestion cluster such as Kafka (Apache Kafka n.d.), a million writes per second is a goal that is achievable with commodity machines and that is something that happens with systems today that have up to 10s of billions of writes per day (e.g. a social media site like Facebook or Linkedin) Here are the node types that would be required in the different cloud providers' offerings.

A million writes per second translates to around 80 billion writes a day, par for the course for data processing systems in industries such as advertising or other verticals such as finance. Given that one would need at least a day's worth of data to be saved on disk for retrieval from batch processing systems one would imagine that the data ingestion cluster would need to have around 80,000,000,000 * 300 = 24,000,000,000,000 = 24 terabytes of storage, given that each event or message size is on average 300 bytes. Add a 3x replication and that amounts to 72TB of storage. So the process of choosing an instance type (EC2 Instance Types—Amazon Web Services (AWS) n.d.) for this cluster in AWS goes something like this.

AWS

- Compute nodes
  - EC2, i2.xlarge

- Storage
  - Elastic Block Storage (EBS) (Amazon Elastic Block Store n.d.) or directly attached SSD. A combination would need to be used given the large amount of data. However with more EBS (Amazon Elastic Block Store n.d.), the access times for log messages goes up. But it depends on how many events need to be available in real-time. Up to a terabyte of data can be available and replicated 3x on the i2.xlarge nodes with 6 nodes that are listed here which should be fine for most systems.
- Networking
  - Enhanced 1Gig or 10Gig. If the data ingestion cluster is to be co-located with the Data processing cluster then a moderate network performance cluster (400–500 Mbits/s) cluster that comes with the i2.xlarge system will suffice. However the needs of the data processing cluster, which requires large amounts of RAM (120GB per machine for a message processing system that can keep latency at 100s milliseconds with a million messages per second) will drive the choice towards r3.4xlarge, which in any case will mean High-Performance (700/800 Mbits/s) networking comes with it. The cost of this cluster would be $7.80 hour just for ingestion. The cost of data processing cluster would be a add-on. One needs to be mindful of the resource contention that might occur if these are in a co-located cluster, where multiple software stacks are running at the same time. If the ingestion cluster were to be separate from the data processing cluster then the choice would be to go with 6 nodes of c3.8xlarge, which is somewhat of an overkill but that is AWS pricing for you. The cost of the above c3.8xlarge cluster would be $10/hr. which amounts to $240/day for annual costs of $86,400, which excludes costs for support, tertiary storage and bandwidth. One could get a yearly discount of 45% if pre-paid. So one could expect annual costs of around $50,000 for just the ingestion cluster.

Let us now consider an on-premise cluster with the same characteristics as above.

On-Premise

In order to do 1 million writes per-second, we expect 6 machines with the following configurations, but again the networking is only 1Gbps so if the Data processing cluster is on a separate cluster then the networking will not suffice and if the data processing cluster is co-located with the messaging cluster then more RAM will be required, at least 120GB of RAM. So a good configuration for a non co-located cluster would be

- Six SSD drives with 1 TB each
- 120GB of RAM
- 10Gbps Ethernet

6.4.2.2 Stream or Data Processing Cluster

Streaming processing engines come in different shapes, sizes and profiles. Examples of these engines include Spark (Apache Spark™–Lightning-Fast Cluster Computing n.d.) and Concord (n.d.).

They work with different models of execution for example micro-batches so their needs are different. However most work in-memory and therefore require large amounts of memory for low-latency processing. For example, an aggregate RAM of 300GB is required to run a benchmark such as spark-perf (Databricks spark-perf n.d.) at a scale of 1.0. Balancing resource usage of a data processing cluster with other clusters is key to building a robust data pipeline.

Hardware Requirements for Data Processing Cluster

So if one were to go for the cheapest option in AWS (Amazon Web Services (AWS) n.d.) which is 20 m^3.xlarge nodes, which offers 300 GB of RAM and about 1.6 TBs of storage across the cluster, then if the data storage cluster is co-located as is being recommended by a lot of vendors, clearly the amount of total storage available is not enough, neither is the RAM which only suffices to run the Data processing cluster. It is also not suitable when changing over to a non-co-located Data processing cluster, since network throughput is only about 800 Mbps.

For a co-located cluster based on RAM, CPU and disk the best configuration in terms of cost/performance tradeoff would be r3.8xlarge nodes primarily because of high RAM and high storage amounting to 244 GB and 640 GB respectively.

For a non co-located cluster the best node configuration would be c3.8xlarge, given that there is 10Gbps network connectivity with that config and 64 GB of RAM per-node so about 5 nodes would be required from a RAM perspective.

6.4.2.3 Data Stores

Apache Cassandra (Cassandra n.d.), Aerospike (Aerospike High Performance NoSQL Database n.d.), ScyllaDB (World Fastest NoSQL Database n.d.), MemSQL (Real-Time Data Warehouse n.d.) and more come out everyday. The best hardware fit for a given cluster depends on the specific software stack but generally veer towards high RAM and large amounts of directly attached SSD. The vendors of such stacks typically overprovision hardware to be on the safe side so configurations such as 32 cores, 244 GB of RAM and 10 TB of storage per node are common. The vendors of such stacks typically recommend co-locating the data processing cluster with the data storage cluster for example Cassandra data nodes being co-located with Spark workers. This results in contention for resources and requires tuning for peak performance. It has also been observed by us while implementing projects at large customers, that performance is dependent on the fit of the data model to the underlying hardware infrastructure. Getting RoI (Return on investment) and keeping TCO (Total cost of ownership) low becomes a challenge for enterprises trying to deploy these clusters. One of the best starting points for companies looking to deploy high performance clusters is to run benchmarking tools and figure out the performance of their hardware in terms of the number of peak operations that their infrastructure can support and then work backwards to profile their applications on smaller PoC (Proof of concept) clusters to characterize the performance in terms of peak operations and get trends. This will help IT teams to better plan, deploy, operate and scale their infrastructure. Good tools are therefore an essential requirement for such an exercise.

6.4.2.4 Indexing and Query-Based Frameworks

There are quite a few of these frameworks available as services in the cloud and or on-premise such as the Elastic Stack (Elastic n.d.), Keen.io (Labs n.d.) and Solr (Apache Lucene™ n.d.). All these frameworks are heavily memory-bound. The number of concurrent queries that can be supported typically is hundreds of thousands with a response time of less than a second, with a cluster of 20 nodes, where each node has at least 64 GB RAM. The challenge is to keep up the performance with more concurrent queries per second. The more the number of replicas the faster the query response will be with higher number of concurrent requests. At some point the number of queries per second does not improve with more memory, at which point, one has to then go ahead and change the caching policies on the cluster to be able to serve queries in near real-time. The key problems that we generally observe for scale and performance are Java garbage collection, cache performance and limitations on the size of heap memory allocated to the Java Virtual Machine (JVM) if present. We see this area as a single most important area of research and improvement.

6.4.2.5 Batch-Processing Frameworks

Batch processing MapReduce frameworks (MapReduce Tutorial n.d.) such as Hadoop have gained in popularity and have become more common. In order to find the right hardware fit for these frameworks one needs to figure out the peak performance, sustained and minimum performance using framework operations per-second that can be executed on the cluster. This can be done by running different types of benchmarks applications which have been categorized into I/O bound such as disk, network, compute or RAM bound jobs in the cluster and observing the operations per-second that can be achieved with different types of workloads. In our tests typical configurations which offer good price-performance tradeoff for running an application such as Terasort (O'Malley 2008) with 1 TB size results in about 12TB of storage being needed for a replication factor of three with five nodes, each of which has eight cores and 64GB of RAM. To maximize cluster usage and drive down RoI time, one needs to have a good job mix in the cluster with a good scheduling mechanism so as to prevent contention for the same resources (memory or network).

6.4.2.6 Interoperability Between Frameworks

In this section we discuss the possibility of running two frameworks together, wherein either both run on the same cluster or on different clusters. We show what to expect with Cassandra-Spark and Kafka-Spark combinations.

Cassandra and Spark Co-Located

The recommendation today from the vendors of Spark and Cassandra namely is to co-locate Spark and Cassandra nodes. To do so let's examine what kind of hardware resources, we will need:

- Both stacks use significant memory, therefore we recommend at least hundreds of Gigabytes of memory on each node.
- Since Cassandra is a data-store, we recommend state-of-the-art SSD or NVME (Jacobi 2015) drives and at least 1 TB storage per node.
- Spark will use significant CPU resources, given it is a processing engine, therefore, we recommend 10s of CPUs per node.
- Based on our experiences, we recommend the peak network bandwidth of at least 1Gbps. This should suffice for the two frameworks to operate in unison given that Spark processing should be restricted as far as possible to data stored on the same node.

Cassandra and Spark located on Different Machines

With this configuration, we are looking at the following requirements:

- Moderate RAM requirements, which could be as low as tens of Gigs of RAM per node
- Storage requirements will still be high for Cassandra nodes and we recommend at least 500 GB of SSD/NVME per node
- Networking requirements in this case are much higher given that data will be continuously transferred between Spark and Cassandra, therefore we recommend a network with peak bandwidth of at least 10 Gbps.
- In this case, we recommend tens of cores for each node on the Spark nodes and about half of those for the Cassandra nodes.

Even with the above ballpark figures it is hard to estimate the exact capacity of such clusters therefore one needs to have ways and means to figure out how to estimate the capacity of such clusters from looking at existing clusters and performing profiling to look for operation/seconds so that capacity can be determined.

Spark and Kafka

If Spark nodes and Kafka nodes are to be deployed together then some of the most important considerations are RAM and the contention for it and the disparate amounts of storage required. Kafka requires moderate amounts of storage (in the terabyte range while streaming events of size 300 bytes at a million messages per second), while Spark typically works with in-memory data sets although some amount of buffering maybe planned to avoid data loss. This data can be used by batch processes or by Spark also with a small delay. There are both receiver and receiver-less approaches to reading data from Kafka into Spark. Receiver-less approaches typically result in lower latencies.

Therefore, Spark nodes and Kafka nodes can be co-located but one needs to monitor RAM usage closely and take appropriate action to start buffering if memory starts running out, so that messages are not lost. If there is sufficient network throughput with link bonding (20-40Gbps) and redundant links then it is best to separate the clusters. One caveat though is monitoring and effective resource utilization so that resource managers do not schedule any network intensive workloads on these clusters. As we will see in following sections on hyper-converged infrastructure that there may be another way of solving this conundrum of co-location of different clusters when planning for high performance.

6.4.2.7 Summary

In Sect. 6.4.2, we have looked at how to get the best hardware fit for your software stack and what stands out is that it is a non-trivial exercise to figure out what hardware to start your deployment with and keep it running for high performance.

We recommend understanding the workings and limitations of each component (ingestion/processing/query) in your data pipeline and specifically identifying the following characteristics:

- Number of events per second or millisecond generated by each component
- Latency between components
- Overall Data volume (number of bytes ingested/stored/processed)

Given the above information, one can move on to the next stage of planning the size of your overall production cluster (more details on how to do so are explained in Sect. 6.5) and determining various co-location strategies.

6.4.3 On-Premise Hardware Configuration and Rack Placement

Based on the discussion so far when designing a cluster for deployment in a datacenter on-premise. The following are the options and the most preferred deployment scenarios for each.

6.4.3.1 On-Premise

1. Indexing and Query engines on one Rack
2. Spark and Cassandra on one rack
3. Kafka on another rack

Racks that are typically laid come with a redundant top-of-rack switch connected by a leaf spine topology in a data center. The top-of-rack switches should have dual connectivity to the servers in the rack which implement storage as well as cluster nodes. Cassandra is rack aware so replicas are typically placed on different racks for redundancy purposes. Enterprise Cassandra vendors such as DataStax (DataStax n.d.-b) advise against using distributed storage so all storage should be local to the racks.

6.4.3.2 On-Premise Converged Infrastructure

Converged infrastructure offerings from companies like Nutanix (Nutanix n.d.) are a new way of providing all infrastructure elements in one chassis. Nutanix implements it's own virtualized infrastructure to provide compute and storage together. The benefits are obvious, however virtualized infrastructure has challenges for high performance computing due to multi-tenancy. The typical deployment for such infrastructure would be to have all data pipeline stacks on one chassis in a rack with inter-chassis replication for fault tolerance in case of rack failures.

6.4.4 Emerging Technologies

6.4.4.1 Software Defined Infrastructure (SDI)

Software Defined Infrastructure (SDI) encompasses technologies such as Software Defined Networking (SDN) (Software-Defined Networking (SDN) Definition n.d.), Software Defined Compute (SDC) (Factsheet–IDC_P10666 2005) and Software

Defined Storage (SDS) (A Guide to Software-Defined Storage n.d.) is a new paradigm of deploying infrastructure, where the hardware is logically separated from the software that manages it and infrastructure can be programmatically provisioned and deployed with a high degree of dynamic control as opposed to statically allocating large sets of resources. This paradigm is perfectly suited for Big Data applications since these technologies can effectively be used to dynamically scale infrastructure based on data volumes or processing needs in an automated manner based on defined SLA's or performance requirements.

Software Defined Networking (SDN)

Software defined Networking is defined by technologies or switches, that allow operators to control and setup networking flows, links or tunnels. A typical use case of SDN is reconstructing a network topology on top of an underlying physical topology to avoid hotspots. Vendors in this space include the Cumulus Virtual Appliance (Better, Faster, Easier Networks n.d.) and Cisco's (n.d.) SDN API's on its routers.

Software Defined Storage (SDS)

Software Defined Storage (SDS) is an approach to data storage in which the programming that controls storage-related tasks is decoupled from the physical storage (Virtual Storage n.d.). What this means in a Big Data context for is the ability to provision storage dynamically as datasets increase. This could include different storage types (such as SSD, HDD's) that reduce hardware and storage management costs. Application performance may or may not be directly affected depending on how the SDS solution is affected but this is definitely a long-term solution for scalable storage solutions. Vendors in this space include EMC's ScaleIO (ScaleIO/Software-defined Block Storage n.d.) solution and HP's StoreVirtual VSA (Virtual Storage n.d.) solution.

Software Defined Compute (SDC)

Software Defined Compute (SDC) is about adding new servers or removing servers in a cluster on-demand as processing goes up or down automatically. This is especially important in Big Data systems as one can start scaling up if we see the compute performance of the cluster is going below desired levels and needs additional resources. SDC can either be achieved by virtualization with vendors such as VMware in a data-center or via cloud-based providers such as AWS EC2.

Since these are emerging technologies, not a whole lot of research has been done to showcase the benefits in Big Data scenarios and we look forward to such case studies going forward.

6.4.4.2 Advanced Hardware

NVME for Storage

Non-Volatile Memory Express or NVME (Jacobi 2015) is a communications interface protocol that enables the SSD to transfer data at very high speed as compared to the traditional SATA or SAS (Serial Attached SCSI). It makes use of the high-bandwidth PCIe bus for communication. With faster speeds and lower latency as compared to SATA (Wilson, R. (2015)), the time taken to access the storage to process and shuffle the data in the Big Data deployment cluster is greatly reduced.

HyperConverged Infrastructure

This involves combining compute, storage and networking into a single chassis. As compared to the traditional infrastructure the performance improves due to the lower latency in transferring data among compute, storage and networking nodes. Nutanix (Nutanix n.d.) and Hypergrid (Hyperdrive Innovation n.d.) are few of the companies that provide hyper-converged infrastructure equipment to enterprises.

6.4.4.3 Intelligent Software for Performance Management

In Sect. 6.4, we have articulated various hardware and software considerations for determining the performance of a Big Data systems and given the sheer complexity and number of options, we at MityLytics find there is a need for intelligent software that:

- Measures and reports the performance of individual Big Data jobs, components and the data pipeline as a whole
- Makes and executes intelligent decisions about how to boost the performance of a data pipeline by optimizing the current infrastructure or by using API's to provision or deprovision SDI (Software Defined Infrastructure) elements given known or the current set of workloads
- Predicts future needs of the infrastructure based on the current performance and resources

We are so convinced that we need this intelligent data driven analytics software that we at MityLytics have been developing software to do exactly the above. To us it makes total sense to use analytics to solve the performance issues of analytics platforms.

6.5 Designing Data Pipelines for High Performance

Now, that we have outlined various considerations for high performance, let us start by seeing about how one can start measuring performance and designing a data pipeline so as to scale with varying dataset sizes (Fig. 6.3).

To do so, we at MityLytics typically go through the following steps in sequence for each component in the pipeline:

1. Identify a series of steps in Big Data application that describes what you typically need to do for that component in the Big Data pipeline; this may be a Big Data streaming or batch job that processes incoming messages or a Hive query that runs across large datasets. Let us call this set of steps as an application job.
2. Deploy a Big Data application on a small cluster size
3. Run an application job with a small dataset and measure time taken and infrastructure statistics (e.g. CPU utilization, memory consumed, bandwidth observed) across various machines in your cluster. At this point, you are just characterising how your application behaves and not whether it meets your end performance goal.
4. If an application is deployed on shared infrastructure, run your job multiple times and record average statistics across all your runs.
5. Increase the size of the dataset and repeat steps 2 and 3.
6. On getting enough data points we can plan infrastructure needs for the application as the dataset is scaled up.

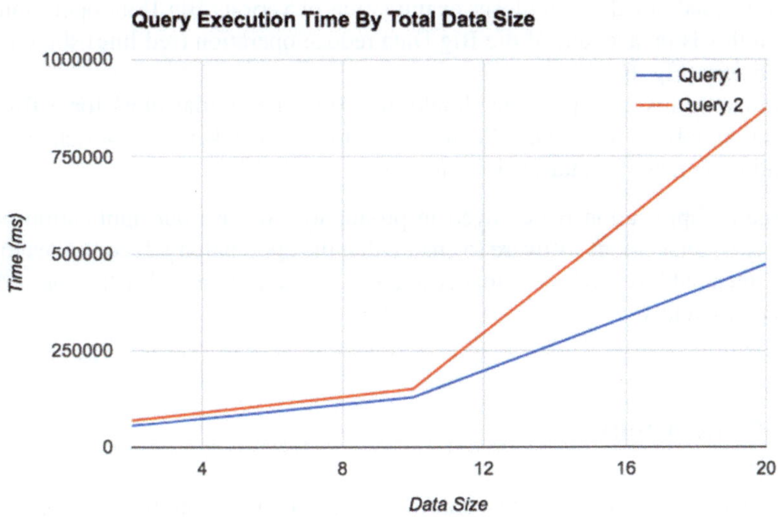

Fig. 6.3 TPC-DS benchmark

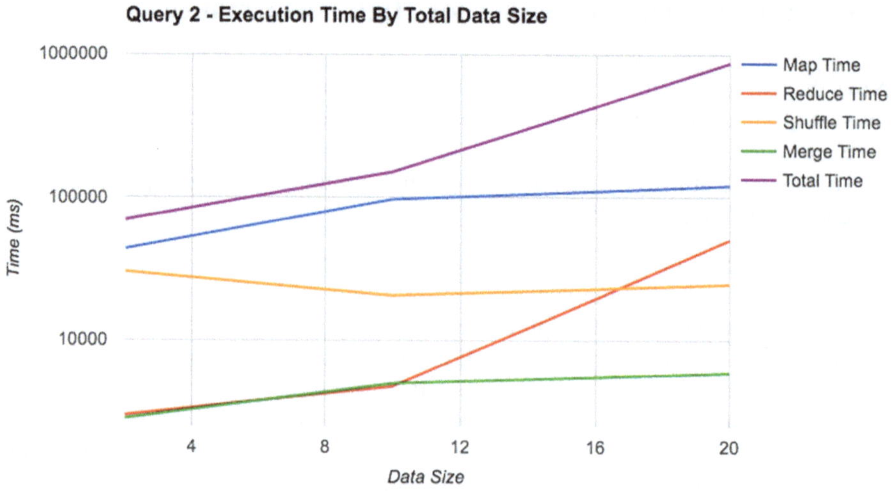

Fig. 6.4 Big Data performance measure

7. For example, in the graph below, we have plotted dataset size by time taken in milliseconds to execute two Big Data queries from the TPC-DS (TPC-DS n.d.) benchmark.

   In this example, we see that initially, the time taken to execute these queries is linear with increase in the data set size but as datasets increase, the time taken increases quadratically. On closer examination of various Big Data operations, we see that this is on account of the Big Data reduce operation (red line) shown in the graph below (Fig. 6.4).

   Similar to this example, one should aim to create similar plots for a Big Data application job so as to be able to show how time taken and system resource utilization increase as dataset sizes increase.

8. Once an application is deployed in production, we use our application performance management software to ensure that the application job continues to meet the intended levels of performance and that the underlying infrastructure is being used efficiently.

6.6 Conclusions

As we have seen above, data analytics is key to the operation and growth of data driven businesses and deploying high performance analytics applications leads to topline growth. Given the importance of data analytics platforms in this environment, it is imperative that data pipelines work in a robust manner with Quality of Service (QoS) guarantees.

It is of the utmost importance then, to plan the data platforms with the right capacity planning tools so that desired levels of performance are met when the platforms are scaled up to handle more data. Once the platforms are deployed, it is imperative that they are continuously monitored using key metrics, which could potentially point to degradation of performance or worse interruptions in service, that could potentially result in loss of revenue. We at MityLytics (MityLytics n.d.) believe that analytics is so important that we use analytics to help drive the performance and hence the viability of analytics platforms.

What is even more desirable however, is to have software and mechanisms to fix problems proactively. When such auto remediation is not possible, support system tickets should be raised which can be manually remediated. The state-of-the-art today is such that operations teams engage in trial-and-error methodologies, which increase the time to RoI (Return on Investment) while increasing the total cost of ownership of the data platform thereby reducing the productivity of development and operations teams. It has also been our experience that consultants from software vendors are tempted to be safe when designing clusters and frequently over-provision resources resulting in increased time to RoI.

One of the most common problems deployments run into is resource contention, since most of the software stacks are designed to be greedy so that they can perform best when working in isolation. However, as we have seen in previous section, in most practical scenarios distributed software systems will be working together and sharing resources such as compute, storage and networking so it is very important that software stacks be evaluated, when multiple components are working in unison. In conclusion, we recommend using software tools to plan, monitor and auto-tune in real-time and perform proactive remediation (prevention). One should operate Big Data clusters with software that will learn, self-heal and use tools to do capacity planning when scaling up. Development processes should also be streamlined to incorporate planning and performance testing for scalability as early in the development cycle as possible to bring HPC like performance to Big Data platforms.

References

A Guide to Software-Defined Storage. (n.d.). Retrieved November 29, 2016, from http://www.computerweekly.com/guides/A-guide-to-software-defined-storage

Aerospike High Performance NoSQL Database. (n.d.). Retrieved November 29, 2016, from http://www.aerospike.com/

Amazon Elastic Block Store (EBS)—Block storage for EC2. (n.d.). Retrieved November 29, 2016, from https://aws.amazon.com/ebs/

Amazon Elastic Compute Cloud. (n.d.). Retrieved May 5, 2017, from https://aws.amazon.com/ec2/

Amazon EMR. (n.d.). Retrieved May 5, 2017, from https://aws.amazon.com/emr/

Amazon Web Services. (n.d.). *What is AWS?* Retrieved November 29, 2016, from https://aws.amazon.com/what-is-aws

Amazon Web Services (AWS). (n.d.). *Cloud Computing Services*. Retrieved November 29, 2016, from https://aws.amazon.com/

Apache Hive. (n.d.). Retrieved November 29, 2016, from https://hive.apache.org/
Apache Kafka. (n.d.). Retrieved November 29, 2016, from http://kafka.apache.org/
Apache Lucene™. (n.d.). *Solr is the popular, blazing-fast, open source enterprise search platform built on Apache Lucene™*. Retrieved November 29, 2016, from http://lucene.apache.org/solr
Apache Spark™—Lightning-Fast Cluster Computing. (n.d.). Retrieved November 29, 2016, from http://spark.apache.org/
Apache Storm. (n.d.). Retrieved November 29, 2016, from http://storm.apache.org/
Better, Faster, Easier Networks. (n.d.). Retrieved November 29, 2016, from https://cumulusnetworks.com/
Cassandra. (n.d.). *Manage massive amounts of data, fast, without losing sleep*. Retrieved November 29, 2016, from http://cassandra.apache.org/.
Cisco. (n.d.). Retrieved November 29, 2016, from http://www.cisco.com/
Concord Documentation. (n.d.). Retrieved November 29, 2016, from http://concord.io/docs/
Data Warehouse. (n.d.). Retrieved November 29, 2016, from https://en.wikipedia.org/wiki/Data_warehouse
Databricks Spark-Perf. (n.d.). Retrieved November 29, 2016, from https://github.com/databricks/spark-perf
Datastax. (n.d.-a). *Case Study: Netflix.*. Retrieved November 29, 2016, from http://www.datastax.com/resources/casestudies/netflix
DataStax. (n.d.-b). Retrieved November 29, 2016, from http://www.datastax.com/
EC2 Instance Types—Amazon Web Services (AWS). (n.d.). Retrieved November 29, 2016, from https://aws.amazon.com/ec2/instance-types/
Elastic. (n.d.). *An introduction to the ELK Stack (Now the Elastic Stack)*. Retrieved November 29, 2016, from https://www.elastic.co/webinars/introduction-elk-stack
Gartner. (n.d.). *Gartner says the internet of things will transform the data center*. Retrieved November 29, 2016, from http://www.gartner.com/newsroom/id/2684915
Google Research Publication. (n.d.). *MapReduce*. Retrieved November 29, 2016, from http://research.google.com/archive/mapreduce.html
Hive. (n.d.). *A Petabyte Scale Data Warehouse using Hadoop–Facebook*. Retrieved November 29, 2016, from https://www.facebook.com/notes/facebook-engineering/hive-a-petabyte-scale-data-warehouse-using-hadoop/89508453919/
Hyperdrive Innovation. (n.d.). Retrieved November 29, 2016, from http://hypergrid.com/
Jacobi, J. L. (2015). *Everything you need to know about NVMe, the insanely fast future for SSDs*. Retrieved November 29, 2016, from http://www.pcworld.com/article/2899351/everything-you-need-to-know-about-nvme.html
Kafka Ecosystem at LinkedIn. (n.d.). Retrieved November 29, 2016, from https://engineering.linkedin.com/blog/2016/04/kafka-ecosystem-at-linkedin
Keen, I. O. (n.d.). Retrieved May 5 2017, from https://keen.io/
Linden, G., Smith, B., & York, J. (2003). Amazon.com recommendations: Item-to-item collaborative filtering. *IEEE Internet Computing, 7*(1), 76–80. doi:10.1109/mic.2003.1167344.
MapReduce Tutorial. (n.d.). Retrieved November 29, 2016, from https://hadoop.apache.org/docs/r2.7.0/hadoop-mapreduce-client/hadoop-mapreduce-client-core/MapReduceTutorial.html
MemSQL. (n.d.). *How pinterest measures real-time user engagement with spark*. Retrieved November 29, 2016, from http://blog.memsql.com/pinterest-apache-spark-use-case/
Microsoft Azure. (n.d.-a). *HDInsight-Hadoop, Spark, and R Solutions for the Cloud/Microsoft Azure*. Retrieved November 29, 2016, from https://azure.microsoft.com/en-us/services/hdinsight
Microsoft Azure. (n.d.-b). Cloud computing platform and services. Retrieved November 29, 2016, from https://azure.microsoft.com/
MityLytics. (n.d.). *High performance analytics at scale*. Retrieved November 29, 2016, from https://mitylytics.com/
Netflix. (n.d.-a). *Kafka inside keystone pipeline*. Retrieved November 29, 2016, from http://techblog.netflix.com/2016/04/kafka-inside-keystone-pipeline.html

Netflix. (n.d.-b). *Netflix Billing Migration to AWS–Part II*. Retrieved November 29, 2016, from http://techblog.netflix.com/2016/07/netflix-billing-migration-to-aws-part-ii.html
Nutanix–The Enterprise Cloud Company. (n.d.). Retrieved November 29, 2016, from http://www.nutanix.com/
O'Malley, O. (2008, May). *TeraByte Sort on Apache Hadoop*. Retrieved November 29, 2016, from http://sortbenchmark.org/YahooHadoop.pdf
Overview/Apache Phoenix. (n.d.). Retrieved November 29, 2016, from http://phoenix.apache.org/
Performance without Compromise/Internap. (n.d.). Retrieved November 29, 2016, from http://www.internap.com/
Platform as a Service. (n.d.). Retrieved November 29, 2016, from https://en.wikipedia.org/wiki/Platform_as_a_service
Premium Bare Metal Servers and Container Hosting–Packet. (n.d.). Retrieved November 29, 2016, from http://www.packet.net/
Real-Time Data Warehouse. (n.d.). Retrieved November 29, 2016, from http://www.memsql.com/
ScaleIO|Software-Defined Block Storage/EMC. (n.d.). Retrieved November 29, 2016, from http://www.emc.com/storage/scaleio/index.htm
SoftLayer|cloud Servers, Storage, Big Data, and more IAAS Solutions. (n.d.). Retrieved November 29, 2016, from http://www.softlayer.com/
Software-Defined Compute–Factsheet–IDC_P10666. (2005). August 31, 2016, https://www.idc.com/getdoc.jsp?containerId=IDC_P10666
Software-Defined Networking (SDN) Definition. (n.d.). Retrieved November 29, 2016, from https://www.opennetworking.org/sdn-resources/sdn-definition
Spark Streaming/Apache Spark. (n.d.). Retrieved November 29, 2016, from https://spark.apache.org/streaming/
TPC-DS–Homepage. (n.d.). Retrieved November 29, 2016, from http://www.tpc.org/tpcds/default.asp
VansonBourne. (2015). The state of big data infrastructure: benchmarking global big data users to drive future performance. Retrieved August 23, 2016, from http://www.ca.com/content/dam/ca/us/files/industry-analyst-report/the-state-of-big-datainfrastructure.pdf
Virtual Storage: Software defined storage array and hyper-converged solutions. (n.d.). Retrieved November 29, 2016, from https://www.hpe.com/us/en/storage/storevirtual.html
Welcome to Apache Flume. (n.d.). Retrieved November 29, 2016, from https://flume.apache.org/.
Welcome to Apache Pig! (n.d.). Retrieved November 29, 2016, from https://pig.apache.org/
Wilson, R. (2015). *Big data needs a new type of non-volatile memory*. Retrieved November 29, 2016, from http://www.electronicsweekly.com/news/big-data-needs-a-new-type-of-non-volatile-memory-2015-10/
World fastest NoSQL Database. (n.d.). Retrieved November 29, 2016, from http://www.scylladb.com/
Xia, F., Lang, L. T., Wang, L., & Vinel, A. (2012). Internet of things. *International Journal of Communication Systems, 25*, 1101–1102. doi:10.1002/dac.2417.

[illegible]

Chapter 7
Managing Uncertainty in Large-Scale Inversions for the Oil and Gas Industry with Big Data

Jiefu Chen, Yueqin Huang, Tommy L. Binford Jr., and Xuqing Wu

7.1 Introduction

Obtaining a holistic view of the interior of the Earth is a very challenging task faced by the oil and gas industry. When direct observation of the Earth's interior is out of reach, we depend on external sensors and measurements to produce a detailed interior map. Inverse problems arise in almost all stages of oil and gas production cycles to address the question of how to reconstruct material properties of the Earth's interior from measurements extracted by various sensors. Inverse theory is a set of mathematical techniques to determine unknown parameters of a postulated model from a set of observed data. Accurate parameter estimation leads to better understanding of the material properties and physical state of the Earth's interior. In general, inverse problems are ill-posed due to the sparsity of the data and incomplete knowledge of all the physical effects that contribute significantly to the data (Menke 2012).

Recently, technological improvements in other fields have enabled the development of more accurate sensors that produce a larger amount of accurate data for well planning and production performance. These data play a critical role in the day-to-day decision-making process. However, the sheer volume of data, and the high-dimensional parameter spaces involved in analyzing those data, means

J. Chen • X. Wu (✉)
University of Houston, Houston, TX 77004, USA
e-mail: jchen84@uh.edu; xwu7@uh.edu

Y. Huang
Cyentech Consulting LLC, Cypress, TX 77479, USA
e-mail: yueqin.duke@gmail.com

Tommy L. Binford Jr.
Weatherford, Houston, TX 77060, USA
e-mail: tommy.binford@weatherford.com

S. Srinivasan (ed.), *Guide to Big Data Applications*, Studies in Big Data 26,
DOI 10.1007/978-3-319-53817-4_7

that established numerical and statistical methods can scarcely keep pace with the demand to deliver information with tangible business value. The objective of this article is to demonstrate some scalable deterministic and statistical algorithms for solving large-scale inverse problems typically encountered during oil exploration and production. Especially, we like to show how these algorithms can be fit into the MapReduce programming model (Dean and Ghemawat 2008) to take advantage of the potential speed up. Both the model and data involved in inverse problems contain uncertainty (Iglesias and Stuart 2014). We will address this fundamental problem by proposing a Bayesian inversion model, which can be used to improve the inversion accuracy in terms of physical state classification by learning from a large dataset.

7.2 Improve Classification Accuracy of Bayesian Inversion Through Big Data Learning

Many solutions of engineering and applied sciences problems involve inversion: to infer values of unknown model parameters through measurement. Accurate parameter estimation leads to better understanding of the physical state of the target. In many cases, these physical states can be categorized into a finite number of classes. In other words, the real objective of the inversion is to differentiate the state of the measured target; and the estimated parameters through inversion act as state indicators or a feature set that can be used to classify the state of the object. For example, well logging plays critical roles in meeting various specific needs of the oil and gas industry (Ellis and Singer 2007). Interpretation of the well logging measurement is a process of inversion. It has many applications for structural mapping, reservoir characterization, sedimentological identification, and well integrity evaluation. The interpretation of the inversion involves correlating inversed parameters with properties of the system and deducing the physical state of the system.

There exist many other situations in engineering and science problems where we need to evaluate the physical state of a target through indirect measurement. For example, scientists who are studying global climate change and its impact on our planet depend on the climate data collected through measurements from various sources, e.g., ocean currents, atmosphere, and speleothems. Researchers categorize ocean eddies through the study and interpretation of satellite oceanography data (Faghmous et al. 2013). All of these critical work involves inversion. In particular, the inversed results can be categorized into a finite number of classes. Past research of inverse problems concentrates on deducing parameters of a model and largely ignores their functionality as a feature set and indicator of the physical state, which is the ultimate objective for many problems. In this article, we suggest a common Bayesian inversion model and an inference framework that is computationally efficient for solving inverse problems. It is in our interests to investigate and provide answers to questions on how to extend the proposed framework to serve more general science and engineering problems that involve categorizing the result of an inversion.

7.2.1 Bayesian Inversion and Measurement Errors

Given a set of measurements, traditionally, parameters are estimated by analyzing discrepancies between the expected and the actual well logging response. The analysis defines the solution as the minimizer of a criterion between two components, the data model matching, and the regularizer. The data model matching defines the distance metric between the observed data and forward model outputs. The regularizer is a smoothing or sparse constraint.

What is missing from this model is the historical data matching, correction, and expertise of the engineer, which play a pivotal role in handling uncertainties in the data or model and minimizing systematic errors. In various fields that involve inversion, historical data keeps accumulating in a large scale. Many of the data sets have been studied and labeled by domain experts. However, only a small part of the data has been studied duo to its volume, variety and variability challenge. Under the explosive increase of the data, the massive unstructured data or so-called Big data brings about opportunities for discovering new values and helps us to gain more in-depth understanding of the hidden characteristics of the system. Our proposed Bayesian inversion model will take advantage of information existed in the large-scale dataset; learn the correlation between inversed parameters and possible physical states; and explore the dependency relationship among direct and indirect observed data, noise/error, and forward model. The target is to improve confidence in justifying and predicting the physical state of the measured target.

Measurement errors pose another challenge to the inverse problem. How best to integrate some knowledge of the measurement error into the context of a Bayesian inversion model has been the subject of much interest (Kaipio and Somersalo 2005). In the ideal cases that are void of any measurement errors, the data y and model parameters u are related through the forward model $y = f(u)$. However, assumptions made by the forward model may not include all factors that affect measurements. The accuracy of observed data is also a consequence of the execution, which adds another layer of uncertainty related to potential human errors. Uncertainty also comes from a lack of direct measurement of the surrounding environment and other coupling factors. The inadequate information will ultimately compromise the accuracy of predictions made by the forward modeling, which needs to be considered during the inverse process.

Although most parameters that characterize the physical properties are continuous, it is the physical state of the subject that is being concerned for solving the inverse problem. For example, properties of the material obtained through the inversion for the well logging are a set of surrogate measures that can be used to characterize the formation by its lithology, pore fluid, porosity, clay content, and water saturation. Therefore, the potential searching area for the parameter space is dramatically reduced when the inversion framework shifts its attention from obtaining an accurate numerical solution for parameters of the model to estimating the possibility of occurrence of a finite number of formation types. One of the objectives here is to propose a unifying approach for improving categorization accuracy and confidence through inversion. Our approach will explore the common mathematical framework that supports the new approach. Furthermore, we will investigate the computational efficiency, scalability and practical implementations of the proposed framework.

7.2.2 Bayesian Graphical Model

Data obtained through indirect measurement of a sensor network often has errors and unknown interferences that can affect the assessment of the relation between environmental factors and the outcome of measuring instruments. The precise mathematical description of the measurement process does not exist. Under the traditional Bayesian inversion framework, a simple way to incorporate surrogate measurement errors into the system is to construct a Berkson error model (Carroll et al. 2006), shown in Fig. 7.1a, if we know how to formulate the error through a rigorous analysis and replace the deterministic f with a stochastic function. A solution of the Bayesian inversion is to find a probability measure of $\tilde{x}$ given the data $\tilde{y}$ and locate the most probable point by searching the pool of points sampled from the posterior distribution. The inversion result does not provide an answer for the classification problem. Instead, it serves as an input for an external classifier. Considering that the parameter space consists of a collection of random variables, the system can be modeled as a stochastic process. Given a big data set collected through the similar data acquisition process, it is possible for us to learn the process autonomously. Alternatively, it is clear that any subset of the parameter space can only represent a finite number of physical states. Following the Bayesian approach, the posterior distribution of model parameters can be deduced given the measurement and prior knowledge of the correlation between parameters and corresponding physical states. In the natural environment, the occurrence of any of those physical states can be random, and they are independent to each other. Since the physical state is classified based on the value of the model parameters, the parameter space can be clusterized. The number of clusters depends on the number of physical states. The center of each cluster is chosen to be typical or

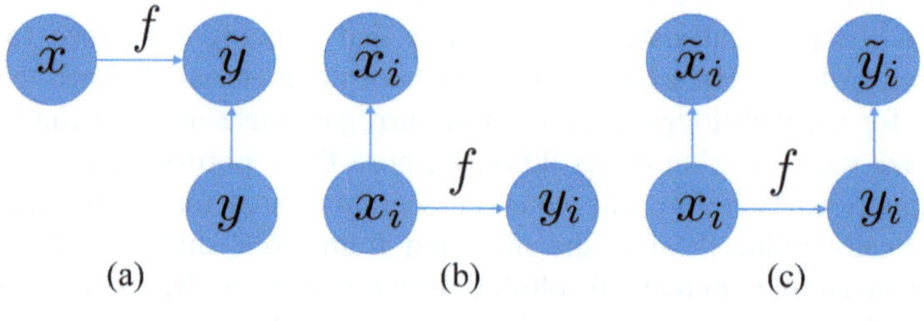

Fig. 7.1 Graphical model (**a**) Berkson error model (**b**) Dependency among $(x_i, \tilde{x}_i, y_i)$, (**c**) Dependency among $(x_i, \tilde{x}_i, y_i, \tilde{y}_i)$

Fig. 7.2 Resistivity vs. neutron-density

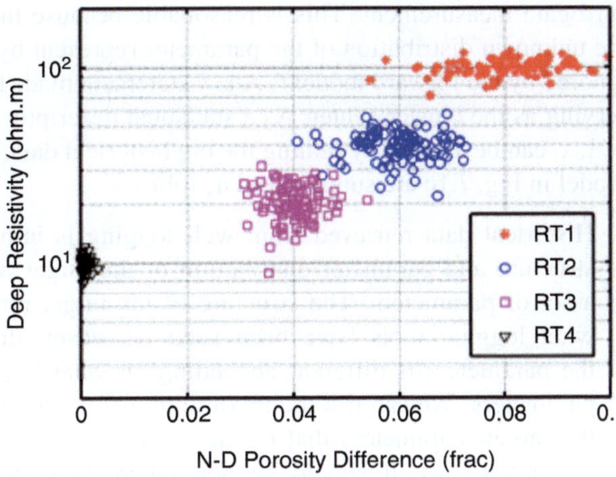

representative of the corresponding physical state. This can be generalized by an additive model or more specifically, a mixture model mathematically represented as $p(x) = \sum_{k=1}^{K} \lambda_k p_k(x)$, with the λ_k being the mixing weights, $\lambda_k > 0$ and $\sum_k \lambda_k = 1$. Figure 7.2 (Xu 2013) shows four rock types (RT1–4) and the distribution of corresponding parameters obtained by the resistivity and neutron-density log. We can use a hierarchical mixture model to formulate the uncertainty involved in identifying the current physical state. The mixture distribution represents the probability distribution of observations in the overall population. The mixture model is a probabilistic model for representing the presence of subpopulations within an overall population (Wikipedia 2016). The mixture is usually multimodal and can facilitate statistical inferences about the properties of the sub-populations given only observations on the pooled population. Rather than modeling all of those noisy details, errors can be estimated by simple statistical measures such as mean and variance.

Let us consider three variables x_i, $\tilde{x}_i$, and y_i for a physical state i. x_i is the parameter (x_i can be a vector) that characterizes the physical state i, y_i is the indirect measurement, and $\tilde{x}_i$ is deduced through an inversion process related to the forward model $y = f(x)$. Conditional independence models will separate known from unknown factors and we could use more general statistical measures to describe the error. In Fig. 7.1b, f is a deterministic function, and $\tilde{x}$ absorbs the system uncertainty due to unknown environmental interferences. The dependency among x_i, $\tilde{x}_i$, and y_i can be formalized as: $[y_i|f(x_i)]$ and $[\tilde{x}_i|x_i]$. The structure of Fig. 7.1b can be enriched by introducing $\tilde{y}$ for the inclusion of surrogate measurement errors, which are the result of the mixture of unknown environmental factors and execution errors. The relationship among $\tilde{x}_i$, x_i, y_i and $\tilde{y}$ can be graphically represented as Fig. 7.1c. The new dependency changes to $[\tilde{y}_i|f(x_i)]$ and $[\tilde{x}_i|x_i]$. The proposed dependency map can handle both forward and inversion errors more robustly since the error model is not explicitly constrained by the forward or inverse modeling process. In the proposed model, it is interesting to see that $\tilde{x}_i$ is conditionally independent of $\tilde{y}_i$, the observed surrogate measurement. This is reasonable because the uncertainty of $\tilde{x}$ is due to the unknown distribution of the parameter represent by x and other factors that are ignored by the forward model f. A key statement made by the new model is that x_i is serving as the cluster center. As a statistical descriptor of the posterior distribution of $\tilde{x}$, x_i can be learned by mining the big historical data. Advantages of the proposed model in Fig. 7.1c are summarized as follows:

1. Historical data retrieved from well logging is important in understanding the structure and parameter distribution of the target, where it is characterized by a set of parameters. The structure of the target remains unchanged regardless what logging tools have been used. However, the perceived distribution of the parameters is different depending on what logs were used to derive those parameters. We can use the mixture model to formulate the parameter space of the target. Parameters that define the mixture model depends on the selected integrated logs. It should be noted here that the conditionally independent relationship in Fig. 7.1c removes the one-over-one mapping constraint between $\tilde{x}_i$ and $\tilde{y}_i$, which makes it possible to learn the mixture model of the parameter space with any selected sets of the logs. There are two advantages brought by the proposed model: first, the information of the distribution of the formation parameters hidden in the historical data is kept intact since the learning process includes all data sources; second, after the learning process, the parameters of the mixture model are re-calibrated to be adaptive to the sensitivity of the new sets of integrated logs while maintaining the discriminative ability for characterizing the target.
2. The model provides much more flexibility in selecting a proper probabilistic model for each of the subpopulation in the mixture model, which can be optimized ahead by learning from the historical data. Just like the traditional Bayesian inversion model, the proposed method does not impose any limitation on the choice of probabilistic approaches towards how to select models for each sub-component in the mixture model. Thus, the richer representation of the probabilistic approach is kept intact in the new model.
3. Since the distribution of the well logging responses $\tilde{y}_i$ are not the target of the learning process. It can be modeled in a nonparametric way. In the case of estimating the joint distribution of multiple log responses, nonparametric statistical inference techniques for measures of location and spread of a distribution on a Riemannian manifold could be considered to capturing global structural features of the high dimensional space (Pelletier 2005; Bengio et al. 2005).
4. According to the Bayesian inversion model, the solution of the inverse problem for x_0 given a new measurement y_0 can be obtained by calculating the MAP:

$$\underset{x_0}{\operatorname{argmax}}\, p(x_0|y_0, u_x, u_y)\,, \tag{7.1}$$

   where p is the pre-learned mixture model with model parameters u_x and u_y. In solving the inverse problem, we try to propagate uncertainty from a selected set of well logging measurements to petrophysical parameters by taking into account uncertainty in the petrophysical forward function and a priori uncertainty in model parameters.

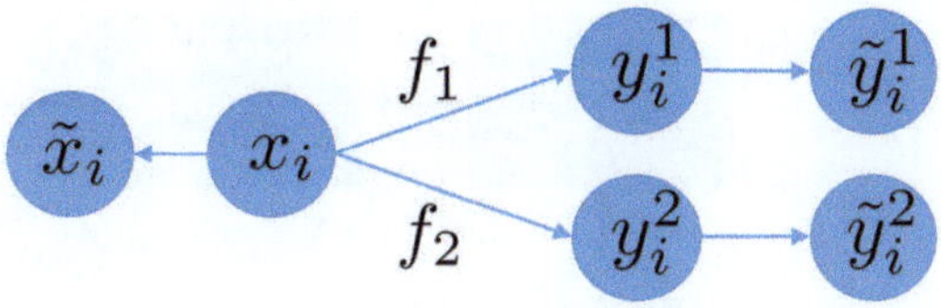

Fig. 7.3 Bayesian inversion with multiple sensors

5. Since the mixture model itself can serve as a classifier, additional classification efforts are not necessary in most cases. And each classification result will be gauged with a probability measure.
6. Expertise and noise patterns hidden in the historical play a key role in optimizing the pre-learned model. In particular, The distribution of the surrogate measurement $\tilde{y}$ directly affects the estimation of the model parameters according to the dependency relationship of the new model. In other words, both unknown environment factors and execution errors were considered when learning model parameters using the historical data.
7. Characterization is an art of multiple log interpretation. Figure 7.3 shows how the proposed model can be easily extended for the multisensory integration, where f_1 and f_2 represent forward models used for different measurement approaches. The dependency relationship remains the same and the likelihood estimation of observed data from different sensory modalities is subject to conditional independence constraints in the additive model.

Figure 7.4a and b highlight the difference between the framework used by the traditional Bayesian inversion and our proposed model optimized by big data learning.

Another advantage inherited by the proposed model is that Bayesian approach allows more sophisticated hierarchical structure as different priors to account for prior knowledge, hypothesis or desired properties about the error and unknown factors. This gives us a large range of flexibilities to regularize smoothness, sparsity, and piecewise continuity for a finite number of states.

7.2.3 Statistical Inference for the Gaussian Mixture Model

The *pdf* of a Gaussian mixture model has the form

$$p(x) = \sum_{k=1}^{K} \pi_k \mathcal{N}(x|\mu_k, \sigma_k), \tag{7.2}$$

where π_k is the mixing coefficient and $\sum_{k=1}^{K} \pi_k = 1$. We augment the model by introducing a latent variable z, which is a K-dimensional binary random variable that has a 1-of-K representation. If x_n, $n \in \{1, \ldots, N\}$ for N observations, belongs

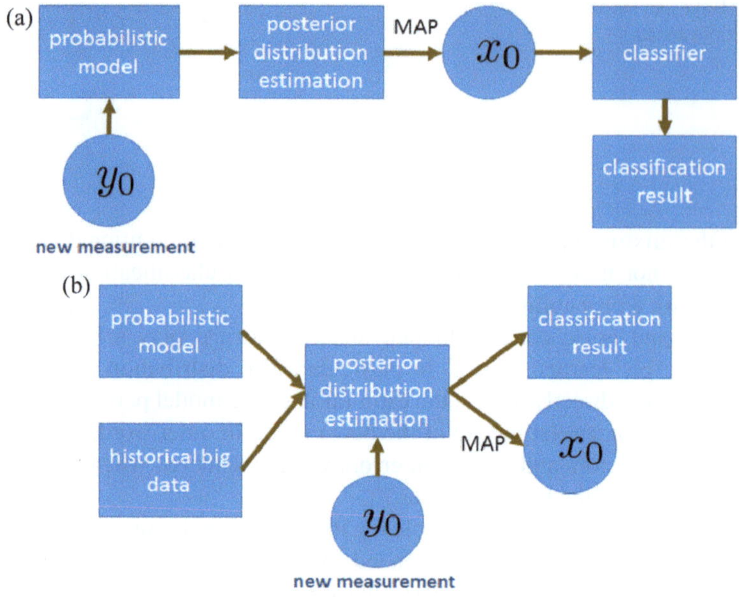

Fig. 7.4 (**a**) Traditional Bayesian inversion. (**b**) Proposed Bayesian inversion

to the kth component of the mixture, then $z_{nk} = 1$. The distribution of z is then in the form $p(z) = \prod_{k=1}^{K} \pi_k^{z_k}$. Conditional distribution of x given z is $p(x|z) = \prod_{k=1}^{K} \mathcal{N}(x|\mu_k, \sigma_k)^{z_k}$. The likelihood of the joint distribution of x and z is then

$$p(X, Z) = \prod_{n=1}^{N} \prod_{k=1}^{K} \pi_k^{z_{nk}} \mathcal{N}(x_n|\mu_k, \sigma_k)^{z_{nk}}. \tag{7.3}$$

According to Bayesian inference rules (Marin et al. 2005), given predefined hyper-parameters ν_0, σ_0, μ_0 and κ_0, the posterior of the conditional distribution of u_k will satisfy

$$u_k \propto \mathcal{N}(\mu_n, \sigma^2/\kappa_n) \cdot \mathcal{L}(\tilde{y}|y), \tag{7.4}$$

$$\mu_n = \frac{\kappa_0 \mu_0 + n\bar{x}}{\kappa_n} \text{ and } \kappa_n = \kappa_0 + n,$$

where $\mathcal{L}$ is the likelihood function. Sampling x is expensive since repeated evaluation of the posterior for many instances means running forward model frequently. A random-walk Metropolis algorithm is shown in Algorithm 1, which summarizes the Metropolis-Hastings sampling steps, an Markov chain Monte Carlo (MCMC) method for obtaining a sequence of random samples from a posterior distribution (Chib and Greenberg 1995).

Algorithm 1: Metroolis-Hastings algorithm for sampling $p(u|\tilde{y})$

input : initial value: $u^{(0)}$, jump function $q(u^{(i)}|u^{(j)})$ (a normal distribution is a popular choice for the jump function q)
output: $u^{(k)}$, where $k \in \{1, 2, \ldots, K\}$
begin
  Initialize with arbitrary value $u^{(0)}$
  while *length of MCMC chain < pre-defined length K* **do**
    Generate $u^{(k)}$ from $q(u^{(k)}|u^{(k-1)})$
    $\alpha(u^{(k)}, u^{(k-1)}) = \min[\frac{p(u^{(k)}|\tilde{y})q(u^{(k-1)}|u^{(k)})}{p(u^{(k-1)}|\tilde{y})q(u^{(k)}|u^{(k-1)})}, 1]$
    Generate α_0 from uniform distribution $\mathcal{U}(0, 1)$
    if $\alpha_0 < \alpha(u^{(k)}, u^{(k-1)})$ **then**
      keep $u^{(k)}$
    else
      $u^{(k)} = u^{(k-1)}$
    end
    save $u^{(k)}$ in the chain
  end
end

7.2.3.1 Distributed Markov Chain Monte Carlo for Big Data

Although the MCMC method guarantees asymptotically exact solution for recovering the posterior distribution, the cost is prohibitively high for training our model with a large-scale and heterogeneous data set. General strategies for parallel MCMC, such as Calderhead (2014) and Song et al. (2014), require full data sets at each node, which is non-practical for Big data. For applications with Big data, multi-machine computing provides scalable memory, disk, and processing power. However, limited storage space and network bandwidth require algorithms in a distributed fashion to minimize communication. Embarrassingly parallel MCMC proposed in Neiswanger et al. (2013) tackles both problems simultaneously. The basic idea is to allocate a portion of the data to each computing node. MCMC is performed independently on each node without communicating. By the end, a combining procedure is deployed to yield asymptotically exact samples from the full-data posterior. This procedure is particularly suitable for use in a MapReduce framework (Dean and Ghemawat 2008).

Embarrassingly parallel MCMC partitions data $x^N = \{x_1, \ldots x_N\}$ into M subsets $\{x^{n_1}, \ldots, x^{n_M}\}$. For each subset $m \in \{1, \ldots, M\}$, sub-posterior is sampled as:

$$p_m(\theta) \propto p(\theta)^{\frac{1}{M}} p(x^{n_m}|\theta) \tag{7.5}$$

The full data posterior is then in proportion to the product of the sub-posterior, i.e. $p_1 \ldots p_M(\theta) \propto p(\theta|x^N)$. When N is large, $p_1 \ldots p_M(\theta)$ can be approximated by $\mathcal{N}_d(\theta|\hat{\mu}_M, \hat{\Sigma}_M)$, where $\hat{\mu}_M$ and $\hat{\Sigma}_M$ are:

$$\hat{\Sigma}_M = \left(\sum_{m=1}^{M} \hat{\Sigma}_m^{-1}\right)^{-1}, \hat{\mu}_M = \hat{\Sigma}_M \left(\sum_{m=1}^{M} \hat{\Sigma}_m^{-1} \hat{\mu}_m\right). \tag{7.6}$$

7.2.4 Tests from Synthetic Well Integrity Logging Data

Well integrity is critical for blocking migration of oil, gas, brine and other detrimental substances to freshwater aquifers and the surface. Any deficiencies in primary cementing tend to affect long-term isolation performance. Wide fluctuations in downhole pressure and temperature can negatively affect cement integrity or cause debonding. Tectonic stresses also can fracture set cement (Newell and Carey 2013). New wells could disturb layers of rock near fragile old wells, the "communication" between the old and the new can create pathways for oil, gas or brine water to contaminate groundwater supplies or to travel up to the surface (Vidic et al. 2013). Regardless of the cause, loss of cement integrity can result in fluid migration and impairment of zonal isolation. In an effort to minimize environmental impacts, advanced sensing technology is urgently needed for continuously monitoring well integrities. However, the precise interpretation of the cement log is a challenging problem since the response of acoustic tools is also related to the acoustic properties of the surrounding environment such as casing, mud, and formation. The quality of the acoustic coupling between the casing, cement, mud, and formation will alter the response as well (Nelson 1990). The analysis of the log requires detailed information concerning the well geometry, formation characteristics, and cement job design to determine the origin of the log response. Therefore, a fair interpretation of an acoustic log can only be made when it is possible to anticipate the log response, which is accomplished through forward modeling.

The goal is to differentiate set cement, contaminated cement and fluid through multiple measurements obtained by different sonic logging tools. If the cement is not properly set, the probability of isolation failure is high in the long term. The key of the research is to estimate the distribution of model parameters under different physical states of the cement. As demonstrated in Fig. 7.5, the red area is marked as contaminated cement, and the green area represents pure cement. The gray area indicates ambiguous cases. Blue curves show the distribution of the density for contaminated cement and pure cement. Numbers on the first row of the axis are the measurement. The second row shows the inversed cement density given the measurement. Accurate estimation of the distribution of the density will improve classification accuracy for those cases with inversed parameters fallen into the gray area.

We have applied the aforementioned framework with simulated data. The simulated training data set is generated to cover a large variety of combinations of different cement status (pure or contaminated) and fluids (water, spacer, mud). The data is generated to simulate five cementing conditions behind the casing: pure cement, contaminated cement, mud, spacer and water. The synthetic log used as the ground-truth is generated with additional artificial noises. Two measurements are used by running the forward-modeling simulator, impedance and flexural-wave attenuation. The setting of the borehole geometry, casing parameters, and formation properties is fixed. During the training stage, 500 samples were generated for each cementing condition and corresponding physical properties. Parameters

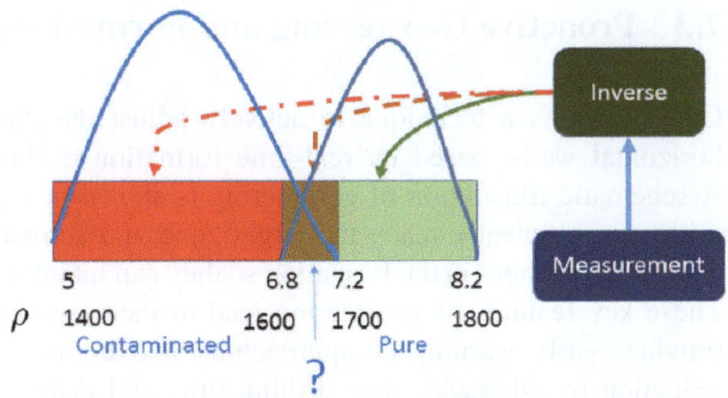

Fig. 7.5 A non-uniform distribution improves classification accuracy

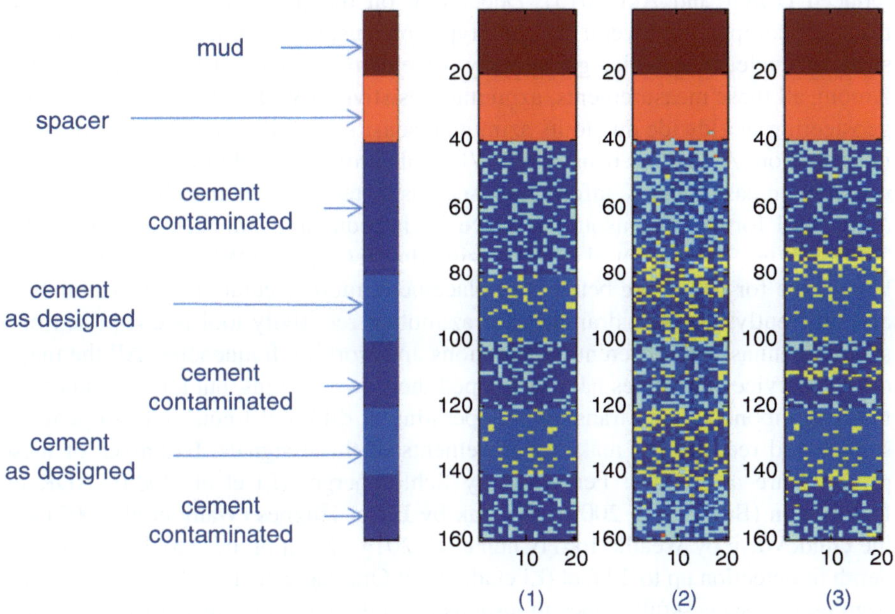

Fig. 7.6 Improve log interpretation with the pre-learned Gaussian mixture model: (**1**) ground truth; (**2**) log interpretation without learning; (**3**) log interpretation with pre-learned Gaussian mixture

for the Gaussian mixture model are learning through MCMC sampling. Inversed parameters, density, compressional and shear velocity, are used to differentiate those five cementing conditions. Test result on the synthetic dataset presented in Fig. 7.6. Figure 7.6(1) is ground truth. Figure 7.6(2) show classification result by running an off-the-shelf classifier on inversed results obtained through traditional inversion tool without learning. Figure 7.6(3) demonstrates the improvement in terms of classification accuracy by applying the pre-learned Gaussian mixture model.

7.3 Proactive Geosteering and Formation Evaluation

Geosteering is a technique to actively adjust the direction of drilling, often in horizontal wells, based on real-time formation evaluation data (Li et al. 2005). A schematic illustration of geosteering is shown in Fig. 7.7. This process enables drillers to efficiently reach the target zone and actively respond while drilling to geological changes in the formation so they can maintain maximal reservoir contact. These key features of geosteering lead to increased production. Geosteering also provides early warning of approaching bed boundaries and faults leading to a reduction in sidetracks, thus drilling time and drilling cost are also significantly reduced (Bittar and Aki 2015). Depending on the properties of formation and reservoir complexity, several different types of sensors can be used for geosteering, such as nuclear, acoustic, gamma ray, or electromagnetic (EM) measurements. Among all these measurements, azimuthal resistivity LWD tools are widely used in geosteering worldwide due to its azimuthal sensitivity and relatively large depth of investigation. Azimuthal resistivity LWD tools provide, in addition to conventional resistivity measurements, information such as distance to bed interface, relative dip angle, and formation anisotropy. Since its introduction into well logging in the 2000s (Bittar 2002; Li et al. 2005), azimuthal resistivity LWD tools have been a key device for delivering better well placement, more accurate reserve estimation, and efficiently reservoir draining. An azimuthal resistivity tool is comprised of a set of antennas with different polarizations and working frequencies. All the major oilfield service companies have developed their own designs, but all tools share the major components, i.e., transmitters operating at different frequencies to generate signals and receivers to make measurements of those signals. Examples of these products are as follows: PeriScope by Schlumberger (Li et al. 2005), ADR by Halliburton (Bittar et al. 2009), AziTrak by Baker Hughes (Wang et al. 2007) and the GuideWave by Weatherford (Chen et al. 2014). Most of these tools can provide depth of detection up to 20 feet (Li et al. 2005; Omeragic et al. 2005), and they have been used to successfully place thousands of wells. We will consider as an example the Weatherford GuideWave azimuthal resistivity tool.

A schematic diagram of the GuideWave Azimuthal Resistivity tool is shown in Fig. 7.8. It has transmitters and receivers both along the tool axis (the Z direction) and perpendicular to it (the X direction). This tool design is fully compensated, i.e., transmitters and receivers are always in pairs and symmetric to the center of the tool.

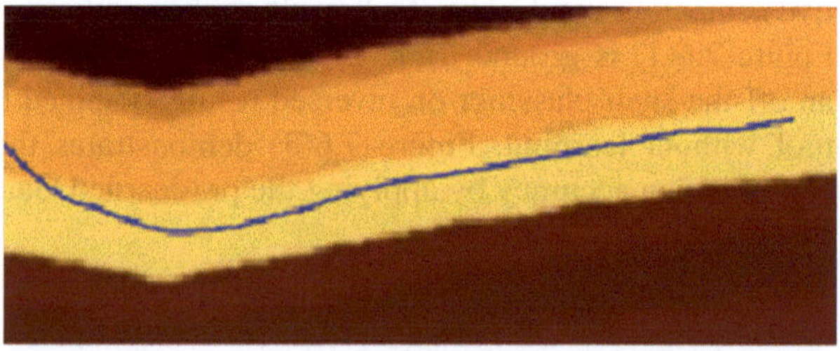

Fig. 7.7 A schematic of geosteering using azimuthal resistivity LWD tool: the center *bright yellow layer* represents reservoir, and the *blue line* denotes the well trajectory

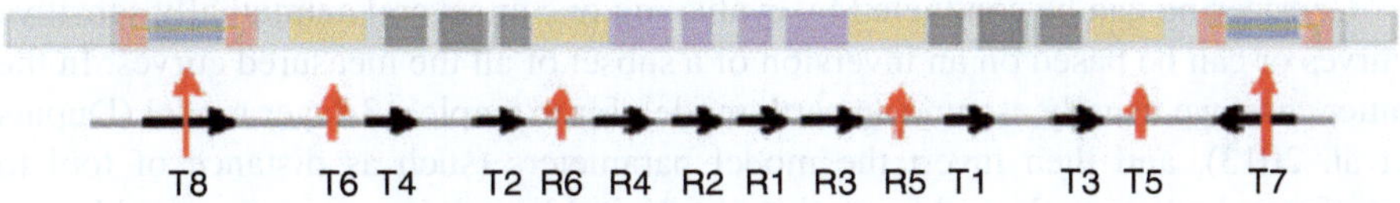

Fig. 7.8 The structure and schematic of an azimuthal resistivity LWD tool. The *black arrows* denote transmitters or receivers with components along the tool axis (the Z direction), and the *red arrows* denote transmitters or receivers with components perpendicular to the tool axis (the X direction)

Fig. 7.9 The full mutual inductance tensors of formation solved from receiver voltages with rotation

$$\begin{bmatrix} xx & xy & xz \\ yx & yy & yz \\ zx & zy & zz \end{bmatrix}$$

While rotating, transmitters are energized and the receivers voltages are measured by electronic components. Processing algorithms in the electronic hardware reduce the measured signals from the receiver antennas to the full mutual inductance tensor (see Fig. 7.9) related to the resistivity tensor of the geological formation occupied by the tool (Dupuis and Denichou 2015). These different inductance tensors are used to generate different types of measurement curves.

Examples of these measurement curves are the ZZ curves of standard resistivity measurement, the azimuthally sensitive ZX curves, and the XX curves with sensitivity to anisotropy. The ZZ curves are obtained by two Z direction transmitters, such as T1 and T2, and two Z direction receivers, such as R1 and R2, thus four separate transmitter-receiver measurement pairs are used to determine a data point by full compensation, which is regarded as very robust to unwanted environmental and temperature variations. The ZZ curves have no directionality: i.e. they cannot distinguish up from down, or left from right. The ZZ curves are used for standard measurement of formation resistivity (Mack et al. 2002). The ZX curves are obtained by transmitters with components along the X direction, such as T7 and T8, and Z direction receivers, such as R3 and R4. Each data point of a ZX curve is calculated after a full 360° of tool rotation with respect to tool direction (the Z axis). The ZX curves have an azimuthal sensitivity, which means they can be used to figure out the azimuthal angle of a bed interface relative to the tool. This property of the ZX curves makes them very suitable for figuring out the azimuth and distance of adjacent boundaries and keeping the tool in the desired bed, or in one word, geosteering. The XX curves are obtained by transmitters with components along the X direction, such as T7 and T8, and X direction receivers, such as R5 and R6. The construction of a data point of a XX curve also requires a full tool rotation around the Z axis. The XX curves are sensitive to formation resistivity anisotropy at any relative dip angle, thus they are the best candidates for anisotropy inversion (Li et al. 2014).

Geosteering can be conducted by evaluating one or several azimuthally sensitive curves or can be based on an inversion of a subset of all the measured curves. In the latter case, we usually assume an earth model, for example a 3-layer model (Dupuis et al. 2013), and then invert the model parameters (such as distance of tool to interfaces, bed azimuth, and formation resistivity) by solving an inverse problem.

In recent years a new generation of EM geosteering tool with a much larger depth of investigation has emerged on the market. This class of devices has longer spacings between sensors compared with the aforementioned tools, and correspondingly the working frequency can be as low as several kilohertz. GeoSphere developed by Schlumberger is such kind of tool claiming a depth of investigation in excess of 100 feet (Seydoux et al. 2014) and the ability of optimized well landing without pilot wells (it can cost tens of millions of dollars to drill a pilot hole) is realized. Another new generation tool recently released by Baker Hughes (Hartmann et al. 2014), commercially named ViziTrak, can also directionally "see" the formation up to 100 feet from the well bore. As the new generation tools can see much further than the previous ones, the associated inverse problem becomes more challenging and the complexity and the uncertainty greatly increase.

7.3.1 Inversion Algorithms

The process of adjusting tool position in real time relies on an inversion method the output of which is used to generate an earth model within the constraints of the measurements and inversion techniques. In real jobs, modeling and inversion of azimuthal resistivity LWD measurements are usually based on a 1D parallel layer model, i.e., all bed interfaces are infinitely large and parallel to each other. Two different inversion schemes, a deterministic inversion method and a stochastic inversion scheme, are introduced below.

7.3.1.1 The Deterministic Inversion Method

The conventional inversion method in geosteering applications is the deterministic approach based on algorithms fitting the model function to measured data. Deterministic inversion methods can produce excellent results for distance-to-boundaries and resolve resistivities of the layers with a good initial guess. Suppose an azimuthal resistivity tool can produce N measurements denoted by $m \in \mathbb{R}^N$. A computational model function $S : \mathbb{R}^M \rightarrow \mathbb{R}^N$ is used to synthesize these same measurements based on M model parameters $x \in \mathbb{R}^M$, where the response of the model to these parameters is denoted $S(x) \in \mathbb{R}^N$. Results are considered of high quality when there is good agreement between the computational model and the measured data. Such agreement is measured by the misfit

$$F(x) = S(x) - m\,, \tag{7.7}$$

where x, S, and m are as defined above. Define the cost function as a sum of squares of nonlinear functions $F_i(x)$, $i = 1, \ldots, N$,

$$f(x) = \sum_{i=1}^{N} F_i^2(x) . \tag{7.8}$$

where, again, N is the number of measurement curves. We consider the optimization problem to iteratively find the value of the variables which minimizes the cost function. This is a unconstrained nonlinear least-squares minimization problem. A family of mature iterative numerical algorithms have been established to solve this least-squares problem, such as the method of gradient descent, Gauss-Newton method, and the Levenberg-Marquardt algorithm (LMA). Here we apply LMA to find the unknown parameters vector x.

Suppose the current value of the parameters vector at the kth iteration is x_k. The goal of successive steps of the iteration is to move x_k in a direction such that the value of the cost function f decreases. Steps h_k in LMA are determined by solving the linear subproblem

$$(J^T(x_k)J(x_k) + \mu I)h_k = -J^T(x_k)(S(x_k) - m) , \tag{7.9}$$

where $J(x_k)$ is the Jacobian matrix evaluated at x_k, I is the identity matrix with the same dimension as $J^T(x_k)J(x_k)$, and μ is a small positive damping parameter for regularization, e.g. $\mu = 0.001 \max(\text{diag}(J^T J))$. Updates to the unknown parameter in this method can be written as

$$x_{k+1} = x_k + h_k, \quad k = 0, 1, 2, \ldots \tag{7.10}$$

During optimization we can stop the iteration when

$$||S(x_k) - m|| < \epsilon_1 , \tag{7.11}$$

which means the data misfit is smaller than threshold ϵ_1 (a small positive number), or

$$||h_i|| < \epsilon_2 , \tag{7.12}$$

which means the increment of parameter vector is smaller than threshold ϵ_2 (also a small positive number), or simply

$$i > i_{max} , \tag{7.13}$$

which means the maximum number of iteration is reached.

In real jobs, it is usually very expensive, if not impossible, to obtain the closed form of Jacobian. Here we use finite difference method to obtain this matrix numerically

$$J_{i,j}(x) = \frac{\partial S_j(x)}{\partial x_i} \approx \frac{S_j(x + \hat{e}_i \delta x_i) - S_j(x)}{\delta x_i} , i = 1, 2, \ldots, N \text{ and } j = 1, 2, \ldots, M . \tag{7.14}$$

The above formulation shows that every iteration during optimization needs to evaluate $2 \times M \times N$ data. This part can be quite expensive requiring significant computational resources. To address this issue we employ Broyden's rank-one update method here for the numerical Jacobian. In Broyden's method the Jacobian is updated at each iteration by

$$J_{k+1} = J_k + \frac{S(x_{k+1}) - S(x_k) - J_k h_k}{\|h_k\|^2} \tag{7.15}$$

Though this kind of gradient-based method is efficient and effective for any number of parameter variables, it has drawbacks that the solution is highly dependent to the initial guess. A poor initial guess can lead the results quickly falling into a local minimum far from the true solution. In our study, to overcome these drawbacks, a scheme of combing the local optimization method with a global searching method is applied for EM geosteering: we make a coarse exponential discretization for the model space and use the global searching method to obtain initial guess for local optimization then choose the LMA method to refine that initial guess.

7.3.1.2 The Statistical Inversion Method

The aforementioned deterministic regularization methods have a few limitations. First, it is sensitive to the choice of the distance metric and the parameter selection of the regularizer. Second, it is hard to quantify the uncertainty of the proposed solutions. Finally, the gradient descent based method is hard to take advantage of the multi-core hardware platform due to computational dependencies in the algorithm. As an alternative method to deal with uncertainty, Bayesian approach has attracted more attentions (Knapik et al. 2011). Unlike the deterministic formulation, the observation model or the so-called likelihood is usually built upon the forward model and some knowledge about the errors (e.g. measurement noise). The desired property and the prior knowledge of the uncertainty of the solution are translated into prior distributions. Bayesian rule is used to obtain the posterior distribution from which the solution is deduced after combining the likelihood and the prior. Parallel sampling is also available for more complicated Bayesian inference cases.

Experiments and observations suggest physical theories, which in turn are used to predict the outcome of experiments. Solving a forward problem is to calculate the output (y) of a physical model (f) given its parameter set u, $y = f(u)$. The inverse relationship can be written as $u = \tilde{f}(\tilde{y})$, where $\tilde{f}$ defines the inverse mapping and $\tilde{y}$ is the observed output. The classical definition of a well-posed problem, due to Hadamard (1923), is that the solution exits is unique and depends continuously on data. On the contrary, most inverse problems are fundamentally underdetermined (ill-posed) because the parameter space is large, measurements are too sparse and different model parameters may be consistent with the same measurement. The problem is usually solved by minimizing

$$\Omega(u) + \lambda R(u), \tag{7.16}$$

where Ω is the objective function and R serves the role of a regularizer with regularization constant $\lambda > 0$. Under the least-squares criterion, the objective function is written as $\|\tilde{y} - f(u)\|_2^2$. This formulation limits its modeling ability with regard to different characteristics of noise. It is also impractical to search for the optimal solution for both u and a proper mapping $\tilde{f}$ when the forward problem is nonlinear.

To overcome the shortcomings of the deterministic method, a stochastic approach generates a set of solutions distributed according to a probability distribution function. In a Bayesian context, the solution is sampled from the posterior distribution of the model parameters u given the measurement y, $p(u|y) \sim p(y|u)p(u)$. $p(u)$ is the prior that describes the dependency between the model parameters and, therefore, constrains the set of possible inverse solutions. The likelihood $p(y|u)$ models the stochastic relationship between the observed data y and parameter set u. The stochastic approach provides more probabilistic arguments for modeling the likelihood and prior, and for interpreting the solution. In particular, it represents a natural way to quantify the uncertainty of the solution via Bayesian inference.

Let's revisit the forward model of the inverse problem. Suppose the noise is additive and comes from external sources, then the relationship between observed outputs and corresponding parameters of the physical system can be represented as:

$$\tilde{y} = f(u) + \epsilon \,, \tag{7.17}$$

where ϵ represents additive noise. In classical Bayesian approaches, ϵ is a zero-mean Gaussian random variable, $\epsilon \sim \mathcal{N}(0, \sigma^2 I)$. Then we have the following statistical model instead, $p(\tilde{y}|u) \sim \mathcal{N}(\tilde{y} - f(u), \sigma^2 I)$. From a Bayesian point of view, suppose the prior distribution of u is governed by a zero-mean isotropic Gaussian such that: $p(u|\beta) \sim \mathcal{N}(0, \beta^2 I)$. By virtue of the Bayes formula, the posterior of x is given by:

$$p(u|\tilde{y}) \sim \mathcal{N}(\tilde{y} - f(u), \sigma^2 I)\mathcal{N}(0, \beta^2 I)) \,. \tag{7.18}$$

It is easy to show that the log likelihood of Eq. (7.18) is equivalent to use Tikhonov regularization method for the ill-posed problems (Bishop 2006). Furthermore, the solution of a lasso estimation can be interpreted as the posterior mode of u in the above Bayesian model under the assumption of a Laplacian prior (Tibshirani 1996).

To search for the Bayesian solution according to Eq. (7.18), we draw samples from the posterior probability density function. The goal is to locate the most likely value or function u is going to be. In other words, we are looking for the most probable point u in the posterior distribution. And usually, the point is attained as the maximum a posteriori (MAP) estimator, namely, the point at which the posterior density is maximized. In the case of a linear mapping function f, the posterior density function can be easily derived when selecting a conjugate prior. However, conducting Bayesian inference according to Eq. (7.18) becomes a challenge when f indicates a non-linear mapping relationship between u and y. Consequently, the analytical solution for the posterior distribution is not always available anymore. In practice, MCMC is one of the most popular sampling approaches to draw *iid* samples from an unknown distribution (Robert and Casella 2004). For solving an inverse problem, the sampling algorithm requires repeated evaluation of the posterior for many instances of u.

7.3.2 Hamiltonian Monte Carlo and MapReduce

The random walk Metropolis algorithm as shown in Algorithm 1 suffers from low acceptance rate and converges slowly with long burning period (Neal 2011). In practice, a burn-in period is needed to avoid starting biases, where an initial set of samples are discarded. It is also hard to determine the length of the Markov chain. Running MCMC with multiple chains is a natural choice when the platform supports parallel computing (Gelman and Rubin 1992). In this section, we like to discuss how the Hamiltonian Monte Carlo and the MapReduce platform can be used to improve the sampling performance.

Hamiltonian Monte Carlo (HMC) (Neal 2011) reduces the correlation between successive sampled states by using a Hamiltonian evolution. In short, let $x \in \mathcal{R}^d$ be a random variable and $p(x) \propto \exp(\mathcal{L}(x))$, where $\mathcal{L}$ is the likelihood function. HMC defines a stationary Markov chain on the augmented state space $\mathcal{X} \times \mathcal{P}$ with distribution $p(x, p) = u(x)k(p)$ (Zhang and Sutton 2011). A Hamiltonian function, the sampler, is defined as

$$H(x,p) = \underbrace{-\mathcal{L}(x)}_{\text{potential energy}} + \underbrace{\frac{1}{2}p^T M^{-1} p}_{\text{kinetic energy}}. \tag{7.19}$$

A new state (x^*, p^*) is set by the Hamiltonian dynamics

$$\dot{x} = \frac{\partial H}{\partial p}, \tag{7.20}$$

$$\dot{p} = \frac{\partial H}{\partial x}. \tag{7.21}$$

Approximation of the above system is obtained with the leapfrog method (Neal 2011). If we consider the kinetic energy part in Eq. (7.19) as a jump function, we could use Newton's method to approximate $\mathcal{L}(x)$ by a quadratic function. In other words, we can use a local approximation to serve as the jump function. Let $H_{\mathcal{L}} = \ddot{\mathcal{L}}$ be the Hessian matrix, then the Hamiltonian function is

$$H(x,p) = -\mathcal{L}(x) + \frac{1}{2}p^T H_{\mathcal{L}}^{-1} p. \tag{7.22}$$

However, since inverse of the Hessian matrix is computationally prohibitive in high-dimensional space, we take the L-BFGS approach (Liu and Nocedal 1989), and the Hamiltonian energy is simply (Zhang and Sutton 2011)

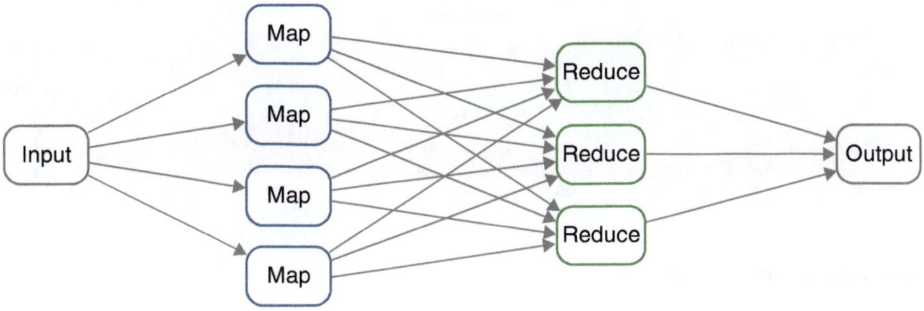

Fig. 7.10 MapReduce programming mode

$$H(x,p) = -\mathcal{L}(x) + \frac{1}{2}p^T H_{BFGS}^{-1} p. \tag{7.23}$$

HMC also receives the benefit from multiple chains and MapReduce is an ideal parallel computing platform for processing and dealing with large data sets on a cluster. A simple MapReduce diagram is shown in Fig. 7.10 (Pal 2016). In the MapReduce mode, a master node is responsible for generating initial starting values and assigning each chain to different mappers. Data generated by each chain is cached locally for the subsequent map-reduce invocations. The master coordinates the mappers and the reducers. After the intermediate data is collected, the master will in turn invoke the reducer to process it and return final results. In the multiple-chain cases, the reducer will calculate the within and between chain variances to evaluate convergence. The mean value of all chains collected by the master will be the final result.

7.3.3 *Examples and Discussions*

In this part, we take the tool design of GuideWave (the first generation azimuthal resistivity LWD tool by Weatherford) for an instance to generate tool measurements and conduct different inversions for demonstration. It is noted that the synthetic data are generated based on analytical EM wave solutions for 1D multi-layer model with dipole transmitters and receivers. The effect of the borehole is considered as negligible in forward modeling here. A field data by this tool will also be shown and processed with these inversion methods.

7.3.3.1 Tests from Synthetic Data

In the first example, we consider a three-layer model as shown in Fig. 7.11. The resistivities of the upper, center and lower layers are $10\,\Omega\,\text{m}$, $50\,\Omega\,\text{m}$, and $1\,\Omega\,\text{m}$, respectively. The central green line indicates the tool navigation trajectory.

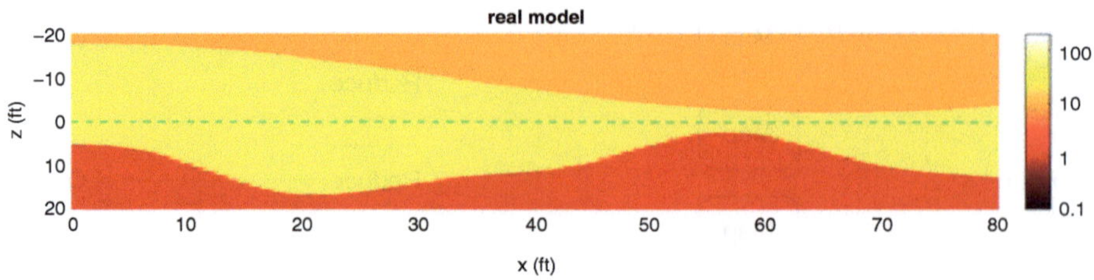

Fig. 7.11 The three-layer model

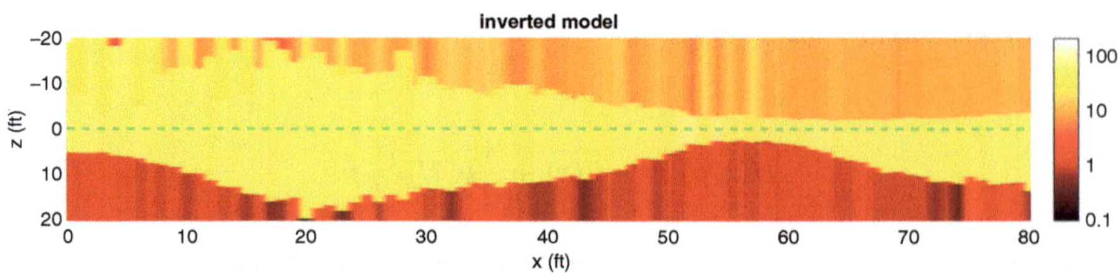

Fig. 7.12 The inverted model by deterministic method

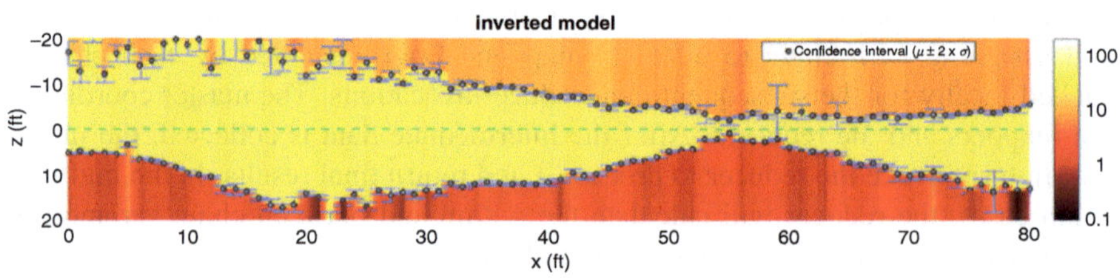

Fig. 7.13 The inverted model by statistical inversion

In this case, we assume the tool dip angle is fixed as 90°. Figures 7.12 and 7.13 show the inverted model by the deterministic and statistical methods. One can see that when the tool is far from the top boundary (see the beginning part), the measurements and the inversion cannot resolve the top boundary accurately. In other words, the uncertainty involved in solving the inverse problem is high. The reason is that the sensitivity of the measurements is relatively poor when the tool is relatively far from a boundary. As the tool moves forward where the top boundary bends downward, the inversion can clearly resolve both the upper and the lower boundaries. Comparing two inverted models obtained through the deterministic and statistical methods in this case, the results are comparable to each other. The statistical method, however, provides quantified interpretation for the uncertainty of the inverse problem (Fig. 7.14).

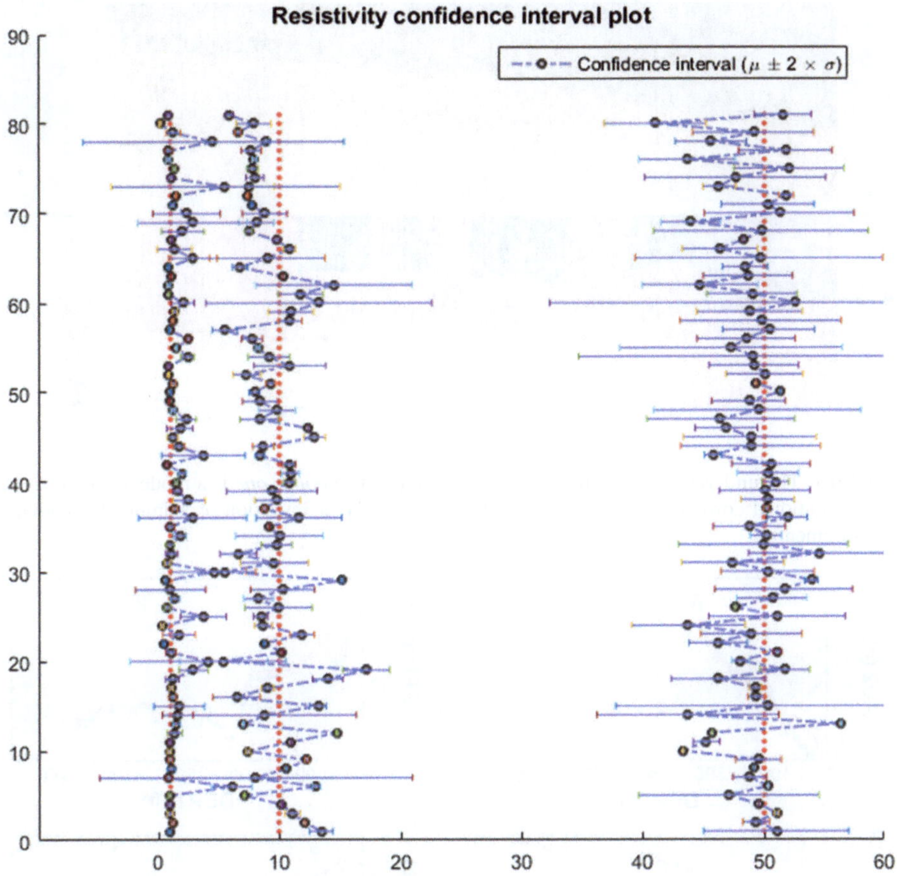

Fig. 7.14 The inverted formation resistivities

7.3.3.2 Test from Field Data

In the end, we will show a real field case. Figure 7.15 shows a model of this well created from the conventional correlation and the inverted 3-layer model based on resistivity measurement. This figure shows that the upper layer has lower resistivity around 1 Ω m, and the lower layer has higher resistivity around 20 Ω m. The tool may penetrate or approach bed boundary three times: the first one is around X050 ft, and second is close to X300 ft, and near X900 ft the tool penetrated into a relatively resistive zone. We can see that this inverted 3-layer model is reasonably consistent with the model created from the conventional correlation, and the good quality of inversion can be also verified by the great agreements between measured and synthetic curve values shown in Fig. 7.16.

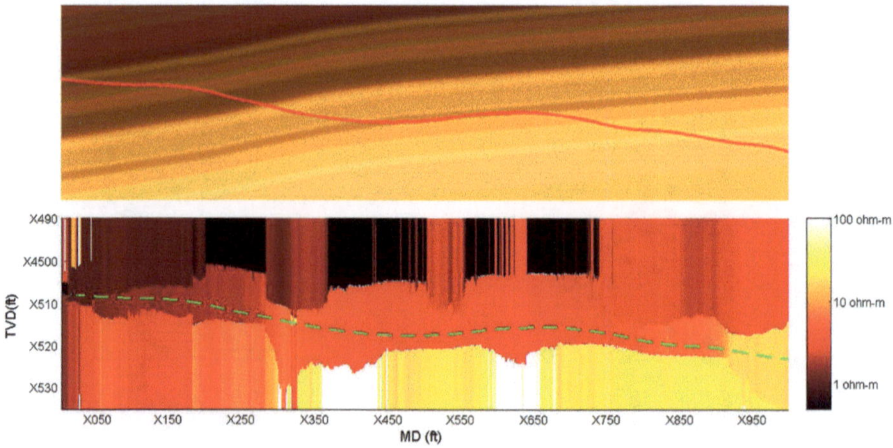

Fig. 7.15 The underground structures of a real field job. The *upper one* is a model created based on conventional correlation, and the *lower one* is by a 3 layer inversion of azimuthal resistivity measurements

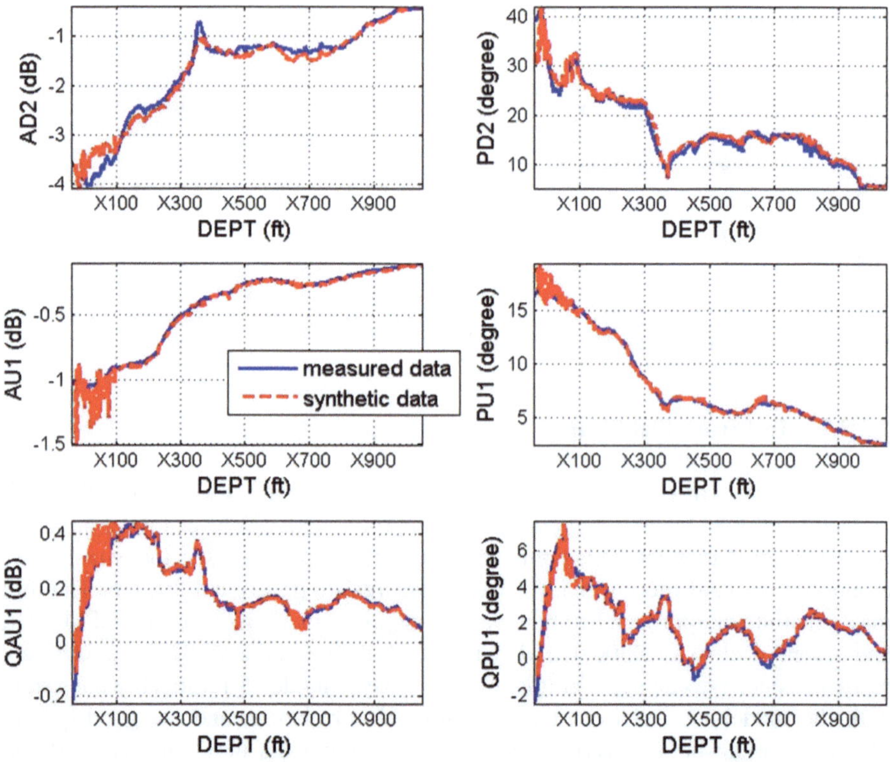

Fig. 7.16 Comparison of measured and synthetic values of six *curves*: AD2 and PD2 are amplitude attenuation and phase difference of a 2 MHz ZZ measurement, AU1 and PU1 are amplitude attenuation and phase difference of a 100 kHz ZZ measurement, QAU1 and QPU1 are amplitude attenuation and phase difference of a 100 kHz ZX measurement. Good agreements can be observed for all the six curves, which indicate a good quality of the inversion for this field example

7.3.4 Conclusion

Most research of inverse problems centers around the development of formulas that yield a description of the system as a function of the measured data, as well as on theoretical properties of such formulas. Bayesian approach is mainly used as a convenient way to estimate the measurement error and model the uncertainty in the system. The proposed work is, foremost, an effort to investigate both the inversion and interpretation process and build an intelligent framework for improving classification results through inversion. A more critical issue is that the deeper knowledge and repeatable patterns that are hidden in the big data accumulated in the past are poorly described and studied. Even if they are used, there is a lack of systematic approach towards inverse problems. In addition, we think the research related to exploration of computationally efficient statistical inference on the large scale is also instrumental in learning and understanding inverse problems. The techniques developed in this research could have a large positive impact on many areas of computing.

References

Bengio, Y., Larochelle, H., & Vincent, P. (2005). Non-local manifold parzen windows. In *Advances in Neural Information Processing Systems* (pp. 115–122).

Bishop, C. M. (2006). *Pattern recognition and machine learning* Berlin, Heidelberg: Springer.

Bittar, M., & Aki, A. (2015). Advancement and economic benefit of geosteering and well-placement technology. *The Leading Edge, 34*(5), 524–528.

Bittar, M. S. (2002, November 5). *Electromagnetic wave resistivity tool having a tilted an- tenna for geosteering within a desired payzone*. Google Patents. (US Patent 6,476,609)

Bittar, M. S., Klein, J. D., Randy B., Hu, G., Wu, M., Pitcher, J. L., et al. (2009). A new azimuthal deep-reading resistivity tool for geosteering and advanced formation evaluation. *SPE Reservoir Evaluation & Engineering, 12*(02), 270–279.

Calderhead, B. (2014). A general construction for parallelizing metropolis- hastings algorithms. *Proceedings of the National Academy of Sciences, 111*(49), 17408–17413.

Carroll, R. J., Ruppert, D., Stefanski, L. A., & Crainiceanu, C. M. (2006). *Measurement error in nonlinear models: a modern perspective*. Boca Raton: CRC Press.

Chen, J., & Yu, Y. (2014). An improved complex image theory for fast 3d resistivity modeling and its application to geosteering in unparallel layers. In *SPE Annual Technical Conference and Exhibition.*

Chib, S., & Greenberg, E. (1995). Understanding the metropolis-hastings algorithm. *The American Statistician, 49*(4), 327–335.

Dean, J., & Ghemawat, S. (2008). Mapreduce: simplified data processing on large clusters. *Communications of the ACM, 51*(1), 107–113.

Dupuis, C., & Denichou, J.-M. (2015). Automatic inversion of deep-directional-resistivity measurements for well placement and reservoir description. *The Leading Edge, 34*(5), 504–512.

Dupuis, C., Omeragic, D., Chen, Y. H., & Habashy, T. (2013). Workflow to image unconformities with deep electromagnetic LWD measurements enables well placement in complex scenarios. In *SPE Annual Technical Conference and Exhibition.*

Ellis, D. V., & Singer, J. M. (2007). *Well logging for earth scientists*. Dordrecht: Springer Science & Business Media.

Faghmous, J. H., Le, M., Uluyol, M., Kumar, V., & Chatterjee, S. (2013). A parameter-free spatio-temporal pattern mining model to catalog global ocean dynamics. In *ICDM* (pp. 151–160).
Gelman, A., & Rubin, D. B. (1992). Inference from iterative simulation using multiple sequences. *Statistical Science*, 457–472.
Hadamard, J. (1923). *Lectures on cauchy's problem in linear differential equations*. London: Yale University Press.
Hartmann, A., Vianna, A., Maurer, H.-M., Sviridov, M., Martakov, S., Lautenschläger, U., et al. (2014). Verification testing of a new extra-deep azimuthal resistivity measurement. In *SPWLA 55th Annual Logging Symposium.*
Iglesias, M., & Stuart, A. M. (2014). Inverse problems and uncertainty quantification. *SIAM News, volume July/August.*
Kaipio, J., & Somersalo, E. (2005). *Statistical and computational inverse problems* New York: Springer.
Knapik, B. T., Vaart, A. W. V. D., & Zanten, J. H. V. (2011). Bayesian inverse problems with Gaussian priors. *The Annals of Statistics, 39*(5), 2626–2657.
Li, Q., Omeragic, D., Chou, L., Yang, L., & Duong, K. (2005). New directional electromagnetic tool for proactive geosteering and accurate formation evaluation while drilling. In *SPWLA 46th Annual Logging Symposium.*
Li, S., Chen, J., & Binford Jr, T. L. (2014). Using new LWD measurements to evaluate formation resistivity anisotropy at any dip angle. In *SPWLA 55th Annual Logging Symposium.*
Liu, D. C., & Nocedal, J. (1989). On the limited memory BFGS method for large scale optimization. *Mathematical Programming, 45*(1–3), 503–528.
Mack, S. G., Wisler, M., & Wu, J. Q. (2002). The design, response, and field test results of a new slim hole LWD multiple frequency resistivity propagation tool. In *SPE Annual Technical Conference and Exhibition.*
Marin, J.-M., Mengersen, K., & Robert, C. P. (2005). Bayesian modelling and inference on mixtures of distributions. In *Handbook of statistics* (Vol. 25, pp. 459–507).
Menke, W. (2012). *Geophysical data analysis: Discrete inverse theory* (Vol. 45). San Diego: Academic Press.
Neal, R. M. (2011). MCMC using Hamiltonian dynamics. In *Handbook of Markov chain Monte Carlo* (Vol. 2, pp. 113–162). Boca Raton: Chapman & Hall/CRC Press.
Neiswanger, W., Wang, C., & Xing, E. (2013). Asymptotically exact, embarrassingly parallel MCMC. arXiv preprint arXiv:1311.4780.
Nelson, E. B. (1990). *Well cementing* (Vol. 28). London: Newnes.
Newell, D. L., & Carey J. W. (2013). Experimental evaluation of wellbore integrity along the cement-rock boundary. *Environmental Science & Technology, 47*(1), 276–282.
Omeragic, D., Li, Q., Chou, L., Yang, L., Duong, K., Smits, J. W., et al. (2005). Deep directional electromagnetic measurements for optimal well placement. In *SPE Annual Technical Conference and Exhibition.*
Pal, K. (2016). *Hadoop key terms, explained.* Retrieved from http://www.kdnuggets.com/2016/05/hadoop-key-terms-explained.html [Online]. Accessed 27 August 2016.
Pelletier, B. (2005). Kernel density estimation on Riemannian manifolds. *Statistics & Probability Letters, 73*(3), 297–304.
Robert, C., & Casella, G. (2004). *Monte carlo statistical methods* (2nd ed.). New York: Springer.
Seydoux, J., Legendre, E., Mirto, E., Dupuis, C., Denichou, J.-M., Bennett, N., et al. (2014). Full 3d deep directional resistivity measurements optimize well placement and provide reservoir-scale imaging while drilling. In *SPWLA 55th Annual Logging Symposium.*
Song, Q., Wu, M., & Liang, F. (2014). Weak convergence rates of population versus single-chain stochastic approximation MCMC algorithms. *Advances in Applied Probability, 46*(4), 1059–1083.
Tibshirani, R. (1996). Regression shrinkage and selection via the lasso. *Journal of the Royal Statistical Society. Series B (Methodological), 1*, 267–288.
Vidic, R. D., Brantley S. L., Vandenbossche, J. M., Yoxtheimer, D., & Abad, J. D. (2013). Impact of shale gas development on regional water quality. *Science, 340* (6134).

Wang, T., Chemali, R., Hart, E., & Cairns, P. (2007). Real-time formation imaging, dip, and azimuth while drilling from compensated deep directional resistivity. In *48th Annual Logging Symposium.*

Wikipedia. (2016). *Mixture model — wikipedia, the free encyclopedia.* Retrieved from https://en.wikipedia.org/w/index.php?title=Mixturemodel&oldid=735277579 [Online]. Accessed 23 August 2016.

Xu, C. (2013). *Reservoir Description with Well-Log-Based and Core-Calibrated Petrophysical Rock Classification* (unpublished doctoral dissertation). University of Texas at Austin.

Zhang, Y., & Sutton, C. A. (2011). Quasi-Newton methods for Markov chain monte carlo. In *Advances in Neural Information Processing Systems* (pp. 2393–2401).

Chapter 8
Big Data in Oil & Gas and Petrophysics

Mark Kerzner and Pierre Jean Daniel

8.1 Introduction

This paper was born as a result of attending O&G and Petrophysics shows, of discussion with multiple smart people involved in the development and operation of petrophysical software and energy development in general and, last but not least, from the personal experience of the authors with oilfield related clients (Bello et al. 2014).

Mark has been involved in petrophysical software early on in his career dealing with advanced log analysis research (Kerzner 1983) and practical software applications (Kerzner 1986). Later, he turned to Big Data software in general, and has been teaching and implementing it for the last eight years for various clients in different countries in the Oil & Gas, high tech companies and legal verticals.

Pierre Jean has been involved in every aspect of the well logging business from tool research creating patents to petrophysics and tool/software integration. He is currently working on the development of a cloud software platform for all oilfield-related activities, including petrophysical interpretation.

When discussing the subject together we arrive at the same conclusion that Big Data is not far away from Petrophysics and it is already there. Multiple companies, large and small, are beginning to implement their first Big Data projects.

As these are their first explorations in the area, they are often faced with the same problems as the designers of current Big Data systems (think Google, Yahoo, Facebook, Twitter) faced early on (Birman 2006). Often, they may not be aware that the

M. Kerzner (✉)
Elephant Scale, Houston, TX, USA
e-mail: mark@elephantscale.com

P.J. Daniel
Antaeus Technologies, Houston, TX, USA
e-mail: pdaniel@antaeus-tech.com

S. Srinivasan (ed.), *Guide to Big Data Applications*, Studies in Big Data 26,
DOI 10.1007/978-3-319-53817-4_8

encountered problems have already been solved in a larger context. This also means that even the implementation itself has already been created and likely open sourced.

Therefore, our intent is to benefit the Big Data work in the industry by describing common problems that we see in implementing Big Data specifically in petrophysics, and by providing guidance for such implementations.

Generally, Mark contributes to the Big Data view of things and Pierre Jean makes sure that it can be practically tied into real life. However, since their experiences cross over, both authors draw on their knowledge of both areas.

Finally, we are well aware that the oilfield industry is going through a deep financial crisis with each company reducing cost to survive the downturn and at the same time preparing for the next cycle. Actual Big Data technology developed through intense innovation for other industry can and must help the oil and gas companies becoming more cost effective through innovation. There has never been so much data and information coming from so many sources without proper software tools to integrate them. There are existing cloud applications but they each cover a very small portion of the oilfield scheme and integration is close to impossible task. Provided that all data information would be interpreted and integrated, higher quality of decisions would then be expected at every level along the decision chain.

8.2 The Value of Big Data for the Petroleum Industry

The oil and gas industry is at the brink of a digital revolution (McGrath and Mahowald 2014) Organizations and measurements scattered across the globe are in dire needs of better environment understanding which can be brought by enhanced calculation capabilities, communication and collaboration within and between companies (Industry Week 2015).

The oilfield industry is left behind the other industries in terms of access to latest cloud software technology (Crabtree 2012). The immediate need is for the creation of a platform in which each user can add plugins, process data, share and visualize information (data, documents, calculation techniques and results) with other users, teams or companies in a very secure and swift process.

This is really the first step to move from an excel-dominated world to the universe of the Big Data analytics. For a platform to become a standard in this industry, four necessary components need to co-exist simultaneously (Fig. 8.1). They are: flexible and simple cost model for all companies adapted to user responsibilities, optimized cloud databases, simplified external plugin integration from all sources, and dashboard accessed by web interface.

Once achieved, this novel platform will bring a new level of work quality and open the way to larger data analytics in the oil and gas industry.

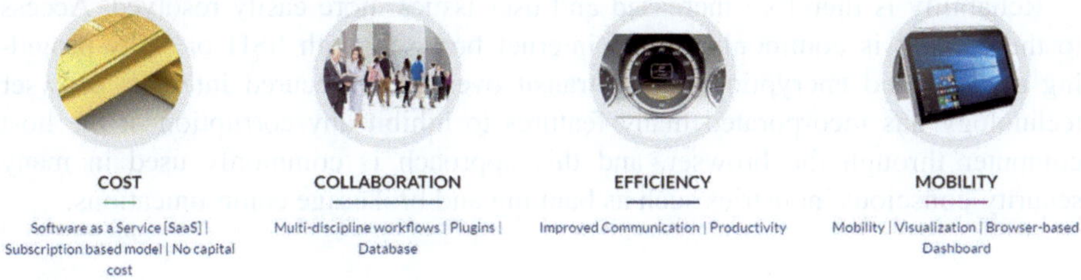

Fig. 8.1 Four components of a petrophysical software platform

8.2.1 *Cost*

The traditional model where software licenses are purchased for and installed on a user's computer is referred to as Product as a Product (PaaP). With the many advantages offered by cloud computing the SaaS model (Wikipedia 2016), where the software provider hosts his software on his own servers, usually in the cloud, and offers a subscription to users of the software, is projected to grow at a rate about five times of the traditional model. The advantages of the SaaS model to the user are: limited capital investment in hardware, many IT functions that are handled by the cloud provider freeing up internal resources, flexibility in changing the system sizing (increasing or decreasing), cost advantages in cloud sourcing, software provider savings in a single-platform development and testing that are passed on to the user, and many other advantages.

8.2.1.1 SaaS (Software as a Service)

SaaS is a business model defined in Wikipedia as Software as a Service (SaaS; pronounced/sæs/) is a software licensing and delivery model in which software is licensed on a subscription basis and is centrally hosted. It is sometimes referred to as "on-demand software". SaaS is typically accessed by users via a web browser. SaaS has become a common delivery model for many business applications, including office and messaging software, payroll processing software, DBMS software, management software, CAD software, development software, gamification, virtualization, accounting, collaboration, customer relationship management (CRM), management information systems (MIS), enterprise resource planning (ERP), invoicing, human resource management (HRM), talent acquisition, content management (CM), antivirus software, and service desk management. [5] SaaS has been incorporated into the strategy of all leading enterprise software companies. See IBM cloud strategy [5]. One of the biggest selling points for these companies is the potential to reduce IT support costs by outsourcing hardware and software maintenance and support to the SaaS provider.

There are significant advantages to a software provider in content delivery via SaaS, primarily that the software always operates on a defined platform without the necessity to design and test the software on all possible hardware and operating systems.

Reliability is therefore increased and user issues more easily resolved. Access to the content is commonly via an internet browser, with SSH or VPN providing security and encryption during transit over the unsecured internet. Browser technology has incorporated many features to inhibit any corruption of the host computer through the browser, and this approach is commonly used in many security-conscious industries such as banking and brokerage communications.

8.2.1.2 Cost Flexibility

One of the main lessons from the drop in oil prices in late 2014 is that companies must be aware of their immutable fixed costs. With the SaaS subscription model (Fig. 8.2), where reassignable seats are provided on a monthly basis with hardware expense included, access to increased resources can be quickly provided and can as quickly be scaled back as markets change. The attempt to make the user interface simpler and more intuitive reduces the time to productive results and mitigates the effects of staff reductions of skilled operators.

8.2.1.3 Reduced IT Support

Many users do not appreciate the hidden costs included in the traditional model of software license purchase. Only very large companies fully account for such essential services as backup, disaster recovery, physical security, proper environmental

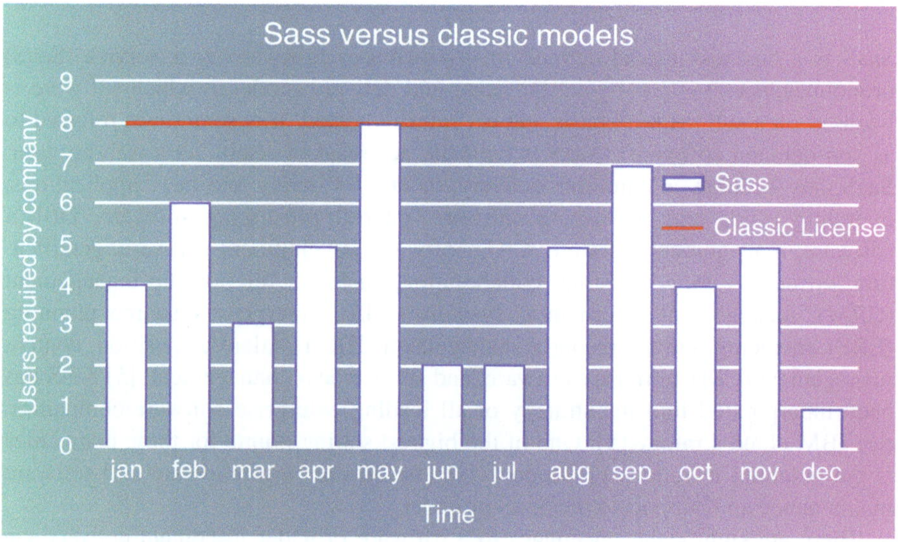

Fig. 8.2 SaaS vs. classic software models

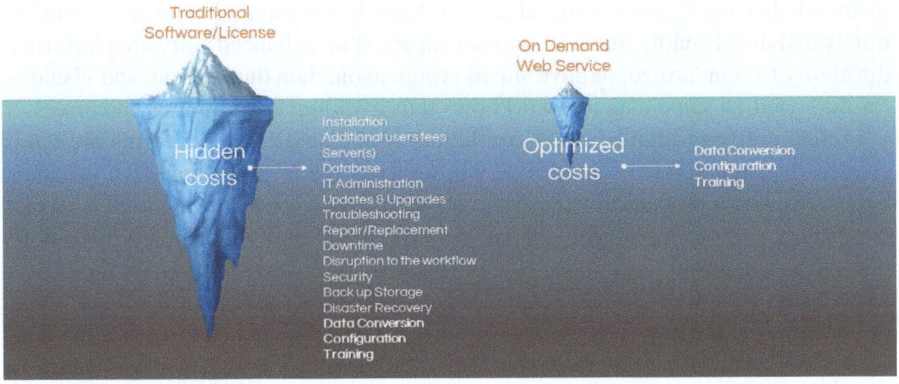

Fig. 8.3 Traditional software licensing vs. on-demand web services

and power conditioning for on-site servers. The planning for hardware resources requires commitments to future use. These are areas that have been recognized as essential to the cloud acceptance and receive significant attention from all cloud providers, allowing IT personnel to concentrate their more immediate tasks (Fig. 8.3). As a cloud software platform provides its software on cloud providers, software updates are provided automatically without threatening other user software or hardware and avoiding the necessity of a lengthy user approval.

8.2.1.4 Efficiency

With the organization of files into a family of searchable databases and the enhanced collaboration in mind, the system increases the efficiency of employees. A cost structure that encourages wider participation in the platform tools encourages the efficiency increase. Direct transfer of log data as inputs to the workflow engine and direct transfer of workflow outputs, together with the automatic identification of all workflow input files and parameters enhance operator efficiency. Software designed for state-of-the art computational efficiency reduces computation time.

8.2.2 Collaboration

Oilfield companies have been cooperating with varying success over the years through joint ventures, shared asset projects, and client/services providers. A very important technical aspect of this collaboration is the transmission itself of the data and having all individuals along the chain of decision to make the right moves to optimize an asset. The ultimate of a browser platform is to add value to the company through a better decision process. To do so, there are different factors such as access to knowledge and response time, ability to share the information between actors in a timely and user friendly manner. Another aspect of an enhanced software platform is the ability to standardize, remove siloed group, avoid data duplication, and visualize results based on same data and workflows thus avoiding uncertainties and errors.

Actual software platforms are conflicting in the sense that platforms are either created by large services companies whose goals are to push their own products, large software companies with prohibitive software cost, or medium software companies without a platform shareable in the cloud and with limited options for internal and external workflows. In this condition, it is very difficult for the oil and gas industry to move to the digital age of Big data analytics. Only a small fraction of the companies, the ones with extensive financial means, can hope to make the leap.

A cloud platform must have the clear objective to correct all these issues by providing a flexible and agile database schema ideal for cooperation, a low cost Software as a Service (SaaS) pricing model widely employed in other industries, an improved cooperation mechanism for external companies, and access to a wide selection of plugins. To allow worldwide collaboration independent of operating system, the secure platform has to be accessible from a secure intuitive browser platform.

8.2.2.1 Ideal Ecosystem

Collaboration is the cooperative effort of more than one person to achieve more progress than the collaborators could achieve individually. It requires shared information and the freedom to manipulate the information to achieve greater knowledge. With cooperative ecosystems as practiced currently transfer of logs and other file types (image, reports) discourage optimal collaboration and division of labor:

- Sources and results of the collaboration are scattered, many intermediate discarded logs are retained and source logs overwritten. The source logs and parameters of workflows are uncertain.
- Reports are scattered on different computers, servers and local memory drives, without easy access by all users.
- High cost of workflow seats and restricted suites of plugins limit the number of collaborators available to work during the cycle of an asset development.
- Communication with customers or consultants requires the ability to share data, workflow, and overall interpretation process for an enhanced collaboration using the same tools.

To reach the Big data analytics stage, the cloud software platform should stimulate collaboration by providing a complete integrated software platform providing cloud databases, calculation plugin engine and browser-based dashboard visualization.

Through the platform, information represented by many file types can then be searched, shared and manipulated. Users have available workflows and plugins for

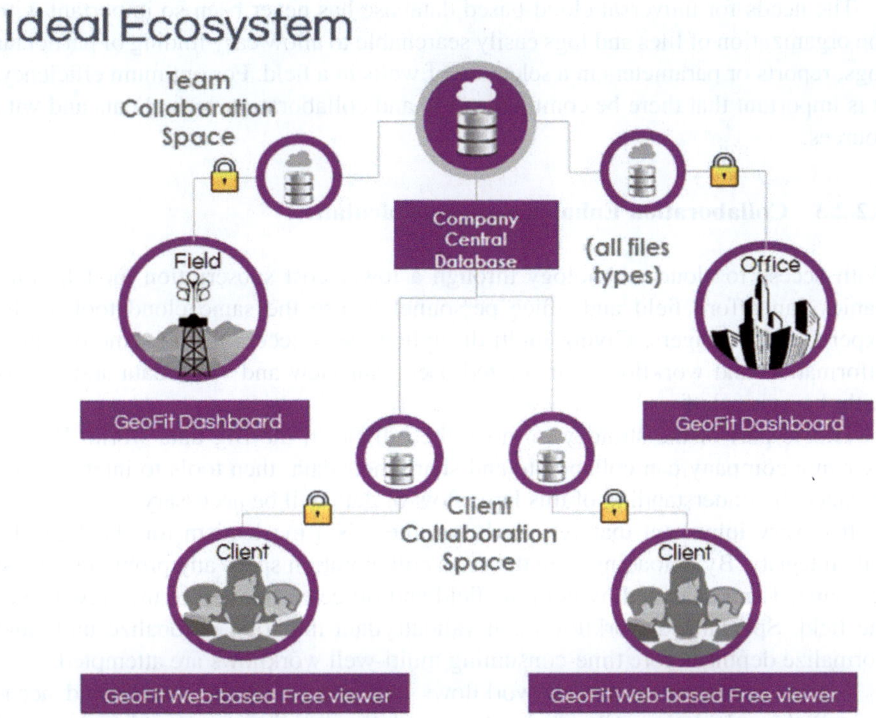

Fig. 8.4 Ideal ecosystem for oilfield actors

a variety of oilfield disciplines under the simplified subscription model. In this ideal ecosystem, each employee in a company is given a specified level of access to particular data (logs, reports, lab results, etc.) to make the best decision.

In this context, workflows developed and shared by other oilfield actors (tool manufacturers, services companies, consultants, academics, oil operators) are accessible by each authorized user, leading to enhanced shared knowledge and higher quality of operation (Fig. 8.4).

8.2.2.2 Collaboration Enhancement by Universal Cloud-based Database Organization

Nothing inhibits evaluation of a well, for instance, more than the frustration of searching for the relevant information which is scattered haphazardly through the computer filing system. Well logs, for instance, may have been obtained at different times, with depths that have been adjusted over time as more information became available. Associated files with relevant information may be in different formats: text, PDF, LAS, DLIS, JPEG, etc.

The needs for universal cloud-based database has never been so important with the organization of files and logs easily searchable to allow easy finding of particular logs, reports or parameters in a selection of wells in a field. For optimum efficiency, It is important that there be communication and collaboration with clients and with sources.

8.2.2.3 Collaboration Enhancement by Calculation

With access to cloud technology through a lower cost subscription model, companies can afford field and office personnel to use the same cloud tool as the experts and managers. Giving multi-discipline users access to the same database information and workflows, authorized users can view and share data and results with the same platform.

This is part of the strategy to move the oilfield to the Big data world. If each user in a company can collaborate and share their data, then tools to interpret and enhance the understanding of this large flow of data will be necessary.

It is very important that very early on, there is a mechanism for checking the data integrity. By uploading field data to a collaboration space any problems can be recognized and analyzed by both the field and office earlier before the crew leaves the field. Specialized workflows can validate data integrity, rationalize units and normalize depths before time-consuming multi-well workflows are attempted.

A very important aspect of workflows and plugins is the quality and accuracy of the calculation. Classical software platform relies on advertisements and conferences to be known and recognized. In the same way as quality is checked by users in many SaaS applications through a rating and comment, this method can be applied to the oilfield with a software platform providing the ability for users to add comments and ratings to workflows developed internally or by oilfield actors (services companies, tool manufacturers, consultants). The gain in quality, efficiency and productivity is immediate since the users will use calculation modules already proven and tested by others.

8.2.2.4 Collaboration Enhancement by Low Cost Subscription Model

If Big Data has penetrated deeper into other industries when compared to the oil and gas, it is also because the actual cost structures of current oilfield software platforms is not adapted to the agility and flexibility required by clients.

Collaboration is inhibited when large immobilized capex and expensive annual subscriptions for each additional module purchased by the company make it prohibitively expensive for many users to share the same access to data, projects and workflows.

SaaS (Software as a Service) has been applied to other sectors such as tourism, travel, accounting, CAD software, development software, gamification, management information systems. This model, through its lower total cost and ability to increase and decrease the commitment, allows wider usage including lab technicians and field operators doing more analysis before involving senior staff.

The new cost structure must start at an inexpensive basic level for those not involved in workflow calculations, such as field personnel who only download field information, management interested in viewing results and accounting or sales personnel. Then more expensive models must be adapted to the job function of the users. For instance a field personnel will not need to run thousands of workflows at the same time while the office expert will.

8.2.2.5 Collaboration Enhanced by Intuitive Browser

Interface System access via a browser allows the platform dashboard to be independent of the operating system or user hardware. Major effort must be expended to make the dashboard very intuitive, thus minimizing the need for lengthy software training. The dashboard can be Google-inspired to intelligently access user profile and preferences, select databases, search for templates views and reports, workflows and reports. The project tree is closer to a data management system than a simple project tree through the ability to browse for existing data, files, templates, reports, or images present in the database.

8.2.3 Knowledge (Efficiency & Mobility)

The efficiency of an organization in which all users can use the same database and software to interpret and/or visualize data and documents is bound to increase. The cloud software platform developed by Antaeus Technologies is seen as an intermediate step between actual software using virtual machines and the ability to use Big Data system for the oilfield arena.

An intuitive interface reduces non-productive time required for training. The full suite of workflows available for each subscription seat allows the sharing of knowledge between all users, while each workflow is optimized for speed of execution and display, reducing non-productive time. Workflows are available from a number of sources and are user-rated with comments to promote workflows found to be very satisfactory by the community of users. The innovation inherited from other industries will minimize the non productive time spent at selecting the right workflow while improving plugin qualities provided (tool manufacturers, oil companies, services companies, consultants, academics).

8.2.3.1 Workflow Variety

The availability of suitable workflows is essential to user efficiency and cloud acceptance by the oilfield industry. There are many sources for workflows and many disciplines for their application. The cloud platform must be provided with essential workflows for many disciplines, and to provide access to the entire suite of workflows to each subscribed seat.

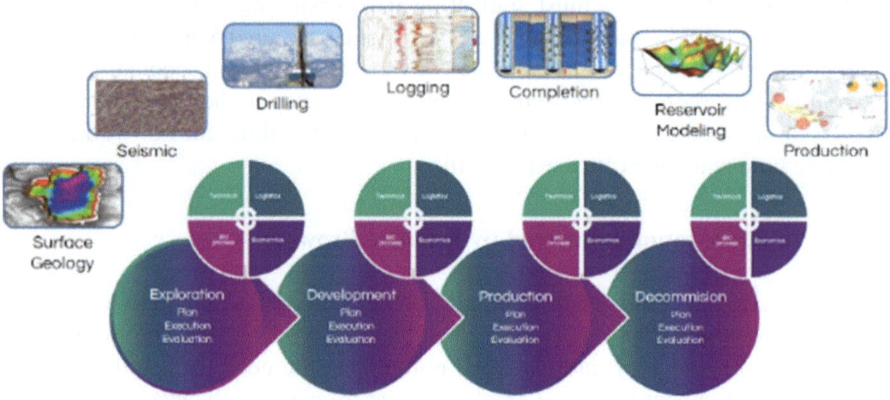

Fig. 8.5 Complete set of oilfield activities

This allows a seat to be shared amongst different disciplines and users. It is also our opinion that many in the community, e.g. academics, students, consultants, would appreciate having a platform for the exercise of their own workflow creations in return for the community recognition and higher visibility of their technical capabilities.

Others who have invested heavily in the development of high end workflows may have a product for which the users would pay additional for their use. For all workflows the mechanism of the workflow has to be hidden from the users for the protection of intellectual property, allowing users to create their own proprietary workflows with limited, controlled dissemination. All oilfield activities can be considered (Fig. 8.5).

8.3 General Explanation of Terms

8.3.1 Big Data

Big Data technically means that you have more data than one computer can hold. This is not a strict definition, and it will depend on the size of this computer, but it is safe to assume that 10 TB already falls in the Big Data category.

However, it would be wrong to limit Big Data just by size. It is also characterized by 3 Vs, that is, volume (we just talked about it), velocity and variety. Velocity means that a lot of data comes to us every second. Hundreds of thousands of transactions per second and even millions are well within reach of Big Data tools. Variety means that the data comes in different formats, and even not formatted at all.

The last aspect of Big Data is its availability. Once engineers got the tools that allowed that to the storage and process of terabytes and petabytes of data, they got into the habit of storing more data. One thing leads to another.

Let us look at a very simple practical example. Imagine that you design a system capable of storing a large number of logs. You do this to help people store each man his own copy on this own computer. That helps, your system gets noticed, and you immediately get these requests: "We heard that you can centrally store a lot of data. Can we please store all of our PDF documents there as well."

Thus, by Big Data we mean all these things. The tools to handle this do exist, and we will be describing them below. Petrophysics is uniquely positioned to benefit from Big Data, because its typical data sizes are big (tens of terabytes) but not too big (petabytes), which would require higher level skills. So, it is a "low hanging fruit".

For a simple introduction to Big Data and Hadoop, please see (Kerzner and Maniyam 2016).

8.3.2 *Cloud*

Cloud is better to be explained as "elastic cloud." People often think that if they store the data in a hosting center and deliver it through the Internet, this is called "cloud-based system." This is a mistake. This is called "internet delivery," and it is as old as the Internet (circa sixties and seventies (Leiner et al. 2016). By contrast, elastic cloud means that your resources are elastic. If you need more storage, you can get it by a simple request, and if you need more processing power, you can get this just as well, all in a matter of minutes. Examples of such clouds are Amazon AWS, Google Cloud, and Microsoft Azure. But any systems that can scale up and down qualifies as a cloud-based system.

What is special about cloud-based system is that they are easily available and very economical. The best, though old, comparison, is by building your own power-generating plant (old style), vs. buying power from a large providers (cloud-based). Because of obvious economies of scale and flexibility and ease of access, once cloud-based system appear, the old systems can't compete. A good example of cloud-based technologies are those behind Uber.

8.3.3 *Browser*

Everybody knows what a browser is, but not everybody appreciates the benefits of delivering the application through it. Browser gives you an universal interface, allows the systems to look the same on any device, and saves developers a lot of work. As modern browser-based systems begin to more and more imitate desktop applications, the distinctions may blur in the users' minds, but the developers should always keep the principle of browser delivery paramount.

In the oil and gas, browser delivery is especially important. Let us remember that security for browser delivery has been under constant development for the last ten years, and it can now be called, "secure browser platform".

The aim of a secure browser platform is to allow the different actors of the oil and gas field to work and collaborate. Current softwares offerings are inadequate to cover all specific needs in a single framework, due to the cost of licensing and the fact that services companies develop software for their own workflows and measurements, thus depriving oil companies from other workflows and measurements as they are not part of the original workflow. The missing link is a software platform independent from main services companies with which companies can interact altogether.

Current companies have different user profiles with very different needs. For instance a field operator will need to run a workflow on a single well at a time whereas an office expert needs to run workflows on hundreds or thousands wells at once, and managers would need reports and history trend.

8.4 Steps to Big Data in the Oilfield

8.4.1 Past Steps

For the last 20 years, software and services companies have been developing PC-based application with large number of calculation modules. The migration to the cloud is bound to be difficult due to the inherent cost of the technical and business legacies, the necessary support of existing systems and the requirement for data continuity. One solution which has been put in place is the use of virtual machine in which the PC-based application is emulated remotely with the user getting a graphic rendering (Angeles 2014). This technology, although facilitating communication between centers, suffers from the absolute requirement of high speed internet since all commands are being sent back and forth to the server with an internet-hungry rendering on the screen. Although a necessary solution for seismic due to large volume of data and cluster-hungry processing, it is also being used for petrophysics, geology and other domain in which this solution is far from ideal. Hence it is extremely sensitive to the network quality and users have been complaining about this technology for domains outside of the seismic arena. Due to these inefficiencies, it does not help the oilfield to migrate to Big Data from the fact that users are not fascinated in this technology.

To reach a point where all data are used efficiently, the IT department has to be able to manage data in databases efficiently and securely, upgrade workflow for every user concurrently, keep time and user versioning of past projects. A central document management is key to make this happen both for structural and non-structural information. On a logistic level, these options are effective with a central platform system and users accessing through the network, provided that the information is passed through the network in a an effective way. The best option is to use user and client machines in the most intelligent way through some artificial intelligence and machine learning.

8.4.2 *Actual Step: Introduce Real Cloud Technology*

The software industry is rapidly accelerating towards cloud computing. Software packages purchased by companies under a Software as a Service (SaaS) model accounted for 10% of software purchased in 2013. This market share increased to $\sim$30% in 2015.

Considering this rapid take-over, companies are developing cloud platforms for multiple industries. Antaeus Technologies has set its course to provide a full cloud collaboration platform to the oilfield industry incorporating various file formats (DLIS, LAS, . . .) and different protocols (WITSMl, PRODML, SCADA). The primary goal is to expand the number of users of its cloud software platform worldwide, then in a second time, connect the office personnel (technicians, accounting, experts, managers . . .) to their instrumented wells and flow stations through the same dashboard. The users will have the ability to run processes to optimize the decision process of the companies. The mid-term goal is to add SCADA and MODBUS protocol integration for intelligent oilfield operations.

Antaeus cloud software platform - GeoFit - is a combination of technologies (Daniel 2016), all inherited from the cloud industry, cross-optimized for speed and user collaboration. It is designed as a cooperation tool inside and between companies. Its workflow engine accepts calculation modules from a variety of sources (oil companies, services companies, consultants, tool manufacturers, academics) supervised by an intelligent input guidance to help users navigate between multiple workflows.

Using Node.js for computational efficiency, it includes multiple wrappers for workflows written in other languages, thus facilitating company migration. The databases accept multiple data types, providing both user and time versioning, and with calculations done at the server for efficiency and security. GeoFit is a full cloud platform where communication and visualization are optimized without requiring any installation, apart from a standard web browser on the user's computer. The cloud platform hosted by Amazon Web Services can also be provided on a portable mini-server allowing the user full mobility without internet connection.

8.4.2.1 What Is GeoFit?

GeoFit is a cloud based software platform providing users in the energy industry access to a cloud database, specialized suite of calculation modules and browser based dashboard to visualize and manipulate well data (Fig. 8.6). The heart of

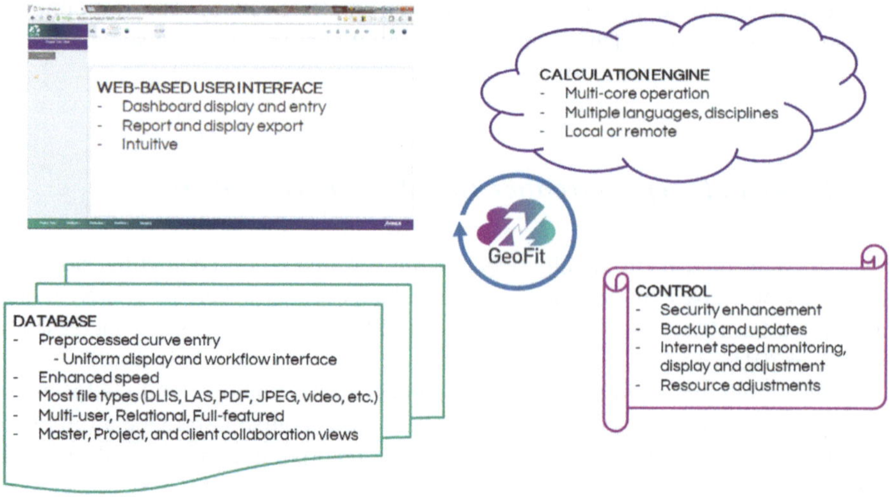

Fig. 8.6 GeoFit architecture

the program is a suite of related databases which are accessed by a computation engine to run workflows and search engines and visualization tools to manipulate and display the contents of the database.

8.4.2.2 Why Use a Universal Cloud-Based Hybrid Database?

GeoFit provides very agile fully multi-user and real-time ready relational database performing at up to 40 Million data entries per second. There is much related information concerning a well. The information can be in a variety of formats, for example: well log formats (DLIS, LAS, CSV, etc); related documents that can be in text, PDF, CSV, etc.; related files (JPEG pictures, audio or video clips, etc.).

The first issue is the access to the files appropriate to the investigation at hand. In normal computer storage relationships are indicated by directories. Should the files related to a log be stored in a file position related to the well, to the field, or to the type of log? The GeoFit database answers such questions by the following properties:

- Searchable: In the database every file and every log has keys indicating searchable properties. Files can be found by searching for those files matching the desired key properties. If you want to find every log in a specified field that is a gamma-ray log created after the year 2000, this is done almost instantaneously. Files and logs could be searched, for example, by a client name or tool manufacturer. Organization is accomplished by the database itself and not by the file location. Spreadsheets such as Excel are not relational databases and do not incorporate this flexible search capability.

- Multiple data types: While specially catering to well log data, the GeoFit database can accommodate multiple file types. This allows retrieving files related to a well without regard to a predefined categorization. PDF, JPEG, Excel, log files and files of many other formats can be stored and referenced.
- Multi-user: Two or more users can access a log at the same time. The system avoids problems that could be created with two users accessing the same file. When the source file is in the Master database the user has read-only access to the data. Any changes or any new files created by a workflow are in a new file that can only be saved to the user's Project database. The system enforces that each new file has a unique name to avoid inadvertent overwriting. Project database files are designed for maximum flexibility in the quest for the best interpretation and users. If users were given delete permissions by the system administrator, they can delete project valuable files; if not given these permissions, they may create confusion or clutter up the project. To alleviate this, project files that are considered worthy of protection can be transferred by the administrator to the Master database to avoid inadvertent destruction while remaining available read only.
- Preprocessed: It would be very inconvenient if the logs in a database were stored in their native format, e.g. some in DLIS and some in LAS. This would require different viewers for each format, would hide inside the formats the information that might be useful in searching, and would require some extraction and formatting of data that might be used in calculations. Before entry of an item into the GeoFit database all the pertinent items in each log are extracted and stored in the database in a uniform proprietary format. The data is stored in a form amenable to rapid calculation and the log information is stored in searchable keys. By this means calculations and searches are sped up by orders of magnitude. The GeoFit viewer can then display any database log without further transformation, and any log can be exported to other users in a standard format by a single export application.
- Calculation access: The GeoFit workflow engine is designed to access the database by its keys so that the selection of workflow inputs is quick and simple. Since the workflow engine knows from the database the units and format of the data no additional specification is necessary, and automatic conversion between user and calculation units are made. With the data in a curve put by the preprocessing into the form for easiest processing, the calculations are orders of magnitude faster. The results can automatically be placed into the database in the GeoFit format for immediate viewing or further manipulation. The identification of the particular workflow version, the identification of the particular source logs and the identification of the input parameters are all stored in the GeoFit entry of every workflow calculation output file.
- Collaboration: GeoFit is supplied with three individual but interactive types of database: a master database with controlled insertion, deletion and modification of masterfiles, allowing their general read-only use; project databases with

shareable files to try different trial interpretations of the data; customer databases which allow secure access by customers to selected files allowing log and report viewing and downloading by the customer.

- Security: An administrator can assign each user an access level which determines privileges, with multiple levels of access available. Secure files can be restricted to only chosen viewers and users requiring very limited access can have access to only those files selected.
- Scalability. The databases scales by building on top of Cassandra. The use of Cassandra is hidden by the API, so to the users it looks like a relational database. Not all the ACID properties are provided though, and they are not needed. The important capability of Cassandra for O&G is it flexible eventual consistency, which allows up to days of offline to be considered as a normal occurrence. If not completely offline, but having very reduced connectivity at time is a norm in O&G.

8.4.2.3 Workflow Engine

The GeoFit workflow engine is designed for optimized performance in conjunction with the GeoFit database.

- Intelligent input guidance: A common problem in the application of workflows is a workflow design that favors inputs from a particular family or tool manufacturer or the application of an inappropriate curve or parameter to a tool. To address this problem, each GeoFit workflow or plugin contains intelligence that directs the user to select from inputs most appropriate to the particular plugin.
- Multidisciplinary library: GeoFit provides at no additional charge an ever increasing selection of multi discipline workflows. GeoFit also provides at no charge third party workflows provided in return for advertisement and attribution. There is a capability for users who develop proprietary workflows to distribute to their customers password enabled versions of those workflows. Third party workflows providing particular value will be provided to users at an additional charge shared with the third party. All workflows are described in the GeoFit form with user comments and rating.
- Computational efficiency: GeoFit-supplied plugins operate multicore in the user's instance and insure the full computer usage and fastest speeds of calculation. Calculations which consume a large percentage of the computer usage do not prohibit the usage of the computer by other operators.
- Easy migration: For ease in migration from other systems, GeoFit can support plugins written in a number of languages: C, Python, Node, Scilab, Matlab.

Software For Reliability, Security and Speed

GeoFit incorporates the latest software techniques. It incorporates Node.js using the V8 engine. Both the display and the workflow interface directly with the database utilizing the specialized structure of the pre-processed log formats to directly and efficiently extract the data points. Wherever possible the data is manipulated in Node.js to accommodate multi core processing. Event loops incorporating single-threaded, non blocking IO allow handling a huge number of simultaneous connections with high throughput. Node uses Google's "V8" JavaScript engine which is also used by Google Chrome. Code optimization and thorough testing is made easier by designing for operation in the defined environment of the cloud, rather than requiring a system adaptable to all operating systems and hardware environments found in traditional sales models. The database is tightly coupled to workflows and displays and was completely rewritten three times before a satisfactory design was arrived at.

8.4.2.4 Cloud Instantiation

As described elsewhere there are many advantages to operation in the cloud, primarily the flexibility in changing the computational and memory resources, the enhanced security and reliability and the cost. For us a major consideration is the opportunity to optimize in a controlled environment allowing rapid problem addressing and a more thorough testing. Traditionally such software was installed on each user's server requiring software compromises to allow operation on many operating systems on different hardware, so testing on each instance must be done on the customer site and problems may be specific to a particular instance.

8.4.2.5 Viewing

Database entries can be viewed in a number of modes: single or multiwell views of multiple logs, cross-plots histograms, etc. (Fig. 8.7) The viewer allows such editing features as formation definition, splicing, depth modification, etc. Searches of the database for the well selection can be shown and other format files selected for viewing. The individual views (2-d, crossplots, etc.) can be extracted in a form suitable for insertion into reports (JPEG, SVG). Plot colors can be easily modified and color preferences (default, company, project, user or customer) can be easily created and utilized. Templates for common groupings are easily created

Communications Remote connectivity is essential in the oil industry. The exchange of information with field offices over the internet is an everyday occurrence. Because many oilfields are in remote places the issue of connectivity over poor or nonexistent internet lines deserves attention. Often connections that are slow or spotty are blamed on faulty software while the true source of the problem is a

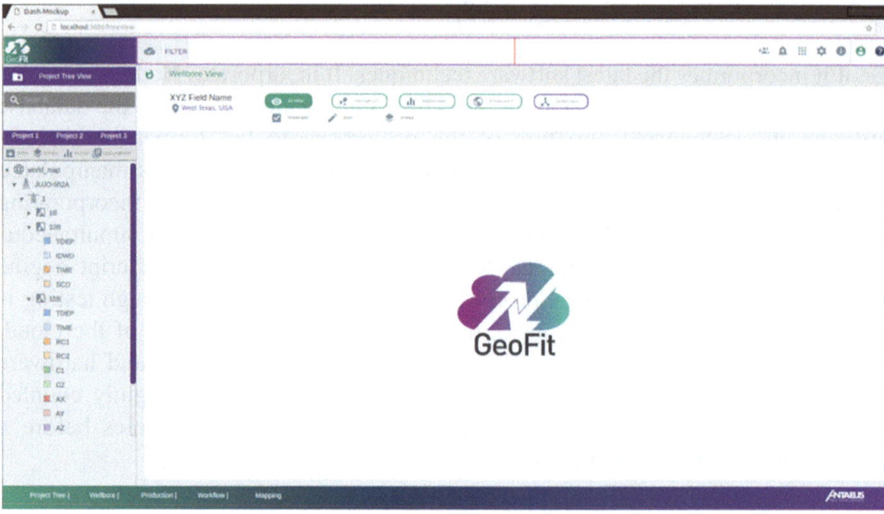

Fig. 8.7 GeoFit main screen

poorly configured firewall or internet connection. In this case effort to improve the internet connection can pay big dividends. GeoFit addresses this issue in a number of ways:

- Internet quality: GeoFit continually monitors the internet quality and displays on the dashboard a measurement of the quality. If there is no connectivity the user is alerted so that corrective actions can be taken and time is not spent trying to fix nonexistent other problems.
- Maximum utilization of bandwidth: If the internet connection is present but the quality is low (e.g. dropped packets requiring multiple retransmissions) GeoFit detects this and automatically takes steps to reduce the transmission of information not necessary for the active display. This makes the best use of the bandwidth available. This adaptive adjustment is not done for high quality connections.
- Portable Server: There are some cases where an internet connection is so unreliable that alternatives are required. The GeoFit platform can be deployed in a portable server providing full functionality (Fig. 8.8). If an internet connection should be occasionally available then the portable server can synchronize with the company database. Selected files can be downloaded to the portable server before deployment and workflows run and the results viewed on those files and any additional files collected in the field. There are additional security measures incorporated into the portable server to protect the code and data stored, but the ability to limit the totality of files initially loaded provides additional security in deployment to high-risk areas

Fig. 8.8 Portable server for limited internet connectivity

8.4.3 Structure Your Data Before You Go Big

8.4.3.1 Project Structure

The project tree is very generic in the sense that it contains all required level for a field study with the option to select worldmap, basin, field, planned and actual well, borehole, activities and container with data created by users or inherited from DLIS/LAS/ASCII files. The project tree has the intrinsic data management ability with the selection of files based on their finality like data, core, plot or document such as reports or videos. Without changes of project tree view, the user can quickly visualize all document types contained at each level of the well hierarchy. This is particularly useful when the user tries to locate the field drilling report or the latest mud content used in wireline operation. It can also be useful for the accountant to understand cost discrepancy between operations. The experts will want to research through a machine learning process if existing operation history could enhance further drilling and completion operation in another field.

Saving all tool parameters saved and stored in the database at each run is the ability to monitor the tool quality, apply preemptive measure to avoid field failure. Having a project tree containing all objects in the database with easy access to tool parameters benefits the quality process of all services companies. It is possible at that point to run Big Data process to fully understand the root of the issues which could very well be linked to the field petrophysical environment or an electronic component reacting to an environment parameter (pressure, temperature, radiation.) over time and over runs. This solution is absolutely impossible to solve if the user is searching at a very limited number of runs and datasets.

For a geoscience expert, having the ability to access reports, images and cores in the same project tree quickly when visualizing and processing data in a wellbore, locating quickly all reports, images or cores from the project tree enhances the quality of the interpretation. This level of integration is necessary for an enhanced decision cycle.

8.4.4 Timestamping in the Field

Data filtering based on time, tool, location, user, version, and other main metadata attached to the project itself. Through the time period selection, the user has the ability to select the period for which he/she would like to visualize the changes in production data, tool run, etc . . . . By selecting a particular date, project tree and all visualization graphics can selectively show data and results at that particular date.

Saving all data and metadata versus time allows propagating the right propagation of parameters along the workflows. For instance an important aspect is the depth of the reservoir. if the kelly bushing (KB) is different from one run to the next, then the measured depth and true vertical depth will be different. if the software platform cannot handle the changes in metadata such as KB, mistakes will be introduced in the processing and wrong depth will be calculated for the reservoir.

The fact that all data and metadata are timestamped will increase the quality of the workflows since the parameters in the workflow will be for the particular run in the well. It also opens the door to the field of data analytics with the ability to monitor specific parameters over runs. For example monitoring Rmf over time provides the petrophysics with information to the well porosity and permeability.

Having metadata linked to runs also increases the chance to fully understand a tool failure or issue in a well since the right environment is known versus the measurement. For instance a tool has failed continuously without the engineering center finding the quality issues. If the problem is due to special environment encountered by the tool while drilling or logging, it is very hard to diagnosticate the main reasons by looking at a single case. With access to past environments for each similar tool, a complete review of all possible reasons become possible through big data analytics processes.

8.5 Future of Tools, Example

For a new software platform to be adopted by users and management at every level in a company, the learning curve must be very fast in the order of a day training. A company like Apple has been successful thanks to the inherent simplicity of its application keeping the complexity in the background of the interface. Advantage of Google and Android is the simplicity to create and share new applications between users creating a whole world of opportunities and access to knowledge previously impossible to apprehend. The software platform developed by Antaeus Technologies is following the same principles of simplicity for the users, extensibility and evolutionary functionalities in the background, incorporating the latest cyber-security for data, workflow and results.

The software design is based on the principle on "for but not with" meaning that the design is not preventing further evolution. Excellent ideas which require additional work are not prevented in the future, while not necessary in the current version for the platform to work properly.

The visualization graphics are being developed with time and real-time data as a major requirement. All views are created through a data-driven graphics language developed for time related information, enhancing the experience and capability of the user to comprehend the flow of data from all origins.

Users in a single team might not be allowed to work directly on the company reference database. In this case, a subset of the database (based on field, well, or zones) can be shared by selected members of a team and publish once approved by the data manager.

To achieve the actual cost efficiency required by the oil and industry to survive during this downturn, the technology has to be extremely optimized. Regarding database, a very strong option to manage speed and security is the use of an database on linux system. Through an optimized process, we obtained a search query in very large and complex file in the order of tens of microseconds per row of data for multiple selected channels. Security has been designed by security experts with years and proven records of data security over browser based platform. In the same line, workflows have the options to have plugins inside the platform, in a separate controller isolated from the platform or run on a remote server.

The scalability of the platform is two-fold, at the database level with the option to select a small computer box as well as using a AWS-type of server, but although at the workflow level with the option to send lengthy processing to high power server for increased efficiency. This is an important factor for a company mid to long term growth. Considering the effort to migrate information, the company has to be certain that there will be no limitation as to the size of the database and the use of the software by its users. The platform developed by Antaeus is being designed with the concept that any users with specific profiles can access and use the platform, in the field without network or connected to localized network, main company intranet network or external network database such as AWS (Amazon web server).

TIME: Overall process time is an important factor to the company effectiveness. Workflows run by users can be lengthy and occupy most of the processing power of a server. To avoid any latency time, Antaeus platform can run a workflow with plugins run on different servers for processing time efficiency. Another reason for loss of time is the amount of data shared on the network. This amount of data can be optimized, lowering the requirement on high network bandwidth. The last main reason to waste time in a project is a lack of collaboration due to system inability to share data, workflow and results. Antaeus is developing a true multi-user, multi-well, realtime architecture in which data, workflows and results can be shared within a group on a team dedicated workspace before publishing to the company database. The same dedicated space can be shared also with partners to accelerate project completion.

Data integration from different origins is a key part of the software platform to achieve digital oilfield through the integration of sensors from any sensors at the rig or the field and wellbore data from logging tools. To be complete, digital oilfield required processing link between data and modeling to ensure adapted model prediction and adapted decisions. Data and metadata are linked through time in the database, integrating these information allows better quality of data, strong decision process and higher return on investment.

The database is collocated at the server level with calculation modules. The user sends request to the server which are handled through an asynchronous, non blocking and concurrent event based language, thus utilizing intrinsically the computational power of the server. The graphics is handled by the client, limiting response delay generated in virtual machines by low network bandwidth. The graphical language is a powerful data driven graphics, developed around the notion of streaming and realtime data. The workflow engine run at the server level can integrate language such as python, Matlab, scilab or C. From a security point of view, plugins can be locally in the server or a separate dedicated server isolated from the software platform with only input/output/parameter accessible.

This overall system creates and optimized cloud server/client platform with operations shared between the server and the client for optimum network usage and client experience.

8.6 Next Step: Big Data Analytics in the Oil Industry

In this section, we set up the major considerations for planning your Big Data implementation.

8.6.1 Planning a Big Data Implementation

8.6.1.1 Planning Storage

The standard platform for Big Data implementations is Hadoop. The description of the technology that eventually became Hadoop was initially published by Google, in 2003–2004.

When I travel and have to explain to my Uber driver what is Hadoop, I usually tell them: "Hadoop is that glue that keep Uber together. Imagine all the millions of drivers and riders, and the size of the computer that does it. So it is not one computer, but many, working together, and Hadoop is that glue that makes sure they work together, registering drivers and riders and all the rest." This explanation always works. Incidentally, Uber has recently made public their architecture stack (Lozinski 2015).

Hadoop gives you two things: unlimited storage and unlimited computation (Fig. 8.9). More precisely, it is limited only by the size of the cluster, that is, by the number of servers that are tied together using the Hadoop software. One better call is Apache Hadoop, because Hadoop is an Apache trademark.

Keep in mind that there are many Big Data projects and products, but Hadoop is the most popular, so it will serve as a base of our first explanation.

Fig. 8.9 Two faces of Hadoop: storage and processing

As an example, let us plan a Hadoop cluster. Practically, here are the steps you would go through. For example, if you plan 1 GB ingestion per day, this means that you need to prepare 3 GB of raw storage per day, since Hadoop replicates all data three times. So, 1 GB becomes 3 GB.

Next, Hadoop wants 25% of storage to be allocated to temporary storage, needed to run the jobs. This turns our 3 GB intro approximately 5 GB. Finally, you never want to run at full storage capacity, and you would want to leave at least 50% of your storage in reserve. This makes 5 GB into 10 GB.

Thus, every GB of data requires 10 GB of raw hard drive storage. This is not a problem, however, since at the current storage prices of about 5 cents per gigabyte, you can store 10 GB for 50 cents. A terabyte (TB) would then be $500. As a consequence, a petabyte (PB) would be $500,000. A petabyte, however, is a very large amount of information, equivalent to a thousand large laptops. An enterprise big enough to need this much data is usually big enough to be able to afford it.

This is illustrated in the Table 8.1 below.

8.6.1.2 Planning CPU Capacity

Usually, as a rule of thumb, once you have enough storage capacity in your cluster, the CPUs that come with the servers hosting the hard drives provide sufficient processing power. In your estimates, processing would include the need to process all the incoming data on time, and accounting for ingest spikes.

Table 8.1 Planning Hadoop storage capacity

Average daily ingest	1 TB
Replication	3×
Daily 'raw data'	3 TB
Node raw storage	24 TB (12 × 2 TB)
Map Reduce temp space	25% 6 TB
'Raw space' available per node	18 TB (raw–mr reserve)
One node can store data for	6 days (18/3)
1 year worth of data (flat growth)	1 TB × 3 × 365 (ingest × replication × days)
1 year (5% growth per month)	1095 TB (1 PB +)
	61 nodes needed (1095/18)
	80+ nodes
1 year (10% growth per month)	110+ nodes

However, you may face a chicken-and-egg problem. If your cluster is very CPU-intensive, like image processing applications, it may require more than the standard processor. You will either know this up-front, or you will see this after you build your cluster and load it. Luckily, clusters are grown by adding, not replacing, the hardware. So if you cluster does not cope, you will increase the size, but not throw it away to buy a new one.

The cost of processing is thus included in the planning for storage above.

8.6.1.3 Planning Memory Requirements

Modern cluster are very memory-hungry, especially with the latest fashion of in-memory computing. The results is that RAM sizes from 128 GB to 256 GB per server are common. Many production clusters that we work with are in the thousands of nodes, with 50–100 TB of RAM total.

The petrophysics clusters for Hadoop and for NoSQL will likely be much more modest, starting from one-machine cluster for development, and extending to 5–10 servers when they go into the first acceptance testing and production.

The real-world petrophysics application will also present a unique problem, not found in the regular Hadoop/Big Data deployments: network may be plentiful in the office, but not so in the field, where it may be limited and unreliable.

The cost of memory is also included in the planning for storage above.

More details on Big Data and Hadoop capacity planning can be found in (Sammer 2012).

Fig. 8.10 Eventual consistency – Starbucks analogy

8.6.2 Eventual Consistency

Consistency is one area that needs to be addressed before we can announce that our O&G application is ready for the real world. Consistency can be strong or eventual.

A good example of eventual consistency is provided by a Starbucks coffee shop line, and this analogy is one that is very often used, see, for example in (Jones 2015). At Starbucks, you submit your order, and the baristas make your coffee, with many people involved. They exchange messages and do not use strongly coupled chain of events. The results is that you may have to wait longer for your coffee than the guy who was after you in line, and getting the coffee itself may require a retry. But eventually you will get yet. This is good enough for you, and it is scalable for Starbucks, and that is exactly the point (Fig. 8.10).

First of all, what is consistency? Databases provide good example of strong consistency: their transaction are atomic (all or nothing), consistent (no inconsistent states are visible to the outside clients), isolated (multiple processes can do updates independently) and durable (data is not lost).

That is all true because of the databases reside on one server. In the world of Big Data, multiple servers are used to store the data, and the ACID properties we just described are not guaranteed. The only one where Big Data systems, such as NoSQL, shine is durability. They don't lose data.

The real question to discuss is consistency. Instead of providing the strong consistency of SQL databases, Big Data systems usually provide what is call eventual consistency. The data will be correct, with the passage of time. How much time? Sometime the data is propagated through the cluster in a second.

In the extreme case, however, the consistency will be preserved if one of the data centers where the data is located is out of circulation for up to ten days. We are talking about Cassandra, of course. Cassandra was designed to be able to tolerate large periods of outage for large parts of it, and then reconcile automatically when the cut-off data center comes back online.

It is exactly this kind of eventual consistency, with very relaxed limitations, that the O&G system will have to provide. In doing so, they can learn from the technologies already developed, tested, and put in practice by the Big Data architect. Cassandra's eventual consistency can be an especially useful example. You do not have to borrow the code, but you certainly study and re-apply the eventual consistency architectural principles. How that can be achieved is addressed in the immediately following section.

8.6.3 Fault Tolerance

Once your system has more than one computer, you have a cluster. And once you have a cluster, some of the computers will fail or will be in the process of failing. Out of 10,000 servers, one fails every day.

This poses two problems. First, if we lose a server, we must make sure that in our unlimited storage there is not definicenty, and that we have not lost any data.

This kind of fault tolerance, the fault tolerance of data, is achieved in Hadoop by replicating all data three times. The replication factor is configurable, first as a default on the cluster, and then as a per-file settings. The replication is going block by block, not file by file, and block are well distributed on the cluster, so that the loss of any server does not lead to the loss of data. Instead, the storage master finds under-replicated blocks and creates more copies of the same data, using good blocks on other servers (Fig. 8.11). Thus the Hadoop storage is self-healing.

In other areas, such as NoSQL, the designers follow the simple principles. For example, in Cassandra data replication is left to each individual server. All servers are imagined as forming a logical ring, and a copy of the data is stored first on the primary server chosen by the hashing algorithm, and then on two next servers on the ring.

This solution is know to work very well, helping with a few aspects of the system design:

- Data loss is prevented: if a server fails, we have more copies of the same data;
- Data locality is preserved: if you have multiple copies of the data, you may be able to find a copy that is local to your server that is running the computations, and not a remote data copy on the network;
- Computations efficiency and good overall throughput. The efficiency is ensured by locality, and good overall throughput by minimizing network traffic.

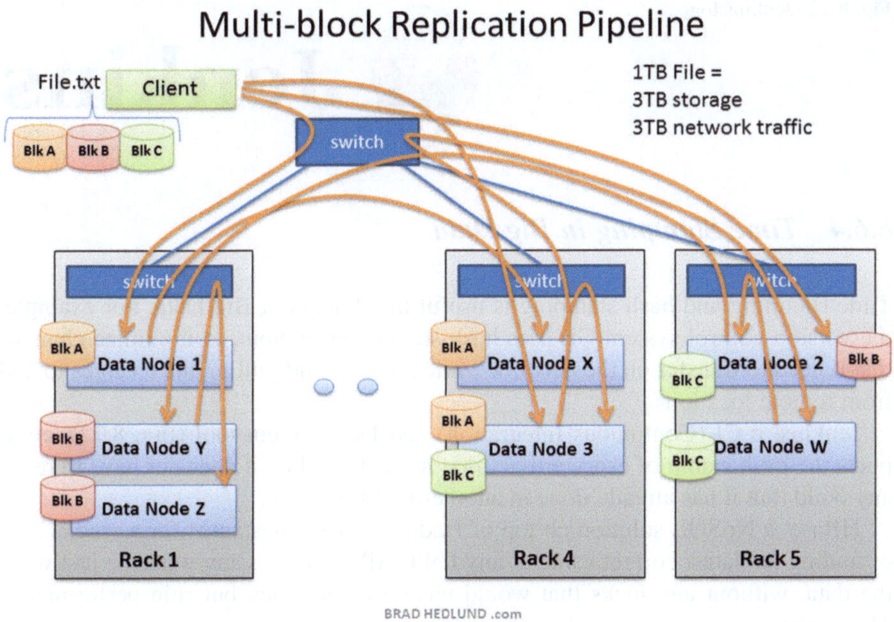

Fig. 8.11 Data replication in HDFS

The designers of Big Data O&G systems, in addition to using tools which already implemented replication, will do well to use the same principles where required by the O&G specifics, and not provided by current tools. How this can be done is explained in the section immediately following.

The second part of fault tolerance implies that your system continues to function even if some of the servers that constitute it fail. This is achieved in Hadoop by having standby, or failover server, which will kick in immediately on the failure of the primary server for a specific function. The function may be storage or processing, or any other part of the workflow.

As Amazon Web Services design courses teach, "Plan for failure, and nothing fails." (Varia 2011). In other words, failure should be considered part of the primary design path. The question is not "if my component fails" but "when this component fails." This is achieved by decoupling the system, using message passing to communicate the tasks, and by having elastic fleet of servers on standby.

All these principle should be used by O&G designers, who are either building on top of the existing Big Data tools, or building their own fault tolerance using the storage and processing fault tolerance ideas developed in Big Data.

Fig. 8.12 Jenkins logo

8.6.4 Time Stamping in Big Data

Time stamping and hash stamping is useful in all areas of Big Data. For example, the version control system Git uses hash stamps throughout, to minimize changes that it need to transfer on the network while pushing and pulling the changes to and from remote locations.

Jenkins, a CI (Continuous Integration) and Deployment tool (Fig. 8.12), computes the hash stamp of every artifact that it creates. Then it does not have to re-do any build that it has already done in another build pipeline.

HBase, a NoSQL solution on top of Hadoop, uses timestamps for a crucial step of finding the latest correct value of any field. HBase allows any writer to just write the data, without any locks that would ensure consistency but ruin performance. Then, when a request for reading the data comes, HBase compares the timestamps on multiple field values written by multiple servers, and simply chooses the latest value as the correct one. Then it updates the other values.

Cassandra is doing the same time comparison, and also fixing the old, incorrect, or corrupt data values. These lessons can be implemented in O&G, and we show in the next section.

8.7 Big Data is the future of O&G

A secure browser platform has intrinsic functionalities in terms of knowledge collaboration once connected to big data database. The collaboration is at multiple level, internally and externally. Internally by creating a workspace shared by a team and linked to the main database through publish/synchronize action, the ability for every user to share the same workflows and results. Externally by allowing companies to interact either at the workflow level or data themselves. A services companies has a strong interest in developing plugins that his clients will use, because the client will be able to make a full use of the data taken in the wellbore. On the other side, an oil company will have access to a larger number of workflows, and thus a larger number of services companies to choose from for a logging work.

In the same way, a tool manufacturer interest is having client hearing about the full potential of a tool, and what better way than promoting it through the full capability of the tool measurements. A consultant strength results in the data processing to get the maximum knowledge out of information from measurements taken on wellbore. Another aspect of the collaboration is the ability to interact faster with the academic work, accelerating the spread of new techniques and measurements.

8.8 In Conclusion

As we have shown above, Big Data is the future of O&G, so now let us summarize the steps that the industry needs to take.

1. Standardize data storage approach. O&G industry is not there yet.
2. With that, you will be able to scale, and move into cloud-based databases.
3. People with a mental block of "my data never goes into the cloud!" thinking will soon be converted. The same situation happened in health and in legal tech, and by now all of them are happily using compute clouds (Kerzner 2015–2016).
4. Scalability will be achieved through cloud-provided scalability.
5. Customer separation will be achieved by each customer working within their own cloud account.
6. Cloud will also allow for multiple clients to inter-operate when this is required.
7. Learning Big Data implementation principles is a must of O&G Big Data system designers.

References

Angeles S. (2014). *Virtualization vs. cloud computing: what's the difference?*. http://www.businessnewsdaily.com/5791-virtualization-vs-cloud-computing.html

Bello O., Srivastava, D., & Smith, D. (2014). *Cloud-based data management in oil and gas fields: advances, challenges, and opportunities*. https://www.onepetro.org/conference-paper/SPE-167882-MS

Birman, K. (2006). *Reliable distributed systems*. London: Springer. http://www.springer.com/us/book/9781447124153

Crabtree, S. (2012). *A new era for energy companies*. https://www.accenture.com/t00010101T000000__w__/fr-fr/_acnmedia/Accenture/Conversion-Assets/DotCom/Documents/Global/PDF/Technology_2/Accenture-New-Era-Energy-Companies-Cloud-Computing-Changes-Game.ashx

Daniel, P. J. (2016). *Secure cloud based oilfield platform*. https://antaeus.cloud/

Industry Week (2015). *The digital oilfield: a new model for optimizing production*. http://www.industryweek.com/connected-enterprise/digital-oilfield-new-model-optimizing-production

Jones, M. (2015), *Eventual consistency: the starbucks problem*. http://www.tother.me.uk/2015/06/29/eventual-consistency-the-starbucks-problem/

Kerzner, M. (1983). Formation dip determination – an artificial intelligence approach. *SPWLA Magazine, 24*(5).

Kerzner, M. (1986). *Image Processing in Well Log Analysis*. IHRDC publishing. Republished by Springer in 2013. https://www.amazon.com/Image-Processing-Well-Log-Analysis/dp/9401085765/.

Kerzner, M. (2015–2016). *Multiple articles on Bloomberg Law*. https://bol.bna.com/author/mkerzner/

Kerzner, M., & Maniyam, S. (2016). *Hadoop Illuminated. Open Source Book about Technology*. http://hadoopilluminated.com/

Leiner, B. M., Cerf, V. G., Clark, D. D., Kahn, R. E., Kleinrock, L., & Lynch, D. C., (2016). *Brief History of the Internet*. http://www.internetsociety.org/internet/what-internet/history-internet/brief-history-internet

Lozinski, L. (2015), *The Uber Engineering Tech Stack.*, https://eng.uber.com/tech-stack-part-one/

McGrath, B., & Mahowald, R. P. (2014). *Worldwide SaaS and Cloud Software 2015–2019 Forecast and 2014 Vendor Shares*. https://www.idc.com/getdoc.jsp?containerId=257397

Sammer, E. (2012). *Hadoop Operations*. Sebastopol, CA: O'Reilly Publishing.

Varia, J. (2011). Architecting for the Cloud: Best Practices. https://media.amazonwebservices.com/AWS_Cloud_Best_Practices.pdf

Wikipedia. (2016). SAAS – Software as a Service. https://en.wikipedia.org/wiki/Software_as_a_service

Chapter 9
Friendship Paradoxes on Quora

Shankar Iyer

9.1 Introduction

The "friendship paradox" is a statistical phenomenon occurring on networks that was first identified in an influential 1991 paper by the sociologist Feld (1991). That paper's title declares that "your friends have more friends than you do," and this pithy and intuitive articulation of the core implication of the friendship paradox is now one of the most popularly recognized discoveries of the field of social network analysis. Despite its name, the "friendship paradox" is not really a paradox. Instead, it is a term for a statistical pattern that may, at first glance, seem surprising or "paradoxical": in many social networks, most individuals' friends have more friends on average than they do. The simplest forms of the friendship paradox can, however, be explained succinctly: compared to the overall population, a person's friends are typically more likely to be people who have a lot of friends rather than people who have few friends.

That explanation of the mechanism underlying the friendship paradox may make the phenomenon seem "obvious," but in the wake of Feld's paper, researchers identified several nontrivial ramifications of the phenomenon for behavior on real-world social networks. Feld himself pointed out that this phenomenon may have psychological consequences: if people tend to determine their own self-worth by comparing their popularity to that of their acquaintances, then the friendship paradox may make many people feel inadequate (Feld 1991). Meanwhile, D. Krackhardt independently identified that the friendship paradox has important implications for marketing approaches where customers are encouraged to recommend products to their friends; essentially, these approaches may be more lucrative than is naively expected, because the friends of the original customers tend to be people who

S. Iyer (✉)
Quora, Inc., Mountain View, CA 94041, USA
e-mail: siyer.shankar@gmail.com

S. Srinivasan (ed.), *Guide to Big Data Applications*, Studies in Big Data 26,
DOI 10.1007/978-3-319-53817-4_9

are better connected and more influential (Krackhardt 1996). This is a line of investigation that continues to be actively pursued in marketing research (Seeman and Singer 2013). The friendship paradox was also found to have implications for designing immunization strategies to combat epidemics spreading in social contact networks. Cohen et al. showed that, with limited immunization resources and incomplete knowledge of the network, it is more effective to immunize randomly chosen acquaintances of randomly chosen people than the randomly chosen people themselves (Cohen et al. 2003). The enhanced effectiveness of such a strategy is again due to the greater connectivity of those acquaintances, and their commensurately greater risk of contracting and transmitting the epidemic. This idea was later turned on its head by N.A. Christakis and J.H. Fowler, who pointed out that those acquaintances make better *sensors* for recognizing the early stages of an outbreak, since they tend to get infected sooner (Christakis and Fowler 2010).

In the past decade, the widespread usage of online social networks and the vast quantities of data that usage generates have enabled the study of the friendship paradox in new contexts and on unprecedented scales. For example, in 2011, researchers at Facebook found that 92.7% of active Facebook users (where an "active" user was defined to be someone who had visited Facebook in a particular 28-day window in the spring of 2011 and who had at least one Facebook friend) had fewer friends than the mean friend count of their friends (Ugander et al. 2011). Similarly, researchers have studied the network of people following one another on Twitter. The Twitter follow network is a directed network, where follow relationships need not be reciprocated. Such a directed network allows for the existence of four different types of degree-based friendship paradoxes:

- Most people have fewer followers than the mean follower count of people whom they follow.
- Most people have fewer followers than the mean follower count of their followers.
- Most people follow fewer people than the mean number followed by people whom they follow.
- Most people follow fewer people than the mean number followed by their followers.

All four of these paradoxes have been shown to occur empirically (Hodas et al. 2013).

Researchers have also extended the core friendship-paradox idea to quantities other than friend count or degree, showing that, in various social networks, the average neighbor of most people in the network scores higher according to some other metric. Hodas et al. have shown, for instance, that for most people on Twitter, the people whom they follow are more active on average and see more viral content on average (Hodas et al. 2013). Very recently, Bollen et al. combined sentiment and network analysis techniques to show that a "happiness" paradox holds in the mutual-follow network on Twitter: if we define a network which contains a link whenever a pair of Twitter users follow one another, then for most people in this network, their neighbors are on average happier than they are, at least according to the sentiment encoded in their tweets (Bollen et al. 2016). These types of phenomena have been called "generalized friendship paradoxes" by Eom and Jo, who identified similar phenomena in academic collaboration networks (Eom and Jo 2014).

Hodas, Kooti, and Lerman have also emphasized the point that, in many social networks, even stronger versions of the friendship paradox and generalized friendship paradoxes occur. For example, not only do most Twitter users have fewer followers than the *mean* follower count of their followers, they also have fewer followers than the *median* follower count of their followers. This stronger phenomenon (which Hodas et al. refer to as the "strong paradox," in contrast to the "weak paradox" in terms of the mean) holds for all four of the degree-based paradoxes in directed networks and also for generalized paradoxes. As some examples of the latter, Hodas et al. show that most people on Twitter are less active and see less diverse content than most of the people whom they follow (Hodas et al. 2014). Similar observations have been made before: the 2011 Facebook study cited above found that 83.6% of active Facebook users had fewer friends than the median friend count of their friends, and indeed, Feld pointed out in this seminal 1991 paper that the strong paradox held empirically in real-world high school friendship networks (Ugander et al. 2011; Feld 1991; Coleman 1961). Therefore, the important contribution of Hodas et al. was not simply to observe that the strong paradox holds in online social networks like Twitter. Instead, their key observation was that, when distributions of quantities over people in a social network follow heavy-tailed distributions (where the mean is higher than the median due to the presence of rare, anomalously large values), the existence of the weak paradox is usually *guaranteed* by the statistical properties of the distribution. The strong paradox, in contrast, is not. Hodas et al. demonstrate this empirically by showing that the weak paradox often survives random reshufflings of the network that destroy correlations between degree and other quantities, but the strong paradox usually disappears. As such, the existence of a strong paradox reveals something about behavioral correlations on the actual network that the weak paradox does not (Hodas et al. 2014).

In the present work, we take these developments in the study of the friendship paradox and examine their implications for Quora, an online knowledge repository whose goal is "to share and grow the world's knowledge." Quora is structured in a question-and-answer format: anyone with a Quora account can add a question about any topic to the product. Quora product features are then designed to route the question to people who have the knowledge required to answer it. This can happen through a question being automatically recommended to an expert in their "homepage feeds," through the question asker or other participants on Quora requesting an answer from a particular individual, or through various other mechanisms. Frequently, these mechanisms lead to questions on physics being answered by physics professors, questions on music theory being answered by professional musicians, questions on politics and international affairs being answered by policy experts and journalists, etc. Processes also exist for merging duplicate questions or questions that are phrased differently but are logically identical. The goal is for there ultimately to be a single page on Quora for each logically distinct question, ideally with one or more high-quality, authoritative answers. Each such page can then serve as a canonical resource for anyone who is interested in that knowledge. People can discover answers that may interest them by reading their homepage feeds (where content is automatically recommended to them), by searching for something they specifically want to learn about, or through several other product features. People can refine Quora's recommendations for them by following topics and questions that interest them, following people whose content they value reading, or simply by providing feedback on answers through "upvoting" or "downvoting." People "upvote" an answer when they consider it to be factually correct, agree with the opinions expressed in the answer, or otherwise find it to be compelling reading; people "downvote" answers that they deem to be factually wrong or low quality. Several of the core interactions on Quora (including following, upvoting, and downvoting) generate rich relationships between people, topics, and questions. Many of these relationships can be represented as networks, and the existence of various strong paradoxes in these networks may reveal important aspects of the Quora ecosystem.

9.1.1 Organization of the Chapter and Summary of Results

The rest of this chapter is devoted to exploring contexts in which variants of the friendship paradox arise on Quora and to exploring the consequences of these paradoxes for the Quora ecosystem. Before diving into the data however, we will briefly review some of the statistics of the friendship paradox in Sect. 9.2, with an emphasis on why strong paradoxes are special and can reveal important aspects of how Quora works.

Then, in Sects. 9.3–9.5, we will explore three different sets of paradoxes on Quora:

- **Section 9.3—Strong Paradoxes in the Quora Follow Network**: We first study the network of people following one another on Quora and show that the strong versions of all four of the directed-network paradoxes occur.
- **Section 9.4—A Strong Paradox in Downvoting**: We next turn our attention to a less traditional setting for the friendship paradox: the network induced by a specific, negative interaction between people during a given time period. More specifically, we will study a network where a directed link exists for each unique "downvoter, downvotee pair" within a given time window; in other words, a link exists from person A to person B if person A downvoted at least one of person B's answers during the time window. We will show that, for most "sufficiently active" downvotees (where "sufficiently active" means that they have written a few answers during the time window), most of their "sufficiently active" downvoters get downvoted more than they do.
- **Section 9.5—A Strong Paradox in Upvoting**: Finally, we will show that, for answerers on Quora who have small numbers of followers, the following property holds: for most of their upvoted answers, most of the upvoters of those answers have more followers than they do.

Each of these incarnations of the friendship paradox is worth studying for different reasons. The paradoxes in the follow network, which we explore in Sect. 9.3, are the "canonical" manifestations of the friendship paradox in directed networks. Measuring and interpreting these paradoxes is a natural precursor to studying paradoxes in less familiar contexts in Sects. 9.4 and 9.5, and it also allows us the opportunity to develop methodologies that are useful in those later analyses. The paradox in downvoting, which we study in Sect. 9.4, is a variant of the friendship paradox in a context that has been rarely explored: a network representing negative interactions. Furthermore, because it occurs in such a context, the downvoting paradox turns the usually demoralizing flavor of the friendship paradox ("your friends are more popular than you are on average") on its head: it may offer some consolation to active writers that the people in their peer group who downvote them typically receive just as much negative feedback themselves. Finally, the paradox in upvoting, which we discuss in Sect. 9.5, has potential practical benefits for the distribution of content that is written by people who have yet to amass a large following. Their content may ultimately win more visibility because this paradox occurs.

After exploring each of these paradoxes and their implications in detail in Sects. 9.3–9.5, we conclude in Sect. 9.6 by recapping our results and indicating interesting avenues for future investigation.

9.2 A Brief Review of the Statistics of Friendship Paradoxes: What are Strong Paradoxes, and Why Should We Measure Them?

Our ultimate goal in this chapter is to understand what various versions of the strong paradox can tell us about the Quora ecosystem. As a precursor to studying any specific paradox however, we will first explore the statistical origins of friendship-paradox phenomena in general. We will begin by reviewing traditional arguments for the friendship paradox in undirected networks. In doing so, we will show how recent work has put Feld's original argumentation on stronger mathematical footing and explained why weak and strong friendship paradoxes in degree are ubiquitous in undirected social networks. We will also explore why weak generalized paradoxes are often inevitable consequences of the distributions of quantities in a social network. In contrast, we will argue that the following types of friendship-paradox phenomena are *not* statistically inevitable:

- Strong degree-based paradoxes in directed networks.
- Strong generalized paradoxes in undirected or directed networks.

The existence of these types of paradoxes depends upon correlations between degree and other quantities, both within individuals and across links in the social network. Thus, these phenomena reveal the impact of these correlations on the network structure, and that network structure can, in turn, reveal nontrivial aspects of the functioning of the Quora ecosystem.

9.2.1 Feld's Mathematical Argument

We now return to Feld's 1991 paper (Feld 1991). In that paper, Feld presented a mathematical argument for the existence of the friendship paradox in which he compared the mean friend count over people in a social network to the mean friend count over neighbors. Suppose a social network contains N nodes (each node represents a person) and has degree distribution $p(k)$. Here, $p(k)$ is the fraction of people in the network with degree k (i.e., with exactly k friends or neighbors - we will use the terms degree, friend count, and neighbor count interchangeably when discussing undirected networks). The mean degree in this network is just:

$$\langle k \rangle = \sum_k kp(k) \tag{9.1}$$

Meanwhile, in the distribution of degree over neighbors, each node with degree k gets represented k times; hence, the degree distribution over neighbors is:

$$p_{\text{nbr}}(k) = \frac{kp(k)}{\sum_{k'} k'p(k')} = \frac{kp(k)}{\langle k \rangle} \tag{9.2}$$

and the mean degree of neighbors is:

$$\langle k_{\text{nbr}} \rangle = \frac{1}{\langle k \rangle} \sum_k k^2 p(k) = \frac{\langle k^2 \rangle}{\langle k \rangle} \tag{9.3}$$

In Eqs. (9.1)–(9.3) above, the sums over k and k' are over all possible degrees in the network; if self-loops and multiple links between individuals are prohibited, then these indices will range from 0 (representing isolated nodes) to $N - 1$ (representing nodes that are connected to all others in an N-node network). From Eq. (9.3), we find that:

$$\langle k_{\text{nbr}} \rangle - \langle k \rangle = \frac{\langle k^2 \rangle - \langle k \rangle^2}{\langle k \rangle} = \frac{Var(k)}{\langle k \rangle} \tag{9.4}$$

This demonstrates that $\langle k_{\text{nbr}} \rangle > \langle k \rangle$ whenever there is non-zero variance (i.e., $Var(k) > 0$) of k in the degree distribution $p(k)$ (Feld 1991).

9.2.2 What Does Feld's Argument Imply?

The argument above is clearly a mathematically valid statement about the two means $\langle k \rangle$ and $\langle k_{\text{nbr}} \rangle$, but it is important to reflect carefully on what it implies. It is, in fact, easy to construct examples of networks where the degree distribution has

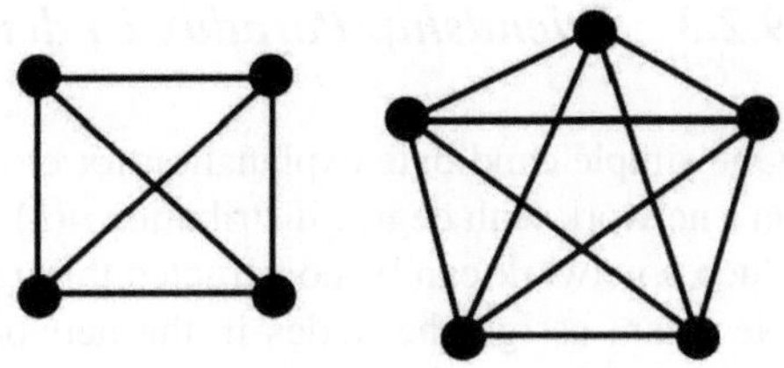

Fig. 9.1 Example network where no one has fewer friends than any of their friends. Inspired by similar examples given by Lattanzi and Singer (2015)

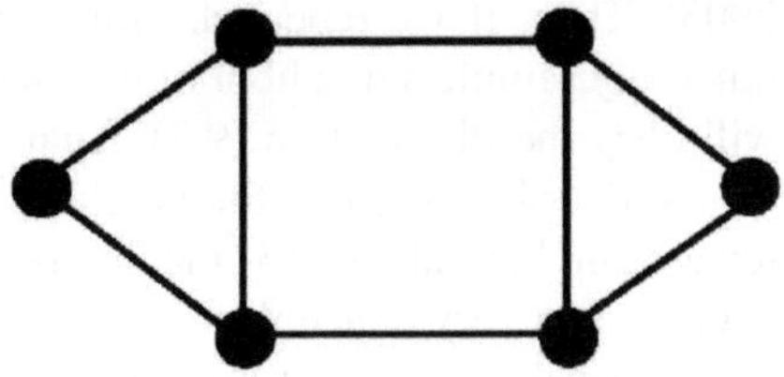

Fig. 9.2 Example network where most people have more friends than the mean number of friends of their friends. Reproduction of an example given by Feld (1991)

non-zero variance, but where no one has a lower degree than the average degree of his or her neighbors. Figure 9.1 provides an example, inspired by similar examples given by Lattanzi and Singer (Lattanzi and Singer 2015). Here, as is mathematically guaranteed, $\langle k_{\text{nbr}} \rangle > \langle k \rangle$ (specifically, $\langle k_{\text{nbr}} \rangle = \frac{29}{8}$ and $\langle k \rangle = \frac{32}{9}$ for the network in Fig. 9.1), but there is no one in the network to whom it would be possible to accurately say, quoting Feld, that "your friends have more friends than you do." Figure 9.2 provides another example, actually provided by Feld himself in his 1991 paper, where most people have *more* friends than the mean friend count of their friends (Feld 1991).

These examples highlight a tension between Feld's argument and the intuitive conception of what the friendship paradox means. This tension exists because Feld's argument actually compares the following two calculations:

- Iterate over each person in the network, write down his or her degree, and take an average over the list.
- Iterate over each friendship pair in the network, write down the degree of each of the people involved, and take an average over the list.

In the process of comparing these two quantities, we actually never directly compare the degree of a person to the degrees of his or her neighbors, and the fact that $\langle k_{\text{nbr}} \rangle > \langle k \rangle$ does not, by itself, put any guarantees on these ratios. Dating back to Feld however, people have empirically measured these ratios and found that most people in many real-world social networks are in the "weak-paradox condition" in terms of degree: in other words, their degree is lower than the mean degree of their neighbors (Feld 1991; Ugander et al. 2011; Hodas et al. 2013). Thus, there is an explanatory gap between Feld's argument and this paradox.

9.2.3 Friendship Paradox Under Random Wiring

One simple candidate explanation for closing this gap is to consider what happens in a network with degree distribution $p(k)$ and purely random wiring between nodes. Such a network can be constructed through the "configuration model" approach of Newman: assign the nodes in the network a number of "stubs" drawn randomly from the degree distribution $p(k)$ and then randomly match the stubs (Newman 2003). Then, if we repeatedly randomly sample a node in this network and then randomly sample a neighbor of that node, the degree distribution over the neighbors will obey the distribution (9.2). Again assuming non-zero variance in the original degree distribution $p(k)$, the neighbor degree distribution $p_{\text{nbr}}(k)$ is clearly shifted towards higher values of k, making both the weak and strong paradoxes in terms of degree seem very plausible.

The weakness in this line of argumentation is that real-world social networks are not randomly wired. Instead, empirical networks exhibit across-link correlations such as *assortativity*, the tendency of people of similar degree to link to one another more often than would be expected by random chance. Assortativity generically tends to reduce the gap in degree between a person and his or her neighbors, and thus, generically tends to weaken the friendship paradox.

9.2.4 Beyond Random-Wiring Assumptions: Why Weak and Strong Friendship Paradoxes are Ubiquitous in Undirected Networks

Very recently, Cao and Ross have proved a relationship between the degree distribution of X, a randomly selected individual from a network, and the degree distribution of Z, a randomly selected neighbor of X, without making any assumptions about random wiring or absence of assortativity. Specifically, these authors have proven that the degree of Z, which we denote by k_Z, is "stochastically larger" than the degree of X, which we denote by k_X. This means that, for any degree k^*, the following property holds (Cao and Ross 2016):

$$P(k_Z \geq k^*) \geq P(k_X \geq k^*) \tag{9.5}$$

If we let k_{med} refer to the median of the overall degree distribution $p(k)$ and set $k^* = k_{\text{med}}$ in Eq. (9.5), then we can see that $P(k_Z \geq k_{\text{med}}) \geq \frac{1}{2}$. Thus, the median of the neighbor degree distribution is at least as large the median of the overall degree distribution.

Returning to Fig. 9.2, Feld explained away this counterexample to the friendship paradox by arguing that it represents a very fine-tuned situation and that there is no reason to expect these improbable situations to be realized in empirical networks. Cao and Ross' work puts this type of argumentation on sounder mathematical footing. In large networks, perhaps there are regions where the local topology of the network happens to align such that some nodes do not experience the friendship paradox; overall though, we can usually expect the weak and strong degree-based paradoxes to prevail in undirected networks.

9.2.5 Weak Generalized Paradoxes are Ubiquitous Too

Hodas et al. have advanced an argument that we should also expect the weak *generalized* paradoxes to occur very generically in undirected and directed social networks. The reason for this is that distributions of quantities in social networks (e.g., content contribution or activity in an online social network) often follow heavy-tailed distributions. These are distributions where the mean is larger than the median due to the presence of rare but anomalously large values that skew the mean. If a person P in the social network has n neighbors, that means the mean of some quantity y over those n neighbors has n opportunities to sample an extreme value in the heavy tail. Meanwhile, P only has one opportunity to sample an extreme value. Therefore, it is statistically likely that the mean of y over neighbors of P will be greater than the value of y for P. As such, we should expect a large number of people in the social network to be in the weak generalized paradox condition, just because of structure of the distribution of y (Hodas et al. 2014).

9.2.6 Strong Degree-Based Paradoxes in Directed Networks and Strong Generalized Paradoxes are Nontrivial

In contrast to the case with weak generalized paradoxes, Hodas et al. show that strong paradoxes are *not* guaranteed by the form of heavy-tailed distributions. These authors make their case empirically by studying real data from two online networking products, Twitter and Digg. First, they show that both weak and strong degree-based paradoxes occur on these products and that weak and strong generalized paradoxes do too. As an example of the degree-based paradoxes, for the majority of Twitter users, most of their followers have more followers than they do; as an example of the generalized paradoxes, for the majority of Twitter users, most of the people whom they follow tweet more frequently than they do. Next, Hodas et al. randomly permute the variable under consideration over the network. In the case of the paradox in follower count, this destroys the correlation between indegree (follower count) and outdegree (followee count). In the case of the paradox in tweet volume, this destroys the correlations between people's degrees and their content contribution. In both cases, the weak paradox survives the scrambling, but the strong paradox disappears: it is no longer the case, for example, that for most people in the scrambled network, most of their followees tweet more frequently than they do. This shows that the strong paradoxes depended upon correlations between degree and other quantities in the network (Hodas et al. 2014).

Another way to see this is to return to our random-wiring assumption and imagine what would happen if we looked for a paradox in a quantity y by sampling a node in our network, sampling a follower of that node, and then comparing the values of y across that link. Under the random-wiring assumption, we would sample followers in proportion to their outdegree. If the joint distribution of outdegree and y in the network is $p(k_{\text{out}}, y)$, we would expect the joint distribution of outdegree and y for the follower to obey:

$$p_{\text{fwr}}(k_{\text{out}}, y) = \frac{k_{\text{out}} p(k_{\text{out}}, y)}{\langle k_{\text{out}} \rangle} \tag{9.6}$$

where $\langle k_{\text{out}} \rangle$ is the mean outdegree in the whole network. If y is a metric that is positively correlated with the number of people an individual follows, then we can expect the marginal distribution of y over followers to shift to higher values compared to the overall distribution of y. In these circumstances, we might expect the typical follower to have a higher value of y, at least within the random-wiring assumption. On the other hand, y could be anti-correlated with outdegree, and this could lead to an "anti-paradox" instead. These possibilities emerge even before we relax the random-wiring assumption and allow correlations across links. These correlations introduce effects like assortativity, which can compete with the within-node correlations between degree and other metrics, enriching and complicating the picture. Thus, the strong paradoxes are not inevitable consequences of the distributions of individual degrees or quantities in the network, and these phenomena can reveal nontrivial features of how the network functions.

9.3 Strong Paradoxes in the Quora Follow Network

9.3.1 Definition of the Network and Core Questions

We now begin our actual analysis of friendship paradoxes on Quora by analyzing the network of people following one another. On Quora, one person (the "follower") follows another person (the "followee") to indicate that he or she is interested in that person's content. These follow relationships then serve as inputs to Quora's recommendation systems, which are designed to surface highly relevant content to people in their homepage feed and digests. The network generated by these follow relationships is a classic example of a directed network, and as we noted in the introduction, a directed network allows for four different degree-based friendship paradoxes. We will now confirm that all four of these paradoxes occur on Quora.

In this analysis, we consider the follow relationships between all people who visited Quora at least once in the 4 weeks preceding June 1, 2016 and who had at least one follower or followee who made a visit during that four-week period. For each of 100,000 randomly chosen people in this group who had at least one follower *and* one followee, we ask the following questions:

- What is the average follower count (i.e., average indegree) of their followers?
- What is the average followee count (i.e., average outdegree) of their followers?
- What is the average follower count (i.e., average indegree) of their followees?
- What is the average followee count (i.e., average outdegree) of their followees?

In all of these cases, the "average" can be either a mean or a median over neighbors, and we compute both to see how they differ, but our claims about the existence of *strong* paradoxes are always on the basis of medians. Note that the followers and followees that we include in each average must also have visited Quora in the 4 weeks preceding June 1, 2016, but need not have both incoming and outgoing links. For the 100,000 randomly chosen people themselves, requiring them to have both followers and followees allows us to actually pose the question of whether they experience the paradox with respect to both types of neighbors.

9.3.2 *All Four Degree-Based Paradoxes Occur in the Quora Follow Network*

In Table 9.1, we report median values of degree over the 100,000 randomly sampled users as well as median values of the averages over their neighbors. The "mean follower" row in Table 9.1 reports the results of the following calculation:

Table 9.1 This table reports statistics for the degrees of 100,000 randomly sampled people in the follow network

Typical values of degree		
	Follower count (indegree)	Followee count (outdegree)
Person	[6.0, **6.0**, 6.0]	[9.0, **9.0**, 9.0]
Mean follower	[35.0, **35.5**. 36.0]	[72.7, **73.5**, 74.2]
Median follower	[17.0, **17.0**, 17.5]	[42.0, **42.0**, 42.5]
Mean followee	[104.7, **106.3**, 108.0]	[63.8, **64.4**, 65.0]
Median followee	[51.0, **52.0**, 52.0]	[32.0, **33.0**, 33.0]

The "person" row shows the median values of indegree (follower count) and outdegree (followee count) over these randomly-sampled people. Meanwhile, the "mean follower" and "mean followee" rows show the "typical" (i.e., median) value of the mean degree of the neighbors of the randomly sampled people. Finally, the "median follower" and "median followee" rows show the "typical" (i.e., median) value of the median degree of the neighbors of the 100,000 randomly sampled people. Since we subsample the full population in these estimates, we also report a 95% confidence interval around each of our estimates, computed using the "distribution-free" method (Hollander et al. 1999). The estimates themselves are in bold

1. For each of the 100,000 randomly sampled users, compute the mean degree (indegree or outdegree) over his or her followers.
2. Compute the median of those 100,000 means. This gives the "typical" value of the mean degree.

The data in Table 9.1 implies that all four directed-network paradoxes occur, because the "typical" values of the averages over neighbors are greater than the typical values for the randomly-chosen people. Table 9.1 is not completely conclusive though, because we have computed statistics over the randomly-chosen people and their neighbors independently, ignoring correlations in degree across links. We can remedy this by computing statistics for *differences* in degree across the links. We do this in Table 9.2, which shows directly that all four variants of directed-network friendship paradoxes occur. For example, a typical followee of a typical individual gets followed by 28 more people and follows 9.5 more people than that individual (note that fractional values like 9.5 are possible, because if someone has an even number of neighbors, it is possible for the computed median to be the midpoint between two integers).

Table 9.2 shows that the typical gap between the degree of a randomly selected person and his or her neighbors is greater when computed in terms of the mean over neighbors (i.e., when measuring a *weak* paradox). As we argued in Sect. 9.2, this is because the mean is more affected by extreme values: when we take a mean over > 1 neighbors, we are giving the mean more opportunities to be inflated by someone with an exceptionally large degree. Consequently, the statement that most people have lower degree than the mean over their neighbors is generally weaker than the statement that most people have lower degree than the median over their neighbors (Hodas et al. 2014). Table 9.2 shows that all four paradoxes survive when we take the median over neighbors and thus measure the "strong" paradox.

Table 9.2 This table reports statistics for the differences in degree between 100,000 randomly sampled people in the follow network and their neighbors

Typical values of differences in degree		
	Follower count (indegree)	Followee count (outdegree)
Mean follower—person	[16.0, **16.4**, 16.7]	[49.4, **50.0**, 50.7]
Median follower—person	[2.0, **2.5**, 2.5]	[20.0, **20.0**, 20.5]
Mean followee—person	[75.0, **76.2**, 77.3]	[35.3, **35.8**, 36.2]
Median followee—person	[27.5, **28.0**, 28.0]	[9.0, **9.5**, 10.0]

The "mean follower—person" and "mean followee—person" rows show the typical (i.e., median) values of the difference between the mean degree of the neighbors of P and the degree of P for each of the randomly sampled people P. Meanwhile, the "median follower—person" and "median followee—person" rows show the typical (i.e., median) values of the difference between the median degree of the neighbors of P and the degree of P for each of the randomly sampled people P. Compared to Table 9.1, averaging over differences better captures correlations in degree across links in the network. Since we subsample the full population in these estimates, we also report a 95% confidence interval around each of our estimates, computed using the "distribution-free" method (Hollander et al. 1999). The estimates themselves are in bold

9.3.3 *Anatomy of a Strong Degree-Based Paradox in Directed Networks*

Before proceeding to look at more exotic paradoxes, we pause to dissect the impact that within-node and across-link correlations have on the paradoxes in the follow network. These paradoxes are ultimately consequences of positive correlations between indegree and outdegree: people who are followed by more people also tend to follow more people. We can verify that these correlations exist by tracking how the distribution of indegrees changes as we condition on people having larger and larger outdegrees. We do this in Fig. 9.3.

To see how across-link correlations impact the paradoxes, we can calculate how strong the paradox would be under a random-wiring assumption, which ignores these correlations. Then, we can compare against reality. To demonstrate this, we focus upon the following paradox: most people have fewer followers than most of their followers. The follow network can be characterized by a joint distribution of indegrees and outdegrees $p(k_{\text{in}}, k_{\text{out}})$, which gives the probability that a person has k_{in} followers and follows k_{out} people. The joint distribution encodes the within-node correlations between indegree and outdegree that we see in Fig. 9.3. Suppose that the wiring *between* nodes is completely random. Now, imagine repeatedly sampling a random person in the network and then sampling a random follower of that

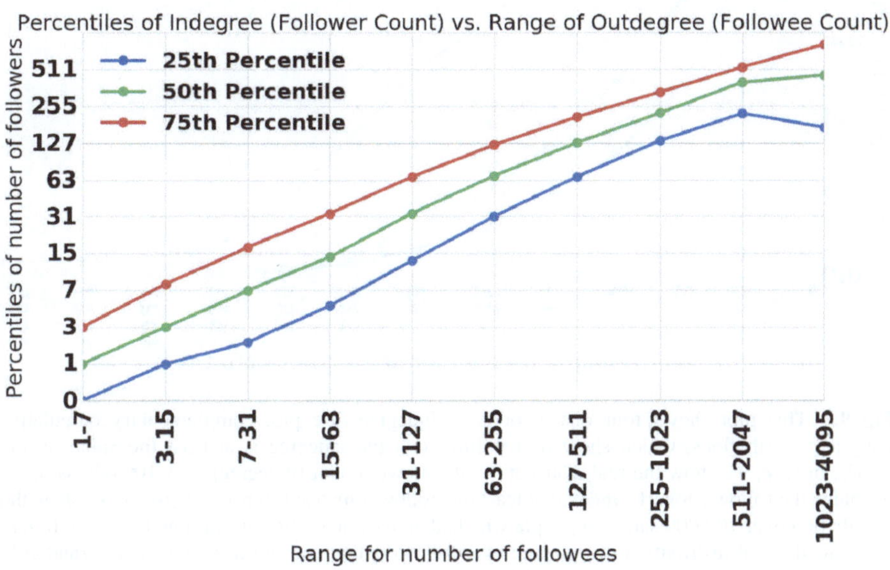

Fig. 9.3 This plot shows percentiles of the overall indegree distribution in the follow network vs. ranges of outdegree. We show the 25th, 50th, and 75th percentiles of the distribution. As we consider people who follow more and more people, the distribution of follower counts shifts to higher and higher values. This reveals strong positive correlations between indegree and outdegree in the follow network

person. Based on the argumentation from Sect. 9.2, we would expect that the joint distribution of indegrees and outdegrees for that person would look like:

$$p_{\text{fwr}}(k_{\text{in}}, k_{\text{out}}) = \frac{k_{\text{out}} p(k_{\text{in}}, k_{\text{out}})}{\langle k_{\text{out}} \rangle} \tag{9.7}$$

Using our empirical distribution $p(k_{\text{in}}, k_{\text{out}})$ and Eq. (9.7), we can compute the expected distribution $p_{\text{fwr}}(k_{\text{in}}, k_{\text{out}})$ under the random-wiring assumption. In practice, we actually calculate the expected marginal distribution over followers of just the variable k_{in} and then calculate the complementary cumulative distribution of this variable:

$$p_{\text{fwr}}(k_{\text{in}} \geq k) = \sum_{k_{\text{out}}} \sum_{k_{\text{in}} \geq k} \frac{k_{\text{out}} p(k_{\text{in}}, k_{\text{out}})}{\langle k_{\text{out}} \rangle} \tag{9.8}$$

In Fig. 9.4, we plot four complementary cumulative distributions of k_{in}:

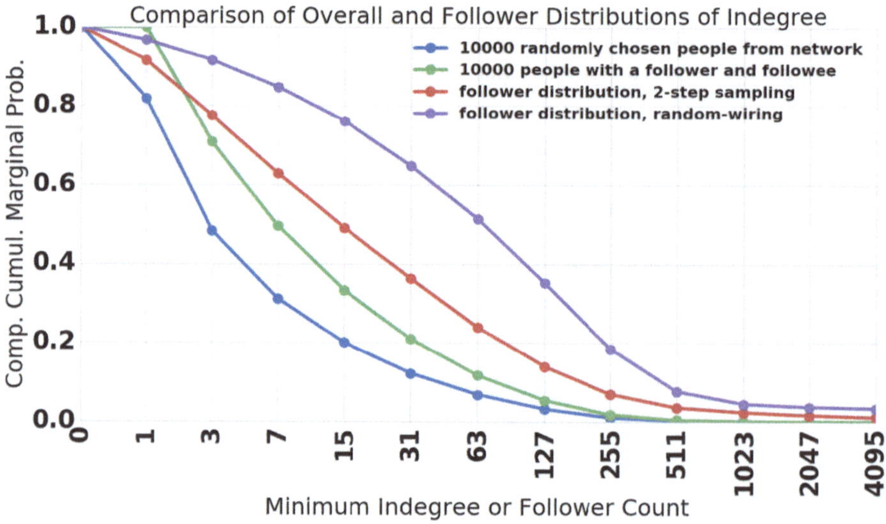

Fig. 9.4 This plot shows four distributions of indegree. We plot complementary cumulative marginal distributions, which show probabilities that the indegree is at least the value on the x-axis. In *blue*, we show the real distribution of follower count (indegree) over 100,000 sampled people in the follow network who had at least one follower **or** one followee. In *green*, we show the distribution over 100,000 sampled people who had at least one follower **and** one followee. In *red*, we show the real distribution of indegree over followers that we find if we repeatedly randomly sample an individual with at least one follower and one followee and then randomly sample one of that person's followers. In *purple*, we show the inferred distribution of indegree over followers that we would expect if we apply the random-wiring assumption in Eq. (9.8) to our empirical data

1. In blue, we plot the distribution over 100,000 randomly sampled people from the network (i.e., people with at least one follower **or** one followee).
2. In green, we plot the distribution over 100,000 randomly sampled people from the network who had at least one follower **and** one followee.
3. In red, we plot the distribution that we find if we adopt a two-step sampling procedure where we repeatedly randomly sample someone with at least one follower and one followee, randomly sample a follower of that person, and measure that follower's indegree.
4. In purple, we measure the distribution implied by the random-wiring assumption from Eqs. (9.7) and (9.8).

The two-step sampling distribution is shifted outward from distribution 1 (i.e., the overall distribution). For all but the lowest indegrees, the two-step sampling distribution is also shifted outward from distribution 2, with the behavior at low indegrees arising from the fact that we have required all people included in distribution 2 to have at least one follower. The fact that the median of the two-step sampling distribution is shifted outward from the median of distribution 2 is consistent with the observation of the paradox in Table 9.2. However, note that the two-step sampling distribution is not shifted out as far as distribution 4, the distribution computed via the random-wiring assumption. This indicates that the paradox is indeed weakened, but not fully destroyed, by correlations in degree across links.

9.3.4 Summary and Implications

In this section, we have established that all four strong degree-based friendship paradoxes occur in the network of people following one another on Quora. We have previously argued that strong friendship paradoxes rely on correlations between quantities in the network. In this case, the relevant correlations are between indegree (i.e., follower count) and outdegree (i.e., followee count). These within-node correlations outcompete across-link correlations (i.e., assortativity) to result in strong friendship paradoxes.

It is worth noting that this need not have worked out this way. It is possible to imagine a product where indegree and outdegree are anti-correlated. Suppose a hypothetical product contains two very different types of users: there are consumers who follow many producers but do not receive many followers themselves, and there are producers who have many followers but do not follow many people themselves. In a situation like this, it is possible to devise scenarios where most people have more followers than most of their followers. Maybe a scenario similar to this is actually realized in online marketplaces, where buyers purchase from various merchants but do not sell to many people themselves and where merchants sell their goods to many buyers without patronizing many other merchants.

Nevertheless, in the Quora context, it is perhaps not very surprising that indegree and outdegree are strongly correlated, as seen in Fig. 9.3. People who attract many followers are typically active writers, who are likely to be sufficiently engaged with Quora that they follow many other writers too. This, in turn, makes the existence of strong paradoxes in the follow network less surprising, but it is still important to examine this set of "canonical" friendship paradoxes before moving on to more exotic examples. Moreover, the technology that we developed in this section to probe the origins of these standard paradoxes will be very useful in dissecting the less familiar paradoxes to come.

9.4 A Strong Paradox in Downvoting

9.4.1 What are Upvotes and Downvotes?

Although the network of people following one another on Quora provides valuable input into the product's recommendation systems, in practice, people end up seeing content originating from outside their follow network as well. This can be because they follow topics or directly follow questions, or because they access content through non-social means like search. In other words, the network of people following one another is not synonymous with the actual network of interactions on Quora. In this section, we show that friendship-paradox phenomena also exist in "induced networks" of real interactions on the product. We focus on a specific interaction, the downvote, for which we identify the special variant of the friendship paradox that we referred to in the Introduction as the "downvoting paradox."

Before proceeding, we should clarify what a downvote is. On any Quora answer, any person who has a Quora account has the opportunity to provide feedback by either "upvoting" or "downvoting" the answer. To understand what a "downvote" represents in the system, it is helpful to first understand what an "upvote" is. An upvote typically signals that the reader identifies the answer as factually correct, agrees with the opinions expressed in the answer, or otherwise finds the answer to be compelling reading. Upvotes are used as one of a large number of signals that decide where the answer gets ranked, relative to other answers to the same question, on the Quora page for that question. Upvotes are also a signal that the reader values the type of content represented by the answer, and therefore, can serve as one of many features for Quora's recommendation systems. Finally, upvotes serve a role in social propagation of content: if someone upvotes an answer (without using Quora's "anonymity" feature), that person's followers are more likely to see that piece of content in their homepage feeds or digests. We will explore this role of the upvote in much greater detail in Sect. 9.5.

In many ways, the "downvote" is the negative action that complements the "upvote." People cast downvotes on answers to indicate that they believe the answer to be factually wrong, that they find the answer to be low quality, etc. Downvotes are used as negative signal in ranking answers on a question page, and they also signal that the reader is *not* interested in seeing further content of this type. Meanwhile, in contrast to the upvote, they do not serve a social distribution function: that is, a follower of someone who downvotes an answer is, naturally, *not* more likely to see that answer in homepage feeds or digest.

There is another way that downvotes differ fundamentally from upvotes. In cases where upvoters do not use Quora's "anonymity" feature, writers can directly see who has upvoted their answers, but the identity of their downvoters is hidden from them. This is true even if the downvoter has not elected to "go anonymous" on the piece of content that he or she is downvoting. This is one of the features that makes downvoting a compelling setting to look for friendship paradoxes: in the ordinary settings in which friendship paradoxes are explored, the public nature of the social ties encoded in the network has implications for the growth of the network. For example, on networks such as Twitter or Quora, information about who is following whom is public, increasing the odds that any particular follow relationship will be reciprocated. The inherently hidden nature of downvoting precludes these types of dynamics, so it is interesting to explore the ramifications of this fundamental difference for friendship paradoxes.

Another reason that downvoting is a compelling setting for exploring this phenomenon is simply that downvoting represents a negative interaction between two individuals. Networks representing negative interactions show several important structural differences with networks representing positive interactions (Harrigan and Yap 2017). Friendship paradoxes, as the term "friendship" itself suggests, have relatively rarely been explored in these types of contexts.

9.4.2 The Downvoting Network and the Core Questions

We are now prepared to define the downvoting network that we study in this section. Our network represents all non-anonymous downvotes cast (and not subsequently removed) on non-anonymous answers written (and not subsequently deleted) in the four-week period preceding June 1, 2016. Note that the conclusions that we reach are not peculiar to this particular four-week window, and we have checked that they hold in other four-week windows as well; we will see one such example of another time window later in this section.

In our network, we draw a directed link for every unique "downvoter, downvotee pair" within the four-week window. In any downvoting interaction, the "downvotee" is the person who wrote the answer that received the downvote, and the "downvoter" is the person who cast that downvote. A directed link exists between two nodes in our network if the downvoter (who is represented by the origin node) downvoted even a single non-anonymous answer written by the downvotee (who is represented by the target node) within the four-week window. In other words, just through the network structure, we cannot tell if a given link represents one or multiple downvotes from a particular downvoter to a particular downvotee. We present a cartoon version of this network in Fig. 9.5.

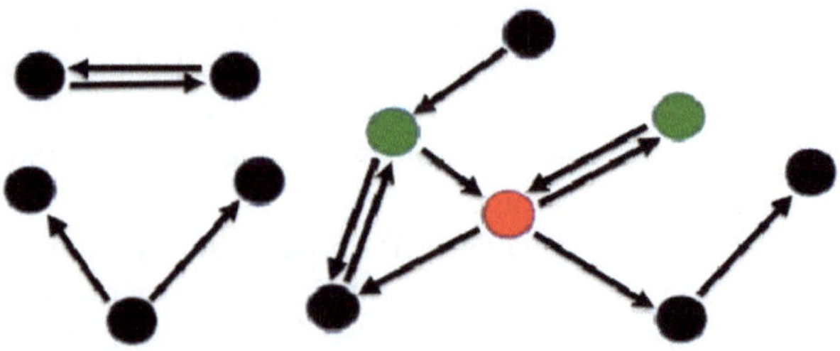

Fig. 9.5 A cartoon illustration of the downvoting network, representing the downvotes within a four-week period on Quora. A directed link exists between two nodes if the person represented by the origin node (the "downvoter") cast at least one downvote on any answer by the person represented by the target node (the downvotee) during the four-week period. In this diagram, the nodes in *green* represent all the unique downvoters of a particular downvotee, who is represented by the node in *red*

Here are the core questions that we pose about this network:

1. **The downvotee → downvoter question**: For most downvotees, did most of their downvoters receive more or fewer downvotes than they did?
2. **The downvoter → downvotee question**: For most downvoters, did most of their downvotees receive more or fewer downvotes than they did?

Note that these questions amount to asking whether two versions of the strong paradox exist in the downvoting network. This is because we ask if *most* of the neighbors of *most* downvotees or downvoters score more highly according to a metric (namely, the total number of downvotes received).

9.4.3 *The Downvoting Paradox is Absent in the Full Downvoting Network*

In Tables 9.3 and 9.4, we report data that addresses the "downvotee → downvoter" and "downvoter → downvotee" questions in the full downvoting network. Table 9.3 reports statistics for the typical values of the number of downvotes received by each party; meanwhile, Table 9.4 reports statistics for the typical values of differences between the number of downvotes received by the average neighbor and the individual in question. As in Sect. 9.3, the statistics reported in Table 9.4 better capture the impact of correlations across links in the downvoting network.

Tables 9.3 and 9.4 reveal that the analog of the strong paradox is present in the answer to the "downvoter → downvotee" question: the typical downvotee of most downvoters gets downvoted more than the downvoter. However, in the "downvotee → downvoter" case, the paradox is absent: for most downvotees, most of their downvoters get downvoted *less* than they do.

Table 9.3 This table reports the typical number of downvotes received by people and their average "neighbors" in the "downvotee → downvoter" and "downvoter → downvotee" questions

Typical values of downvotes received		
	Downvotee → downvoter	Downvoter → downvotee
Mean for downvotee	1.0	40.0
Mean for downvoter	0.2	0.0
Median for downvotee	1.0	30.0
Median for downvoter	0.0	0.0

Consider the "downvotee → downvoter" column. The "mean over downvotee" and "median over downvotee" rows here are identical because they correspond to iterating over each downvotee in the network, taking a mean or median over a single value (the number of downvotes received by that downvotee), and then taking a median of those values over the downvotee in the network. Meanwhile, the "mean for downvoter" and "median for downvoter" rows are different because they correspond to iterating over each downvotee in the network, taking a mean or median over each downvotee's downvoters, and then taking a median over downvotees to compute a typical value of the average over downvoters. The statistics in the "downvoter → downvotee" are analogous, with the roles of downvotees and downvoters in the computation reversed. Note that these are population values over the entire downvoting network

Table 9.4 This table reports the typical differences in the number of downvotes received by people and their average "neighbors" in the "downvotee → downvoter" and "downvoter → downvotee" questions

Typical values of differences in downvotes received		
	Downvotee → downvoter	Downvoter → downvotee
Mean downvoter–downvotee	−1.0	−39.0
Median downvoter–downvotee	−1.0	−29.0

Consider the "downvotee → downvoter" column. The "mean downvoter–downvotee" and "median downvoter–downvotee" rows correspond to the following calculations: (1) iterate over each downvotee in the network, (2) compute the mean or median number of downvotes received by the downvotee's downvoters, (3) subtract the number of downvotes received by the downvotee, and (4) take a median of the difference from step 3 over all downvotees. The statistics in the "downvoter → downvotee" are analogous, with the roles of downvotees and downvoters in the computation reversed except in the order in which we compute the difference. In other words, we continue to subtract the number of downvotes received by downvotees from the number received by downvoters, rather than reversing the order of the subtraction. Note that these are population values over the entire downvoting network

From one perspective, the fact that the paradox does not occur in the "downvotee → downvoter" case may be unsurprising. It may be the case that most downvotees get downvoted for understandable reasons (e.g., they write controversial or factually incorrect content). Consequently, we may *expect* them to get downvoted more frequently than their downvoters. However, it is useful to think about what it would mean if the analog of this paradox was absent in the follow network. In that context, the analogous situation would be if, for most people, most of their followers have fewer followers than they do. As we saw in the previous section, the strong positive correlations between indegree and outdegree actually produce the opposite trend:

we showed that most people who have both followers and followees have fewer followers than most of their followers. We now examine how the correlations between downvoting and being downvoted produce a different outcome in the downvoting network.

Consider the joint distribution over people in the downvoting network of four variables:

- k_{in}: the number of unique downvoters of the person (i.e., that person's indegree in the downvoting network).
- d_{in}: the number of downvotes the person received, which should respect $d_{\mathrm{in}} \geq k_{\mathrm{in}}$.
- k_{out}: the number of unique downvotees of the person (i.e., that person's outdegree in the downvoting network).
- d_{out}: the total number of downvotes the person cast, which should respect $d_{\mathrm{out}} \geq k_{\mathrm{out}}$.

We call this joint distribution $p(k_{\mathrm{in}}, d_{\mathrm{in}}, k_{\mathrm{out}}, d_{\mathrm{out}})$. Now, imagine choosing a downvotee and then following a random incoming link out to a downvoter of that downvotee. If we adopt a random-wiring assumption, then we ought to reach downvoters in proportion to the number of distinct people whom they downvoted, which is k_{out}. We expect the distribution of the four variables over the randomly sampled downvoter to follow:

$$p_{\mathrm{dvr}}(k_{\mathrm{in}}, d_{\mathrm{in}}, k_{\mathrm{out}}, d_{\mathrm{out}}) = \frac{k_{\mathrm{out}} p(k_{\mathrm{in}}, d_{\mathrm{in}}, k_{\mathrm{out}}, d_{\mathrm{out}})}{\langle k_{\mathrm{out}} \rangle} \tag{9.9}$$

This shifts the distribution to higher values of k_{out}. If k_{out} and d_{in} are positively correlated, we might expect the typical downvoter of most downvotees to get downvoted more than the downvotee.

We probe correlations between k_{out} and d_{in} in Fig. 9.6 by bucketing downvoters by their values of k_{out} and then plotting percentiles of the distribution of d_{in}. The plot shows that, over a large range of k_{out}, the majority of downvoters actually receive no downvotes at all (i.e., they have $d_{\mathrm{in}} = 0$ and are "undownvoted downvoters"). This reveals a type of anti-correlation that is at play in the downvoting network: a typical person who has $k_{\mathrm{out}} > 0$ (i.e., they have downvoted someone) is actually more likely to have $k_{\mathrm{in}} = d_{\mathrm{in}} = 0$ than to have $k_{\mathrm{in}} > 0$. We actually could have anticipated this from Table 9.3 above: the typical downvoter of most downvotees is an undownvoted downvoter, which trivially means that this downvoter gets downvoted less than the downvotee.

In Fig. 9.7, we follow the logic that led to Fig. 9.4 to plot four complementary cumulative distributions of d_{in}:

1. In blue, we plot the overall distribution of d_{in} in the downvoting network, including all downvoters and downvotees.
2. In green, we plot the distribution of d_{in} over downvotees in the network.

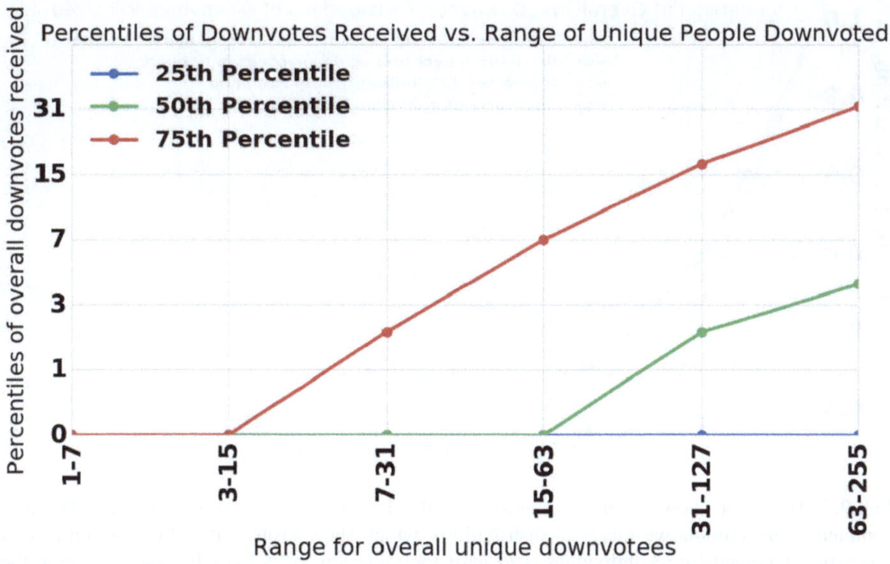

Fig. 9.6 This plot shows percentiles of the number of downvotes an individual received vs. ranges of the number of unique people that individual downvoted. We show the 25th, 50th, and 75th percentiles of the distribution. This plot shows that, over a large range of unique downvotee counts, the median number of downvotes received is zero. In other words, the distribution is strongly affected by the presence of "undownvoted downvoters"

3. In red, we plot the distribution that arises from repeatedly randomly sampling a downvotee, randomly sampling a downvoter of that downvotee, and then measuring d_{in} for the downvoter.
4. In purple, we plot the distribution of d_{in} over downvoters that we would expect from the random-wiring assumption in Eq. (9.9).

Here, the random-wiring assumption actually predicts an *inward* shift of the median of the distribution of d_{in} over downvoters (relative to the overall distribution 1). This inward shift survives in distribution 3 (which takes into account correlations across links through applying the two-step sampling procedure to the actual network). This, in turn, helps to explain the strong anti-paradox that we observe in Table 9.4.

9.4.4 *The Downvoting Paradox Occurs When The Downvotee and Downvoter are Active Contributors*

Let us now ask *why*, in terms of real behavior on the product, it may be more likely for the typical downvoter of most downvotees to not receive any downvotes. One possible explanation is that there is a barrier to receiving downvotes that does

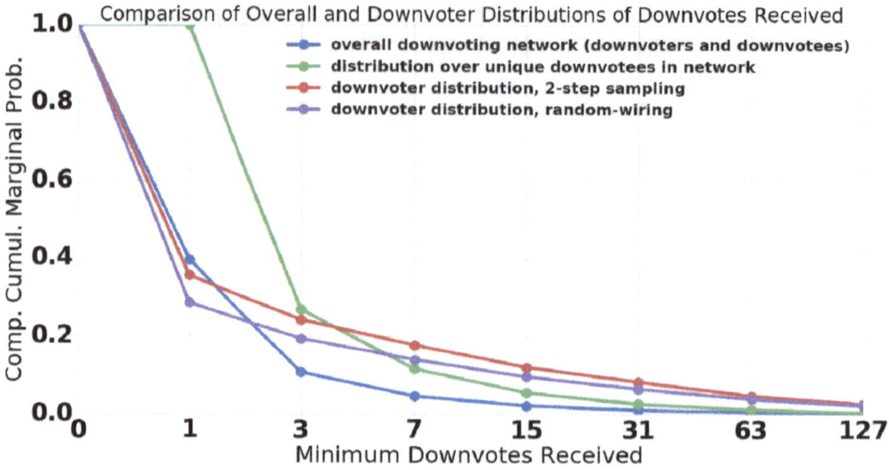

Fig. 9.7 This plot shows four distributions of the number of downvotes received. We plot complementary cumulative marginal distributions, which show probabilities that the number of downvotes received by an individual is at least the value on the x-axis. In *blue*, we show the real distribution of downvotes received over all people in the downvoting network, including both downvoters and downvotees. In *green*, we show the real distribution of downvotes received over people who received at least one downvote (i.e., over all downvotees). In *red*, we show the real distribution of downvotes received over downvoters that we find if we repeatedly randomly sample a downvotee and then a randomly sample a downvoter of that downvotee. In *purple*, we show the inferred distribution of downvotes received over downvoters that we would expect if we apply the random-wiring assumption in Eq. (9.9) to our empirical data

not exist for casting downvotes: in particular, to receive downvotes, it is necessary to actually write answers. This motivates posing the "downvotee → downvoter" *and* "downvoter → downvotee" questions again, but just for people who are active answer writers.

We now pursue this line of investigation and show that, when we condition on content contribution, both sides of the downvoting paradox hold. In particular, we revise the two questions as follows:

1. **The downvotee → downvoter question**: For most people who wrote at least n answers and who received downvotes from people who also wrote at least n answers, did most of those downvoters receive more or fewer *total* downvotes than they did?
2. **The downvoter → downvotee question**: For most people who wrote at least n answers and who downvoted people who also wrote at least n answers, did most of those downvotees receive more or fewer *total* downvotes than they did?

Note that, as before, we are always referring to non-anonymous answers and non-anonymous downvotes here, even if sometimes omit the adjectives for convenience. In Tables 9.5 and 9.6, we fix $n = 3$ and results for the revised questions. These results reveal that both sides of the downvoting paradox now hold.

Table 9.5 In this table, we report statistics that we obtain when we repeat the calculations that led to Table 9.3 but restrict our attention to downvotees and downvoters who contributed at least $n = 3$ non-anonymous answers during the four-week window that our downvoting network represents

Typical values of downvotes received		
	Downvotee → downvoter	Downvoter → downvotee
Mean for downvotee	4.0	54.7
Mean for downvoter	13.8	2.0
Median for downvotee	4.0	30.0
Median for downvoter	8.0	2.0

These are population values over all downvoting pairs in the downvoting network that satisfy the content-contribution condition. Note that the variable that we compare between downvoters and downvotees is still *total* downvotes received, not just downvotes received from active contributors

Table 9.6 In this table, we report statistics that we obtain when we repeat the calculations that led to Table 9.4 but restrict our attention to downvotees and downvoters who contributed at least $n = 3$ non-anonymous answers during the four-week window that our downvoting network represents

Typical values of differences in downvotes received		
	Downvotee → downvoter	Downvoter → downvotee
Mean downvoter—downvotee	5.0	−44.5
Median downvoter—downvotee	2.0	−23.0

These are population values over all downvoting pairs in the downvoting network that satisfy the content-contribution condition. Note that the variable that we compare between downvoters and downvotees is still *total* downvotes received, not just downvotes received from active contributors

We now study the "anatomy" of the "downvotee → downvoter" side of the paradox, under the content-contribution condition. Note first that the content-contribution condition motivates revising the definitions of the four variables in Eq. (9.9):

- k_{in}: the number of unique downvoters of the person who have written at least n answers.
- d_{in}: the number of downvotes the person received, which should still respect $d_{in} \geq k_{in}$.
- k_{out}: the number of unique downvotees of the person who have written at least n answers.
- d_{out}: the total number of downvotes the person cast, which should still respect $d_{out} \geq k_{out}$.

If we study correlations between k_{out} and d_{in} for just those people with $n \geq 3$, we find that, consistent with the existence of the strong paradox in the "downvotee → downvoter" analysis, a strong positive correlation is evident. We plot this in Fig. 9.8. Note that k_{out} for each node now just refers to the number of people whom the downvoter downvoted who wrote at least three answers. We can now probe the impact that the correlations in Fig. 9.8 have upon the downvoter distribution of d_{in} by plotting the analog of Fig. 9.7, but with the content-contribution condition. We do this in Fig. 9.9 and observe that the distribution of d_{in} over downvoters

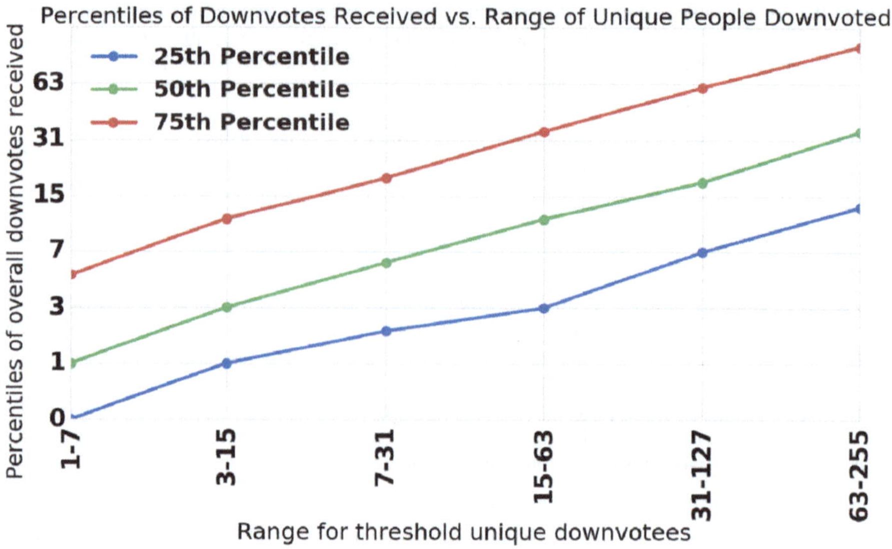

Fig. 9.8 This plot, like Fig. 9.7, shows percentiles of the number of downvotes an individual received vs. ranges of the number of unique people that individual downvoted. The difference with respect to Fig. 9.7 is that we have imposed the content-contribution threshold that we discuss in the text. This means that all people considered for this plot contributed at least $n = 3$ non-anonymous answers during the four-week window represented by the downvoting network. Furthermore, the number of "threshold unique downvotees" for each individual only counts those downvotees who also satisfy the content-contribution criteria. Meanwhile, the number of "overall downvotes received" still includes all downvotes received from any downvoter, not just those who satisfy the content-contribution threshold

now shifts outward as expected. Moreover, we observe that the random-wiring assumption works extremely well in this context: this means that the outward shift of the distribution is approximately what we would expect from the within-node correlations seen in Fig. 9.8, with correlations across links playing a minor role.

9.4.5 Does a "Content-Contribution Paradox" Explain the Downvoting Paradox?

We now consider a possible explanation for the "downvotee → downvoter" side of the downvoting paradox: that the typical downvoter of the typical downvotee contributes more content. If this is the case, then maybe the downvoter generally gives himself or herself more opportunities to be downvoted, and consequently, we should not be surprised that downvoter typically gets downvoted more. Table 9.7 shows that this "content-contribution paradox" actually occurs: for most downvotees who wrote at least three recent answers and got downvoted by someone who also

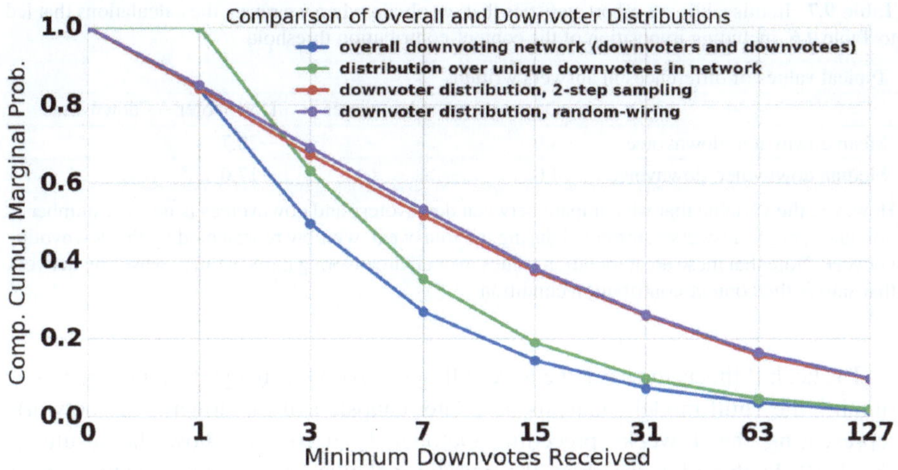

Fig. 9.9 This plot, like Fig. 9.7, shows four distributions of the number of downvotes received. We plot complementary cumulative marginal distributions, which show probabilities that the number of downvotes received by an individual is at least the value on the x-axis. In *blue*, we show the real distribution of downvotes received over all people in the downvoting network, including both downvoters and downvotees. In *green*, we show the real distribution of downvotes received over people who received at least one downvote (i.e., over all downvotees). In red, we show the real distribution of downvotes received over downvoters that we find if we repeatedly randomly sample a downvotee and then a randomly sample a downvoter of that downvotee. In *purple*, we show the inferred distribution of downvotes received over downvoters that we would expect if we apply the random-wiring assumption in Eq. (9.9) to our empirical data. The difference with respect to Fig. 9.7 is that we have imposed the content-contribution threshold that we discuss in the text. Thus, the distributions are computed over people who contributed at least $n = 3$ non-anonymous answers during the four-week window represented by the downvoting network. Furthermore, when we randomly sample a downvotee and then a downvoter, we require that both parties satisfy the threshold. However, the number of downvotes received still includes all downvotes received from any downvoter, not just those who satisfy the content-contribution threshold

wrote at least three recent answers, most of those downvoters wrote at least four more recent answers than they did. Moreover, comparing Tables 9.6 and 9.7, we see that the ratio of downvotes to recent answers may actually be lower for the downvoters.

Does this "content-contribution paradox" fully account for the downvoting paradox? To determine this, we can see if the downvoting paradox survives the following procedure: in each unique downvoting pair of the dataset, in place of the actual number of downvotes received by the downvoter, assign the number of downvotes received by a randomly chosen person who wrote the same number of recent public answers. Then, we can redo the calculations for the "downvotee → downvoter" side of the downvoting paradox in this "null model" and check if the paradox still occurs. If so, then that would support the argument that the content-contribution paradox explains the downvoting paradox.

Table 9.7 In this table, we report statistics that we obtain when we repeat the calculations that led to Table 9.6, including imposition of the content-contribution threshold

Typical values of differences in answers written		
	Downvotee → downvoter	Downvoter → downvotee
Mean downvoter–downvotee	8.0	−28.7
Median downvoter–downvotee	4.0	−17.0

However, the variable that we compare between downvoters and downvotees is now the number of non-anonymous answers contributed during the four-week window represented by the downvoting network. Note that these are population values over all downvoting pairs in the downvoting network that satisfy the content-contribution condition

To check if this is the case, we have followed up on the analysis in this section by running the "null-model" analysis on a later snapshot of the downvoting network, representing the 4 weeks preceding October 1, 2016. We show the results in Fig. 9.10. In this plot, we show the paradox calculation for various values of the content-contribution threshold n in the real downvoting network, as well as in the null model described above. This analysis reveals three things:

1. The "downvotee → downvoter" side of the downvoting paradox is not a peculiarity of a particular snapshot of the downvoting network, since it occurs in this later snapshot as well. The "downvoter → downvotee" side of the paradox also occurs in this later snapshot, although that data is not represented in this plot.
2. The "downvotee → downvoter" side of the downvoting paradox is not a peculiarity of a specific content-contribution threshold n. In the plot, the paradox gets stronger as n grows.
3. More to the current point, the content-contribution paradox in terms of recent answers composed does not fully account for the downvoting paradox, since the paradox disappears under the null model.

A follow-up suggestion might be that, if recent content-contribution volume cannot account for the paradox, then maybe *historical* content-contribution volume does: namely, maybe the downvoters in the "downvotee → downvoter" side of the paradox receive more downvotes because they have more answers over their entire history on Quora. Figure 9.10 provides evidence against this possibility too, by presenting data for a null model with respect to historical answers.

One weakness of the null-model analysis above might be that, for high values of content-contribution volume, we may not have sufficient examples of people who contributed precisely that number of answers. This could make it difficult to get a good estimate of the distribution of number of downvotes received over people who contribute a specific number of answers, and that could, in turn, compromise the randomization process employed by the null model. To check if this is a problem, we have rerun the null models while treating everyone with more than n_t recent or historical answers equivalently. In other words, in any downvoter, downvotee pair where the downvoter has at least n_t answers, we assign that person a number of downvotes received by someone else who also wrote at least n_t answers, irrespective

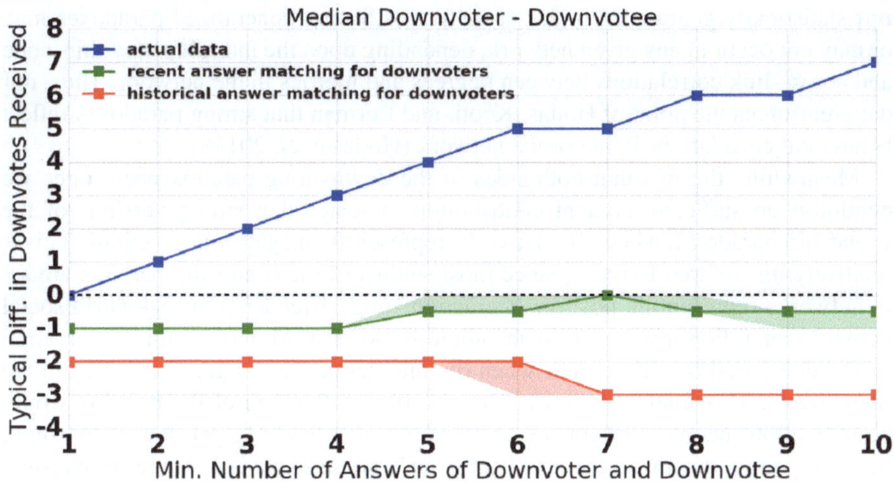

Fig. 9.10 This plot compares the "downvotee → downvoter" side of the downvoting paradox (the *blue line*) to two null models. In the *red line*, for each downvotee, downvoter pair in the downvoting network, we match the downvoter to someone who contributed the same number of recent public answers and use that person's downvote count instead. In the *green line*, we do the same type of matching, but with someone who has the same number of historical public answers. This plot shows that recent and historical content-contribution volume alone cannot explain the downvoting paradox. The plot also shows that the downvoting paradox is not a peculiarity of a particular choice of the content contribution threshold n. Note that this plot, unlike all the others in this section, is for a later snapshot of the downvoting network (representing the 4 weeks preceding October 1, 2016) and shows that the paradox is not a peculiarity of any one time window. Two points about the null-model data: (1) It may be surprising that fractional values appear. This can happen because, if a downvotee has an even number of downvoters, then the median downvote count may fall between two integers. (2) The error bars on the null-model data are computed by repeating the null model analysis 100 times and then taking 2.5th and 97.5th percentiles of those 100 samples; more sampling would have been ideal, but this is a slow calculation, and the current level of sampling already shows that the actual observations (i.e., the *blue line*) are unlikely to be explained by either null model

of the precise number of answers. The paradox disappears for various choices of n_t, so the conclusions that we draw from Fig. 9.10 appear to be robust to this issue.

Thus, content-contribution volume alone does not seem to account for the downvoting paradox, despite the fact that a "content-contribution paradox" also occurs in the downvoting network.

9.4.6 Summary and Implications

We now discuss the implications of the observations in this section. First of all, the absence of the "downvotee → downvoter" side of the downvoting paradox in the full downvoting network provides an example of why the strong generalized paradox is

not statistically guaranteed in social networks. Strong generalized paradoxes may or may not occur in any given network, depending upon the interplay of within-node and across-link correlations between degrees and metrics in the network. Thus, our data reinforces the point of Hodas, Kooti, and Lerman that strong paradoxes reflect behavioral correlations in the social network (Hodas et al. 2014).

Meanwhile, the fact that both sides of the downvoting paradox occur once we condition on sufficient content contribution indicates that strong versions of the friendship paradox can occur in networks representing negative interactions. This is relatively unexplored territory, since most studies of the friendship paradox are set in networks representing positive interactions (e.g., friendship in real-world social networks or follow relationships in online networks). Moreover, this observation shows that paradoxes also occur when the interactions are largely hidden from the participants. This matters because one possible explanation of the paradox would be retaliation: people who downvote often are also likely to get downvoted more because people downvote them to "get even." These explanations are implausible in the Quora context, because the identity of downvoters is hidden. Meanwhile, explanations in terms of more content contribution on the part of downvoters are also insufficient, based on our argumentation in this section. This leaves the intriguing possibility that the actual explanation lies in a correlation between downvoting and composing controversial content. A deeper natural-language study may be needed to assess whether such a correlation exists, whether it accounts for the downvoting paradox, or whether the actual explanation is something altogether different (e.g., more subtle explanations in terms of content contribution may still work, an example being a situation where being a downvoter is correlated with getting more views on your historical content).

Finally, we note that the setting of these strong paradoxes in a network representing negative interactions reverses the usually demoralizing nature of the friendship paradox: Feld's 1991 paper announced that "your friends have more friends than you do," and he noted that this statistical pattern could have demoralizing consequences for many people (Feld 1991). In contrast, when analogs of the paradox occur in networks representing negative interactions, that means that the people who interact negatively with you are usually just as likely to be the recipients of negative interactions themselves. This may provide some comfort to participants in online social communities.

9.5 A Strong Paradox in Upvoting

9.5.1 Content Dynamics in the Follow Network

In this section, we turn our attention to upvoting and, in particular, to its role as a social distribution mechanism. When a person on Quora casts an upvote on an answer, that answer then has an increased chance to be seen by that person's followers in their homepage feeds, digest emails, and certain other social distribution channels. As such, a plausible route for a piece of content to be discovered by a sequence of people on Quora is the one depicted in Fig. 9.11. In this network diagram, each node represents a person on Quora, and the links represent follow relationships. Recall that one person (the "follower") follows another (the "followee") because the follower is interested in the content contributed by the followee. In Fig. 9.11, we depict a situation in which the person represented by the node with an "A" has written an answer. A follower of this author reads the answer in homepage feed and upvotes it. This first upvoter's followers then have an increased chance of encountering the answer for social reasons, and one of them finds the answer in a digest email and upvotes it. This allows the answer to propagate to the second upvoter's followers, and a third person finds the answer in feed and upvotes it. This is a purely social or viral pathway for content to propagate through Quora.

We take this opportunity to introduce the notion of an "upvote distance": since the first upvoter in Fig. 9.11 is a direct follower of the answer author, once that upvote is cast, the answer reaches distance $d = 1$ from the author. Until then, we say it is at distance $d = 0$ (i.e., still with the author). After the third upvote is cast, because it would take three hops along directed links to get back to the answer author, we say that the answer is at distance $d = 3$. Since this is a purely social or viral pathway for propagation, the answer sequentially moves from distance $d = 0$ to 1 to 2 to 3.

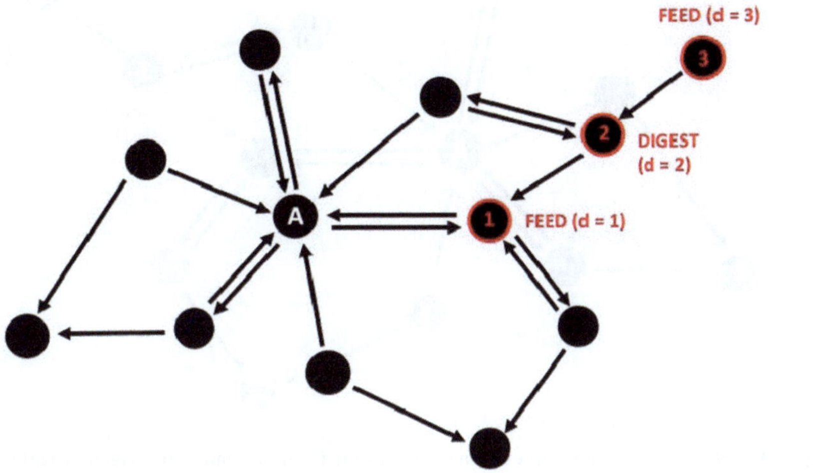

Fig. 9.11 Network diagram representing, in cartoon form, a fully social way for content to propagate in the Quora follow network. Each node represents a person, and each *arrow* represents a follow relationship. The person represented by the center node (indicated with an "A") has written a non-anonymous answer and then three people read and upvote the answer for social reasons. The upvote plays a social distribution role at each step of this process. See the text for further discussion

But content on Quora also has the opportunity to be discovered via non-social channels. A prominent one is search: people can find answers on Quora by issuing a relevant query to an external search engine like Google or Bing or by using Quora's internal search engine. Moreover, people can be shown content in their homepage feeds and digest emails for non-social reasons (e.g., because they follow a relevant topic), so these parts of the product are themselves only partially social. Yet another non-social channel for content discovery is to navigate directly to a topic page of interest and read content specifically about that topic.

Figure 9.12 illustrates how these non-social channels may impact the dynamics of content. As before, the person indicated with an "A" has written an answer. In this case however, the first person to discover and upvote the answer does so via internal search. This first upvoter happens to follow someone who directly follows the answer author, but does not directly follow the answer author. As such, the answer hops out to distance $d = 2$ without ever having been at distance $d = 1$; this "leapfrogging" is possible because internal search is not a social means of discovery and does not rely upon the follow graph. On the basis of the first upvoter's action though, that person's followers are more likely to see the answer, and a second person finds the answer in homepage feed for this social reason and upvotes it. Thus, the answer reaches a distance of $d = 3$ through a combination of social and non-social means. The third upvoter in this cartoon scenario is a direct follower of the answerer who encounters the answer in a digest email; this person's upvote is

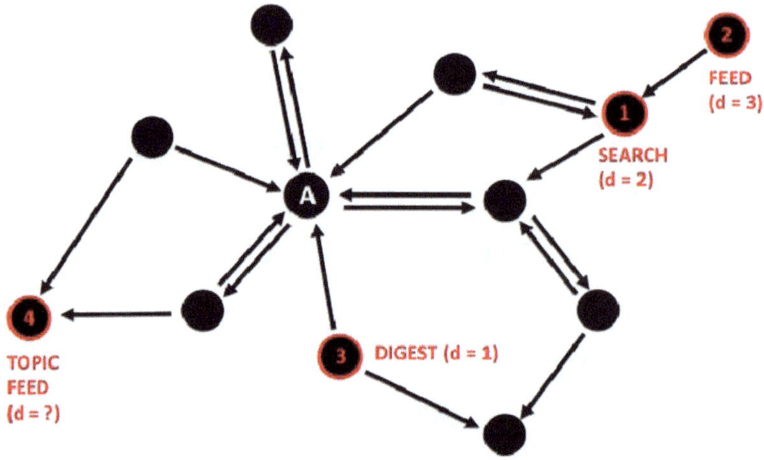

Fig. 9.12 Network diagram representing, in cartoon form, a content propagation pathway that mixes social and non-social channels. Each node represents a person, and each arrow represents a follow relationship. The person represented by the center node (indicated with an "A") has written a non-anonymous answer. A person discovers the answer via search. This person happens to be a second-degree follower of the author (i.e., a follower of a direct follower of the author). Thus, the answer hops out to distance $d = 2$ and does so without ever having been at distance $d = 1$, since search is a non-social pathway for propagation. See the text for further discussion

at $d = 1$, since he or she directly follows the author. The fourth and final upvoter finds the answer on a topic page. There actually is no path back from this person to the answer author, so it is less clear what distance to assign to this upvote. Nevertheless, this fact highlights an important aspect of the non-social channels of content discovery: through them, answer authors can reach audiences that they could *never* reach through purely social distribution.

9.5.2 Core Questions and Methodology

The fundamental question that we want to address in this section is the following: when an answer receives an upvote, does the upvoter who casts that upvote typically have more or fewer followers than the answer author? Furthermore, how does the answer to this question vary with the number of followers of the answer author and with the network distance of the upvote? We will see below that, in many practically relevant cases, the following property holds: for most answer authors, for most of their non-anonymous answers that get non-anonymous upvotes, most of the non-anonymous upvoters have more followers than they do. This is a strong variant of the friendship paradox that can have very important consequences for how content dynamics play out in the Quora ecosystem.

To address the questions above, we will track all answers written on Quora in January 2015 and track all upvotes cast on these answers until near the end of October 2016. In this section, as in Sect. 9.4), we always mean that we look at data for non-anonymous, non-deleted answers and non-anonymous upvotes (that were not removed after being cast) on those answers, but we sometimes drop the adjectives for brevity. To each answer, we attach the number of followers the answer author had when the answer was composed, and to each upvote, we attach the number of followers the upvoter had when the upvote was cast. Furthermore, to each upvote, we attach the network distance from the upvoter to answer author when the upvote was cast. Now, it is important to note that the underlying network changed significantly between January 2015 and October 2016, with many more people joining Quora and many new follow relationships being added. We take these dynamics seriously and update the underlying network as we assign distances / follower counts to each upvote. Our overall procedure is as follows:

1. Construct the follow network including people who signed up before January 1, 2015 and the links between them.
2. Iterate day-by-day from the start of the answer cohort (January 1, 205) to the end of the observation period. For each day, perform two actions:

    (a) Update the network by adding a node for new people who signed up and adding a link for new follow relationships.
    (b) Compute the follower count for the upvoter and the distance between the upvoter and answer author for each upvote that was cast during that day (and was not subsequently removed) on the January 2015 cohort of answers.

3. Look at all upvotes or at some relevant subset of upvotes (e.g., those cast at distance *d*) and ask:
   (a) For each answer, what is the median follower count of the upvoters?
   (b) For each answer author, what is the median over answers of the median upvoter follower count?
   (c) For each answer author, what is the ratio of the answer to question (b) to the author's follower count when the answer was composed?
   (d) What is the median over the ratios in question (c)? If this median ratio exceeds 1, then the strong paradox holds.

We implement these calculations using the NetworkX Python package (Hagberg et al. 2008).

There are a number of subtleties to consider in the procedure outlined above. Here are the most important ones:

- **What links should we include in the network?** In Sect. 9.3 above, we kept only links between people who had been active on Quora within a four-week window. Similarly, in this setting, we may want to avoid including all links, so as to not compute artificially short paths through people who do not actively engage with the product. In our analysis, we only add a link to the follow network if the person being followed has added a non-deleted, non-anonymous answer or non-anonymously upvoted a non-deleted answer since joining Quora. If that is not the case, then we defer adding the link until that person takes one of those actions. The interested reader can refer to our blog post, where we check how pruning links affects the results of this analysis; the high-level takeaway is that it does not dramatically affect the results (Iyer 2015).
- **Within a given day of the calculation (see step 2 in the procedure above), should we update the network or compute distances first?** Both options introduce some amount of error. To see why, consider a scenario where someone finds an answer by an author that he or she has not yet followed and upvotes it. In fact, the answer is so compelling that the reader also follows the answerer. Later in the day, the reader visits feed, sees new content by the person whom he or she just followed, and upvotes that too. Suppose we treat this scenario within the "update-first" protocol where we update the graph before computing distances. In this case, we would miss the fact that the original upvote happened when the network distance between upvoter and answerer was greater than 1, and possibly substantially greater. We end up underestimating the upvote distance. Alternatively, suppose we use the "compute-first" protocol where we compute distances before updating the graph. In this case, we miss out on the fact that the second upvote likely happened because the reader was a first-degree follower of the author. We end up overestimating the upvote distance. In the calculations reported in this chapter, we always use the "update-first" protocol, but we check robustness to changing the protocol in the blog post (Iyer 2015).
- **What should we do about link removal?** On Quora, people can stop following others whose content no longer interests them. Our analysis only includes links

that survived when the analysis was performed for the last time (in November 2016) and does not consider links that existed in the past that were subsequently removed. Link removal is relatively rare compared to link addition (a very rough estimate is that it happens about 4–5% as often) and substantially harder to track in our dataset. Furthermore, we consider it very unlikely that link removal would qualitatively change the results because of the robustness of the findings to other structural modifications to the network, such as the two discussed above. For these reasons, we have not checked robustness to this issue directly; however, we cannot completely exclude the possibility that people who take the time to curate their networks through link removal represent unusually important nodes in the network.

9.5.3 Demonstration of the Existence of the Paradox

We first quote results for the paradox occurring over all upvotes. For most answer authors from January 2015, for most of their answers written during that month that subsequently received upvotes, most of those upvotes came from people who had at least 2 more followers than the answer author. This means that a strong paradox occurs, but this paradox may not seem very dramatic. To see why this paradox may actually have practical implications for content dynamics, we need to look deeper, and in particular, we need to look at the strength of the paradox for answer authors who had few followers themselves. For example, for answer authors who had 1–9 followers when composing their answers, for most of their upvoted answers, the median upvoter on those answers had around five times as many followers as they did. This suggests that these answers could typically be exposed to larger audiences from these upvotes.

The potential impact is made clearer by incorporating network structure into the analysis: in Fig. 9.13, we plot the output of the procedure from Sect. 9.5.3, broken down by order-of-magnitude of the follower count of the answerer and by network distance from the upvoter to the answerer. The plot shows that, for answer authors who had 1–9 followers, when their answers were upvoted by people at distance $d = 1$ on the network, the median follower count of the upvoters on those answers was around 6.9 times their own. The effect is actually more dramatic at $d = 2$, with the median follower count of the $d = 2$ upvoters on most answers being around 29.6 times the follower count of the answer authors. For these answer authors, the typical ratio remains above 1 all the way out to $d = 6$, which is the largest distance that we study. Meanwhile, the paradox also holds for authors with tens of followers out to distance $d = 3$, again peaking at $d = 2$ with a typical ratio of around 3.4.

The practical consequence of this version of the strong friendship paradox is the following: answers by people with low follower counts initially face a stark funnel for social distribution. However, if these answers can get upvoted through this stark funnel or circumvent the funnel through non-social means, then future social propagation may be easier. This may help content written by these people to reach

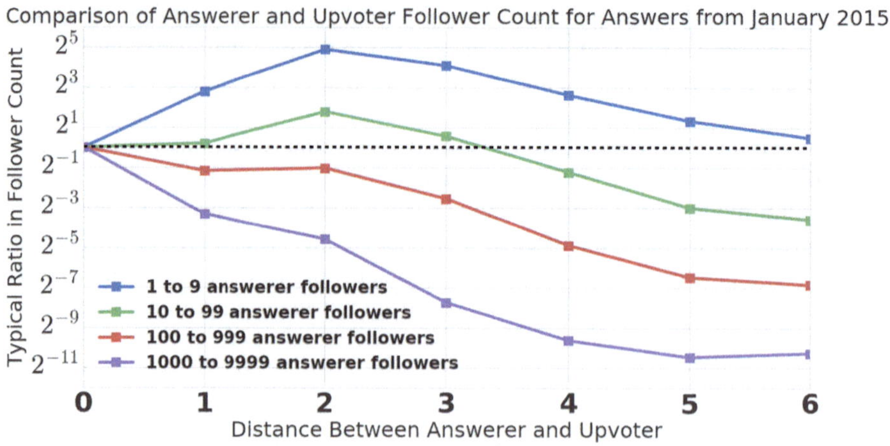

Fig. 9.13 For the January 2015 cohort of answers, this plot tracks "typical" ratio of the follower count of the upvoter to the follower count of the answer author vs. the network distance between the upvoter and the answer author at the time of the upvote. Here, "typical" refers to the median over answers of the median within each answer; see the procedure outlined in Sect. 9.5.2 for more details. This plot demonstrates the existence of the strong paradox for answer authors with low follower counts, which has potentially beneficial consequences for distribution of content by these people

readers who are further from them in the follow network. These benefits are possible because the upvoters in these situations are *typically* much better followed than the answer author. We can make the claim about typicality because we have measured a *strong* paradox. We emphasize again that strong paradoxes are not bound to occur: it could have been the case that the typical upvoter of an author with few followers has a comparable or lower number of followers than the answer author. This would curtail the answer's opportunities for social propagation, even if it was discovered by some other means.

We should briefly comment on the behavior in Fig. 9.13 for answers by people with many followers. In these cases, we see a sharp drop off of the ratio at each network distance. This makes sense, because the distance of an upvoter from a highly-connected author tells us something about the upvoter. If someone is several hops away from a highly-connected author that means that they have not followed that author, they have not followed anyone who follows that author, etc. This means that they are unlikely to be very active participants in the online community, and therefore, are unlikely to have many followers themselves. This selection effect gets more dramatic with each step away from a highly-connected author, so the sharp decay of the red and purple curves in Fig. 9.13 is completely expected.

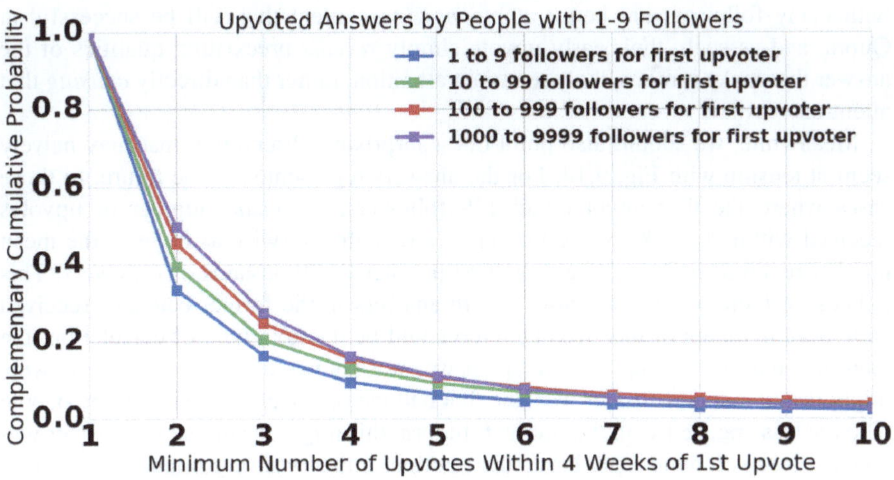

Fig. 9.14 For answers written in January 2015 by answer authors with 1–9 followers at the time of answer composition, this plot shows the distribution of upvotes received in the first 4 weeks after the answer received its first upvote, broken down by the follower count of the first upvoter. The plot includes those answers that received upvotes from people who had follower counts in the stated ranges. We plot complementary cumulative distributions, which show the fractions of answers that received at least the number of upvotes on the x-axis. The outward shift of the distribution with the order-of-magnitude of follower count of the first upvoter suggests that the upvoting paradox may assist in answer distribution, but we point out important caveats in the text

9.5.4 Can We Measure the Impact of the Paradox?

We now ask if we can measure the potential impact of the paradox reported in this section. One way that we might probe this is the following:

1. Consider all answers written in January 2015 by people with 1–9 followers at the time of answer composition.
2. For those answers that received upvotes (by 4 weeks before the cutoff time of the analysis, which was November 1, 2016), record the follower count of the first upvoter at the time the upvote was cast.
3. Measure how many upvotes the answer received in the 4 weeks beginning at the time of that first upvote.
4. Look at the distribution of number of upvotes received, broken down by the order-of-magnitude of followers of the first upvoter.

We report the results of this analysis in Fig. 9.14.

As the order-of-magnitude of the follower count of the first upvoter grows in Fig. 9.14, the distribution shifts outward. This suggests that the upvoting paradox could actually boost distribution of answers from people with few followers, as described above. However, it is important to treat this observation with care: there are many confounding effects here. For instance, it might be the case that people with many followers are better at *recongizing* content that will be successful at Quora, and as such, their early upvote simply *reveals* preexisting qualities of the answer that make it more likely to attract attention, rather than directly *causing* that attention.

Meanwhile, we should also point out a surprising observation that may naively seem at tension with Fig. 9.14. For the answers represented in the figure, in those cases where the first upvoter had 1–9 followers, the *mean* number of upvotes received within 4 weeks of the first upvote was almost twice as large as the mean number received in cases where the first upvoter had thousands of followers. This is because there were more cases where answers in the former category received very large numbers of upvotes. This too could be due to a type of correlation: the mere fact that the first upvoter on an answer was someone with very few followers might reveal that this answer was one that had the opportunity to be seen by people who are less engaged with the product. In turn, this might mean that the answer was one that received wide topic-based distribution or that was very popular via search. In these cases, maybe we should not be surprised that the answer went on to receive many upvotes.

We cannot completely remove these confounding effects via the type of observational analysis that we have done in this section. Nevertheless, the outward shift of the distributions in Fig. 9.14 should serve as further motivation for trying to ascertain the actual impact of the upvoting paradox, perhaps through an appropriately-designed controlled experiment.

9.5.5 Summary and Implications

In this section, we have shown that the following variant of the strong friendship paradox holds: for most answer authors with low follower counts, for most of their answers, most of their distance d upvoters have more followers than they do. This holds for a large range of d for people with the lowest follower counts. We have demonstrated this friendship paradox for the January 2015 cohort of answers in Fig. 9.13, but our findings are not peculiar to this group of answers. The interested reader can find relevant data for other sets of answers in a related blog post entitled "Upvote Dynamics on the Quora Network" (Iyer 2015).

This variant of the paradox is special for a number of reasons. First, the analysis takes into account dynamics of the underlying network, whereas most studies of the friendship paradox focus upon static snapshots of a social network. Secondly, the paradox does not fit neatly into either the standard or generalized paradox categories. The metric being compared is a form of degree (specifically, indegree or follower count), but the links of the network that get considered in the comparison are subject to a condition: that an upvote happened across the link. Furthermore, the same link can count multiple times if one person upvotes another many times. Finally, this paradox generalizes the notion of a "link" itself to consider followers of followers (i.e., $d = 2$ upvoters), followers of followers of followers (i.e., $d = 3$ upvoters), etc. Studies of the friendship paradox for anything other than first-degree neighbors have been rather rare, although Lattanzi and Singer have recently touched upon this subject (Lattanzi and Singer 2015).

Most importantly though, this variant of the friendship paradox highlights how the friendship paradox, or variants thereof, can have practical ramifications for the fate of content on online social networks. People with low follower counts would have quantitatively lower opportunity for their content to win visibility if this paradox did not occur. This issue could actually be more acute in social networks other than Quora, where people do not have the opportunity to win visibility for non-social reasons. In products where distribution is purely social, the existence of this type of paradox (or something similar) may be vital for new participants to be able to attract an audience. Therefore, we hope that this study will inspire inquiry into similar phenomena in other online social products.

9.6 Conclusion

The "friendship paradox," the statistical pattern where the average neighbor of a typical person in a social network is better connected than that person, is one of the most celebrated findings of social network analysis. Variants of this phenomenon have been observed in real-world social networks over the past 25 years, dating back to the work of Feld (Feld 1991). In recent years, the availability of large volumes of data collected from online social networks has ushered in a new era of theoretical and empirical developments on this phenomenon. There has been theoretical work aimed at clarifying when and how certain variants of the friendship paradox occur, putting the original work of Feld on stronger mathematical footing (Cao and Ross 2016; Lattanzi and Singer 2015). There has also been empirical work that points out that the friendship paradox occurs for metrics other than friend count or degree, so that the average neighbor of most individuals in many social networks scores higher according to several metrics, for example activity or content contribution in that network (Eom and Jo 2014; Hodas et al. 2013; Bollen et al. 2016). Finally, there has been work that clarifies that, when analyzing friendship paradoxes, not all averages are created equally (Hodas et al. 2014). The so-called *strong* paradox (where the median neighbor of most individuals in a network scores higher on some metric) can often teach us much more about the functioning of the network than the *weak* paradox (where only the mean neighbor of most individuals scores higher on that metric).

In this chapter, we have applied these recent developments to the study of various realizations of the friendship paradox on Quora, an online knowledge-sharing platform that is structured in a question-and-answer format. We have identified three different incarnations of the strong paradox in networks that represent core parts of the Quora ecosystem. First, in Sect. 9.3, we have analyzed the network of people following one another on Quora. We have confirmed that the four "canonical" degree-based paradoxes in directed social networks all occur in the follow network. Next, in Sect. 9.4, we studied the network induced by people downvoting one another during a four-week period and found that, for most sufficiently active writers who got downvoted, most of the sufficiently-active writers who downvoted them got downvoted just as frequently. Finally, in Sect. 9.5, we found that, for writers with low follow counts, most of their upvoters have many more followers than they do. We noted the potential benefits that this phenomenon has for the distribution of content written by people who have yet to amass a large following on Quora.

Our results in Sect. 9.3 represent the first published measurements of the standard degree-based paradoxes on Quora, and investigating these paradoxes is a natural and necessary precursor to examining more exotic variants of the phenomenon. However, it is the more exotic paradoxes that we study in Sects. 9.4 and 9.5 that, we believe, point the way to important future studies. As we have mentioned above, the "downvoting paradox" in Sect. 9.5 occurs in a context that is relatively rarely examined in research on the friendship paradox: a network representing adversarial or negative interactions. Our analysis of the downvoting paradox motivates many follow-up questions. For example, to what extent is the downvoting paradox explained by an increased tendency of downvoters to produce controversial content themselves? Furthermore, downvoting on Quora represents a very particular type of negative interaction. The identity of downvoters is hidden from downvotees and this can have important consequences for the behavior of these parties: downvoters may feel freer to give negative feedback if they are not publicly identified, and the downvotees cannot retaliate against any specific individual if they believe that they have been downvoted. Does something like the downvoting paradox survive if the underlying product principles are different (e.g., if the identity of downvoters is public), or would such a situation fundamentally alter the dynamics? We may be able to address these questions by analyzing friendship paradoxes in networks representing other types of negative interactions in online or real-world social systems.

Meanwhile, we have noted that the paradox in upvoting that we demonstrate in Sect. 9.5 can have direct practical consequences for the fate of content on Quora. This underscores why the friendship paradox should not be thought of as *merely* a sampling bias. It actually matters that the typical upvoter of a typical piece of content by a relatively undiscovered writer is *not* a typical person from the full population. The fact that this typical upvoter is more highly followed than that typical person may help new writers be discovered and win influence in the network. We hope that this study motivates researchers to study the role that strong friendship paradoxes play in content propagation on online social networks. There has been recent work on how various versions of the friendship paradox can influence opinion spreading and adoption on social networks, but as far as we know, the role that friendship paradoxes play in the discovery of individual pieces of content is relatively unexplored (Jackson 2016; Lerman et al. 2016). As we have mentioned above, the role that these phenomena play may be especially important in products that employ purely social means of content distribution and discovery.

Acknowledgements Some of the data that is presented in this chapter was initially shared publicly through two posts on Quora's data blog, entitled *Upvote Dynamics on the Quora Network* and *Friendship Paradoxes and the Quora Downvoting Paradox* (Iyer 2015; Iyer and Cashore 2016a). M. Cashore, who was an intern on Quora's data team during the winter of 2015, collaborated with the present author on early stages of the research for *Friendship Paradoxes and the Quora Downvoting Paradox* and coauthored a forthcoming article summarizing some of the findings from Sects. 9.3 and 9.4 in the 2016 Proceedings of the American Statistical Association's Joint Statistical Meeting (Iyer and Cashore 2016b). The present author would also like to thank K. Lerman and Y. Singer for stimulating discussions about their work and B. Golub for informing him about the recent work of Cao and Ross (Cao and Ross 2016). Finally, the author thanks the Quora data team for reviewing the work reported in this chapter: within the team, the author is particularly indebted to W. Chen for reviewing important parts of the code for Sect. 9.4 and to W. Chen, O. Angiuli, and Z. Kao for introducing him to the "distribution-free" method for computing confidence intervals on percentile metrics, which was useful in Sect. 9.3 (Hollander et al. 1999).

References

Bollen, J., Gonççalves, B., van de Leemput, I., & Ruan, G. (2016). The happiness paradox: your friends are happier than you. arXiv preprint arXiv:1602.02665.

Cao, Y., & Ross, S. M. (2016). The friendship paradox. *Mathematical Scientist, 41*, (1).

Christakis, N. A., & Fowler, J. H. (2010). Social network sensors for early detection of contagious outbreaks. *PLOS One, 5*(9), e12948.

Cohen, R., Havlin, S., & Ben-Avraham, D. (2003). Efficient immunization strategies for computer networks and populations. *Physical Review Letters, 91*(24), 247901.

Coleman, J. S. (1961). *The adolescent society*. New York: Free Press.

Eom, Y.-H., & Jo, H.-H. (2014). Generalized friendship paradox in complex networks: The case of scientific collaboration. *Scientific Reports, 4*, 4603.

Feld, S. L. (1991). Why your friends have more friends than you do. *American Journal of Sociology, 96*(6), 1464–1477.

Hagberg, A. A., Schult, D. A., & Swart, P. J. (2008, August). Exploring network structure, dynamics, and function using NetworkX. In *Proceedings of the 7th Python in Science Conference (SCIPY2008)*, Pasadena, CA, USA (pp. 11–15).

Harrigan, N., & Yap, J. (2017). Avoidance in negative ties: Inhibiting closure, reciprocity and homophily. *Social Networks, 48*, 126–141.

Hodas, N., Kooti, F., & Lerman, K. (2013). Friendship paradox redux: Your friends are more interesting than you. arXiv preprint arXiv:1304.3480.

Hodas, N., Kooti, F., & Lerman, K. (2014). Network weirdness: Exploring the origins of network paradoxes. In *Proceedings of the International Conference on Web and Social Media* (pp. 8–10).

Hollander, M., Wolfe, D., & Chicken, E. (1999). Wiley series in probability and statistics. *Nonparametric Statistical Methods* (3rd ed., pp. 821–828).

Iyer, S. (2015). *Upvote dynamics on the quora network.* Data @ Quora (data.quora.com).

Iyer S., & Cashore, M. (2016a). *Friendship paradoxes and the quora downvoting paradox.* Data@Quora (data.quora.com).

Iyer, S., & Cashore, M. (2016). Friendship paradoxes and the quora downvoting paradox. In *JSM Proceedings, Section on Statistics in Marketing* (pp. 52–67). Alexandria: American Statistical Association.

Jackson, M. O. (2016). The friendship paradox and systematic biases in perceptions and social norms. Available at SSRN.

Krackhardt, D. (1996). Structural leverage in marketing. In *Networks in marketing* (pp. 50–59). Thousand Oaks: Sage.

Lattanzi, S., & Singer Y. (2015). The power of random neighbors in social networks. In *Proceedings of the Eighth ACM International Conference on Web Search and Data Mining* (pp. 77–86).

Lerman, K., Yan, X., & Wu, X.-Z. (2016). The "majority illusion" in social networks. *PLOS One, 11*(2), e0147617.

Newman, M. E. (2003). The structure and function of complex networks. *SIAM Review, 45*(2), 167–256.

Seeman, L., & Singer, Y. (2013). Adaptive seeding in social networks. In *2013 IEEE 54th Annual Symposium on Foundations of Computer Science (FOCS)* (pp. 459–468).

Ugander, J., Karrer, B., Backstrom, L., & Marlow, C. (2011). The anatomy of the facebook social graph. arXiv preprint arXiv:1111.4503.

Chapter 10
Deduplication Practices for Multimedia Data in the Cloud

Fatema Rashid and Ali Miri

10.1 Context and Motivation

With the advent of cloud computing and its digital storage services, the growth of digital content has become irrepressible at both enterprise and individual levels. According to the EMC Digital Universe Study (Gantz and Reinsel 2010), the global data supply had already reached 2.8 trillion Giga Bytes (GB) in 2012, with the expectation that volumes of data are projected to reach about 5247 GB per person by 2020. Due to this explosive growth of digital data, there is a clear demand from CSPs for more cost effective use of their storage and network bandwidth for data transfer.

A recent survey (Gantz and Reinsel 2010) revealed that only 25% of the data in data warehouses are unique. According to a survey (Anderson 2017), currently only 25 GB of the total data for each individual user are unique and the remainder are similar shared data among various users. At the enterprise level (Adshead 2017), it was reported that businesses hold an average of three to five copies of files, with 15–25% of these organizations having more than ten. This information points to massive storage savings that can be gained, if CSPs store only a single copy of duplicate data.

A major issue hindering the acceptance of cloud storage services by users has been the data privacy issue, as the user no longer have physical control of their data that now reside at the cloud. Although some CSPs such as Spider Oak, Tresorit, and Mozy allow users to encrypt the data using your own keys, before uploading it to the cloud, many other CSPs, such as popular Dropbox require access to the plaintext data, in order to enable deduplication services. Although many of these service providers, who do not allow encryption of data at the client site using users'

F. Rashid • A. Miri (✉)
Department of Computer Science, Ryerson University, Toronto, ON, Canada
e-mail: fatema.rashid; Ali.Miri@ryerson.ca

S. Srinivasan (ed.), *Guide to Big Data Applications*, Studies in Big Data 26,
DOI 10.1007/978-3-319-53817-4_10

keys, do encrypt the data themselves during transmission, this data either would be accessible at the server end, exposing the data to both internal and external attacks (Kovach 2017).

In the next section, we will discuss some of technical design issues pertaining to data deduplication.

10.2 Data Deduplication Technical Design Issues

In this section, we will briefly discuss common types of deduplications used in practice, based on methods used or where and how they take place.

10.2.1 Types of Data Deduplication

- Hash-Based: Hash-based data deduplication methods use a hashing algorithm to identify duplicate chunks of data after a data chunking process.
- Content Aware: Content aware data deduplication methods rely on the structure or common patterns of data used by applications. Content aware technologies are also called byte-level deduplication or delta-differencing deduplication. A key element of the content-aware approach is that it uses a higher level of abstraction when analyzing the data (Keerthyrajan 2017). In this approach , the deduplication server sees the actual objects (files, database objects etc.) and divides data into larger segments of typical sizes from 8 to 100 MB. Since it sees the content of the data, it finds segments that are similar and stores only bytes that have between objects. That is why it is called a byte-level comparison.
- HyperFactor: HyperFactor is a patent pending data deduplication technology that is included in the IBM System Storage ProtecTIER Enterprise Edition V2.1 software (IBM Corporation 2017). According to Osuna et al. (2016), a new data stream is sent to the ProtecTIER server, where it is first received and analyzed by HyperFactor. For each data element in the new data stream, HyperFactor searches the Memory Resident Index in ProtecTIER to locate the data in the repository that is most similar to the data element. The similar data from the repository is read. A binary differential between the new data element and the data from the repository is performed, resulting in the delta difference which is stored with corresponding pointers.

10.2.2 Deduplication Level

Another aspect to be considered when dealing with any deduplication system is the level of deduplication. Data deduplication can take place at two levels, file level or block level. In block level deduplication, a file is first broken down into blocks (often called *chunks*) and then the blocks are compared with other blocks to find duplicates. In some systems, only complete files are compared, which is called *Single Instance Storage (SIS)*. This is not as efficient as block level deduplication as entire files have to be stored again as a result of any minor modification to it.

10.2.3 Inline vs Post-Processing Data Deduplication

Inline data deduplication refers to the situation when deduplication is performed as the data is written to the storage system. With inline deduplication, the entire hash catalog is usually placed into system memory to facilitate fast object comparisons. The advantage of inline deduplication is that it does not require the duplicate data to actually be written to the disk. If the priority is high-speed data backups with optimal space conservation, inline deduplication is probably the best option. *Post-processing deduplication* refers to the situation when deduplication is performed after the data is written to the storage system. With post-processing, deduplication can be performed at a more leisurely pace, and it typically does not require heavy utilization of system resources. The disadvantage of post processing is that all duplicate data must first be written to the storage system, requiring additional, although temporary, physical space on the system.

10.2.4 Client- vs Server-Side Deduplication

Client side deduplication refers to the comparison of data objects at the source before they are sent to a destination (usually a data backup destination). A benefit of client side deduplication is that less data is required to be transmitted and stored at the destination point. A disadvantage is that the deduplication catalog and indexing components are dispersed over the network so that deduplication potentially becomes more difficult to administer. If the main objective is to reduce the amount of network traffic when copying files, client deduplication is the only feasible option. On the other hand, *server-side deduplication* refers to the comparison of data objects after they arrive at the server/destination point. A benefit of server deduplication is that all the deduplication management components are centralized. A disadvantage is that the whole data object must be transmitted over the network before deduplication occurs. If the goal is to simplify the management of the deduplication process server side deduplication is preferred. Among many popular vendors such as DropBox, SpiderOak, Microsoft Sky Drive, Amazon S3, Apple iCloud and Google Drive, only SpiderOak performs server-side deduplication (Xu et al. 2013).

10.2.5 Single-User vs Cross-User Deduplication

Single User: Deduplication can be done by a single user, where the redundancy among his/her data is identified and removed, but single-user data deduplication is not very practical and does not yield maximum space saving. To maximize the benefits of data deduplication, cross user deduplication is used in practice. This technique consists in identifying the redundant data among different users and then removing the redundancy by saving a single copy of the data. According to (Anderson 2017), 60% of data can be deduplicated on average for individual users by using cross-user deduplication techniques. In order to save bandwidth, CSPs and users of their services are inclined to apply client-side deduplication where similar data is identified at the client side before being transmitted to cloud storage. However, the potential benefit of data deduplication in terms of storage space and storage cost can also have some associated drawbacks. In February 2012, DropBox disabled client-side, cross-user deduplication due to security concerns (Xu et al. 2013).

10.3 Chapter Highlights

The highlights of this chapter can be summarized as follows.

- We will introduce a secure image deduplication scheme in cloud storage environments through image compression, that consists of a partial encryption and a unique image hashing scheme embedded into the SPIHT compression algorithm. This work initially appeared in part in Rashid et al. (2014).
- We will introduce a secure video deduplication scheme in cloud environments using H.264 compression, which consists of embedding a partially convergent encryption along with a unique signature generation scheme into a H.264 video compression scheme. This work initially appeared in part in Rashid et al. (2015).

10.4 Secure Image Deduplication Through Image Compression

In this chapter, a novel compression scheme that achieves secure deduplication of images in the cloud storages is proposed (Rashid et al. 2014). Its design consists of embedding a partial encryption along with a unique image hashing into the SPIHT compression algorithm. The partial encryption scheme is meant to ensure the security of the proposed scheme against a semi-honest CSP whereas the unique image hashing scheme is meant to enable classification of the identical compressed and encrypted images so that deduplication can be performed on them, resulting in a secure deduplication strategy with no extra computational overhead incurred for image encryption, hashing and deduplication.

10.5 Background

The scheme proposed in this chapter is compromised of three components, namely, image compression, partial image encryption, and image hashing. For our scheme, we have used the SPHIT algorithm (Said and Pearlman 2017) for image compression, which is a variant of the EZW algorithm. The EZW algorithm (Shapiro 1993) has the property that the bits in the bit stream are generated in order of importance, yielding a fully embedded code. For partial encryption of the image, we need to utilize data produced during the compression process. Cheng and Li (2000) indicates some of the specific data as significant information which is vital for decompressing the image. Hence by encrypting only this partial information can improve security of the compressed images. We, therefore, use some of the basic concepts presented in Cheng and Li (2000). In Yang and Chen (2005), Yang and Chen introduced an image hash scheme which is based on the knowledge of the locations of the significant wavelet coefficients of the SPIHT algorithm. In our work, we have adapted this scheme, by utilizing the wavelet coefficients of the SPIHT compression algorithm to generate the image hash.

10.6 Proposed Image Deduplication Scheme

The design of the proposed approach for cross-user client side secure deduplication of images in the cloud involves three components, namely, an image compression scheme, a partial encryption scheme, and a hashing scheme as shown in Fig. 10.1. Typically, on the user's side, the user will process the image by applying image compression, partial encryption, and will calculate the hash signature of the image, in that order. Next, he/she will only send the hash signature to the CSP. On the CSP side, the CSP will compare the received image hash against all the signatures already present in the cloud storage. If a match is not found, the CSP will instruct the user to upload the image. Otherwise, the CSP will update the image metadata and then will deduplicate the image by saving only a single, unique copy.

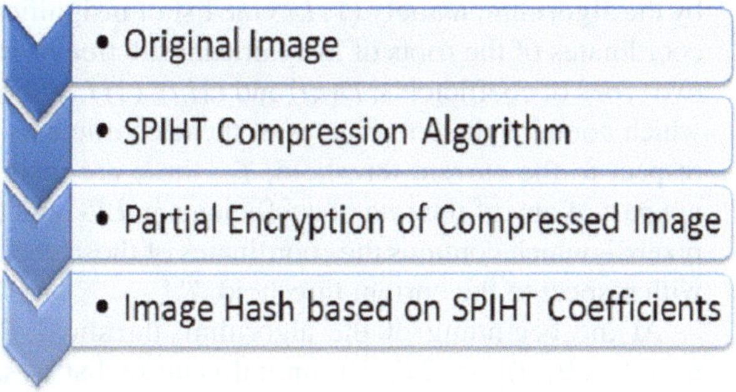

Fig. 10.1 Image deduplication process

10.6.1 Image Compression

We propose to utilize image compression in order to achieve image deduplication. The reason for doing so is that the images are compressed anyways for efficient storage and transmission purposes. In fact, applying the compression first and encrypting the compressed information next, one can save computation time and resources. As discussed above, the SPIHT algorithm (Said and Pearlman 2017) is used for image compression since it produces an embedded bit stream from which the best images can be reconstructed (Singh and Singh 2011). In addition, it uses an embedded coding method which makes it a suitable scheme for progressive optimal transmission that produces fine compression ratios. It also uses significant information sets to determine tree structures, and its execution relies upon the structure of the zero trees of the EZW algorithm. The SPIHT compression algorithm is independent of any keys or other form of user input. Hence, it is suitable for deduplication of identical images compressed by different users, as these identical images should appear identical in the compressed or encrypted form to the CSPs. Furthermore, this algorithm involves the use of chaos theory or random permutation from the users side, adding to the security of the system.

The SPIHT algorithm is based on the fact that there is a correlation between the coefficients that are in different levels of the hierarchy pyramid (bands) of the underlying structure. It maintains this information in the zero trees by grouping insignificant coefficients together. Typically, each 2×2 block of coefficients at the root level of this tree structure corresponds to further trees of coefficients. Basically, the SPIHT algorithm can be in three phases, namely, initialization, sorting, and refinement phases.

- Initialization phase: Let *Coef(i,j)* denote the coefficient at node *(i,j)*. For *Coef(i,j)*, three coefficient sets are further defined: *O(i,j)* the set of coordinates of the children of node *(i,j)* when this node has children; *D(i,j)* the set of coordinates of the descendants of the node *(i,j)*; and the set of coordinates of all coefficients at the root level. Then $L(i,j) = D(i,j) - O(i,j)$ is the set of descendants of the tree node, except for its four direct offspring. Starting at threshold *T*, any set of coefficients *S* is said to be significant (with respect to that threshold) if there is a coefficient in *S* that has a magnitude at least equal to *T*. Three lists are maintained by the algorithm, namely (1) *LIS* the list of insignificant sets—which contains the coordinates of the roots of insignificant sub trees—this list is further divided into two types of coefficients: *D(i,j)* and *L(i,j)*; (2) *LIP* the list of insignificant pixels—which contains the coordinates of those coefficients which are insignificant with respect to the current threshold, *T*—these are insignificant coefficients that are not part of any of the sets of coefficients in *LIS*; and (3) *LSP* the list of significant pixels—which contains the coordinates of those coefficients which are significant with respect to the current threshold, *T*.

  At the beginning of the algorithm, threshold *T* is selected such that $T \leq \max_{i,j} |\mathrm{Coef}(i,j)| < 2T$. The initial state of list is set in a specific manner, i.e. *LIP* maintains *H* the set of coordinates of tree roots, *LIS* maintains the set *D(i,j)*, where *(i,j)* are coordinates with descendants in *H*, and *LSP* is empty.

- Sorting phase: The algorithm identifies the significant coefficients by dividing the *D(i,j)* sets into two subsets, namely the set *L(i,j)* and the set *O(i,j)* of individual coefficients, and by putting each significant coefficient into the *LSP* list with respect to the current threshold level.
- Refinement phase: This step consists of the refinement of the significant coefficients of the list *LSP* from the previous pass in a binary search fashion. After each pass, the threshold is decreased by a factor of 2 and the whole process starts again. The algorithm ends when the desired bit rate is achieved.

A description of the SPIHT algorithm is available in Said and Pearlman (1996).

10.6.2 Partial Encryption of the Compressed Image

We propose to partially encrypt the compressed image (obtained from the image compression step) before uploading it to cloud storage. The reason for doing so is to ensure the security of the data from the CSP or any malicious user. Encrypting only a part of the compressed image will reduce the amount of data to be encrypted, thereby reducing the computational time and resources (Cheng and Li 2000). In order to satisfy the basic requirement of cross user deduplication, i.e. identical images compressed by different users should appear identical in the compressed/encrypted form, we propose to use convergent encryption (Thwel and Thein 2009) to encrypt the coefficients generated by the SPIHT algorithm since such a scheme will allow the CSP to classify the identical compressed and encrypted images.

Typically, the output of the SPIHT compression algorithm is a stream of encoded wavelet coefficients along with the zero trees for the structure of the coefficients. It contains the sign bits, the refinement bits, the significance of the pixels, and the significance of the sets. Thus, to correctly decompress the image, the decompression algorithm must infer the significant bits accurately. In this regard, it was suggested in Cheng and Li (2000) that only the significant information be encrypted. This information is determined by the significant information of the pixels in the highest two levels of the pyramid as well as in the initial threshold for the SPIHT algorithm. As an example, if the root is of dimension 8×8, then the *(i,j)* coordinates are encrypted if and only if $0 \leq i, j < 16$, and there is no need to encrypt the significant information belonging to subsequent levels, i.e. the pixels in the third, fourth, or sixth level of the pyramid. This is due to the fact that the coefficients in those levels will be reconstructed with the help of the information obtained from the coefficients belonging to the highest two levels of the pyramid.

Using the information obtained from the coefficients at the highest two levels of the pyramid, the states of the above-mentioned lists (i.e. *LIS*, *LIP*, *LSP*) will also be initialized. From these levels, the states of these lists will constantly be changed, depending upon the subsequent levels. But if the initial information is not correctly derived, these lists will be pruned to errors and the image will not be decompressed accurately. Hence, by having the user encrypt only the significant information of the highest two levels of the pyramid of the compressed image generated by the SPIHT algorithm before uploading it to the cloud storage, the image will be prevented from being decompressed by the CSP. The reason is that without the knowledge of the significant information from the highest two levels, the CSP will not be able to know the initial states of the above-mentioned lists, hence will not be able to decompress the image.

The convergent encryption algorithm (Thwel and Thein 2009) uses the content of the image to generate the key to be utilized by the users for encryption purpose, thus it is expected that two identical images will generate identical keys, thus identical ciphertext or cipher images. More precisely, the user constructs a SHA-3 hash of the significant information *Sig*_Info of the coefficients belonging to the highest two levels of the pyramid and use it as the key for encryption, i.e.

$$\text{Image\_Key}(IK) = \text{Hash}(Sig\_\text{Info}) \tag{10.1}$$

Using Eq. (10.1), identical encrypted significant information of the first two levels of the pyramid (hence identical ciphertext) will be produced from two identical compressed images, irrespective of the fact that these images have been encrypted by different users. Using this key, the users will perform symmetric encryption on the significant information as shown in Eq. (10.2)

$$Sig\_\text{Info}' = (IK, Sig\_\text{Info}) \tag{10.2}$$

This way, the CSP will be able to determine the match between identical images from different users without having to process the images in plaintext form.

10.6.3 Image Hashing from the Compressed Image

In previous steps, the image has been compressed (using the SPIHT algorithm) and only the significant information of the highest two levels of the pyramid has been encrypted. In order for the CSP to perform client side deduplication on the image, there has to be a unique identity (referred to as image hash) for each image. This step consists in generating such image hash. The image hash is generated by the user and sent to the CSP for performing the client side deduplication. In case of a redundant image, the CSP will only set the pointers of the image ownership to the new user and will not request the image to be sent again. Using this hash, the CSP will be able to identify and classify the identical images without the necessity to possess the original image in plaintext format. Indeed, the CSP will need to scan through the already stored hashes in the cloud storage to find the matches rather than scanning the entire images. It should be noted that by doing so, the CSP will be able to remove the redundant images and store only a single unique copy of the image. The CSP only has the knowledge of the compressed image and its partially encrypted version, and the original image cannot be reconstructed from these elements in any case. It is worth mentioning that the image hash generation will not involve any additional computational or storage overhead since the coefficients generated by the SPIHT algorithm are known. In addition, the image deduplication performed in this way will be secured against the CSP since all the significant information are encrypted.

Typically, the sorting pass of the SPIHT algorithm identifies the significant map, i.e. the locations of the significant coefficients with respect to a threshold. The binary hash sequence (where 1 denotes a significant coefficient and 0 denotes a non significant coefficient) is then designed based on this significance map and on the fact that the first four highest pyramid levels under the first three thresholds are more than enough to be used as the signature of the image (Yang and Chen 2005). In addition, the use of the convergent encryption is meant to ensure that the same ciphertext is obtained for the same coefficients in the plaintext format.

10.6.4 Experimental Results

Experiments were conducted by keeping in mind the following three requirements: (1) using the SPIHT compression algorithm, the deduplication of the image should be performed accurately, i.e. should not be faulty and the proposed scheme should not identify two different images as identical, and even minor alterations should be trapped; a failure to this requirement will enable the CSP to eliminate one of the images and keep only one single copy of the two images which appear to be identical but are actually not; (2) the proposed scheme should be secure enough against the semi-honest CSP since the CSP is not given access to the compressed image and thus cannot decompress it; the CSP can identify the identical images from different users only through the significant maps of the image generated by the SPIHT algorithm;and (3) the proposed scheme should be efficient in terms of computation, complexity and storage overheads.

10.6.4.1 Experimental Settings

First, some altered images from one single image are generated. For this purpose, six classic images are considered: Lena, bridge, mandrill, goldhill, pepper, and clock, all represented in gray scale format. Ten different kinds of altered images are introduced for each single classic image and 10 versions of that particular classic image are generated, yielding a total of 126 images (i.e. 120 altered images added to the six original images). The reason for using a large number of altered images is to demonstrate the affects of compression on different kinds of alterations. The first, second, and third altered images are obtained by changing the brightness, contrast, and gamma levels of the original image respectively. The fourth altered image is obtained by compressing the original image first (using a JPEG compression algorithm), then decompressing it. The fifth altered image is obtained by applying the rotations of the original image on the standard angular positions of 90, 180 and 270° respectively. The sixth altered image is obtained by applying a 2D median filtration on the original image. The seventh altered image is obtained by introducing a Poisson noise in the original image. The Poisson noise is a type of noise which is generated from the content of the data itself rather than by adding some artificial noise. The eighth altered image is constructed by rescaling the original image, i.e. by converting the unit eight pixel values into double data type. Rescaling can also be done to convert the image pixels into an integer of 16 bits, single type or double type. The ninth altered image is obtained by applying a rectangular shear on both sides of the original image; and the tenth altered image is obtained by removing the noise from the original image using the Weiner filter.

Next, the Computer Vision Toolbox and Image Processing Toolbox of MATLAB are used for applying the above-mentioned alterations to the original images. For our experiments, we have also developed a MATLAB implementation of the SPIHT compression algorithm described in Said and Pearlman (2017).

10.6.4.2 Deduplication Analysis

As stated earlier, one of the main requirements of the proposed scheme is that it should be robust enough to identify the minor changes between two images even if they are in the compressed form, i.e. it is desirable that the signature of the image generated from the SPIHT compression algorithm be different even when the original image has been slightly altered (hence, producing a different image). The results showing the percentage of dissimilarity between the signature of the original image and that of the altered image (obtained as described above) are captured in Tables 10.1 and 10.2, for images data coded at a rate of 0.50 bpp and 0.80 bpp respectively, on two dimensions, namely 256×256 and 512×512. In Table 10.1, it is observed that the difference between the signatures is lesser for the commonly used alterations (namely brightness, gamma correction, contrast and rotation changes) compared to the uncommon and severe alterations. These alterations are often performed by the users on the images but they do result in an entirely different appearance of the image. Although the images do not appear to change to a great extend for a human eye, the uncommon alterations (i.e. median filtration, noise insertion and removal, shearing and rescaling) are causing significant changes in the signature of the image. As shown in Table 10.1, the percentage difference between the original and altered image increases significantly for the severe alterations. As far as the dimensions are concerned, the percentage change between the signatures tends to decrease slightly as the image dimension increases. The same trend is maintained in the case the images data are coded at a rate of 0.80 bpp as shown in Table 10.2. The rest of the 63 images have been tested on 0.80 bpp to ensure that the images exhibit the same patterns when generating the signatures at different rates as well. As shown in Table 10.2, the percentage of change for the commonly used alterations are less than that observed for the severe alterations, including at higher compression rates.

Table 10.1 % difference between significant maps of the original and altered images at 0.05 bpp

Alteration	Clock (256×256)	Clock (512×512)	Pepper (256×256)	Pepper (512×512)	Lena (256×256)	Lena (512×512)
Brightness	6.8	5.7	6.4	4.8	5.3	4.2
Gamma	5.5	3.9	6.3	4.5	5.7	4.8
Contrast	6.3	6.2	6.0	4.9	6.7	4.6
Rotation	6.4	4.4	5.1	4.3	7.1	5.7
JPEG	0	0	0	0	0	0
Median filtration	13.3	9.6	29.6	19.8	26.2	30.8
Noise removal	11.4	8.2	30.5	24.66	26.3	30.1
Noise insertion	29.1	24.2	28.8	24.18	27.1	30.9
Shearing	29.8	15.1	46.8	33.0	48.1	35.2
Rescaling	29.2	19.1	28.8	18.75	26.5	36.7

Table 10.2 % difference between significant maps of the original and altered images at 0.08 bpp

Alteration	Bridge (256×256)	Bridge (512×512)	Goldhill (256×256)	Goldhill (512×512)	Mandrill (256×256)	Mandrill (512×512)
Brightness	7.5	7.1	7.4	6.0	9.5	6.2
Gamma	8.2	7.8	9.15	6.7	10.4	6.3
Contrast	7.5	7.7	7.7	5.3	10.6	5.1
Rotation	7.12	6.42	6.9	6.25	8.0	6.2
JPEG	0	0	0	0	0	0
Median filtration	19.6	7.5	28.8	19.0	20.2	11.5
Noise removal	23.0	7.6	28.7	15.4	23.4	7.2
Noise insertion	14.4	7.8	28.4	19.9	15.6	6.8
Shearing	29.8	17.3	30.8	21.3	30.1	18.5
Rescaling	14.4	7.8	27.5	19.7	15.5	7.1

The alterations that have been introduced into the original images are also analyzed to determine how much these alterations affect the images' signatures. The results are captured in Fig. 10.2. In Fig. 10.2, it is observed that the shearing process has drastically changed the mathematical content of the image, producing a compressed image data significantly different from the original one. The same trend prevails in terms of difference between the compressed image data and the original data, when the other alteration techniques are utilized, namely noise insertion, rescaling, noise removal, and median filtration, in this order. It is observed that when alterations such as rotation, gamma, and contrast are used, their impact on the images' signatures are not that significant. In the case of JPEG compression, it is

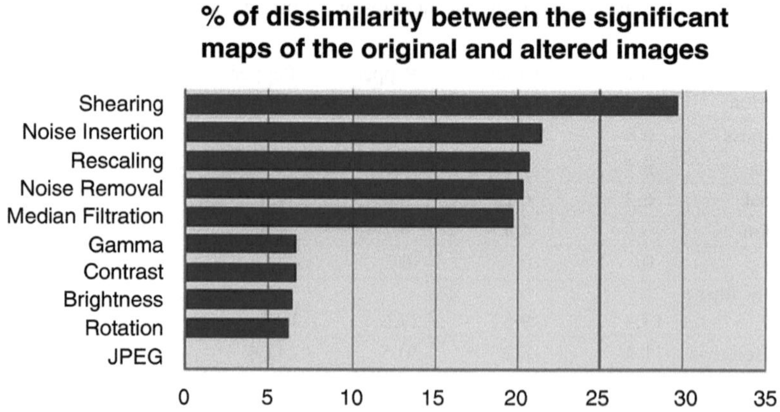

Fig. 10.2 Percentage of dissimilarity caused by the different alterations

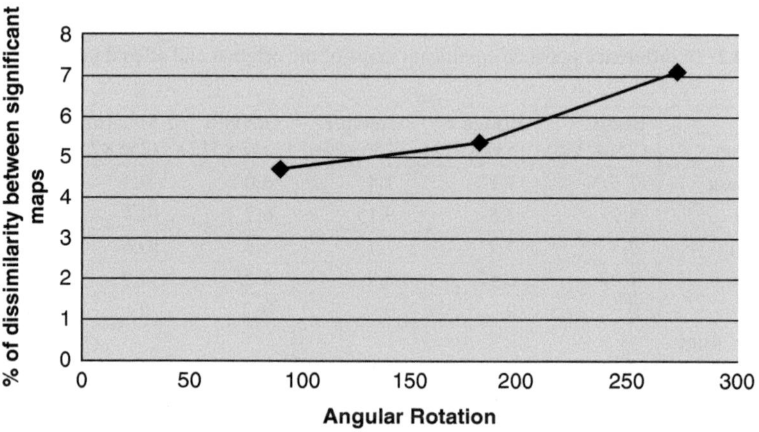

Fig. 10.3 Percentage of dissimilarity caused by the different rotations

observed that the significant maps of the original images and that of the compressed images are similar. This is attributed to the fact that the significant maps of the original image are produced by a compression algorithm as well. Thus, in all the trials with different dimensions and compression rates, the percentage of change in the case of the JPEG compression is 0.

Angular rotations are among the most commonly performed alterations performed by the users on the images before uploading them in the cloud storage (Yang and Chen 2005). In this work, angular rotations are performed and their impact on the images' signatures are captured in Fig. 10.3. In Fig. 10.3, it can be observed that as the image moved away from its initial reference point, the images' signatures start to change progressively, then increases drastically, resulting to a percentage of dissimilarity between the signature of the original image and that of the altered image of about 4.6% (resp. 5.4 and 7.1%) when the angular position of 90° (resp. 180° and 270°) are applied. This confirms a normal behavior of the signatures since the image itself changes in appearance as it is rotated.

10.6.4.3 Performance Analysis

The deduplication process of our proposed scheme also deserves some analysis to determine its performance in terms of computational overheads generated from the users' part. Due to the fact that the compressed images' data have been used to produce the unique hashes of the images for deduplication purpose, there is no other algorithm than the SPIHT compression algorithm that has been involved in the proposed scheme for generating the images' signatures. The SPIHT compression algorithm is invoked by the users to generate their compressed images before uploading them to the cloud storage. For doing so, the users needed to calculate the images' signatures from the lists *LIS*, *LIP*, and *LSP*) that have been generated during the compression process under the first three threshold levels. The performance of the deduplication process can be judged by quantifying the time taken for the signatures of the images generation versus the time taken for the generation of the above-mentioned lists.

Let T_1 be the time (measured in seconds) taken by the user to calculate the images' signatures from the lists; let T_2 (also measured in seconds) be the time taken by the SPIHT algorithm to generate the three lists under all the threshold levels, not including the time taken to perform the binary encoding and decoding steps.

In Tables 10.3 and 10.4, T1 is the time consumed in calculating the signatures, T2 is the time taken by the compression algorithm to calculate the three lists under all the threshold levels and The third column shows how much percent of T2 is T1. T2 is not the total time taken by the compression algorithm since that includes

Table 10.3 Performance in terms of time for 0.80 bpp

Alterations	Pepper (256×256)			Pepper (512×512)		
	T1 (s)	T2 (s)	T1% of T2	T1 (s)	T2 (s)	T1% of T2
Original	0.094	0.58	16.21	0.035	2.137	1.64
Brightness	0.093	0.593	15.68	0.062	2.418	2.56
Gamma	0.046	0.608	7.57	0.031	2.496	1.24
Contrast	0.078	0.562	13.88	0.046	2.465	1.87
JPEG	0.062	0.546	11.36	0.031	2.122	1.46
Rotation	0.078	0.546	14.29	0.035	2.106	1.66
Median filtration	0.015	0.39	3.85	0.015	1.544	0.97
Noise insertion	0.031	0.421	7.36	0.078	1.763	4.42
Noise removal	0.031	0.406	7.64	0.062	1.607	3.86
Shearing	0.031	0.53	5.85	0.036	1.934	1.86
Rescaling	0.031	0.421	7.36	0.031	1.591	1.95

Table 10.4 Performance in terms of time for 0.50 bpp

	Goldhill (256×256)			Goldhill (512×512)		
Alterations	T1 (s)	T2 (s)	T1% of T2	T1 (s)	T2 (s)	T1% of T2
Original	0.031	0.953	3.25	0.031	3.198	0.96
Brightness	0.031	0.952	3.25	0.093	3.494	2.66
Gamma	0.031	0.998	3.10	0.031	3.635	0.85
Contrast	0.031	0.936	3.31	0.031	3.588	0.86
JPEG	0.046	0.936	4.91	0.031	3.167	0.97
Rotation	0.015	0.952	1.57	0.031	3.182	0.97
Median filtration	0.062	0.593	10.45	0.015	2.184	0.68
Noise insertion	0.015	0.593	2.52	0.046	2.168	2.12
Noise removal	0.015	0.53	2.83	0.015	2.163	0.69
Shearing	0.031	0.577	5.37	0.062	2.434	2.54
Rescaling	0.109	0.562	19.39	0.046	2.168	2.12

some binary encoding and decoding as well. Tables 10.3 and 10.4 shows that the time taken to calculate the signatures is significantly low compared to that taken by the SPIHT algorithm to generate the data for the signatures. This was expected since the work carried out by the user on his/her own machine, i.e. performing the compression and extracting the signatures already generated by the SPIHT algorithm, is expected to last for a relative small amount of time (or CPU cycles). Thereby, some storage space is saved by performing the image deduplication. At a compression rate of 0.50 bpp, it is also be noticed that for all the images of 256×256 dimensions (resp. 512 × 512 dimensions), T_2 is more than 10 times (resp. 90 times) the value of $T1$ in average. Therefore, as the dimension of the image increases, the time T_1 tends to decrease to further refine the performance of the reduplication process. The same observations prevail when the images data are compressed at a rate of 0.80 bpp. In this case, T_2 is more than 10 times (resp. 56 times) the value of $T1$ in average for the images of 256 × 256 dimensions (resp. 512 × 512 dimensions). As far as the compression efficiency is concerned, the SPIHT compression algorithm is not affected by the encryption and image hashing steps since these steps are performed only after the image compression has been completed. This way, the image deduplication is achieved with the help of the image compression.

10.6.5 Security Analysis

The setup of the proposed scheme allows the user to only upload the images data generated during the image compression step (it should be noted that a part of it is encrypted by the user) and the unique hash of the image calculated by the user. The CSP should not be able to decompress the image from the received compressed image data.

The security of the proposed scheme depends upon the partial encryption of the wavelet coefficients. The significant information in encrypted form is required in order to correctly decompress the image. The decoder of the SPIHT algorithm keeps track of the order of execution of the encoder; the lists and image mask are both required for the most recent iteration of the encoder. From there on, the SPIHT algorithm starts to decode/decompress the image in the exactly reverse order of what the encoder has done until this process terminates.

In Cheng and Li (2000), it has been proved that the size of the important part is at least 160 bits if at least two sorting passes are performed by the EZT compression algorithm. Therefore, for an exhaustive search, an attacker would need at least 2^{160} tries before heading to a conclusion. For example, considering the Lena image with 512×512 dimension, only the significant coefficients in the first two largest sub bands are required to be encrypted. These are the coefficients in the 16×16 sub band with 8×8 be the root level according (Cheng and Li 2000). We have calculated the size of the indicated data to be encrypted. The size of the two lists to be encrypted comes to 768 bytes in total. In addition, the mask of the image indicating if the pixels in the 16×16 sub bands are significant or not also needs to be encrypted. The size of the mask is 575 bytes compared to that of the complete mask for 512×512 pixels, which is 26000 bytes. Therefore, the partial encryption is indeed very efficient since the encrypted part is significantly less compared to the entire image data.

As far as convergent encryption is concerned, each user has his own key for the decryption of the image, which is kept secret from the CSP. Due to the deduplication requirements, the users cannot be allowed to choose different keys (using a public or private key model) for encryption purpose. The keys generated for two different users for the same image through convergent encryption are exactly the same. Hence, we assume that the users of the same cloud storage will not collude with the CSP, otherwise the security of the proposed scheme may be compromised.

10.7 Secure Video Deduplication Scheme in Cloud Storage Environment Using H.264 Compression

In this section, a secure scheme is proposed that achieves cross user client side video deduplication in cloud storage environments. Since in Venter and Stein (2012), it is claimed that a major part of the digital data these days is composed of videos and pictures, we are focusing on the video deduplication in this section. Its design consists of embedding a partial convergent encryption along with a unique signature generation scheme into a H.264 video compression scheme. The partial convergent encryption scheme is meant to ensure that the proposed scheme is secured against a semi-honest CSP; the unique signature generation scheme is meant to enable a classification of the encrypted compressed video data in such a way that the deduplication can be efficiently performed on them. Experimental results are provided, showing the effectiveness and security of our proposed schemes.

10.8 Background

H.264 is an object-based algorithm that makes use of local processing power to recreate sounds and images (Richardson 2011). The H.264 algorithm (Al Muhit 2017) is a block-based motion compensation codec most widely used for HD videos. Its working is based on frames, which can be further grouped into GOP and thus can yield high deduplication ratios for your scheme. A selective encryption of the H.264 compressed video is proposed in Zhao and Zhuo (2012) and it works on the GOP level and on various types of videos. These characteristics makes this selective encryption scheme very suitable for our video deduplication scheme. The signature generation is inspired by the method proposed in Saadi et al. (2009), but this method is modified to fit our requirements. In Saadi et al. (2009), the scheme proposed is used for watermarking which is not useful for deduplication, therefore, it is used only for signature extraction purposes in our scheme.

10.9 Proposed Video Deduplication Scheme

The design of the proposed approach for secure deduplication of videos in the cloud storage involves three components: H.264 video compression scheme, signature generation from the compressed videos, and selective encryption of the compressed videos as shown in Fig. 10.4.

First, the user compresses the original video using the H.264 compression algorithm. Second, he/she calculates the signatures based on the GOP from the output bit stream. Third, he/she encrypt the important parts of the DCT coefficients and motion vectors according to the type of the videos. After these processing steps on the original video, it will be uploaded in the cloud storage. The CSP will then check for the identical GOPS with the help of the signatures and encrypted data. If identical GOPs are detected, the CSP will delete the new data and update the metadata for this particular data already in the cloud storage. In this way, the CSP will save huge space by performing cross-user video deduplication in the cloud storage. The detailed description of the proposed scheme follows.

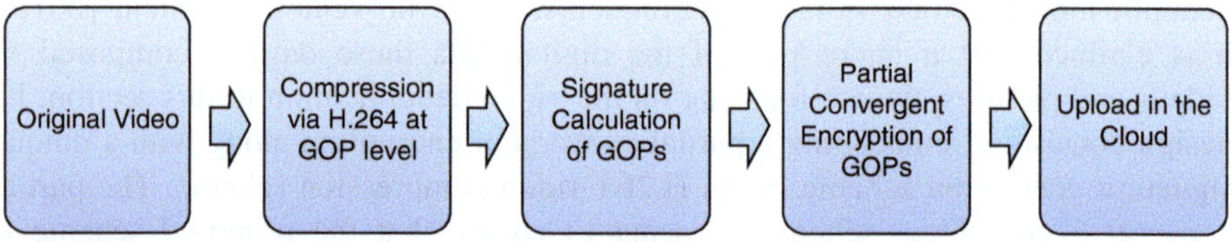

Fig. 10.4 Proposed secure video deduplication scheme

10.9.1 H.264 Video Compression

The first part of our video deduplication scheme is to compress the original video sequence. The compression algorithm should be strong and efficient enough to produce highly compressed, retrievable and smaller in size version of the video. We use the H.264 compression algorithm (Richardson 2011). This algorithm includes a prediction step, a transform step, and an entropy coding step.

- The prediction step: The coded video sequence generated by the H.264 algorithm is made of a sequence of coded pictures. Each picture is divided into fixed size macro blocks. It should be noted that macro blocks are the basic building blocks in the H.264 algorithm. Each macro block is a rectangular picture area made of 16×16 samples for the luma component and the associated 8×8 sample regions for the two chroma components. The luma and chroma samples of a macro block are used for prediction purpose. They are either spatially predicted or temporally predicted by the algorithm and the resulting residual is divided into blocks. These blocks are then transformed, quantized and encoded in the final stages.

  The macro blocks of the picture are further grouped into slices, which themselves divide the above regions of the picture that are eligible to be decoded independently. Each slice is a sequence of macro blocks. It is processed following a raster scan, i.e. from top-left to bottom-right. The slices can be used for error resilience since the partitioning of the picture allows a spatial concealment within the picture and the start of each slice provides a resynchronization point at which the decoding process can be reinitialized (Richardson 2011). The slices can also be used for parallel processing since each frame can be encoded and decoded independently of the other frames of the picture. The robustness of the algorithm can be further strengthen by separating the more important data (such as macro block types and motion vector values) from the less important ones (such as inter residual transform coefficient values).

  The two types of fundamental frames used are, namely (1) the I slice—i.e. a frame in which all macro blocks are coded using the intra prediction. It is the most important type of frame used by all versions of the algorithm; (2) the P frame—i.e. a frame in which the macro blocks can also be coded using the inter prediction with at most one MCP signal per block;

  It should be noted that the intra-picture prediction can be performed in two types of intra coding modes: the intra 4×4 and the intra 16×16 modes. In the intra 4×4 mode, each 4×4 luma block is predicted from spatially neighboring samples. In the intra 16×16 mode, the whole 16×16 luma component of the macro block is predicted with four prediction modes, namely, vertical, horizontal, DC, and plane.

  The inter-picture prediction in P frames is another prediction method that can be utilized by the H.264 algorithm. It involves the use of *P* macro blocks with luma block sizes of 16×16, 16×8, 8×16, and 8×8 samples. For each 8×8 partition, this method can be used to decide on whether or not that partition should be further partitioned into smaller sizes of 8×4, 4×8, or 4×4 luma samples and corresponding chroma samples.

- The transform step: This includes the transform, scaling, and quantization sub-steps. In the H.264 algorithm, an integer transform such as the 4×4 DCT is applied to 4×4 blocks instead of larger 8×8 blocks as used in previous standard methods. The inverse transform of H.264 uses simple exact integer operations so that mismatches are avoided and the decoding complexity is minimized. For the luma component in the intra 16×16 mode and the chroma components in all intra macro blocks, the DC coefficients of the 4×4 transform blocks undergo a second transform, so that the lowest-frequency transform basis functions cover the entire macro block (Richardson 2011). A quantization parameter (QP) is then used for determining the quantization of the transform coefficients. The quantization step size is controlled logarithmically by this QP in such a way that the decoding complexity is reduced and the bit rate control capability is enhanced (Stutz and Uhl 2012).
- The entropy coding: In the H.264 algorithm, two entropy coding are supported: the CAVLC and the CABAC. The CABAC has better coding efficiency compared to the CAVLC, but yields a higher complexity. In both of these modes, the syntax elements are coded using a single infinite-extent codeword set (referred to as Exp-Golomb code).

In our proposed video deduplication scheme, the videos are divided into GOP based on similarity; where each GOP is made of I frames and P frames. The H.264 algorithm is used with the following specifications: the GOP size is set to 15 frames, where the first frame is always the I frame used as reference for the subsequent 14 frames, which are all P frames.

10.9.2 Signature Generation from the Compressed Videos

The generated signature captures the content dependent robust bits from the macro blocks generated by the H.264 Compression algorithm. These bits are then used to authenticate or identify one GOP from another, hence, they are treated as the signature of the GOP. The signature calculation is done in parallel with the compression algorithm as the GOPs are being generated by the H.264 algorithm.

The signature generation is carried out in the compressed domain, and the signatures are generated from the information produced in the transform domain of the H.264 compression algorithm. The content dependent robust bits are extracted from the macro blocks and are further used as the signature for authenticating the compressed video. It should be noted that in our use of the H.264 algorithm, the I and P frames are the minimum types of the frames that should be present for the algorithm to work.

Indeed, the video is first broken down into GOPs, which are authenticated individually by hashing the features extracted from their I frames and P frames. The hash is then considered as the digest digital signature for all the frames in the GOP. The digital signature is composed of the features extracted from the intra 16×16, intra 4 × 4 and inter 4 × 4 MBs. The I slices are composed of intra coded MBs in which each intra 16×16 and intra 4×4 luma regions are predicted from the previous 16 × 16 and 4 × 4 coded MBs of the same I frame.

For the intra 4 × 4 and inter 4 × 4, the quantized DC coefficients and the first two quantized AC coefficients in the zig zag scan order (surrounding the only DC coefficient value) for every 4 × 4 MB are pull out as the signature data. Then, for every intra 16 × 16, all the nonzero quantized *Hadamard* transform coefficients and the first two quantized AC coefficients in the zig zag scan order (also surrounding the only DC coefficient value) for every 16 × 16 MB are pull out as the signature data. The signature is calculated for each MB in a frame until the end of the GOP is reached. Meanwhile, these signatures are saved in a buffer and when the algorithm signals IDR, the signatures for all the MBs for all the frames in the GOP are hashed by using *SHA2-256*, producing a 256 bit long signature for each GOP. Finally, the signatures for all the GOPs are calculated and transmitted along with the video. The CSP will compare these signatures against the ones already stored in the cloud storage in order to identify any possible duplicated parts of the videos. The deduplication at the GOP level will further increase the deduplication ratios since it is less likely that identical entire videos by different users will be uploaded in the cloud as opposed to parts of the videos.

10.9.3 Selective Encryption of the Compressed Videos

Assuming that the compressed videos have been produced by the H.264 algorithm, and the signatures have been generated by the users, the next step consists of encrypting the compressed videos so that it will not be possible for the CSP to access the plain videos.

We have used the partial encryption scheme proposed in Zhao and Zhuo (2012), with some modifications applied on it in order to fulfill the requirements of cross-user deduplication. The partial encryption is carried out in the compressed domain and therefore is well in line with the signature generation process since that is also performed in the compressed domain. Typically, since the video is split into GOPs, the encryption is performed at the GOP level. This encryption scheme is content-based since user dependent variables are not involved in its process. Content-based encryption ensures that the same encrypted videos will be obtained for the same plain videos by different users, which is the basic requirement for cross-user deduplication.

The encryption process starts by first generating the compressed bit stream for the video. First, the user classifies the video into six different categories, namely high, medium and low intensity motion (for complex texture) and high, medium and low intensity motion (for non-complex texture) by utilizing the information generated in the intra prediction mode, DCT coefficients, and motion vectors (as described earlier). It is assumed that if the I frame of a GOP is complex in texture, the corresponding P frames will also be complex in texture. For the motion intensity, the motion vectors of the P frames are taken into account. Second, the user organizes the GOPs of the above mentioned six classes. Basically, the user calculates the average of nonzero coefficients and the average number of intra 4×4 prediction for the I frames. For the P frames, the average of the suffix length of the MV keywords and the corresponding variance are calculated. These values are then compared against the threshold values given by the CSP to all the users of that cloud. The users will use these threshold values to classify their videos and start the encryption process. The details of the classification of the videos are beyond the scope of our current research, but can be found in Zhao and Zhuo (2012). The partial encryption then follows after the video classification step.

The compressed video has already been broken into GOPs for the purpose of signature generation. This same set of GOPs is used for partial encryption as follows. The DCT coefficients and intra prediction modes in every GOP are encrypted according to the texture of the video. In doing so, the I frames are considered as crucial for the decoding of the P frames in the sense if errors occurred in the I frames or if these frames are tampered (even slightly), the decoding of the P frame will be affected for any GOP. Therefore, all the intra prediction modes are encrypted, no matter what the texture or motion intensity of the video is. For complex texture videos, all the nonzero DCT coefficients, except for the trailing ones are encrypted because of the fact that the videos will have a large number of high band DCT coefficients. For non-complex texture videos, only the first three low band DCT coefficients are encrypted. For motion intensity purpose, the user will encrypt the motion vectors of the P frames. In case of the complex texture video, the user will encrypt the motion vector difference in all the P frames, i.e. the first 70% and then the first 30% of the P frames respectively for high, medium and low intensity videos for each GOP. In case of the non-complex texture video, the user will encrypt the motion vector difference in all the P frames, i.e. the first 70% and then the first 30% of the P frames respectively for high, medium and low intensity videos for each GOP.

In our scheme, the convergent encryption (Wang et al. 2010) is employed to derive the key for partial encryption from the content of the compressed video rather than getting the key chosen by the users individually. Therefore, for the same content, the same key will be generated without the users knowing each other. Thus, different users will have the same key as well as the same encrypted videos, irrespective of the knowledge of each other keys. This will make it easier for the CSP to compare the encrypted parts of the videos and perform deduplication in case of duplicated videos without actually decrypting or decompressing the video data.

The convergent encryption steps are described as follows. Let us assume that user partially encrypts the video data as described earlier and the video is complex textured, with medium motion intensity. Then three data sets are to be encrypted, namely, $INTRA - PRED = (All\ the\ intra\ prediction\ modes\ in\ a\ GOP)$, $NZ - DCT = (All\ non\ zero\ DCT\ coefficients\ in\ a\ GOP)$ and $PER - MVD = (\%\ of\ the\ MVDs\ in\ a\ GOP)$. The user will calculate the keys $K1$, $K2$, and $K3$ as follows: $K1 = SHA2\ (INTRA - PRED)$, $K2 = SHA2\ (NZ - DCT)$ and $K3 = SHA2(PER - MVD)$; then will perform the following encryption: $Enc(INTRA-PRED) = AES(K1, INTRA-PRED)$, $Enc(NZ_{D}CT) = AES(K2, NZ_{D}CT)$ and $Enc(PER-MVD) = AES(K3, PER-MVD)$. In this way, for every vector and for every MB of any two identical GOPs, the encrypted data will appear to be similar. The CSP will then be able to compare the GOPs without breaching the privacy and security of the uploaded data while storage space will be maximized through cross-user deduplication. Once the video and the signatures are uploaded by the user to the cloud, the CSP will compare the signatures of the compressed video for deduplication purposes. If the signatures match, the new copy is discarded and the metadata is updated. Depending upon the resource and security requirements, the CSP can compare the encrypted parts of the videos after the signature matches, in order to ensure that the two GOPs are identical before deleting the data. Since convergent encryption is applied, the same plain videos would result is the same cipher videos, thus enabling cross-user deduplication in the presence of a semi-honest CSP.

10.9.4 Experimental Results

The experiments have been conducted by keeping in mind the following requirements: (1) some digital space must be saved by applying the H.264 compression algorithm and deduplication; (2) the compression algorithm must be efficient in terms of computation, complexity and storage overheads; (3) the signature generation step must be robust enough to identify the GOPs for deduplication; and (4) attacks from external users are much more straightforward than those from internal users. Therefore, our goal is to show that our proposed scheme is secure against a semi-honest CSP. The use of partial encryption and signature generation done at the user end is therefore analyzed thoroughly for security and efficiency purposes.

In order to analyze the security and efficiency of our proposed scheme, we tested it on six different video sequences, namely Akiyo, Foreman, Claire, Grandma, and Highway. The specifications of all of these video sequences are well known and are presented in Table 10.5. We have chosen these videos for our testing because they belong to different classes of videos (Thomas et al. 2007). Foreman and Highway

Table 10.5 Video sequence information

Video sequence	Size of video (MB)	Total GOPs	Avg size of GOP (KB)
Akiyo	1.17	20	53.33
Grandma	3.07	58	54.9
Mobile	2.80	20	154.66
Foreman	0.944	14	60.74
Claire	1.33	33	42.10
Highway	6.14	134	49.45

Table 10.6 Deduplication performance of the video sequences

Video sequence	% of space saved for 25% of GOPs	% of Space saved for 50% of GOPs	% of space saved for 75% of GOPs	% of space saved for 100% of GOPs
Akiyo	26.65	53.33	79.99	100
Grandma	25.92	51.85	77.78	100
Mobile	27.61	55.23	82.85	100
Foreman	22.5	45.02	67.53	100
Claire	27.06	51.12	75.17	100
Highway	26.98	53.96	80.94	100

are classified as non-complex textured and medium intensity videos. Grandma, Akiyo and Claire are classified as non-complex textured and low intensity motion videos. Each video sequence is first encoded into the QCIF, and then the frames are extracted. The videos are then input into the H.264 compression algorithm with a GOP size of 15. The first frame is the I frame and the rest of the frames are P frames, which implies an IPPP format. The QP value is set to 28. The algorithm works on 4×4, 16×16 MB modes for I and P frames. The rate of compression for H.264 is set to 30 Hz. The details of the algorithm can be found at (Al Muhit 2017). The system configuration for the experiment is an Intel Core i3 processor with 2.40 GHz frequency and 4 GB RAM. The underlying operating system is 64 bits. We implemented our scheme in MATLAB version R2014a. Finally, different methods, all implemented in MATLAB, were used for the extraction and processing of QCIF videos, as well as the extraction of the frames.

Video deduplication is the most important requirement which we need to fulfill. The deduplication aspect of our scheme is captured in Table 10.6. We have calculated the amount of space saved in cloud storage for the case where the CSP practices cross-user deduplication at the GOP level for 25, 50 , 75 and 100% of GOPs in our experiments. From the results shown in Table 10.6, it can be observed that for complex textured videos such as mobile, the percentage of digital space saved is higher than that of the non-complex textured video. This is attributed to the fact that these videos are larger in size (i.e. there is more information for complex textured videos) on the disk than the others. As more and more GOPs are replicated, the space savings increase at the cloud storage end. The CSP will simply need to update some metadata for the GOP, indicating that this particular GOP had also belonged to this user in addition to some other users. The size of the metadata is very nominal compared to the space saved by simply storing a single copy of the GOP rather than a copy for each user. Moreover, for CSPs with large storage, the metadata space is negligible because of its textual nature. For videos that are large in size and high motion intensity, it can be seen that there are substantial space savings as in case of the Highway video. The percentage of space saved increases as more and more GOPs are identical. When the entire video is the same, 100% of the space is saved. It can also be noted that the amount of space saved increases as the size of the video increases, which is very favourable for a cloud computing setup since users try to upload large videos at times. Videos uploaded to the cloud by different users are often not exactly identical, but at times cropped at the beginning or at the end. Unlike images, the brightness and colour of videos are not often modified by users, but videos can be changed in length by simply cropping some parts of them. This explains why our scheme is based on GOPs so that the deduplication ratios can be improved as much as possible.

The second requirement of our scheme is that the inclusion of the deduplication process should not incur any extra overhead on the process of uploading the videos in the cloud storage from the user's end, i.e. the proposed scheme should be computationally cost-effective in terms of resources at the user's end. We are proposing to use the H.264 compression algorithm as the basis of all our further computations. Since videos are compressed anyways before being uploaded in the cloud, the computational overhead in reduced to a great extend. The user is calculating the signatures from the information produced from the compression algorithm. The encryption is also carried out on the information generated by the compression algorithm. Therefore, the performance efficiency should be determined in terms of how much time the user spent on these two actions. From Table 10.7, it can be observed that the average time to encode the videos is much higher than that to calculate the signature. In case of the Highway video sequence, the average time to compress the video is 165.65 s and the average time to calculate the signature at the GOP level is 0.0603 s, which is nominal compared to the time to encode. It can also be noticed that as the size of the video increases (from foreman to highway), the number of GOPs naturally increases, but the time taken to encode and the PNSR also depends on the nature of the video. The mobile video sequence is complex in texture and smaller in size, but took the highest time to encode. The time taken (and consequently the number of CPU cycles used) to compare the signatures is also insignificant in the case of large cloud storages since the size of the signature is merely 256 bits. For instance, as can be seen from Table 10.7, the size of the actual signature for the grandma video is 1568745 bytes, which has been reduced

Table 10.7 Compression performance of the video sequences

Video sequence	Avg time to encode (s)	Avg PNSR	Avg time to calculate hash of sig (s)	Avg length of sig (Bytes)
Akiyo	146.685	39.9	0.0524	1536480
Grandma	173.846	39.3	0.0480	1568745
Mobile	220.54	35.21	0.0632	1743569
Foreman	160.66	37.7	0.0832	1812960
Claire	166.38	41.5	0.068	1463256
Highway	165.65	39.4	0.0603	1722373

to 256 bits by the SHA hash. These signatures will be transmitted along with the compressed videos, but because of their size, they do not incur any overhead in terms of bandwidth consumption.

For the grandma video, a total of 58 signatures need to be transmitted with an approximate size of 14848 bits, which again is negligible for the user. The signatures are also calculated at the GOPs level, i.e. for each GOP, a 256 bits signature is generated, which is very small compared to the original size of the GOP. Depending upon the computational capacity of the cloud and security and efficiency tradeoff, the signatures can be calculated for each frame rather than at the GOP level, and each frame signature can be checked for deduplication. The time taken to calculate those signatures is also negligible, even at the user's end. For the CSP, the benefit comes when he/she has to compare the signatures (256 bits) rather than the GOPs themselves (e.g. 53.33 KB). The size of the video to be stored also gets reduced after compression, which is also a benefit for the CSP in terms of storage savings. The proposed scheme is designed in such a way that the performance of the H.264 compression algorithm is not affected by the signature calculation and the partial encryption because these methods are applied on top of the information generated by the H.264 compression algorithm. The performance of the H.264 compression algorithm is shown in Table 10.7 in terms of PSNR. The performance of our proposed scheme can be judged from the results depicted in Tables 10.6 and 10.7.

We have also quantified the time consumed in the partial encryption of each GOP. In our proposed video deduplication scheme, the partial encryption presented in Zhao and Zhuo (2012) is adopted and adjusted to meet the requirements of cross-user deduplication. Therefore, we have considered the same experimental settings used in Zhao and Zhuo (2012), in terms of number of frames in GOPs, compression rate, and QP factor. We conducted the experiment on a few samples and obtained almost similar results as the ones shown in Zhao and Zhuo (2012) in terms of time consumed for partial encryption. Indeed, the Foreman video took 0.44019 s, the Highway video took 2.26 s, the Akiyo video took 0.354 s and the mobile video took 1.308 s. From these results, it is clear that the time taken to perform partial encryption increases as the size of the video increases (the Highway video being the largest of all) and this time tends to increase for videos with complex texture since they have more information to encrypt. Overall, the time taken to encrypt the partial compressed information is negligible compared to the time taken to encode the video. Moreover, considering the tradeoff between security and efficiency, this amount of time can be overlooked when it comes to cloud computing.

10.9.5 Security Analysis

The setup of the proposed scheme allows the user to only upload the video data generated during the H.264 compression step, and at the same time, to partially encrypt the video and generate the unique hash of each GOP. The CSP should not be able to decompress the video from the received compressed data, which is partially encrypted. The security of the proposed scheme depends upon the partial encryption of the DCT coefficients and motion vectors. We have used the convergent encryption, which utilizes the 256 bits hash of the data as the key for the AES encryption. The possibility of AES encryption being compromised is very seldom since it requires a computational complexity of $2^{126.1}$ to recover the key. The security of the SHA-256 bits is very strong since it requires a computational complexity of minimum 2^{178} to be attacked in case of differential attacks. The perceptual quality of the videos is too bad if decrypted by the CSP since he does not have access to the required video compression information in order to accurately recover the video. Due to the deduplication requirement, the users cannot be allowed to choose different keys or a public/ private key model for the encryption step. The keys generated for two different users for the same GOP through convergent encryption are exactly the same. Hence, it is assumed that the users of the same cloud storage will not collude with the CSP, otherwise the security of the proposed scheme could be compromised.

10.10 Chapter Summary

In the first part of this chapter, we have presented a novel secure image deduplication scheme through image compression for cloud storage services purpose. The proposed scheme is composed of three parts, namely: SPIHT compression algorithm, partial encryption, and hashing. Experimental results have shown that (1) the proposed scheme is robust enough to identify minor changes between two images even if they are in the compressed form; (2) the proposed scheme is secured against the semi-honest CSP since the CSP does not have access to the compressed images, but can identify the identical images from different users only through the significant maps of these compressed images; (3) the proposed POR and POW schemes are efficient.

In the second half, we have proposed a secure video deduplication scheme through video compression in cloud storage environments. The proposed scheme is made of three components: H.264 video compression scheme, signature generation from the compressed videos, and selective encryption of the compressed videos. The compressed video obtained from the H.264 algorithm is partially encrypted through convergent encryption, in such a way that the semi-honest CSP or any malicious user cannot have access to it in plain, hence ensuring the security of the video data from the CSP or any malicious user. Experimental results have shown that: (1) For complex textured videos, the percentage of digital storage space saved by the CSP practicing cross-user deduplication using our scheme is higher than that of the non-complex textured video; (2) For videos that are large in size and high motion intensity, there are substantial space savings for the CSP practicing cross-user deduplication using our scheme; (3) the average time taken to encode the videos is much higher than that taken to calculate the signature; (4) the proposed scheme is secured against the semi-honest CSP since the CSP does not have access to the video compression information required for the recovery of the video.

References

Adshead, A. (2017). A guide to data de-duplication. http://www.computerweekly.com/feature/A-guide-to-data-de-duplication. Accessed 24 August 2016.

Al Muhit, A. (2017). H.264 baseline codec v2. http://www.mathworks.com/matlabcentral/fileexchange/40359-h-264-baseline-codec-v2. Accessed 24 August 2016.

Anderson, R. (2017). Can you compress and dedupe? it depends. http://storagesavvy.com. Accessed 23 August 2016.

Cheng, H., & Li, X. (2000). Partial encryption of compressed images and videos. *IEEE Transactions on Signal Processing, 48*(8), 2439–2451.

Gantz, J., & Reinsel, D. (May 2010). The digital universe decade - are you ready? https://www.emc.com/collateral/analyst-reports/idc-digital-universe-are-you-ready.pdf. Accessed 24 August 2016.

IBM Corporation (2017). IBM protectier deduplication. http://www-03.ibm.com/systems/storage/tape/protectier/index.html. Accessed 24 August 2016.

Keerthyrajan, G. (2017). Deduplication internals. https://pibytes.wordpress.com/2013/02/17/deduplication-internals-content-aware-deduplication-part-3/. Accessed February 2016.

Kovach, S. (2017). Dropbox hacked. http://www.businessinsider.com/dropbox-hacked-2014-10. Accessed 24 August 2016.

Osuna, A., Balogh, E., Ramos, A., de Carvalho, G., Javier, R. F., & Mann, Z. (2011). Implementing IBM storage data deduplication solutions. http://www.redbooks.ibm.com/redbooks/pdfs/sg247888.pdf. Accessed 24 August 2016.

Rashid, F., Miri, A., & Woungang, I. (2014). Proof of retrieval and ownership protocols for images through SPIHT compression. In *Proceedings of The 2014 IEEE 6th International Symposium on Cyberspace Safety and Security (CSS'14)*. New York: IEEE.

Rashid, F., Miri, A., & Woungang, I. (March 2015). A secure video deduplication in cloud storage environments using h.264 compression. In *Proceedings of the First IEEE International Conference on Big Data Computing Service and Applications San Francisco Bay, USA*. New York, IEEE.

Richardson, I. E. (2011). *The H. 264 advanced video compression standard*. New York: Wiley.

Saadi, K., Bouridane, A., & Guessoum, A. (2009). Combined fragile watermark and digital signature for H. 264/AVC video authentication. In *Proceedings of The 17th European signal Processing Conference (EUSIPCO 2009)*.

Said, A., & Pearlman, W. (2017). SPIHT image compression. http://www.cipr.rpi.edu/research/SPIHT/spiht1.html. Accessed 23 August 2016.

Said, A., & Pearlman, W. A. (1996). A new, fast, and efficient image codec based on set partitioning in hierarchical trees. *IEEE Transactions on Circuits and Systems for Video Technology, 6*(3), 243–250.

Shapiro, J. M. (1993). Embedded image coding using zerotrees of wavelet coefficients. *IEEE Transactions on Signal Processing, 41*(12), 3445–3462.

Singh, P., & Singh, P. (2011). Design and implementation of EZW and SPIHT image coder for virtual images. *International Journal of Computer Science and Security (IJCSS), 5*(5), 433.

Stutz, T., & Uhl, A. (2012). A survey of H. 264 AVC/SVC encryption. *IEEE Transactions on Circuits and Systems for Video Technology, 22*(3), 325–339.

Thomas, N. M., Lefol, D., Bull, D. R., & Redmill, D. (2007). A novel secure H. 264 transcoder using selective encryption. In *Proceedings of The IEEE International Conference on Image Processing (ICIP 2007)* (Vol. 4, pp. 85–88). New York: IEEE.

Thwel, T. T., & Thein, N. L. (December 2009). An efficient indexing mechanism for data deduplication. In *Proceedings of The 2009 International Conference on the Current Trends in Information Technology (CTIT)* (pp. 1–5).

Venter, F., & Stein, A. (2012). Images & videos: really big data. http://analytics-magazine.org/images-a-videos-really-big-data/. Accessed 24 August 2016.

Wang, C., Qin, Z. g., & Peng, J., & Wang, J. (July 2010). A novel encryption scheme for data deduplication system. In *Proceedings of The 2010 International Conference on Communications, Circuits and Systems (ICCCAS)* (pp. 265–269).

Xu, J., Chang, E.-C., & Zhou, J. (2013). Weak leakage-resilient client-side deduplication of encrypted data in cloud storage. In *Proceedings of the 8th ACM SIGSAC Symposium on Information, Computer and Communications Security (ASIA CCS '13)* (pp. 195–206). New York: ACM.

Yang, S.-H., & Chen, C.-F. (2005). Robust image hashing based on SPIHT. In *Proceedings of The 3rd International Conference on Information Technology: Research and Education (ITRE 2005)* (pp. 110–114). New York: IEEE.

Zhao, Y., & Zhuo, L. (2012). A content-based encryption scheme for wireless H. 264 compressed videos. In *Proceedings of The 2012 International Conference on Wireless Communications and Signal Processing (WCSP)* (pp. 1–6). New York: IEEE.

Chapter 11
Privacy-Aware Search and Computation Over Encrypted Data Stores

Hoi Ting Poon and Ali Miri

11.1 Introduction

It is only recently, with the rapid development of Internet technologies, the emergence of Internet of things and the appetite for multimedia that Big Data began to capture the attention of companies and researchers alike. Amidst the constant reminder of the importance of Big Data and the role of data scientist would have on our future, there is also a growing awareness that the availability of data in all forms and in large scale would constitute an unprecedented breach of security and privacy.

While much progress were made in the past decade, a practical and secure big data solution remains elusive. In response to the privacy concerns, anonymization techniques, through removal of personally identifiable information such as names and identification numbers from customer data, have been in use by many companies such as Facebook and Netflix. Yet, many studies have shown that they do not provide reliable protection. For instance, a study by MIT (*How hard is it to 'de-anonymize' cellphone data?* n.d.) showed that knowing a person's location four times in a year is enough to uniquely identify 95% of users in a set of 1.5 million cellphone usage records. In genomics, short subsequences of chromosomes were found to be enough to identify individuals with high probability (Gymrek et al. 2013). The anonymized Netflix Prize dataset was famously deanonymized using publicly available information (Narayanan and Shmatikov 2008). In all cases, the conclusion seems to be that reliable anonymization could well be infeasible since information about individuals is so widely available and easily accessible largely due to the Internet.

H.T. Poon • A. Miri (✉)
Department of Computer Science, Ryerson University, Toronto, ON, Canada
e-mail: hoiting.poon@ryerson.ca; Ali.Miri@ryerson.ca

S. Srinivasan (ed.), *Guide to Big Data Applications*, Studies in Big Data 26,
DOI 10.1007/978-3-319-53817-4_11

As an alternative to anonymization, encryption has well defined security properties and had endured under the scrutiny of academics and security professionals. Rather than maintaining seemingly non-identifying information in plain, all data is encrypted with mechanisms in place to perform the required functionalities. Much of the difficulty in securing distributed computation and data storage is due to the fact that strong encryption tends to require significant computations, which in turn reduces the throughput of the system. Nonetheless, there have been a growing body of work on the topic of processing of encrypted data, that holds promise on what a secure encrypted big data system may one day be.

In this chapter, we will look at some recent works on enabling search over encrypted data. Their common objective is to enable data to be stored and queried in encrypted form at numerous facilities, which may not be in the organizations control since data centers tend to be expensive to maintain and cloud solutions are fast becoming the standard. We will also briefly present the topic of homomorphic encryption, which has been a very active area of research due to the breakthrough work on the first fully homomorphic encryption scheme. The ability to perform computations in encrypted form has significant ramification in secure distributed computing, increasing end node security and privacy.

11.2 Searchable Encryption Models

Since Google popularized MapReduce in the early 2000s, search has been recognized as a central function of many big data systems. Similarly, research into encrypted data processing also began with search. Searchable encryption schemes generally involve up to three parties:

- Data Owner
- Users
- Storage Provider

The Data Owner has access to the secret/private key to decrypt the data set and is considered to be trusted. Users are parties other than the data owner that add material to the encrypted data set or that search over it. Users do not have the ability to decrypt or obtain information without obtaining Data Owner's authorization.

The encrypted data set is stored in an untrusted storage server, such as a cloud service provider (e.g. Amazon EC2, Microsoft Azure). While it is accepted that legitimate service providers will perform the required protocols and encryptions without fault to ensure their business operation, it is also conceivable that they may wish to perform analysis on the information available to them, such as the search patterns, to acquire additional knowledge on their clients that may benefit them. This is termed the honest but curious model, where a non-malicious entity follows the required protocols but desires to learn more on the protected data set. In addition to the storage provider, users are also generally considered to be honest but curious. More complicated schemes have also been proposed which consider the malicious settings where the storage provider or users may deliberately manipulate the protocols and data to compromise the data security. Other models also exist where researchers propose solutions to mitigate information leakage resulting from colluding users and cloud operators. For our discussion, we will restrict to the honest but curious model for cloud storage providers and users.

Searchable encryption schemes can be classified into four categories, denoted by contributors/searchers:

- Private/Private: A private keyword search scheme that allows the data owner possessing the secret key to search over an encrypted data set placed by the data owner without compromising or decrypting the data.
- Public/Private: A keyword search scheme that allows the data owner possessing the secret key to search over an encrypted data set consisting of content submitted by various users without compromising or decrypting the data.
- Private/Public: A keyword search scheme that allows any authorized user to search over an encrypted data set placed by the data owner without compromising or decrypting the data.
- Public/Public: A public keyword search scheme that allows any authorized user to search over an encrypted corpus consisting of content submitted by various users without compromising or decrypting the data.

In a completely private scenario, the data owner is the sole party performing searches and providing the encrypted data set. As the security risk is limited to the storage provider, this setting also entails the most efficient solutions using symmetric encryption. In a private/public setting, where other users may search over a private collection, an extension of the completely private solutions may be used. Where public contribution to the encrypted data set is required, asymmetric encryption is used to maintain secrecy.

To provide a better understanding of searchable encryption, this chapter will begin by describing some classic and recent works in the context of textual information for the aforementioned categories.

11.3 Text

Many sensitive and confidential information are stored in texts. Documents containing medical records, financial spreadsheets, business transactions, credit card records and customer information are among the most cited that require privacy protections. One of the central needs of text processing systems is search.

For our discussions, we will focus on conjunctive keyword search, which deals with queries for multiple keywords linked with an AND relationship. While English is assumed in our discussions, the techniques can easily be extended to other natural languages.

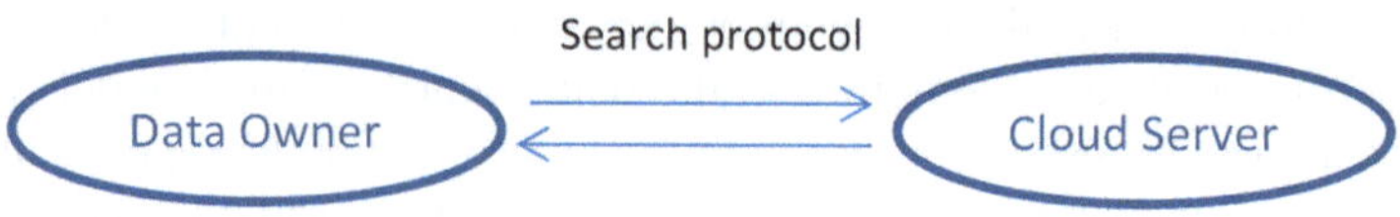

Fig. 11.1 Private/private search scheme

11.3.1 Private/Private Search Scheme: Cloud Document Storage

The simplest scenario is the case where the data owner uploads data to a third party server and wishes to selectively access the data while hiding the content from the server. A typical example would be a student uploading his assignments and notes to a cloud storage host such as DropBox so that he may easily access the documents at home or at school. Another example would be an executive uploading its merger and acquisition records to their cloud hosted systems. In both cases, the data owner is to be the only person searching and generating the encrypted data (Fig. 11.1).

11.3.1.1 Encrypted Indexes

Indexing has been one of the most efficient approaches to search over data. The technique can also be extended to encrypted data.

An Index works by first parsing a data set for keywords and then generating a table that maps the keywords to the data. Consider a document set with three books with the following keywords:

Book A	'Horror', 'Fiction'
Book B	'World War', 'Biography'
Book C	'World War', 'Pandemic', 'Fiction'

Parsing the document set would result in the following index:

'Horror'	A
'Fiction'	A,C
'World War'	B,C
'Biography'	B
'Pandemic'	C

Extending the approach to encrypted data consists simply of hashing and encrypting the keys and entries in a manner that is consistent with the index structure. The data set itself is symmetrically encrypted using a separate secret key.

H_k('Horror')	$E_k(A)$
H_k('Fiction')	$E_k(A, C)$
H_k('World War')	$E_k(B, C)$
H_k('Biography')	$E_k(B)$
H_k('Pandemic')	$E_k(C)$

Suppose a user wishes to upload a document collection, $D = \{D_1, D_2, \ldots, D_n\}$. It is first parsed for a list of keywords, kw_j, which may include document content or meta-data such as date and department. An index is generated mapping keywords to documents such that $I(kw_j) = \{d_a, d_b, \ldots, d_n\}$, where $d_i = 1$ if kw_j is a keyword for the ith document and $d_i = 0$ otherwise. The index is then encrypted and uploaded to the cloud server:

$$I(H_K(kw_j)) = \{E_K(d_a, d_b, \ldots, d_n)\}, \tag{11.1}$$

where $H_K()$ is a cryptographic hash function and $E_k()$ is a symmetric encryption algorithm such as AES. Briefly, cryptographic hash functions are mapping $H(x)$: $C \to D$, where $x \in C$, $|C| \geq |D|$ and where it is computationally infeasible to determine any information about x given $H(x)$. Common cryptographic hash functions include SHA-2, SHA-3 and BLAKE.

For the discussed example, the encrypted index would be

H_k('Horror')	$E_k(100)$
H_k('Fiction')	$E_k(101)$
H_k('World War')	$E_k(011)$
H_k('Biography')	$E_k(010)$
H_k('Pandemic')	$E_k(001)$

To perform a search for a set of keywords $kw' = \{kw_1, kw_2, \ldots, kw_q\}$, the data owner computes their hashes, $H_K(kw')$, using the secret key and sends them to the cloud server. The cloud server looks up entries in the index tables corresponding to $H_K(kw')$ and return the encrypted index entries to the data owner. The data owner then decrypts and finds the intersection of index entries and identifies the matching documents:

$$D_K(I(H_K(kw_1))) \,\&\, D_K(I(H_K(kw_2))) \cdots \,\&\, D_K(I(H_K(kw_q))), \tag{11.2}$$

where & is a bitwise AND operation. Suppose a query was made for all biographies from World war veterans, a search for 'World War' and 'Biography' would require H_k('World War') and H_k('Biography') to be sent to the cloud server. $E_k(011)$ and $E_k(010)$ would respectively be returned to the data owner, who identifies B as the matching results from $011\&010 = 010$.

11.3.1.2 Bloom Filters

While indexes provide a reliable and familiar approach to searching encrypted data, the need for decryption and encryption during search can be computationally expensive for certain applications. As an alternative, Bloom filters offer similar level of performance without the need for decryption and requires only one round of communication, but, unlike indexing, results can contain false positives. While generally undesirable, false positives can provide some level of privacy protection (Goh 2003).

Bloom filters are space-efficient probabilistic data structure used to test whether an element is a member of a set. A Bloom filter contains m bits and μ hash functions, $H_i(x)$, are used to map elements to the m-bits in the filter. All bits in the filter are initially set to zeros. To add an element, a, to the filter, we compute $H_i(a)$ for $i = 1$ to μ, and set the corresponding positions in the filter to 1. For example, for $\mu = 2$ and $m = 5$, to add 'Happy' to the filter, we compute H_1('Happy')= 1 and H_2('Happy')= 4. Setting the position 1 and 4, the Bloom filter becomes $1, 0, 0, 1, 0$. To test for membership of an element, b, in a sample Bloom filter, we compute $H_i(b)$ for $i = 1$ to μ, the element is determined to be a member if all corresponding positions of the sample Bloom filter is set to 1. For example, 'Happy' would be a member of the Bloom filter, $1, 1, 0, 1, 1$.

While Bloom filters have no false-negatives, it can falsely identify an element as member of a set. Given μ hash functions, n items inserted and m bits used in the filter, the probability of false positives is approximately $p = (1 - e^{-\mu n/m})^{\mu}$.

Applying Bloom filters for search consists of viewing the keywords associated with a document as a set and individual keywords as its members. Using the same example as in previous section, Book A would need to add 'Horror' and 'Fiction' to its filter. Suppose $\mu = 2$, $m = 5$, H_1('Horror')= 1, H_2('Horror')= 4, H_1('Fiction')= 2 and H_2('Fiction')= 4, Book A's keyword filter would be $1, 1, 0, 1, 0$. Proceeding similarly for the remaining documents yield the following Bloom filters analogous to the index table in previous section:

Book A	11010
Book B	01101
Book C	11011

To search for 'World War' and 'Biography', we would construct a query filter where H_1('World War'), H_2('World War'), H_1('Biography') and H_2('Biography') are set and send it to the server. Suppose, the query filter is 01101, the server identifies all filters with the 2nd, 3rd and 5th bits set and returns Book B as the result.

Using Bloom filters for encrypted data proceeds in the same manner except members of filters consist of encrypted keywords. That is, to add 'Happy' to a filter, we first compute its keyed cryptographic hash, H_k('Happy'). Then, we hash the result and set the filter bits as before using $H_1(H_k(\text{'Happy'}))$ and $H_2(H_k(\text{'Happy'}))$. To perform a search, we construct a query using the cryptographic hash of keywords under search as members. Since the cloud server does not have access to k, it cannot perform searches without data owner's authorization.

Note that if the file names also require privacy, a small lookup table matching numerical identifiers to file names can be stored privately by the data owner. The matching numerical identifiers can then be used in place of file names on the cloud server. A sample file name to identifier table is as follows:

Book A	83546
Book B	15378
Book C	43879

When compared to the encrypted indexes approach, the use of Bloom filters will generally lead to a much smaller storage requirement at the cost of having false positives.

11.3.2 *Private/Public Search Scheme: Cloud Document Storage Extended*

Suppose you wish to allow certain users to search your documents (Fig. 11.2), the previous solutions can be extended to achieve this goal .

Consider the encrypted indexes based search scheme, a user wishing to search data owner's documents can simply send the queried keywords to the data owner. The data owner computes the encrypted keywords and either forward them to the server or return them to the user. The server then processes the query as in the private scenario and return the results to the user. Figure 11.3 illustrates the technique.

The Bloom filter based scheme can also be extended by having the data owner process all user query requests prior to sending the query filter to the server. Note that the encrypted keyword set or the query filter generated from encrypted keywords constitute a trapdoor, i.e. data that can be used to search the data set but reveal no information on the keywords. This hides the keywords being searched for from the cloud server.

Fig. 11.2 Private/public search scheme

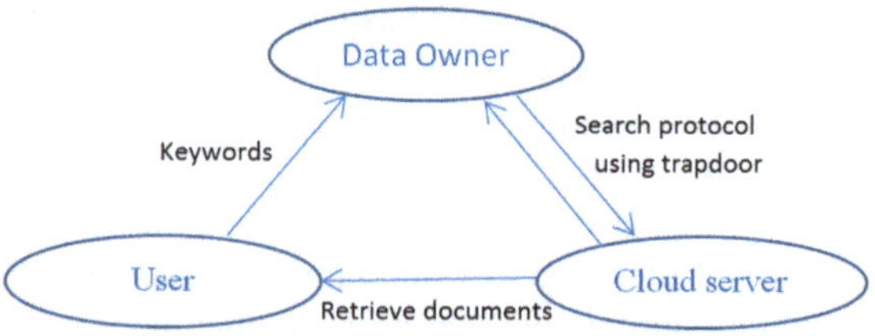

Fig. 11.3 Extension of an encrypted private storage to allow third party searches

Fig. 11.4 Public/private search scheme

One advantage of this setup is that it allows an access control mechanism to be in place where the data owner can authorize and revoke user access to the data set. Consider a multi-national corporation with offices in many nations, but needs to share documents between its employees. The corporation's master server(s) may be considered data owners and its employees are users. Depending on their roles, they may not be able to search for certain keywords. The simple extension discussed in this section would allow the data owner to process all query requests as opposed to a non-interactive model.

11.3.3 Public/Private Search Scheme: Email Filtering System

At a time when email was the primary method of communication online, Boneh et al. (2004) proposed a system that would allow emails to be securely stored on an untrusted email server while allowing selective retrieval of messages based on keywords. In this setting, the users are entities sending emails to the data owner and the cloud server is the untrusted email server. Since there are many users, we have public contribution of encrypted data and private search by the data owner (Fig. 11.4).

The usefulness of the system, shown in Fig. 11.5, is demonstrated in a scenario where certain emails sent by various people may be urgent and required immediate attention from the recipient. Hence, rather than waiting for a email retrieval request, the recipient may be immediately alerted to the urgent matter, all while maintaining the secrecy of the email contents.

The construction is based on public key encryption, which facilitates the key management in the considered scenario, where a single public key may be used by all users wishing to encrypt and send emails to the recipient. In particular, the authors noted that the scenario implies that an Identity-Based Encryption is required.

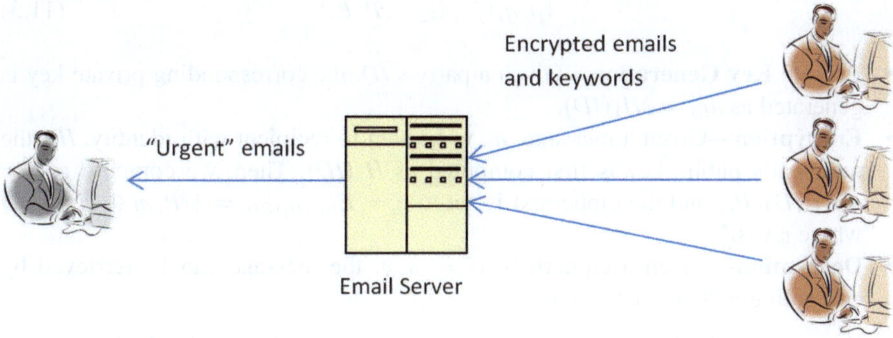

Fig. 11.5 Email filtering system

11.3.3.1 Identity-Based Encryption (IBE)

Identity-Based Encryption schemes allow any arbitrary strings to be used to generate a public key. With traditional public key encryption, different parties must go to trusted certificate authorities to obtain the public key of the intended recipient to encrypt a message. The main advantage of IBE is that parties can effectively forego this process by using the recipient's identity, such as his email address, as the public key to encrypt messages. To decrypt the messages, the recipient must obtain the corresponding private key from a trusted authority. Note that the trusted authority in IBE is a key escrow, that is, it can generate private keys for any users in the system and must be considered highly trusted. It is interesting to note, however, that in a system with a finite number of users, the key escrow can be removed once all users have obtained their private keys. Furthermore, the ability to generate public keys from arbitrary strings also enable the use of attributes in public key, where access to private key may depend upon. For example, emails sent to support@abc.com may be encrypted using 'support@abc.com,lvl=2,Exp=09Oct2016' to signify that the email is valid only for level 2 support staff and until October 9th. If the party requesting the private key fails to meet either condition, the private key would not be issued.

The most efficient IBE schemes today are based on bilinear pairings over elliptic curves (Boneh and Franklin 2001). Due to its importance in the literature as the basis of many solutions on searchable encryption, we'll include the basic scheme here to illustrate the operations. Interested reader is encouraged to refer to Boneh and Franklin (2001) for detailed description.

An Identity-Based Encryption based on bilinear pairings, generally Weil or Tate pairings, consists of the following four algorithms:

- **Setup**—Select two large primes, p and q, two groups, $\mathbb{G}_1$ and $\mathbb{G}_2$ of order q and a generator, $P \in \mathbb{G}_1$. Two cryptographic hash functions, $H(a) : \{0, 1\}^* \rightarrow \mathbb{G}_1$ and $H(b) : \mathbb{G}_2 \rightarrow \{0, 1\}^*$ and a bilinear map $e : \mathbb{G}_1 \times \mathbb{G}_1 \rightarrow \mathbb{G}_2$ are also selected. A master secret $s \in \mathbb{Z}_q$ is held by the authority responsible for generating private keys. The public key, P_s is set to sP. The public parameters of the scheme are:

$$\{p, q, \mathbb{G}_1, \mathbb{G}_2, e, P, P_s\}. \tag{11.3}$$

- **Private Key Generation**—Given a party's *ID*, the corresponding private key is generated as $d_{ID} = sH_1(ID)$.
- **Encryption**—Given a message, m, to be sent to recipient with identity, *ID*, the recipient's public key is first computed as $H_1(ID)$. Then, we compute $g_{ID} = e(H_1(ID), P_s)$ and the ciphertext is set to $c = E_{IBE,H_1(ID)} = \{rP, m \oplus H_2(g_{ID}^r)\}$ where $r \in \mathbb{Z}_q^*$.
- **Decryption**—Given a ciphertext, $c = u, v$, the message can be retrieved by computing $v \oplus H_2(e(d_{ID}, u))$.

Setup and **Private Key Generation** are generally performed by the key escrow authority. **Encryption** is performed by message senders and **Decryption** is performed by recipients. IBE is a probabilistic encryption algorithm. The use of randomness allows different encryption of the same plaintext to result in different ciphertexts. The security of IBE is based on the discrete log problem in elliptic curves and the Bilinear Diffie-Hellman Assumption. The former states that, given P and sP, it is computationally infeasible to determine s. The latter states that, given P, sP, it is computationally infeasible to determine $e(P, P)^s$.

11.3.3.2 An IBE-Based Secure Email Filtering System

We first provide a high level description of the email filtering system, which also involves four algorithms:

- **KeyGen**—Generates public and private key, A_{pub} and A_{priv}.
- **PEKS**(A_{pub},W)—Computes a searchable encryption of the keyword, W, for the recipient with public key, A_{pub}.
- **Trapdoor**(A_{priv},W)—Computes a trapdoor for the keyword, W, for the recipient with private key, A_{priv}.
- **Test**(A_{pub}, S, T_W)—Tests if the keywords used to generate the searchable encryption, $S = PEKS(A_{pub}, W')$, and the trapdoor, $T_W = Trapdoor(A_{priv}, W)$ match.

Suppose a user wishes to filter for the keyword 'urgent', it generates the public and private keys using **KeyGen**. Then, to allow filtering for the keyword, the user computes **Trapdoor**(A_{priv}, **'urgent'**), and place the result on the mail server. Emails sent to the user would contain the encrypted email and a series of keywords encrypted using the recipient's public key, $\{E(email)\|PEKS(A_{pub}, W_1)\| PEKS(A_{pub}, W_2)\ldots\}$. Suppose W_2 = 'urgent', the mail server would first compute **Test**(A_{pub}, $PEKS(A_{pub}, W_1)$, $T_{\text{'urgent'}}$) to find that $W_1 \neq$ 'urgent'. Then, **Test**(A_{pub}, $PEKS(A_{pub}, W_2)$, $T_{\text{'urgent'}}$) would reveal that W_2 = 'urgent' and that the email is urgent while protecting the content of the email and the non-matched keyword W_1. One of the advantages of this approach is that it's non-interactive. Once the recipient has generated the trapdoor for the keyword filters, the incoming emails may be filtered even if the recipient is offline.

To implement the filtering system using IBE, the various parameters are chosen as follows:

- $A_{priv} = a$, where $a \in \mathbb{Z}_q^*$ is randomly chosen, and $A_{pub} = \{P, aP\}$
- **PEKS**(A_{pub},W) $= \{rP, H_2(e(H_1(W), r(aP)))\}$, where $r \in \mathbb{Z}_q^*$ is randomly chosen
- **Trapdoor**(A_{priv},W) $= T_W = aH_1(W)$
- **Test**(A_{pub}, S, T_W) = 'Match' if $H_2(e(T_W, A)) = B$, where $S = \{A, B\}$

As seen, the recipient acts as the key escrow authority using the master secret, a, to generate private keys for identities that are keywords he wish the email server to filter for. The email server is given these private keys. Any users wishing to send emails to the recipient can encrypt using the public keys assigned to the keywords. **PEKS**(A_{pub},W) is equivalent to an IBE encryption of the message, $m = 0$. Given the private keys for the specified keywords, the email server can decrypt only *PEKS*() corresponding those keywords, allowing matching. Since the content of the email is encrypted separately, the security of the scheme is achieved.

11.3.4 Public/Public Search Scheme: Delegated Investigation of Secured Audit Logs

Waters et al. (2004) considered a distributed system where multiple servers, belonging to the same organization, are generating audit logs while in operation (Fig. 11.7). The servers are off-site, e.g. in the cloud, and raise privacy and security concerns. Third parties, such as a consulting firm's investigators, wishing to access the audit log may do so but only records that are relevant to their investigation. Hence, we have public generation of encrypted data, i.e. the audit logs, by the off-site servers destined for the data owner, which is the organization. We also have public search of the encrypted data by the investigators, the users. Figure 11.6 illustrates the scenario.

Two solutions, one based on symmetric encryption and one based on public key encryption, are described, combining some of the techniques from previous sections.

Fig. 11.6 Public/public search scheme

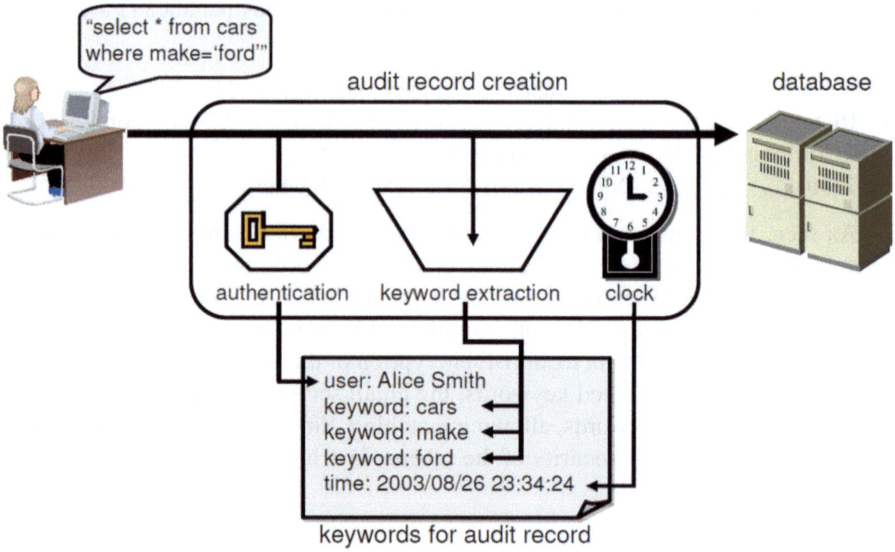

Fig. 11.7 Audit records, with keywords and meta-data such as user and time (Waters et al. 2004)

11.3.4.1 Symmetric Scheme

The symmetric scheme provides an efficient and practical solution for searching and delegating audit log queries. Symmetric encryption has well established security properties and is generally very efficient. In addition, cryptographic hash functions provide an equally efficient tool for one-way mapping of data.

Suppose there are n servers generating audit logs, each server holds a secret key, S_i and encrypts its log entries, m, using symmetric encryption algorithm, $E_k(m)$, where k is randomly generated for each entry. For each keyword, w_i, the server computes

$$a_i = H_S(w_i), b_i = H_{a_i}(r), c_i = b_i \oplus (K|CRC(K)), \tag{11.4}$$

where $H_{k'}()$ is a keyed cryptographic hash functions with the key k'. That is, we compute the hash of the keyword using the server's secret key. Then, the result is used as the key to hash a random value, r. The final result is XOR'ed with the symmetric encryption key used on the log entry. $CRC(K)$ provides a mean to verify if the decrypted symmetric key is correct, i.e. the keywords match. The encrypted log entry is stored as:

$$\{E_K(m), r, c_1 \dots c_n\}. \tag{11.5}$$

To delegate log investigations, a search capability ticket, d_w, must be issued, as shown in Fig. 11.8. Suppose all logs of the class 'Security' is required, the organization computes $d_{w,i} = H_{S_i}(\text{'}Security\text{'})$ for each audit log server and gives

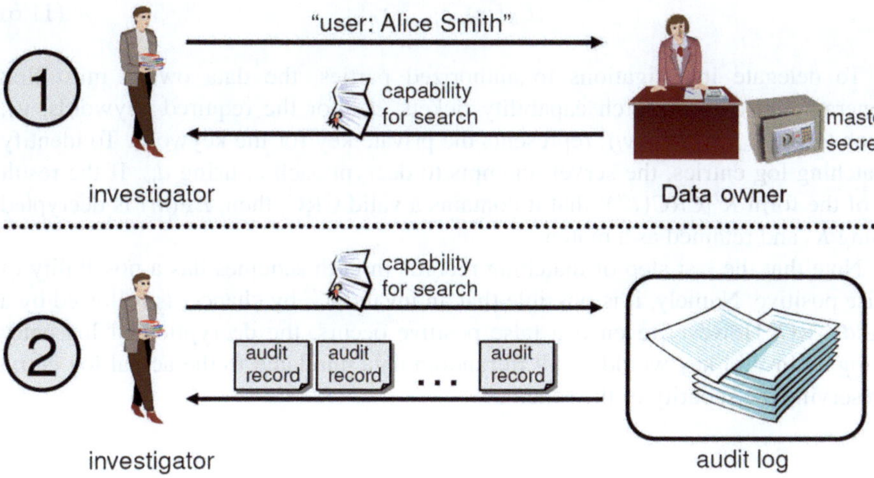

Fig. 11.8 Delegation of log auditing to investigators through search capability tickets (Waters et al. 2004)

them to the investigator. To obtain the log entries with the specified keyword at server i, the server computes $p = H_{d_{w,i}}(r)$ where r is the random value stored with each encrypted entry. Then, it computes $p \oplus c_i$ for each c_i listed. If the result of the form $K'|H$ is such that $H = CRC(K')$, a match is found and the encrypted log entry, $E_K(m)$, is decrypted using K'.

Despite the scheme's efficiency, it can be problematic in the event of a compromise of a server secret. Since each individual log server (encrypted data generator) maintain its own secret key, S_J, to provide keyword search capability for any keywords, compromise of any server's secret would allow an attacker to gain search capability of any keywords he chooses for that server.

11.3.4.2 Asymmetric Scheme

The asymmetric solution addresses this issue by using IBE to provide search capability similar to Sect. 11.3.3. In addition to simplifying key management, asymmetric encryption facilitates the protection of private keys by reducing the number of parties that must learn them. However, asymmetric encryption are also computationally more expensive than symmetric encryption.

Recall from Sect. 11.3.3.1, a master secret, s, is held by the trusted private key generating authority in IBE. In our scenario, this would be the organization, i.e. the data owner. To encrypt a log entry, a server chooses a random secret symmetric encryption key, k, and encrypts the entry to produce, $E_k(m)$. For each keyword, w_i, the server computes the IBE encryption of $K|CRC(K)$ using $H_1(w_i)$ as the public key to produce $c_i = E_{IBE,H_1(w_i)}(m = K|CRC(K))$. The encrypted log entry is stored as:

$$\{E_K(m), c_1 \ldots c_n\}. \tag{11.6}$$

To delegate investigations to authorized parties, the data owner must first generate and assign search capability tickets, d_w, for the required keywords, w_i. Each ticket, $d_w = sH_1(w_i)$, represents the private key for the keyword. To identify matching log entries, the server attempts to decrypt each c_i using d_w. If the result is of the form $K'|CRC(K')$, that it contains a valid CRC, then, $E_K(m)$ is decrypted using K' and retained as a match.

Note that the last step of matching records in both schemes has a possibility of false positive. Namely, it is possible that an invalid K', by chance, is followed by a valid CRC. However, even if a false positive occurs, the decryption of log entry using an invalid key would result in random data unrelated to the actual log entry, preserving the security of the schemes.

11.4 Range Queries

Suppose we wish to retrieve audit logs during a time period in which we believe an online attack had occurred, that is we wish to do "Select * from SecLogs where time > 2014/04/10 and time < 2014/04/13" where the time element may be as in Fig. 11.7. It would have been impractically inefficient to do a keyword search for every single possible time value in the interval using the scheme in Sect. 11.3.4. A recent proposal, however, generated significant interest by showing that it is possible to preserve the ordering after encryption, thus enabling the possibility of performing numerical value comparisons of plaintexts in encrypted form.

11.4.1 Order Preserving Encryption

Order preserving encryption (OPE) (Agrawal et al. 2004) centers on the idea that, given a plaintext space $\mathbb{P}$ and a ciphertext space $\mathbb{C}$,

$$p_1 > p_2 \implies c_1 > c_2 \tag{11.7}$$

where

$$\forall p_1, p_2 \in \mathbb{P} \text{ and } c_1 = E(p_1), c_2 = E(p_2) \in \mathbb{C}. \tag{11.8}$$

The encryption scheme does not rely on any mathematical hard problem, but instead generates a random mapping from plaintext to ciphertext through the use of a pseudo-random function. Boldyreva et al. (2009) later improves the efficiency by providing a technique to generate the encryption mappings on-the-fly without having to reproduce the entire mapping from the key each time.

The security guarantee is that an attacker would not learn the plaintexts p_i besides the order. While there has been significant interest and adaptation of order preserving encryption, it should be noted that the scheme has been shown to leak information, up to half the bits of plaintexts, under certain conditions (Boldyreva et al. 2011).

Adapting OPE to range queries is straightforward. For the example of audit log in Fig. 11.7, encrypt all time elements using OPE and all keywords using the scheme in Sect. 11.3.4. To search for logs on 2014/04/10, we must search for time > 2014/04/09 and time < 2014/04/11. For $A = E(2014/04/09)$ and $B = E(2014/04/11)$ and an encrypted log record, $\{E(Log), E(keywords), E(time)\}$, a match is identified if

$$B > E(time) > A. \tag{11.9}$$

11.5 Media

The growing importance of media cannot be understated. It is estimated that, by 2019, 80% of the world's bandwidth would be consumed by media. Currently, at the end of 2015, the number sits at 70%, with Netflix and Youtube combining for over 50% of data sent over the Internet. Naturally, the need to process videos, audios or images in a secure and privacy-aware manner will be of growing interest in coming years. While studies in encrypted media processing is still at its infancy, there exists some interesting and working solutions. As with texts, investigations began with search.

11.5.1 Keyword Based Media Search

The simplest approach to adapt existing text-based searchable encryption to media is through a meta-data only media search. Consider a set of images, $I = \{I_1, I_2, \ldots, I_n\}$, we first extract a list of keywords for each image. This extraction process can be manual, i.e. a person assigning keywords such as 'Man', 'bird', 'table', 'HighRes', or it may be done through artificial intelligence (AI) and image recognition systems.

Once all images have been assigned keywords, all text-based solutions discussed in Sect. 11.3 apply as is by considering the image set as the document set. This is achievable because the search mechanism for the text-based solutions are all based on extracted keywords and the documents are encrypted separately from the search mechanism, be it an index, Bloom filters or IBE encrypted keywords. The reason for the separation is because searching data content on-the-fly is computationally expensive, even when unencrypted, since each file must be scanned as a whole if no pre-processing was performed. Due to the security guarantees of standard

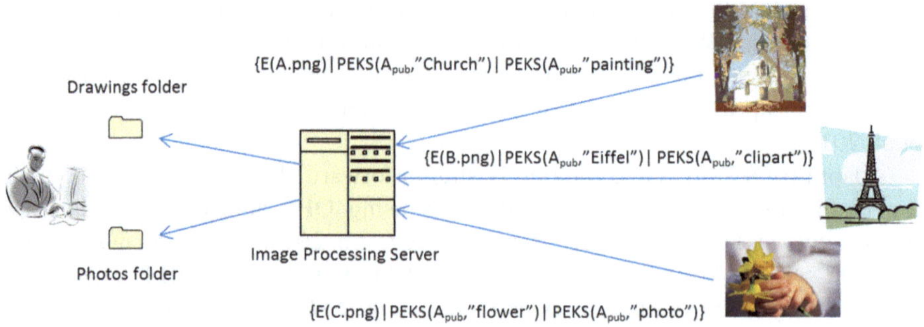

Fig. 11.9 Keywords-based image filtering system based on IBE

symmetric and asymmetric encryption algorithm, processing of encrypted data is often impossible or very computationally intensive.

Figure 11.9 shows a media filtering system based on the Email filtering system in Sect. 11.3.3. The system depicts a image bank that receives images from various sources. Occasionally, some images may contain sensitive or personally identifiable information that requires blurring before being placed in the image bank. The data owner would like to maintain privacy and prevent the host server from learning the image contents. However, due to the quantity of images received, the data owner would like an automatic sorting system that would facilitate post-processing upon decryption. The depicted system achieves this by requiring each user encrypts their images using a symmetric encryption algorithm and encrypts a series of keywords using IBE to describe the image.

11.5.2 Content Based Media Search

Searching for images based on its content is a far more challenging task. Generally, images are first processed to produce feature vectors, analogous to keywords in documents, using algorithms such as SIFT (Lowe 1999). Then, the Euclidean distance between feature vectors provides a mean to measure the similarity between different images. However, doing so in encrypted domain is generally impossible with standard encryption algorithms. We'll describe two recently proposed solutions, one based on OPE and one based on homomorphic encryption.

11.5.2.1 Media Search Using OPE

Nister and Stewenius (2006) introduced a highly cited technique for searching through a database of images. The idea was that feature vectors extracted using popular extraction algorithms can be arranged into clusters and each cluster can be thought of as a visual word. This opens up the possibility of using well studied text-based solutions for content-based image search. In particular, it was shown that using an vocabulary tree can efficiently process searches in extremely large database.

In Lu et al. (2009), Lu described an adaptation of the technique to enable secure searches on encrypted images. The solution considers the visual words as keywords for images and builds a visual word to image index. Unlike with text, the goal of the index is not provide a simple mapping to all images containing a visual word, but rather to enable comparison between a query image and the images in the database.

Suppose we are given a database of n images, feature vectors are first extracted and using k-means clustering to separate into sets of visual words. Figure 11.10 illustrates the process. Then, the feature cluster frequencies are counted to produce an index mapping visual words to image along with the word's frequency, as shown in Fig. 11.11.

All visual word frequency values are encrypted using OPE, $E(w_i)$. Then, borrowing the notion of inverse document frequency (IDF), each encrypted value is further scaled by

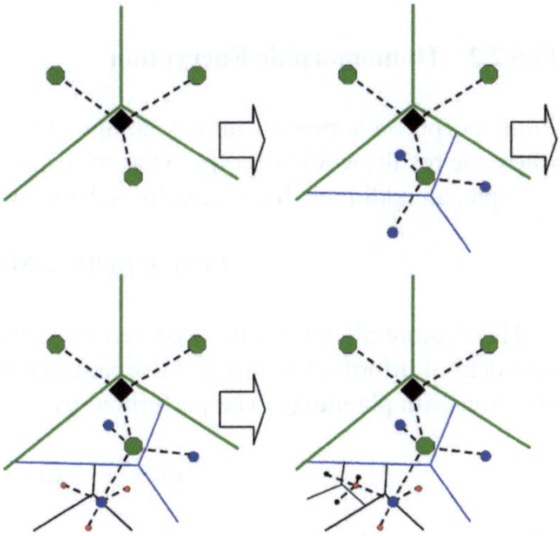

Fig. 11.10 Vocabulary tree generation using k=3 means clustering

Fig. 11.11 Visual word to image index with frequencies

Word ID	i			
Image ID	I_1	I_2	$\cdots$	I_{N_i}
Word frequency	w_1	w_2	$\cdots$	w_{N_i}

$$IDF = log(\frac{M}{N_i}), \tag{11.10}$$

where M is the total number of images in the database and N_i is the number of images containing the visual word, i. The scaling factor, IDF, skews the frequency values such that less common and more distinctive visual words carry more weight in determining similarity matches. The resulting index is stored on the cloud server.

The similarity of a query image and an image in the database is determined by the similarity between their visual word frequency patterns. Given a visual word frequency list of a query image, we encrypt the values using OPE, $E(Q_1), E(Q_2), \ldots E(Q_V)$, and scale the values by IDF. Note that V represents the total number of visual keywords in the database. To compare against the word frequency of a database image, $E(D_1), E(D_2), \ldots E(D_V)$, the Jaccard similarity is computed:

$$Sim(Q, D) = \frac{E(Q) \cap E(D)}{E(Q) \cup E(D)} = \frac{\sum_{i=1}^{V} min(E(Q_i), E(D_i))}{\sum_{i=1}^{V} max(E(Q_i), E(D_i))} \tag{11.11}$$

Due to the use of OPE, the order of the encrypted frequencies and thus the results of $min()$ and $max()$ functions are preserved. The computed similarity reflects that computed over plaintext. Therefore, the matching can be performed server side. In addition to protecting the frequency values, the use of OPE also protects their distribution, which may reveal information on the image content, by flattening it.

11.5.2.2 Homomorphic Encryption

Homomorphic encryption allows computations to be carried out on ciphertexts, where the results would decrypt to the corresponding computation on plaintexts. For example, an additively homomorphic scheme would have the following property:

$$E(A) + E(B) = E(A + B) \tag{11.12}$$

This feature allows for third parties to perform computations without exposing confidential information. An additive homomorphic scheme may also allow multiplication with plaintexts to be performed by

$$E(m_1)^{m_2} = E(m_1 m_2), \tag{11.13}$$

where m_2 is in plain. Until recently, most homomorphic encryption algorithms are either additive or multiplicative, but not both. Gentry (2009), in 2009, described the first fully homomorphic encryption algorithm which supports both addition and multiplication over ciphertext, opening the door to many applications and a dramatic increase in interest on computing over encrypted data. After many years of improvements, fully homomorphic encryption algorithms can now run in relatively reasonable time. However, its computational cost remains much higher than all popular encryption algorithms, limiting its use in practise.

Recall that content-based image search typically relies on euclidean distances between feature vectors as a measure of similarity. That is to compute

$$\|F_Q - F_D\| = \sqrt{\sum_{i=1}^{n}(F_{Q,i} - F_{D_j,i})^2} \tag{11.14}$$

Since the square root operator applies to all values, it is easy to see that using the squared Euclidean distance is equally valid as a distance measure. Naturally, computing the summation is possible using fully homomorphic encryption although at a high computational cost. To do so, the data owner encrypts the features $F_{D,i}$ for each image D_j and uploads to server. To perform a query for image, Q, the data owner encrypts all features, $F_{Q,i}$, of the query image and uploads to the server. The server computes $dist_j = \sum_{i=1}^{n}(F_{Q,i} - F_{D_j,i})^2$ for all images in the database and returns $dist_j$'s to data owner. The smallest distance values are identified as matches by decrypting the results.

Additive homomorphic encryption, such as Paillier cryptosystem (Paillier 1999), can also be used to query an image in plaintext against an encrypted image database, by exploiting the following property:

$$\sum_{i=1}^{n}(F_{Q,i} - F_{D_j,i})^2 = \sum_{i=1}^{n} F_{Q,i}^2 - 2\sum_{i=1}^{n} F_{Q,i}F_{D_j,i} + \sum_{i=1}^{n} F_{D_j,i}^2 \tag{11.15}$$

Since $F_{Q,i}$ is in plain, the first term is available. With $E(F_{D_j,i})$ as ciphertext, the second term can be computed using Eq. (11.13). The third term can be uploaded with the features during setup. The server can then compute the encrypted distance without interacting with data owner. The encrypted distance must then be sent back to the data owner for decryption. It should be noted, however, that the ability to perform similarity search using images in plain allows the server to infer that the matched encrypted image retrieved is close to the query image presented.

11.6 Other Applications

While conjunctive keyword searches remain the central functionality required for many secure data outsourcing applications, researchers have also investigated more specialized searches and functionalities. Examples include:

- Ranking of search results
- Fuzzy search
- Phrase search

- Similarity search for texts
- Privacy-protected recommender system
- Copyright management
- Signal Processing

The ability to rank search results can increase the relevance and accuracy of matches particularly in large databases. Fuzzy searches offer a more user friendly and intelligent search environment. Phrase search deals with sequential data processing, which may find use in genomics, where individuals' DNA's may post a privacy risk. Even virus detection (Poon and Miri 2016) has been suggested to identify the increasing amount of malware that may remain dormant in cold storage.

Aside from search, the ability to compute over encrypted data presented by homomorphic encryption could lead to secure outsourcing of computation. In addition to enabling a remote server to store and search over encrypted data without revealing their content, it may even manipulate the data in meaningful ways without learning its content. A sample application would be privacy-protected recommender system. Suppose a recommender system has identified that users belonging to certain clusters are interested in certain products, a customer may send his feature sets, which may include previously view products and set preferences, in encrypted form and have the recommendation be computed by the server and sent back in encrypted form without learning what products he had viewed or his preferences.

Homomorphic encryption has also been suggested as a way to protect user privacy while managing digital rights of encrypted media. In particular, the use of watermarks and a simple detection algorithm consisting of the computation of correlation between the watermark and an image can be performed in a similar manner as described in Sect. 11.5.2.2.

11.7 Conclusions

In this chapter, we gave an overview of the most important works in searchable encryption along with some of the more recent works and the relationships between them. Classic keyword search techniques such as encrypted indexes, IBE and Bloom filters continue to form the basis of many works in literature today. Recent years have also seen the development of order preserving encryption and fully homomorphic encryption, which generated significant interest in the research community. While the sample applications discussed in this chapter were described in simple scenarios, it is not difficult to imagine the applications in larger scale. For example, sentimental analysis on private Twitter feeds could well be encrypted with searchable keyword tags. Recommender systems could provide recommendations without revealing users' interests, by using homomorphic encryption.

As the world continues to become more connected and the amount of data generated continues to climb, the need for privacy and security will equally become more important. Much of the difficulties in adapting cryptographic techniques to

Big Data lies in their relatively high computational cost for today's technology. As our computational power increases and algorithms improve, the cryptographic techniques developed today may well form the basis of a secure big data system in the future.

References

Agrawal, R., Kiernan, J., Srikant, R., & Xu, Y. (2004). Order preserving encryption for numeric data. In *Proceedings of the 2004 ACM Sigmod International Conference on Management of Data* (pp. 563–574).

Boldyreva, A., Chenette, N., Lee, Y., & O'Neill, A. (2009). Order-preserving symmetric encryption. In *Proceedings of the 28th Annual International Conference on Advances in Cryptology: The Theory and Applications of Cryptographic Techniques* (pp. 224–241).

Boldyreva, A., Chenette, N., & O'Neill, A. (2011). Order-preserving encryption revisited: improved security analysis and alternative solutions. In *Proceedings of the 31st Annual Conference on Advances in Cryptology* (pp. 578–595).

Boneh, D., & Franklin, M. (2001). Identity-based encryption from the weil pairing. In *Advances in Cryptology - Crypto 2001: 21st Annual International Cryptology Conference, Santa Barbara, California, USA, August 19–23, 2001 Proceedings* (pp. 213–229).

Boneh, D., Crescenzo, G. D., Ostrovsky, R., & Persiano, G. (2004). Public key encryption with keyword search. In *Proceedings of Eurocrypt* (pp. 506–522).

Gentry, C. (2009). Fully homomorphic encryption using ideal lattices. In *Proceedings of the Forty-First Annual ACM Symposium on Theory of Computing* (pp. 169–178).

Goh, E.-J. (2003). Secure indexes. Cryptology ePrint Archive, Report 2003/216.

Gymrek, M., McGuire, A. L., Golan, D., Halperin, E., & Erlich, Y. (2013). Identifying personal genomes by surname inference. *Science, 339*(6117), 321–324.

How hard is it to 'de-anonymize' cellphone data? (n.d.). http://news.mit.edu/2013/how-hard-it-de-anonymize-cellphone-data. Accessed 10 September 2016.

Lowe, D. G. (1999). Object recognition from local scale-invariant features. In *Proceedings of the International Conference on Computer Vision.*

Lu, W., Swaminathan, A., Varna, A. L., & Wu, M. (2009). Enabling search over encrypted multimedia databases. In *Proceedings of SPIE, Media Forensics and Security* (pp. 7254–7318).

Narayanan, A., & Shmatikov, V. (2008). Robust de-anonymization of large sparse data set. In *2008 IEEE Symposium on Security and Privacy* (pp. 111–125).

Nister, D., & Stewenius, H. (2006). Scalable recognition with a vocabulary tree. In *Proceedings of the 2006 IEEE Computer Society Conference on Computer Vision and Pattern Recognition* (pp. 2161–2168).

Paillier, P. (1999). Public-key cryptosystems based on composite degree residuosity classes. *Lecture Notes in Computer Science, 1592*, 223–238.

Poon, H., & Miri, A. (2016). Scanning for viruses on encrypted cloud storage. In *IEEE Conferences on Ubiquitous Intelligence & Computing, Advanced and Trusted Computing, Scalable Computing and Communications, Cloud and Big Data Computing, Internet of People, and Smart World Congress* (pp. 954–959).

Waters, B., Balfanz, D., Durfee, G., & Smetters, D. K. (2004). Building an encrypted and searchable audit log. In *Network and Distributed System Security Symposium* (pp. 215–224).

Chapter 12
Civil Infrastructure Serviceability Evaluation Based on Big Data

Yu Liang, Dalei Wu, Dryver Huston, Guirong Liu, Yaohang Li, Cuilan Gao, and Zhongguo John Ma

12.1 Introduction

12.1.1 Motivations of Big-Data Enabled Structural Health Monitoring

Civil infrastructure, such as bridges and pipelines, is the mainstay of economic growth and sustainable development; however, the safety and economic viability of civil infrastructure is becoming an increasingly critical issue. According to a report of USA's Federal National Bridge Inventory, the average age of the nation's 607,380

Y. Liang (✉) • D. Wu
Department of Computer Science and Engineering, University of Tennessee at Chattanooga, Chattanooga, TN 37403, USA
e-mail: yu-liang@utc.edu; dalei-wu@utc.edu

D. Huston
Department of Mechanical Engineering, University of Vermont, Burlington, VT 05405, USA

G. Liu
Department of Aerospace Engineering & Engineering Mechanics, University of Cincinnati, Cincinnati, OH 45221, USA

Y. Li
Department of Computer Science, Old Dominion University, Norfolk, VA 23529, USA

C. Gao
Department of Mathematics, University of Tennessee at Chattanooga, Chattanooga, TN 37403, USA

Z.J. Ma
Department of Civil and Environmental Engineering, University of Tennessee, Knoxville, TN 37996, USA

S. Srinivasan (ed.), *Guide to Big Data Applications*, Studies in Big Data 26,
DOI 10.1007/978-3-319-53817-4_12

bridges is currently 42 years old. One in nine of those bridges is rated as structurally deficient. According to the report of United States Department of Transportation, the U.S. oil and gas pipeline mileages has reached 158,329 miles till 2014. A new analysis of oil and gas pipeline safety in the United States reveals that nearly 300 incidents have occurred per year since 1986.

On the other hand, a prompt and accurate detection of infrastructure deficiency, such as bridge collapse and pipeline leaking, is extremely challenging. In this work, a multiscale structural health monitoring and measuring system (Catbas et al. 2012; Ye et al. 2012) based on Hadoop Ecosystem (Landset et al. 2015), denoted as MS-SHM-Hadoop for simplicity, is proposed. By integrating sensor technology, advanced wireless network, data-mining based on big-data platform, and structural mechanics modeling and simulation, MS-SHM-Hadoop is equipped with the following functions: (1) real-time sensory data acquisition, integration, and analysis (Liang and Wu 2014, 2016; Liang et al. 2013a,b); (2) quantitative measurement of the deficiency of nation-wide civil infrastructure; (3) identification of civil infrastructure's structural fault and quantitative prediction of their life expectancy according to the long-term surveillance about the dynamics behavior of civil infrastructure.

To monitor and manage civil infrastructure effectively, wireless sensor networks (WSNs) have received extensive attention. A WSN is generally comprised of multiple wireless smart sensor nodes (WSSNs) and a base station which can be a computer server with ample computation and storage resources. Featured with low cost in installation and maintenance and high scalability, the WSSNs have been deployed on the Golden Gate Bridge by UC Berkeley in 2006 (Kim et al. 2007) and recently on Jindo Bridge in Korea through a collaborative research among Korea, US, and Japan (Jang et al. 2010). Researchers have also demonstrated enthusiasm using wireless smart sensors to monitor full-scale civil bridge structures in Lynch et al. (2006) and Pakzad (2008). Full-scale deployment of WSN on real bridge structure is transformative because the employment of wired sensor network still dominates SHM projects. Challenges lay the availability of power supply and mature damage monitoring algorithms.

Heterogeneous and multi-modal data about the structural health of civil infrastructures has been collected. Although there have been some work that adopted data management systems and machine learning techniques for structural monitoring, few platforms have been investigated to integrate full spectrum input data seamlessly. In Sofge (1994) neural network based techniques are used for modeling and analyzing dynamic structural information for recognizing structural defects. In Guo et al. (2014), to avoid the need of a large amount of labeled real-world data as training data, a large amount of unlabeled data are used to train a feature extractor based on the sparse coding algorithm. Features learned from sparse coding are then used to train a neural network classifier to distinguish different statuses of a bridge. The work in Roshandeh et al. (2014) presents a layered big data and a real-time decision-making framework for bridge data management as well as health monitoring. In Nick et al. (2015), both supervised and unsupervised learning techniques for structural health monitoring are investigated by considering acoustic emission signals. A data management civil infrastructure based on NoSQL

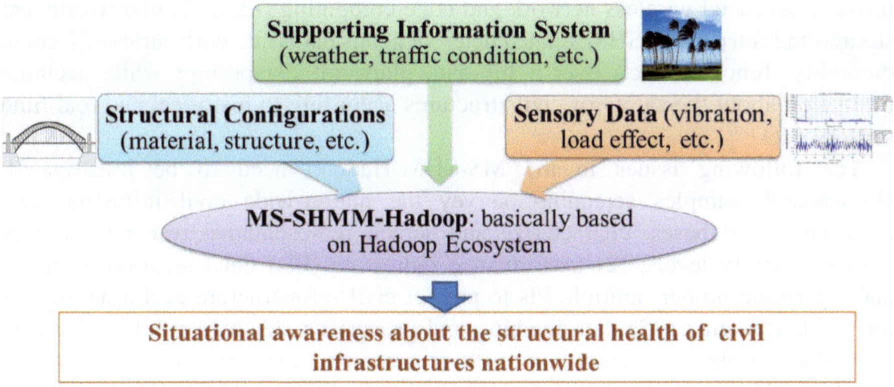

Fig. 12.1 Three major inputs for MS-SHM-Hadoop

database technologies for bridge monitoring applications was proposed in Jeong et al. (2016b). Cloud service platform is also deployed to enhance scalability, flexibility and accessibility of the data management system (Jeong et al. 2016a).

12.1.2 Overview of the Proposed MS-SHM-Hadoop

The three major inputs for MS-SHM-Hadoop is shown in Fig. 12.1. Sensory data includes the cyclic external load and structural response, and surrounding environmental conditions. Supporting information refers to all civil infrastructure related information, such as bridge configuration database (National Bridge Inventory), pipeline safety information (the Pipeline and Hazardous Materials Safety Administration Inventory), transportation status (National Transit Database), and weather conditions (National Climatic Data Center). Structural configurations include the geometric formulation of civil infrastructure and construction material description.

The joint consideration of big-data and sensor-oriented structural health monitoring and measuring is based on the following considerations: (1) Many critical aspects of civil infrastructure performance are not well understood. The reasons for this include the extreme diversity of the civil infrastructure, the widely varying conditions under which civil infrastructure serve, and the lack of reliable data needed to understand performance and loads. Meanwhile, as sensors for civil infrastructure structural health monitoring are increasingly employed across the country, massive information-rich data from different kinds of sensors are acquired and transmitted to the racks of civil infrastructure management administration database. (2) There exists high-degree correlations among civil infrastructure's data, which can be effectively discovered by data mining over big-data platform.

The MS-SHM-Hadoop has the following capabilities: (1) real-time processing and integration of structure-related sensory data derived from heterogeneous sensors through advanced wireless network and edge computing (EC); (2) highly efficient storage and retrieval of SHM-related heterogeneous data (i.e., with varies of format, durability, function, etc.) over a big-data platform; (3) prompt while accurate evaluation about the safety of civil structures according to historical and real-time sensory data.

The following issues in the MS-SHM-Hadoop need to be investigated: (1) research samples screening: survey the nation-wide civil infrastructure's information databases, characterize and screen representative research samples with low safety levels; (2) performance indicators (PIs) determination: evaluate and determine proper multiple PIs to predict civil infrastructure performance in a quantitative manner; (3) data fetching and processing: fetch relevant sensor data from Hadoop platform, according to PIs requirement, and process raw sensor data into load effects and load spectrum (Sohn et al. 2000) through edge computing technology; (4) multi-scale structural dynamic modeling and simulation: based on historical data of sample civil infrastructure, establish finite element (FE) and particle models for global structural analysis and local component fatigue analysis (Zhu et al. 2011); (5) evaluation of the impact of innovative civil infrastructures construction methods on infrastructure performance by instrumenting two new bridges in Tennessee. Civil infrastructure construction, design, and materials have changed over time, and these changes may affect civil infrastructure performance. For example, accelerated bridge construction (ABC) is a new process in bridge construction and may affect bridge performance (He et al. 2012). These two new bridges can also serve as a testing bed for the proposed activities in this project. (6) Civil infrastructure performance evaluation: assess civil infrastructure's performance by PIs of global structure and local critical components (Frangopol et al. 2008).

The implementation of MS-SHM-Hadoop involves the following cutting-edge technologies: (1) machine learning including classification, clustering, regression, and predictive analysis, based on general civil infrastructure information (e.g., age, maintenance management, and weather conditions, etc.), sensory data, and structural configurations (e.g., infrastructure material, length, etc.), Bayesian network and stochastic analysis; (2) structural dynamic analysis; (3) signal processing for external load and structure response; (4) multi-scale strategy ranging from nationwide civil infrastructure survey to specific components' structural reliability analysis; and (5) Hadoop ecosystem deployed in edge computing server to achieve high-scalability including acquisition, fusion, normalization of heterogeneous sensory data, and highly scalable and robust data analysis and information query.

12.1.3 Organization of This Chapter

The remainder of this chapter is organized as follows: Sect. 12.2 describes the architecture and flowchart of MS-SHM-Hadoop; Sect. 12.3 introduces the acquisition of sensory data and the integration of structure-related data; Sect. 12.4 presents nationwide bridge survey; Sect. 12.5 investigates the global structural integrity of civil infrastructure according to structural vibration; Sect. 12.6 investigates the reliability analysis about localized critical component; Sect. 12.7 employs Bayesian network to investigate civil infrastructure's global integrity according to components' reliability, which is obtained in Sect. 12.6; and Sect. 12.8 concludes the paper.

12.2 Implementation Framework About MS-SHM-Hadoop

12.2.1 Infrastructure of MS-SHM-Hadoop

Figure 12.2 shows the infrastructure of MS-SHM-Hadoop, which consists of the following three modules: the sensor grid (SG) module, the data processing and management (DPM) module based on Hadoop platform, and the reliability evaluation (RE) module based on structural dynamics modeling and simulation. A more detailed description about each module is given below.

The sensor grid (SG) module mainly acquires, pre-processes the raw sensory data and then transmits it to the data processing and management (DPM) module. Mobile computing gateway (denoted as MC for simplicity) coordinates with each other through a wireless network. SensorCtrl is the control-module that tunes the sensor's configurations for better observation of the area-of-interest, which is located through structural analysis (RE module).

The Hadoop-enabled data processing and management (DPM) module mainly integrates, transforms, classifies, and stores the data with high fault-tolerance

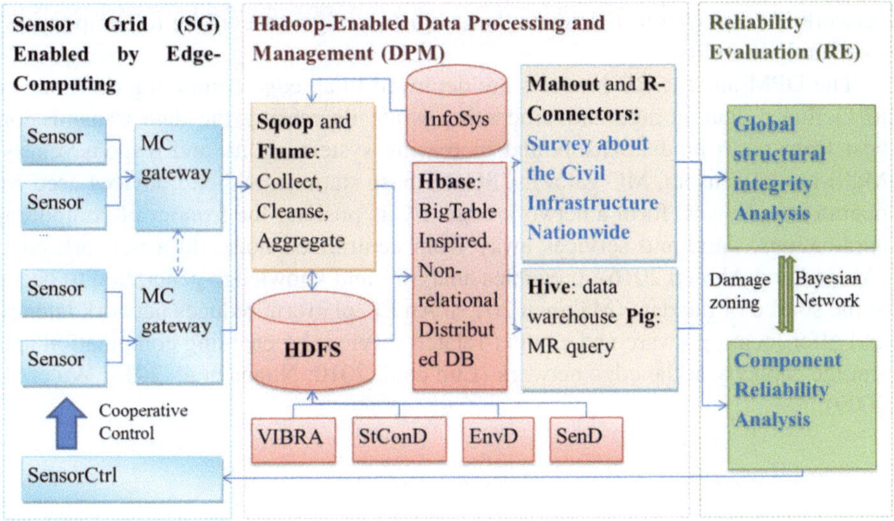

Fig. 12.2 Infrastructure of MS-SHM-Hadoop

and scalability. Based on Hadoop Distributed File System (HDFS) and MapReduce high-performance parallel data processing paradigm (Landset et al. 2015), R-Connector and Mahout (Landset et al. 2015) provide powerful statistics and machine learning capability; Inspired by big-table techniques (including row-key, column-key, and time-stamp), HBase (Landset et al. 2015) efficiently accesses large-scale heterogeneous real-time or historical data; Flume (Landset et al. 2015) collects, aggregates, and moves large amounts of streaming data (i.e., the sensory data about civil infrastructure status) into Hadoop from variety of sources; Hive (Landset et al. 2015) provides a data warehouse infrastructure to manage all the data corresponding to civil infrastructure's serviceability; Pig (Landset et al. 2015) offers MapReduce-enabled query and processing; Sqoop (Landset et al. 2015) supports the ingestion of log data, which is related to civil infrastructure design and operation such as civil infrastructure configuration (e.g., National Bridge Inventory, the Pipeline and Hazardous Materials Safety Administration Inventory), transportation status (e.g., National Transit Database), and weather conditions (e.g., NOAA's National Climatic Data Center). In this work, InfoSys manages the external log data. VIBRA stores the cyclic external force load (or vibration signals), which is applied by the wind or vehicles, and the corresponding structural response. StConD component stores the structure configuration (i.e., geometry configuration and mesh) of civil structure. EnvD (Environmental data component) keeps circumstance parameters such as temperature, moisture, etc. SenD is a database component that keeps the configurations (e.g., location, brand, mechanism, maintenance schedule, etc.) of sensors attached to the civil infrastructure.

Based on structural dynamics theory and signal processing techniques, the RE module mainly uses historical or real-time sensory data to identify the global (or infrastructure-wise) or component-wise structural faults. In addition, Bayesian network is employed to formulate the integrity analysis according to components' structural reliability.

The DPM and the RE modules are deployed in an edge computing (EC) server to facilitate data storage and processing while decreasing the data transmission cost from multi-modal sensors and increasing system agility and responsiveness. Multi-modal sensors, MC gateway, 3G/4G base stations, and end devices used by operators/users will form a network edge. EC is pushing the frontier of computing applications, data, and services away from centralized nodes to a network edge (Ahmed and Ahmed 2016). It enables analytics and knowledge generation to occur at the source of the data (Mäkinen 2015). An EC platform reduces network latency and offer location-aware and context-related services by enabling computation and storage capacity at the edge network (Luo et al. 2010; Nunna et al. 2015; Wu et al. 2009).

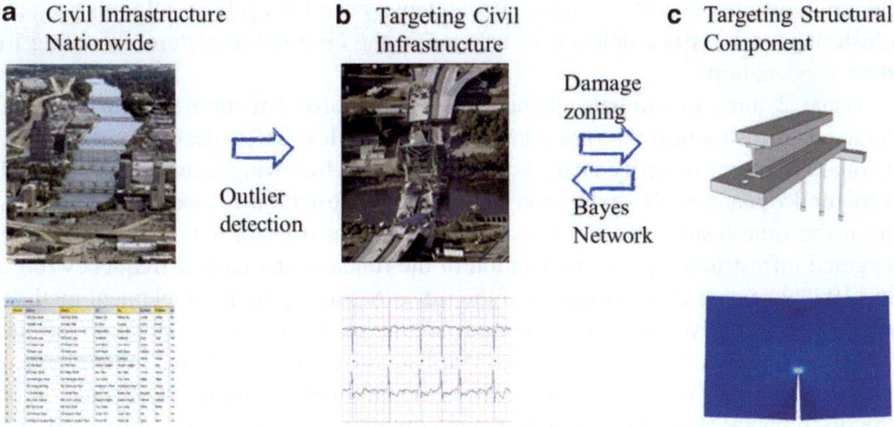

Fig. 12.3 Flowchart of multiscale structural health evaluation: (**a**) nationwide civil infrastructure survey; (**b**) global structural integrity analysis; (**c**) localized structural component reliability analysis (The bridge pictures are derived from mnpoliticalroundtable.com)

12.2.2 Flowchart of the MS-SHM-Hadoop

Figure 12.3 shows the systematic approach of the implementation of MS-SHM-Hadoop system. Based on the acquired sensory data and civil infrastructure related log data, a multiscale structural health monitoring and measurement consist of the following stages: (Stage 1) nationwide civil infrastructure database survey using machine learning techniques; (Stage 2) global structural integrity analysis using signal processing, and structure dynamics; and (Stage 3) localized structural component reliability analysis using stochastic methods, or multiscale modeling and simulation.

With reference to Fig. 12.2, it can be observed that: Stage 1 is implemented in the Sensor Grid (SG) module and partially in the Data Processing and Management (DPM) module; Stage 2 is implemented in the DPM module; and Stage 3 is implemented in the Structure Evaluation module.

By surveying the nation-wide civil infrastructure's status on a big-data platform or other information systems, Stage 1 aims to obtain a preliminary characterization of the safety level of major bridges in United States from the National Bridge Inventory (NBI) database and pipelines from the Pipeline and Hazardous Materials Safety Administration Inventory. NBI involves dimensions, location, type, design criteria, traffic, structural and functional condition, and lots of other information. A general screening and prioritization analysis based on weighting/loading consideration can be performed to determine relatively low safety level aging civil infrastructure. The serviceability of a civil infrastructure is qualitatively determined by a number of overall factors, such as, the year-of-build, structure configuration, construction material, weather conditions, traffic flow intensity and life cycle cost. In this project, clustering analysis is employed to categorize the civil infrastructure according to their serviceability.

Stage 2 aims to evaluate quantitatively the global structural health status of targeted infrastructure that are characterized with low safety level from Stage 1. Global structural integrity analysis consists of the following intensive data-based structural dynamics: (1) extraction the measured structural resonance frequencies from the time-history sensory data via Fast Fourier transformation (FFT) for the targeted infrastructure; (2) computation of the fundamental natural frequency (e.g., the 10 lowest natural frequencies) of the infrastructure using finite-element method (FEM), which gives the upper bound of the solution; (3) computation of the fundamental natural frequency of the infrastructure using node-based finite-element method (NS-FEM), which gives the lower bound of the solution; (4) evaluation of the discrepancy about fundamental natural frequencies between the measured and computed ones; (5) establishment of the relationship between the discrepancy of the fundamental natural frequencies and the healthy status of the infrastructure; and (6) based on the distribution of discrepancy obtained using sufficient large number of sensors deployed over the span of infrastructure, the possible zones with heavy damages and degradations are identified.

Following the time-domain or frequency-domain algorithm, Stage 3 aims to obtain a precise description about the serviceability of local components in the heavily damaged zones identified in Stage 2. This is to provide the remaining service life of the infrastructure, as well as prepare possible strategies for life-prolongation. With the load effects from sensors and computational value from FE analysis, structural performance indicators can be calculated respectively in local scale and global scale. Proper assessment theory, such as neuro-fuzzy hybrid method (Kawamura and Miyamoto 2003) or DER&U method (Zhao and Chen 2001), can be evaluated and utilized. Finally the structural performance evaluation results can be updated to the management system of structural administration to provide professional support for decision making (Catbas et al. 2008).

12.3 Acquisition of Sensory Data and Integration of Structure-Related Data

Table 12.1 lists representative sensors in the proposed system needed to acquire the following information: external load; structure's response to external load; and environmental circumstance parameters. To provide a localized monitoring data analysis we adopt a mobile computing (MC) gateway that collects the raw sensory data, pre-processes and sends them to the DPM module via a wired or wireless network. The MC provides real-time analysis on the situation at a specific location on the infrastructure. The MC is carried by a robot or unmanned aerial vehicles (UAVs) to collect the acquired data from the sensors covering a specified area on

Table 12.1 List of sensors

Monitoring data category	Sensor type	Data to be collected
External loading and structural response	Accelerometer	Proper acceleration
	Displacement transducer	Structural displacement
	Strain gage	Strain of the structure
	Laser doppler vibrometer	Vibration amplitude and frequency
	GPS station	Location of structure and time synchronization
Environmental conditions	Thermometer	Temperature and humidity
	Anemometer and wind-vane	Wind speed and direction
Traffic flow	CCD camera	Vehicle type, throughput, velocity
	Weight in motion	Weight of the still/moving vehicles

the infrastructure. The MC also communicates with the DPM module where further extensive analysis of the collected data is performed. For large-scale monitoring, multiple MCs can be deployed based on the structure of a infrastructure and communicate with each other to acquire more data from sensors and broaden the monitoring analysis.

Wireless sensor networks (WSNs) play a big role in monitoring the infrastructure health, where data is collected and sent to the data processing management module (Wu et al. 2011, 2014, 2015a). Despite the benefits that WSNs can provide, such as, high scalability, high deployment flexibility of deployment, and low maintenance cost, sensors suffer from computational and energy limitations, which need to be taken into consideration for extended, reliable and robust monitoring.

Energy-efficient sensors are crucial for accurate long-duration monitoring in SHM systems. On the one hand, to accurately capture the random process of structural mechanics and detect the potential damage of complex structures in real time, both long-term and real-time monitoring of these structures by sensor networks are needed. On the other hand, sensors usually have very limited energy supply, for example, powered by battery, which is consumed by different modules in the sensors, including the sensing module, the on-board data processing and storage module, and the communication module. Therefore, development of methods and strategies for the optimization of sensors' energy consumption is imperative. Also, the proposed system may incorporate various energy-harvesting devices which can capture and generate power from ambient energy sources, such as vibration, strain, wind, solar, and thermal. Civil infrastructure are ideally suited to harvest such types of energy (Gupta et al. 2014). For example, sensors with piezoelectric materials can be mounted to infrastructure based on their structural information to harvest vibrational energy.

In the proposed SHM system, the parameters to be monitored are heterogeneous, such as temperature, wind, acceleration, displacement, corrosion, strain, traffic, etc. These parameters have different spatial and temporal properties, for example, different variation speeds and locations. Depending on the nature of the monitored parameters, some sensors may work continuously while others may work in the trigger mode. Based on these observations, sampling rate in data acquisition and duty cycles (Ye et al. 2002) in wireless networking is optimized in different types of sensors.

12.4 Nationwide Civil Infrastructure Survey

As the major task of the data processing and management (DPM) module, nationwide civil infrastructure survey is dedicated to classifying the nationwide civil infrastructure according to their life-expectancy. Hadoop Ecosystem (Landset et al. 2015) and deep learning (LeCun et al. 2010) are two enabling techniques for nationwide civil infrastructure survey.

12.4.1 The Features Used in Nationwide Civil Infrastructure Survey

A variety of accurate features can be used in nationwide civil infrastructure survey. Material erosion, cyclic and random external loads, and the corresponding structural responses are the major causes of civil infrastructure's aging. A quantitative investigation about the dynamic behavior of civil infrastructure helps to extract the features for structural health. The following governing equation (Hulbert 1992), Petyt (2015) shows the linear dynamics of infrastructure:

$$[M]\{\ddot{u}\} + [C]\{\dot{u}\} + [K]\{u\} = \{L_{traffic}\} + \{L_{wind}\} + \{L_{self\_weight}\}. \qquad (12.1)$$

where $[M]$, $[C]$, and $[K]$ are mass, damping and stiffness matrices respectively ($[C] = \alpha[M] + \beta[K]$); $\{\ddot{u}\}$, $\{\dot{u}\}$ and $\{u\}$ are acceleration, velocity, and displacement vectors, respectively; external load effects $\{L_{selfweight}\}$, $\{L_{traffic}\}$ and $\{L_{wind}\}$ are self-weight of bridge, traffic load, and aerodynamic load incurred by wind, respectively. Load effects are stochastic due to random variations in space and time. The Turkstra load combination (add up the peak values) (Naess and Rúyset 2000) and Ferry Borges-Castanheta load combination (time-scale) (Thoft-Christensen and Baker 1982) are two applicable strategies to model the uncertainty combination of load.

For small or medium scale civil infrastructure, traffic load ($\{L_{traffic}\}$), which is determined by the traffic velocity, density, and vehicle/load weight, often dominates the external load effects. For large-scale long span civil infrastructure like suspension bridges and cable-stayed bridges, wind load ($\{L_{wind}\}$) dominates the external loads. $\{L_{wind}\}$ consists of two degree-of-freedom: lift (or drag) force and moment.

They are formulated due to the applied force and the corresponding aerodynamic derivatives (Simiu and Scanlan 1986):

$$
\begin{aligned}
L_{wind-lift} &= \tfrac{1}{2}\rho U^2(2B)[KH_1^*\tfrac{\dot{h}}{U} + KH_2^*\tfrac{B\dot{\alpha}}{U} + K^2H_3^*\alpha + K^2H_4^*\tfrac{h}{B}] \\
L_{wind-moment} &= \tfrac{1}{2}\rho U^2(2B^2)[KA_1^*\tfrac{\dot{h}}{U} + KA_2^*\tfrac{B\dot{\alpha}}{U} + K^2A_3^*\alpha + K^2A_4^*\tfrac{h}{B}]
\end{aligned} \tag{12.2}
$$

In the above equations, ρ is air mass density; U is mean wind velocity; B is the width of bridge's girder; $k = B\omega/U$ is the reduced non-dimensional frequency where ω is the circular frequency; h and $\dot{h}$ are the wind-induced displacement and its derivative; α and $\dot{\alpha}$ are structure's rotation and its derivative; A_i^* and $H_i^*\,(i = 1, 2, 3, 4)$ are the aerodynamic derivatives, which are computed from the results obtained by the wind tunnel experiments (Simiu and Scanlan 1986).

The above discussion about the wind-induced load only focuses vortex induced response (Simiu and Scanlan 1986). Those uncommon aerodynamic phenomenon such as buffeting response (i.e., random forces in turbulent wind) and flutter (i.e., self-excited forces in smooth turbulent wind) (Simiu and Scanlan 1986) are not investigated in this work. The dynamic behavior of civil infrastructure caused by extreme weather or environmental conditions is not covered in this work either.

Figure 12.4 shows the features to be used to measure civil infrastructure' life-expectancy. Structural dynamics features include civil infrastructure' structural configuration (e.g., mass, damping and stiffness matrices), and cyclic external load-effect/structural response (derived from in-house sensors or National Transit Database). The weather information can be derived from NOAA's National Climatic Data Center. The accessory civil infrastructure' information such as age, maintenance policy, and construction budgets can be found in related databases, such as, the National Bridge Inventory (NBI) database. Particularly, Nationwide Bridge Sufficiency rating provides training data (fttps://www.fhwa.dot.gov/bridge/).

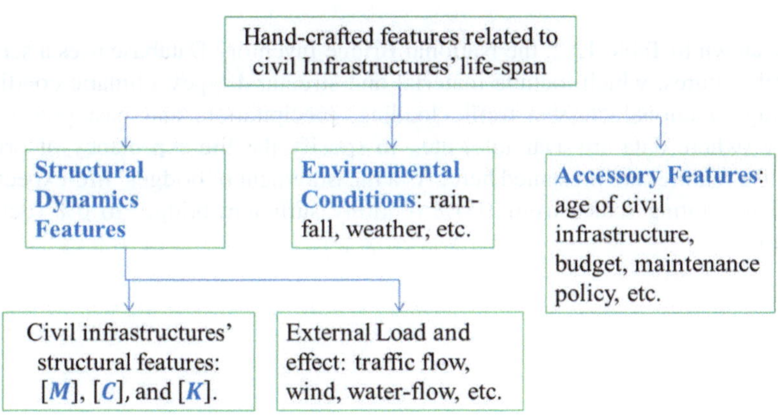

Fig. 12.4 Classification of features involved in nationwide civil infrastructure survey

Table 12.2 Sample bridge data from the National Bridge Inventory Database (updated by 2012)

Year built	Structure	Material	ADT	Status	SR
1914	Stringer/multi-beam or girder	Steel	660	Structurally deficient	6.5
1940	Tee beam	Concrete	210	Structurally deficient	47
1965	Stringer/multi-beam or girder	Steel	170	Structurally deficient	23.6
1941	Stringer/multi-beam or girder	Steel	1320	Structurally deficient	61.3
1975	Stringer/multi-beam or girder	Wood	80	Structurally deficient	29.4
1952	Stringer/multi-beam or girder	Concrete	1080	Functionally obsolete	69.8
1984	Culvert	Concrete	50	Functionally obsolete	87.5
1940	Stringer/multi-beam or girder	Steel	1530	Functionally obsolete	50.8
1950	Stringer/multi-beam or girder	Steel	650	Functionally obsolete	78.6
1946	Tee beam	Concrete	4350	Functionally obsolete	43.9
1982	Stringer/multi-beam or girder	Prest. concrete	1010	Good condition	87.4
1999	Box beam or girders	Prest. concrete	420	Excellent condition	88
1993	Culvert	Concrete	1020	Not applicable	78.8
1988	Culvert	Concrete	1670	Not applicable	99.4
1970	Culvert	Concrete	990	Not applicable	99.4

ADT (ton/day) average daily traffic, *SR* Sufficiency rate

As shown in Table 12.2, the National Bridge Inventory Database uses a series of general features, which include material and structural types, climatic conditions, highway functional classes, traffic loading, precipitation, and past preservation history (where data are available) etc., to specify the life-expectancy of bridges. Only five features are presented here. As a measurement of bridges' life-expectancy, sufficiency rating scales from 100% (entirely sufficient bridge) to 0% (deficient bridge).

12.4.2 *Estimation of the Life-Expectancy of Nationwide Civil Infrastructure Using Machine Learning Method*

12.4.2.1 Overview of Machine-Learning-Enabled Life-Expectancy Estimation

The goal of the statistical analysis of the Life-expectancy of civil infrastructure using empirical models (statistical evidence based) is to identify our target civil infrastructure that are in risk of short service life. To estimate the life expectancy based on statistical analysis, the following three tasks need to be completed at first:

(1) Definition of the end of functioning life. Various definitions of end-of-life may be applied. For instance, National Bridge Inventory Database uses Sufficiency Rating, which takes four rationales. If we use sufficiency rating as the definition of end-of-life, we may use the threshold of the end-of-life as sufficiency rating first drops to or below 50% on a scale from 100% (entirely sufficient bridge) to 0% (deficient bridge).

(2) Selection of general approaches. Three general life estimation approaches that are common in the current literature are: (a) condition based approach, (b) age-based approach, and (c) hybrid approach. As we will have a large amount of time series data from sensors, a condition-based approach may be more appropriate. Civil infrastructure are periodically monitored with respect to their condition. As such, deterioration models can be readily developed. For instance, if a performance threshold is set at which point a bridge "failure" is considered to occur, then the lifetime is the time from construction or last reconstruction to the time when the threshold is first crossed. Combining two approaches may be preferred as we have large amount of all different types of data.

(3) Model Selection. From the literature review (Melhem and Cheng 2003), three potential models may be applied in this project: (a) Linear or non-linear regression models, which fits continuous, deterministic model type with direct interpretations of model fit and parameter strength. These regression models are commonly used due to their simplicity, clarity of results, and ability to be calibrated with widely available statistical software such as SAS and R. Linear regression methods are only appropriate when the dependent variable has linear explanatory variables, which may not necessarily be the case for highway asset performance behavior over time. Furthermore, such models are deterministic estimate that may not reflect the true value of condition or service life that could be expected. On the other hand, for non-linear models, it is difficult to develop a set of significant independent variables, while providing a higher good of fit metric such as R^2. (b) Markov chain-based model fits a continuous, probabilistic, fully-parametric survival curve to a Markov Chain estimate. A Markov chain is a transition probability based solely on the present state not on the past state of civil infrastructure, stochastic process with a finite, integer number of possible, no-negative states that is used to predict the probability

of being in any state after a period of time. (c) Artificial neural networks are non-linear, adaptive models that can predict conditions based on what it has "learned" from the historical data. The approach updates posterior means by applying weighted averages based on previous estimates. Typically the weights are based on the number of observations. Such models have been found work well with noisy data and relatively quick [53]. (d) Deep learning models, which provides a more accurate and powerful formulation about hypothesis functions.

12.4.2.2 Weibull Linear Regression Model

The use of a Weibull model is justified by past findings in the literature, comparing model fit statistics of alternative distributions and validating the prediction against the non-parametric methods such as Kaplan-Meier estimate. The lifetime factors to be investigated may include material and structural types, climatic conditions, highway functional classes, traffic loading, precipitation and past preservation history where data are available. Table 12.3 provides a demo output of estimated average life times with 90% confidence interval for the Sufficiency Rate ranging from 40 to 80.

As our preliminary results, Table 12.2 provides a sample of data of 15 bridges located on the highways of Tennessee from the National Bridges Inventory Database. Estimation of the lifetime function is a regression analysis of life-expectancy T on the covariates (factors). Only six variables are presented here. A parametric fitting with Weibull distribution will be carried out using SAS procedure "lifereg". This procedure handles both left-censored observations and right-censored observations and includes checks on the significance of the estimated parameters. Based on the significance (P-value or associated t-statistics) associated with each covariate (lifetime factor) and the estimated parameters, a final model of lifetime can be determined after the variable selection procedure. Using the final lifetime model, we can obtain the targeted bridges with short lifetime via prediction of life-expectancy of the new bridges with sensor data. Table 12.3 provides a demo output of estimated average life times with 90% confidence interval for Sufficiency Rate ranging from 40 to 80.

Table 12.3 Weibull regression models predictions of bridge life

Sufficiency rate	Expected life (years)	%90 CI	*Weibull model parameters and statistics*
40	42	[15, 65]	Scale factor $\alpha = 57.04$
50	58	[22, 96]	Shape factor $\beta = 2.73$
60	64	[27, 84]	*Model statistics*
70	71	[34, 123]	Log-likelihood function at convergence $= -1043.6$
80	83	[56, 183]	Restricted log-likelihood function $= -1903.5$

End-of-life is age when sufficiency rating drops to or below 50%

12.4.2.3 Markov Chain Models

Markov chains can be used to model condition ratings based on the data from large numbers of infrastructure systems using transitional probabilities. Jiang (2010) used Markov chains to model the condition of bridge substructures in Indiana. Table 12.3 shows the transitional probabilities for concrete bridge substructures. In this case, the transitional probabilities change as the bridge ages. Each entity in the table indicates the probability that a bridge that is currently in a certain state will remain in that state next year.

Using the transition matrix, the Markov models were then calibrated by minimizing the Root Mean Sum of Errors (RMSE) while holding the median life prediction constant and changing the baseline ancillary survival factors. Advantages of Markov-based models include a probabilistic estimate, sole dependence on current conditions (i.e., minimal data needed if the transition probabilities known), flexibility in modifying state duration, and efficiency for dealing with larger networks.

12.4.2.4 Neural Network Models

Neural networks are a class of parametric models that can accommodate a wider variety of nonlinear relationships between a set of predictors and a target variable. Being far more complicated than regression models or a decision tree, neural network models have much stronger prediction and classification capability.

Building a neural network model involves two main phases: (1) configuring the network model and (2) iteratively training the model.

12.4.2.5 Deep Learning Models

The goal of nationwide civil infrastructure survey is to identify those target infrastructure systems that are in risk of short service life. Most of previous work adopted supervised machine learning methods (Agrawal and Kawaguchi 2009; Morcous 2006; Robelin and Madanat 2007) such as linear and nonlinear regression, Markov Chain (Jiang 2010), and Support Vector Machine (SVM) (Zhou and Yang 2013) to estimate civil infrastructure's life expectancy according to hand-crafted features. This work emphatically investigates a deep learning algorithm (LeCun et al. 2015), which automatically formulates constructive features without supervision from raw input data.

Figure 12.5 shows a flowchart of deep learning enabled nationwide civil infrastructure survey. Compared with many other classifiers, a deep learning algorithm has the following advantages: (1) little or no human supervision is needed; (2) some un-interpretable yet constructive features (or intermediate representation) can be directly derived from raw data; (3) less training data is required (this advantage is very important in the addressed project because the archived real world sensory

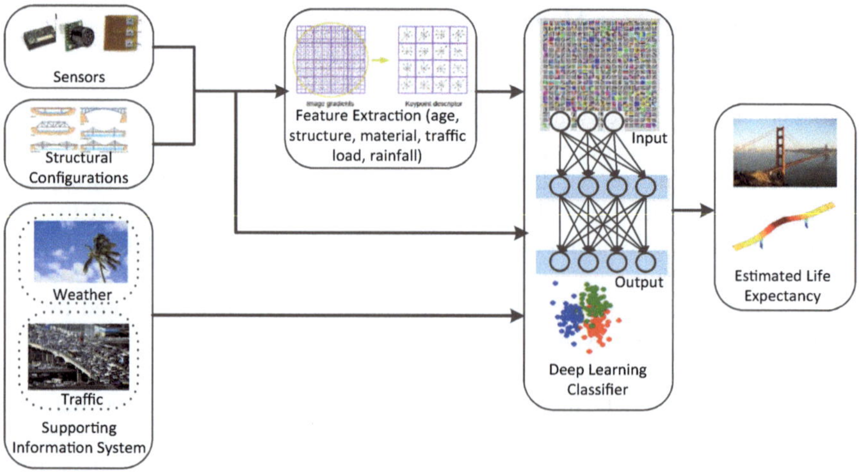

Fig. 12.5 Flow-chart of deep-learning-centric nation-wide infrastructure survey (the original pictures referred in the figure are derived from images.google.com)

data for highly deficient civil infrastructure is limited); (4) the mid layers of deep networks can be repurposed from one application to another, and this advantage is the motivation for using hybrid deep learning method (HDL) (Wang et al. 2009; Wu et al. 2015b) that arises by merging multiple different deep learning algorithms to handle heterogeneous raw input data.

To efficiently and accurately classify the observed civil infrastructure, a hybrid deep-learning (HDL) algorithm is investigated in this work. HDL is featured with the following techniques: (1) Multiple data with heterogeneous modalities, such as raw streams of sensory data, i.e., audio/video data, images, textual information like operational data, city-open-data, environment factors, and other hand-crafted data, is exploited so as to give a panoramic and full-spectrum description about the status of targeted civil infrastructure. (2) HDL is equipped with different deep-learning algorithms, at least at the lower levels, to learn the features from multiple input data with heterogeneous modality. Convolutional neural network (CNN) (LeCun et al. 2010) is used to learn spatial features from visual media such as video and images because it demonstrates superior performance (high accuracy and fast training speed) on matrix-oriented feature-learning. Recurrent Neural Network (RNN) (Sak et al. 2014) is employed to learn temporal features from streaming data such as acoustic signal or vibration signals because RNN exhibits dynamic temporal behavior (enabled by the directed cycle inside RNN). Deep Boltzmann machine (DBM) (Salakhutdinov and Hinton 2009) specializes on learning the high-level features from textual information such as weather conditions, traffic status, and maintenance policy, etc. (3) Deep learning algorithms always learn the upper-level features from lower ones (LeCun et al. 2015), and the input data with heterogeneous modality eventually fuse at upper layers with somewhat homogeneous modality. Therefore, the HDL can use a unified deep learning algorithm such as DBM in the feature-learning of upper levels.

12.4.2.6 Champion Model Selection

In real-world applications, variance of models can be jointly used to predict civil infrastructure' life expectancy. First, we will select a champion model that, according to an evaluation criterion, performs best in the validation data; second, we will employ the champion model to score new data.

12.4.3 Techniques to Boost Nationwide Civil Infrastructure Survey

To boost the performance of the nationwide civil infrastructure survey, various techniques such as missing data handling, variable transformation, data management optimization, and dimensionality reduction, etc. are employed in this work.

12.4.3.1 Imputation and Transformation of Variables

Most of the software packages such as "WeibullReg" in R or "lifereg" in SAS can handle missing data. However if there is a relatively large amount of missing data in the input data of statistical model, some data imputations are required so that all observed value can be explored during model training.

Besides imputation, Variable transformation techniques can improve the quality of data.

12.4.3.2 Optimization of Data Management

The proposed project employs discrete Hash-tables to formulate the correlation among data, control the data partitioning to optimize data placement, and use in-memory technology (Robelin and Madanat 2007).

Using word-count problems as a benchmark and Amazon EC2 as the computing platform, Table 12.4 demonstrates that Apache Spark, an implementation of Resilient Distributed Datasets (RDD) that enables users to explicitly persist intermediate results in memory and control their partitioning to optimize data placement, is 40 times faster than Hadoop.

12.4.3.3 Dimensionality Reduction

The data involved in sensor-oriented structural analysis is always extremely high-dimensional (Tran et al. 2013). As one of our preliminary achievements, a rank revealing randomized singular value decomposition (R^3SVD) (Ji et al. 2016) was

Table 12.4 Time cost of Spark (with in-memory) vs. Hadoop (without in-memory) on word-count problem

	Time on spark (s)			Time on Hadoop (s)		
File size (B)	1st count	2nd count	3rd count	1st count	2nd count	3rd count
1486.	0.7923488	0.268379	0.5591409	35.	35.	37.
214144.	0.7908732	1.040815	0.8181214	40.	38.	39.
491430.	0.5838947	0.570067	0.5644267	40.	41.	41.
674570.	0.7753005	1.310154	0.6197348	40.	41.	40.
1573150.	1.076778	0.9380654	1.096354	43.	40.	41.

proposed to reduce the dimensionality of dataset by adaptive sampling. As a variance of primary component analysis (PCA), R^3SVD uses local statistical errors to estimate global approximation error.

The preliminary investigations (Ji et al. 2013; Liu and Han 2004) demonstrated that R^3SVD scales well to extremely big matrices and us efficient with minimal sacrifices in accuracy due to the following reasons: (1) R^3SVD is based on statistical sampling, which is also applicable to incomplete or noisy data. (2) R^3SVD is able to obtain low-accuracy approximation quickly, which is particularly suitable for many applications where high-accuracy solutions are not necessary but fast decision making is, on the other hand, of most importance. (3) R^3SVD is trivially naturally parallelizable.

As illustrated in Fig. 12.6, R^3SVD algorithm with adaptive sampling dominates its rivals such as general randomized SVD algorithm [57–59] from the point of view of stability and timing performance [60] [61] [55] [62]. Next step we intend to transplant R^3SVD into sensor-oriented structural analysis for civil infrastructure.

12.5 Global Structural Integrity Analysis

The global structural integrity analysis module aims to provide further structural integrity analysis of the deficient civil infrastructure identified in Sect. 12.5. The objectives are itemized as follows: (1) to apply the big-data and perform quantitative analysis of global structural integrity of targeted bridges; (2) to provide guidelines for more intensive and predictive examination of the civil infrastructure at component level to be carried out at Sect. 12.7; and (3) to feed back to the database with integrity analysis results for future uses.

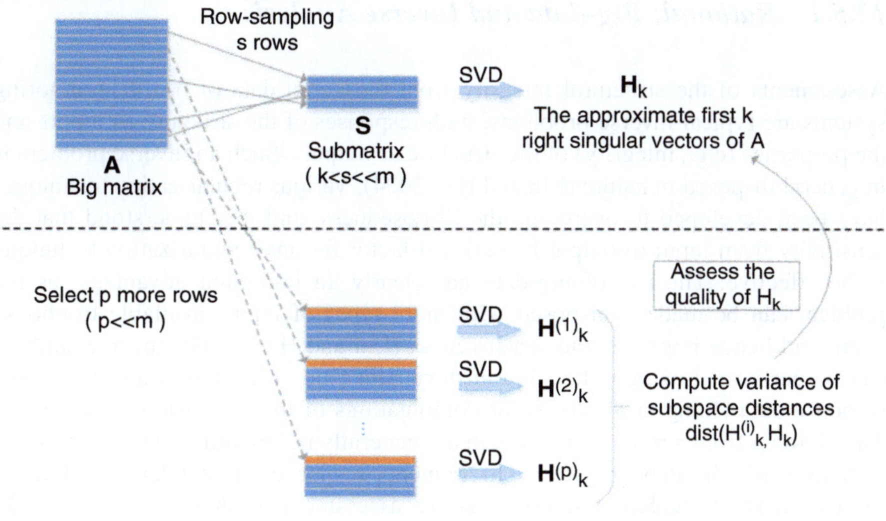

(a) Original

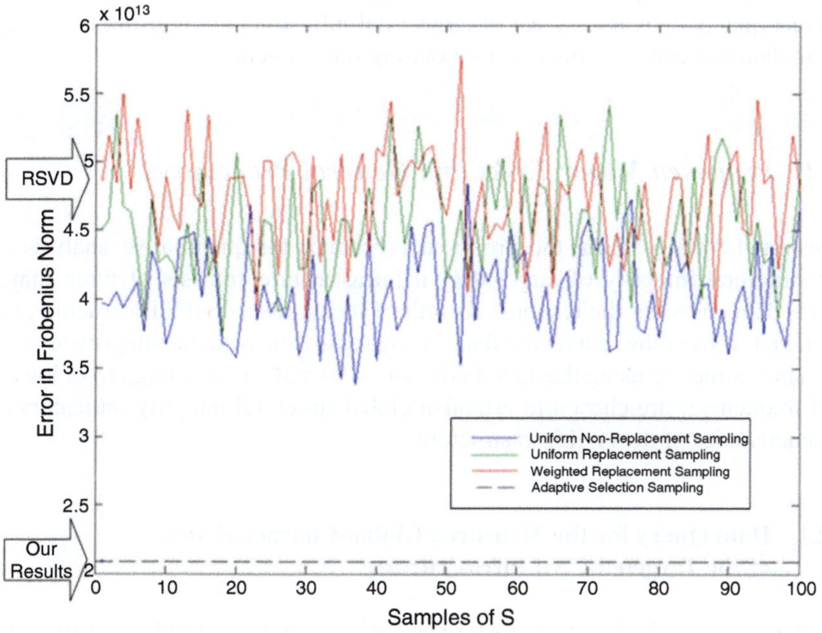

(b) Proposed

Fig. 12.6 (**a**) Rank-Revealing Randomized SVD using Adaptive Sampling. Additional p rows are selected to ensure the quality of the approximated top-k right singular vectors H_k by computing the variance of subspace distances. (**b**) Adaptive sampling leads to more stable and accurate approximation of top-512 singular vectors. A single run of adaptive sampling with a few additional testing rows yields significantly smaller errors than multiple runs of sampling with different strategies (uniform non-replacement, uniform-replacement, and weighted-replacement [55, 62])

12.5.1 Rational: Big-data and Inverse Analysis

Assessments of the structural integrity from measured data of health monitoring systems are typical inverse problems, with responses of the structure as inputs and the properties (e.g., integrity) of the structure as outputs. Such an inverse problem is in general ill-posed in nature (Liu and Han 2004). Various regularization techniques have been developed to overcome the ill-posedness, and it is understood that the sensitivity from input to output is a critical factor for any regularization technique to be effective. The use of big-data has clearly an important advantage, as the problem can be made over-posed with more types of inputs available to choose from, and hence improves the sensitiveness (Liu and Han 2004). Inverse analysis can be performed using either time-history data (Liu et al. 2002), and frequency response data (Jiang et al. 2008), or combinations of the two (Ishak et al. 2001). The big-data from a monitoring system are generally rich in time-history records of responses, which can be transferred to frequency responses via standard Fast Fourier Transform (FFT) techniques. For effective assessment of slender structures like bridge, the global structural integrity relates well to the lowest natural frequencies or to the frequency responses in the lower frequency range. Therefore, we propose to conduct quantitative assessment of target civil infrastructure by using frequency response data that can be extracted from our big-data system.

12.5.2 Proposed Major Tasks and General Procedures

As illustrated by Fig. 12.7a, the proposed procedure for quantitative analysis of global structural integrity of targeted civil infrastructure consists of three major tasks: (1) data query for the response records of the targeted civil infrastructure, (2) computer analysis of the rich record data, and (3) assessment on the integrity level of the civil infrastructure using the data. In this study, a small number (e.g., 6) of lowest natural frequencies are chosen to establish global structural integrity indicators of the characteristics of the civil infrastructure.

12.5.2.1 Data Query for the Measured Global Characteristics of the Targeted Civil Infrastructure

To obtain the actual global characteristics of the targeted civil infrastructure, the following analysis is performed. (1) Based on the data made available for querying in Sect. 12.5, civil infrastructure systems under monitoring are selected for qualitatively integrity assessments. (2) A query is then made to the database for civil infrastructures that have high possibility of short life, and a list of targeted civil infrastructure is created, in the order of urgency. (3) For each targeted civil infrastructure system, a query is next made for the major excitation events that may

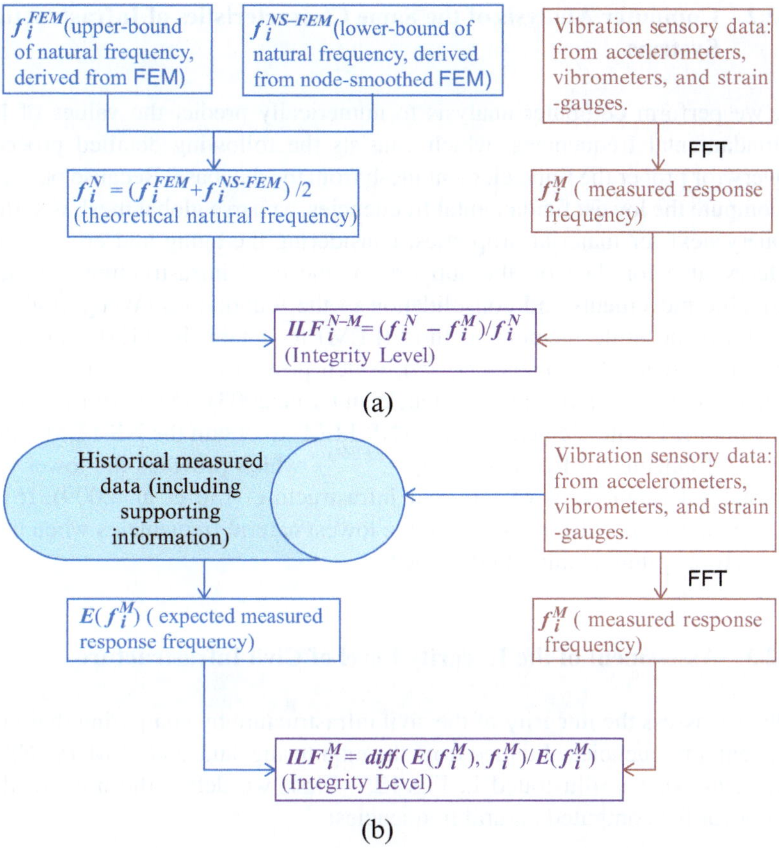

Fig. 12.7 Global structural integrity analysis with reference to: (**a**) theoretical response frequency; and (**b**) historical measured response frequency

have happened to the civil infrastructure system. Such events include earthquake, wind storms, and major traffic loading at levels closest to the level used in the design of the civil infrastructure system. (4) For each targeted infrastructure system and each major event, the detailed health monitoring data related to the global responses and behavior are extracted. The data includes the time-history of accelerometers, vibrometers, and strain gauges installed at various locations on the civil infrastructure system. (5) Fast Fourier Transform (FFT) is next performed to the time-history data to obtained frequency response data (f_i^M, if this is not readily available in the database). (6) Estimate the lowest few fundamental frequencies of the civil infrastructure system from the frequency response data.

12.5.2.2 Computer Analysis of the Same Characteristics of Infrastructure Systems

Next, we perform computer analysis to numerically predict the values of lowest few fundamental frequencies, which consists the following detailed procedures. (1) Query for proper the finite element mesh from the database. Because our purpose is to compute the lowest fundamental frequencies, a coarse global mesh is sufficient. (2) Query next for material properties, considering the aging and erosion effects. (3) Query also for data on the supports of the civil infrastructure, considering the possible movements and consolidation of the foundations (Wang et al. 2002). (4) Perform the finite element method (FEM) to obtain the FEM values of the lowest fundamental frequencies (f_i^{FEM}), which provides the upper bounds of the natural frequencies of the bridge (Quek and Liu 2003). (5) Perform the node-based smoothed finite element method (NS-FEM) to obtain the NS-FEM values of the lowest fundamental frequencies ($f_i^{(NS-FEM)}$), which provides the lower bounds of the natural frequencies of the civil infrastructure (Liu et al. 2009). (6) As a reference, query may also be made for the lowest natural frequencies when the civil infrastructure system is initially designed.

12.5.2.3 Assessment of the Integrity Level of Civil Infrastructure

Finally, we assess the integrity of the civil infrastructure by comparing these lowest fundamental frequencies obtained from the monitoring data, and FEM and NS-FEM analyses, as what is illustrated in Fig. 12.7. First, we define the numerical error indicator for the computed natural frequencies:

$$Error_i = f_i^{FEM} - f_i^{(NS-FEM)} \tag{12.3}$$

which gives a good indication on how accurate the numerical value is. Note that the numerical error can be reduced if finer mesh is used. Therefore, if the error is too big we can use a fine mesh to reduce the error gap. In general, the average of both FEM and NS-FEM value gives a good approximation (Liu and Trung 2010)

$$f_i^N = (f_i^{FEM} + f_i^{(NS-FEM)})/2 \tag{12.4}$$

where the superscript N denotes the numerical natural frequency. The integrity level in terms of the rate of frequency reduction (ILF) is defined as:

$$ILF_i^{(N-M)} = (f_i^N - f_i^M)/f_i^N \tag{12.5}$$

where the superscript M denotes the measured natural frequency. We know that a degradation of a bridge structure may lead to a reduction of some fundamental frequencies. In addition, we have a general understanding that the frequency is related to the square-root of the stiffness of the structure of the bridge. The integrity level in terms of the rate of stiffness reduction (ILK) indicators can then be defined as:

$$ILK_i^{(N-M)} = (\sqrt{f_i^N} - \sqrt{f_i^M})/\sqrt{f_i^N}. \tag{12.6}$$

In the end, a criterion (e.g., 10% reduction) can be set to categorize the civil infrastructure into the list of civil infrastructure to be further studied in detail in Sect. 12.7.

As illustrated in Fig. 12.7b, integrity analysis can be made by comparing the newly observed response signals with the historical response signals.

12.6 Localized Critical Component Reliability Analysis

Different from Sect. 12.6, this section mainly focuses on the measurement of structural component deterioration.

12.6.1 Deep-Learning-Enabled Component Reliability Analysis

Figure 12.8 shows the infrastructure of component reliability analysis. Just as the nationwide civil infrastructure survey, the deep learning technique is employed to digest the input data with heterogeneous modality so as to obtain the reliability of structural components. Component reliability involves two strategies: structural reliability analysis and observation-oriented method. The former is derived from the probabilistic evaluation of load-effect (denoted as S) resistance (denoted as R). The

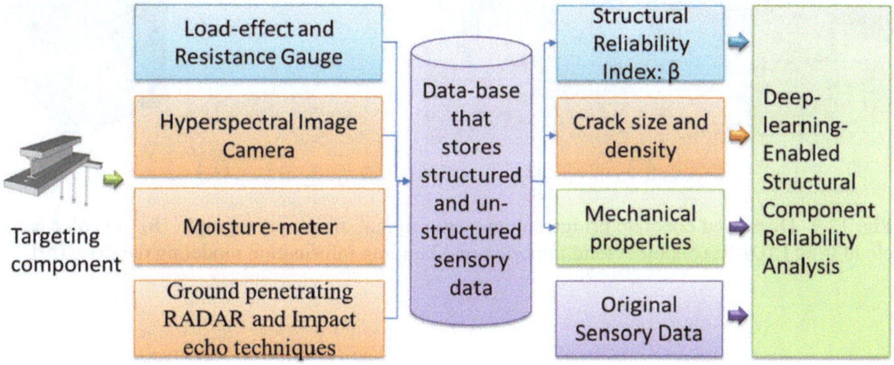

Fig. 12.8 Infrastructure for component reliability analysis

latter is derived from the direct observation about component using optical-electro sensors (e.g., hyper-spectral image cameras and moisture meters).

Structural reliability is conventionally measured by reliability index, β, which is determined by the limit state function $Z = R - S$. Structural component failure occurs whenever $Z < 0$. If R and S follows Gaussian distributions, the reliability index (Hasofer 1974) is a function of the mean and standard deviation of Z, namely, $\beta = \frac{\bar{Z}}{\sigma_Z} = (\bar{R} - \bar{S})/\sqrt{(\sigma_R^2 + \sigma_S^2)}$. Commonly used numerical methods to calculate reliability index include Monte Carlo simulation (Jirutitijaroen and Singh 2008) (random sampling to artificially simulate a large number of experiments and observe the results), first-order reliability method (Cizelj et al. 1994) (approximating limit-state function with a first-order function), response surface method (Huh 2000) (approximating the unknown explicit limit state functions by a polynomial function), Latin hypercube simulation (Jirutitijaroen and Singh 2008), genetic search algorithm (Samaan and Singh 2002), and subset simulation algorithm (Au and Beck 2001).

The crack density and size inside or outside the structural component are also an index to evaluate the reliability of structural component. Hyper-spectral and 3D point cloud image processing (Fernandes et al. 2010) technique and concrete moisture measurement are commonly used techniques to probe crack size and density.

As an example, Fig. 12.9a, b shows a distressed concrete bridge column, first as a digital 2-D image and then as a 3-D color point cloud. Figure 12.9c shows how the point cloud imports into a Building Information Modeling (BIM) database, which can then help with an automated concrete repair decision-making tool.

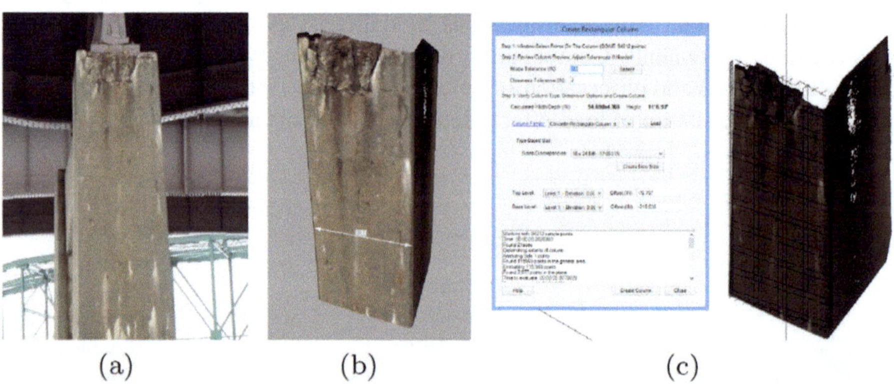

Fig. 12.9 Distressed concrete bridge column: (**a**) original digital photograph; (**b**) 3D color point cloud from LIDAR; (**c**) point cloud imported into building information modeling database (MD)

12.6.2 Probe Prolongation Strategies via Simulating Crack Initialization and Growth

Fatigue failure is a complex and progressive form of local damage which is significantly influenced by many factors such as magnitude and frequency of the loads causing fluctuating stress, temperature, environment, geometrical complexities, material imperfections and discontinuities (Aygul 2012). Durability of the civil infrastructure is mainly dominated by the fatigue behavior of those critical components of the civil infrastructure system.

In this work, both time-domain and frequency-domain finite-element-based (FEM) (Liang et al. 2013c, 2002; Liang 2013) fatigue analyses are investigated to measure the life expectancy of bridge component under random cyclic external load. The former is implemented by formulating the transient solution to the dynamics of structure. The latter formulates the random cyclic load and structural response using Power Spectral Density (PSD) (Halfpenny 1999). Numerically, frequency-domain approaches are more efficient because they do not need to solve the dynamics equation at each time step. However, frequency domain approaches are not applicable for cyclic loads that are too irregular.

Crack generation and crack growth give us a more in-depth understanding about the fatigue behavior of material. Figure 12.10a–c shows our preliminary results in crack generation and growth using extended FEM (X-FEM) (Liang et al. 2008), molecular dynamics (MD) (Mohan et al. 2012), and smoothed particle methods (Monaghan 1992). Our future work will focus on the application of generalized smoothed particle methods in the modeling and simulation of component fatigue, based on which a potential life prolongation strategy will be discussed.

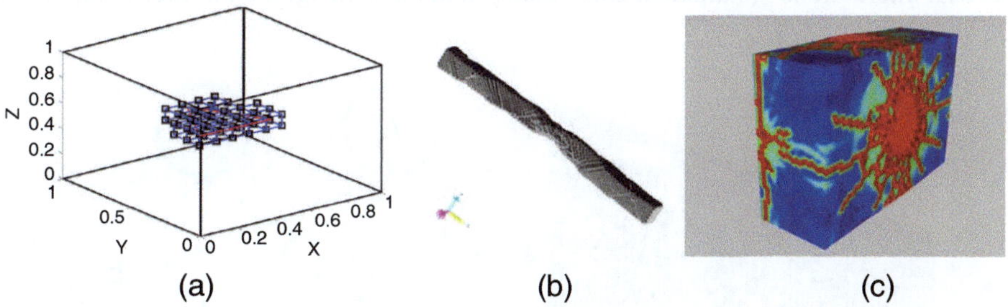

Fig. 12.10 Modeling and simulation of crack generation and growth: (**a**) growth of planar crack (X-FEM); (**b**) deformation of nano-wire (MD); (**c**) crack generation in concrete block (smooth particle method)

12.7 Civil Infrastructure's Reliability Analysis Based on Bayesian Network

Bayesian network (Torres-Toledano and Sucar 1998) is a probabilistic graphical model that represents a set of random variables (nodes) and their conditional dependencies (arcs) via a directed acyclic graph (DAG). As one of the major contributions of this work, Bayesian network is employed to formulate the reliability of civil infrastructure systems according to components' reliability examined in the previous section (or Sect. 12.7).

As illustrated in Fig. 12.11a, b, Bayesian network for civil infrastructure has the following features: (1) Each node represents a structural component and takes discrete value to describe the serviceability (e.g. whether or not the component still functions, or the life expectancy of component, etc.). (2) The topology of Bayesian network is determined according to components' qualitative relationship. Two nodes should be connected directly if one affects or causes the other, with the arc indicating the direction of effect. (3) Once the topology of Bayesian network is specified, the inter-component dependency can be quantified. As its creative contribution, the inter-component interactions are jointly formulated according to mechanical interaction (e.g., pin and hanger) and statistical correlation (e.g., two pins not directly related).

It is extremely computationally costly to construct the Bayesian network of a civil infrastructure system constituted of tens of thousands of components. Multiple techniques are introduced to reduce the computing complexity. For example, Bayesian Network nodes are classified into essential and non-essential components, only those essential ones will be considered in integrity analysis. The inter-component dependency is either derived from mechanical interaction or "inferred causal interactions" (statistical correlation), and those insignificant inter-component

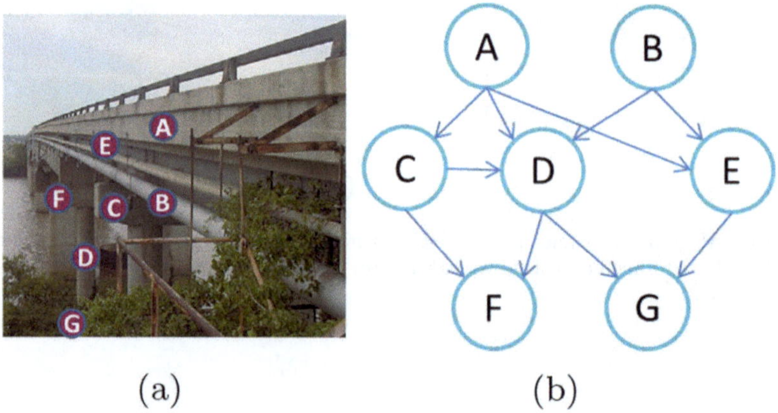

(a) (b)

Fig. 12.11 (**a**) Civil infrastructure's components; (**b**) Bayesian network for the civil infrastructure system correlation are ignored. In addition, sub-system, a self-contained system within a larger one, can be considered to formulate the Bayesian network into hierarchy structure.

12.8 Conclusion and Future Work

This work developed a framework to construct a multi-scale structural health monitoring system based on Hadoop Ecosystem (MS-SHM-Hadoop) to monitor and evaluate the serviceability of civil infrastructure. MS-SHM-Hadoop is a multi-scale reliability analysis system, which ranges from nationwide civil infrastructure survey, global structural integrity analysis, to structural components' reliability analysis. As one of its major technical contribution, this system employs Bayesian network to formulate the integral serviceability of a civil infrastructure system according to components' serviceability and inter-component correlations. Enabled by deep learning and Hadoop techniques, a full-spectrum, sustainable, and effective evaluation can be made to cover nationwide civil infrastructure.

Acknowledgements This work is jointly sponsored by the National Science Foundation (NSF) with proposal number 1240734 and UTC THEC/CEACSE 2016 Grant Program.

References

Agrawal, A. K. & Kawaguchi, A. (2009). *Bridge Element Deterioration Rates: Final Report.* Albany, NY: New York State Department of Transportation.

Ahmed, A. & Ahmed, E. (2016). A survey on mobile edge computing. In *Proceedings of the 10th IEEE International Conference on Intelligent Systems and Control.*

Au, S. & Beck, J. L. (2001). Estimation of small failure probabilities in high dimensions by subset simulation. *Probabilistic Engineering Mechanics, 16*, 263–277.

Aygul, M. (2012). *Fatigue analysis of welded structures using the finite element method.* Gothenburg: Chalmers University Of Technology.

Catbas, F. N., Gul, M., Zaurin, R., Gokce, H. B., Maier, D., & Terrell, T. (2008, June). Structural health monitoring for life cycle management of bridges. In *Proceedings the International Symposium on Life-Cycle Civil Engineering*, Varenna, Lake Como, Italy (pp. 613–618).

Catbas, F. N., Gokce, H. B., & Gul, M. (2012). Nonparametric analysis of structural health monitoring data for identification and localization of changes: Concept, lab, and real-life studies. *Structural Health Monitoring, 11*(5), 613–626.

Cizelj, L., Mavko, B., & Riesch-Oppermann, H. (1994). Application of first and second order reliability methods in the safety assessment of cracked steam generator tubing. *Nuclear Engineering and Design, 147*, 1–10.

Fernandes, S., Liang, Y., Sritharan, S., Wei, X., & Kandiah, R. (2010, July). Real time detection of improvised explosive devices using hyperspectral image analysis. In *Proceeding of the 2010 IEEE National Aerospace and Electronics Conference*, Dayton, OH, USA.

Frangopol, D. M., Strauss, A., & Kim, S. (2008). Bridge reliability assessment based on monitoring. *Journal of Bridge Engineering, 13*(3), 258–270.

Guo, J., Xie, X., Bie, R., & Sun, L. (2014). Structural health monitoring by using a sparse coding-based deep learning algorithm with wireless sensor networks. *Personal and Ubiquitous Computing, 18*(8), 1977–1987.

Gupta, M. N., Suman, & Yadav, S. (2014). Electricity generation due to vibration of moving vehicles using piezoelectric effect. *Advance in Electronic and Electric Engineering, 4*(3), 313–318.

Halfpenny, A. (1999). A frequency domain approach for fatigue life estimation from finite element analysis. *LAP LAMBERT Academic Publishing, 167*, 401–410.

Hasofer, A. M. (1974). Reliability index and failure probability. *Journal of Structural Mechanics, 3*(1), 25–27.

He, Z.-Q., Ma, Z. J., Chapman, C. E., & Liu, Z. (2012). Longitudinal joints with accelerated construction features in decked bulb-tee girder bridges: Strut-and-tie model and design guidelines. *Journal of Bridge Engineering, 18*(5), 372–379.

Huh, J. (2000). Reliability analysis of nonlinear structural systems using response surface method. *KSCE Journal of Civil Engineering, 4*(3), 135–143.

Hulbert, G. M. (1992). Time finite element methods for structural dynamics. *Internal Journal for Numerical Methods in Engineering, 33*, 307–331.

Ishak, S., Liu, G., Lim, S., & Shang, H. (2001). Experimental study on employing flexural wave measurement to characterize delamination in beams. *Experimental Mechanics, 41*(2), 57–164.

Jang, S., Jo, H., Cho, S., et al. (2010). Structural health monitoring of a cable-stayed bridge using smart sensor technology: Deployment and evaluation. *Smart Structures and Systems, 6*(5–6), 439–459.

Jeong, S., Zhang, Y., Hou, R., Lynch, J. P., Sohn, H., & Law, K. H. (2016a, April). A cloud based information repository for bridge monitoring applications. In *Proceedings of the SPIE Smart Structures/NDE Conference*, Baltimore, Maryland, USA (pp. 1–14).

Jeong, S., Zhang, Y., O'Connor, S., Lynch, J. P., Sohn, H., & Law, K. H. (2016b). A nosql data management infrastructure for bridge monitoring. *Smart Structures and Systems, 17*(4), 669–690.

Ji, H., Mascagni, M., & Li, Y. (2013). Convergence analysis of markov chain monte carlo linear solvers using ulam–von neumann algorithm. *SIAM Journal on Numerical Analysis, 51*, 2107–2122.

Ji, H., Yu, W., & Li, Y. (2016). A rank revealing randomized singular value decomposition (r3svd) algorithm for low-rank matrix approximations. *Computing Research Repository*, 1–10. arXiv:1605.08134.

Jiang, Y. (2010). Application and comparison of regression and markov chain methods in bridge condition prediction and system benefit optimization. *Journal of the Transportation Research Forum, 49*(2), 91–110.

Jiang, C., Liu, G., & Han, X. (2008). A novel method for uncertainty inverse problems and application to material characterization of composites. *Experimental Mechanics, 48*(4), 539–548.

Jirutitijaroen, P. & Singh, C. (2008). Comparison of simulation methods for power system reliability indexes and their distributions. *IEEE Transactions on Power Systems, 23*(2), 486–493.

Kawamura, K. & Miyamoto, A. (2003). Condition state evaluation of existing reinforced concrete bridges using neuro-fuzzy hybrid system. *Computers & Structures, 81*(18–19), 1931–1940.

Kim, S., Pakzad, S., Culler, D., Demmel, J., Fenves, G., Glaser, S., et al. (2007, April). Health monitoring of civil infrastructures using wireless sensor networks. In *Proceedings of 6th International Symposium on Information Processing in Sensor Networks*, Cambridge, MA, USA.

Landset, S., Khoshgoftaar, T. M., RichterEmail, A. N., & Hasanin, T. (2015). A survey of open source tools for machine learning with big data in the hadoop ecosystem. *Journal of Big Data, 2*(24), 1–50.

LeCun, Y., Kavukcuoglu, K., & Farabet, C. (2010, May 30–June 2). Convolutional networks and applications in vision. In *Proceedings of 2010 IEEE International Symposium on Circuits and Systems*, Paris, France (pp. 253–256).

LeCun, Y., Bengio, Y., & Hinton, G. (2015). Deep learning. *Nature, 521*, 436–444.

Liang, Y. (2013). *The use of parallel polynomial preconditioners in the solution of systems of linear equations*. Saarbrücken: LAP LAMBERT Academic Publishing.

Liang, Y. & Wu, C. (2014, June). A sensor-oriented information system based on hadoop cluster. In *Proceedings of international Conference on Internet Computing and Big Data*, Las Vegas, NV (pp. 1–5).

Liang, Y. & Wu, C. (2016). A hadoop-enabled sensor-oriented information system for knowledge discovery about target-of-interest. *Internet of Things - Special Issue of FUEE Scientific Journal, 29*(3), 437–450.

Liang, Y., Weston, J., & Szularz, M. (2002). Generalized least-squares polynomial preconditioners for symmetric indefinite linear equations. *Parallel computing, 28*(2), 323–341.

Liang, Y., Waisman, H., Shi, J., Liu, P., & Lua, J. (2008, July). Pre-processing toolkit for three-dimensional x-fem. In *Proceedings of IEEE National Aerospace and Electronics Conference*, Dayton, OH, USA (pp. 265–272).

Liang, Y., Henderson, M., Fernandes, S., & Sanderson, J. (2013a). Vehicle tracking and analysis within a city. In *Proceedings of SPIE Defense, Security, and Sensing*, Baltimore, Maryland (pp. 1–15).

Liang, Y., Melvin, W., Sritharan, S., Fernandes, S., & Barker, B. (2013b). A crowd motion analysis framework based on analog heat-transfer model. *American Journal of Science and Engineering, 2*(1), 33–43.

Liang, Y., Szularz, M., & Yang, L. T. (2013c). Finite-element-wise domain decomposition iterative solvers with polynomial preconditioning. *Mathematical and Computer Modelling, 58*(1–2), 421–437.

Liu, G.-R. & Han, X. (2004). *Computational inverse techniques in nondestructive evaluation*. Boca Raton: CRC.

Liu, G.-R. & Trung, N. T. (2010). *Smoothed finite element methods*. Boca Raton: CRC.

Liu, G., Han, X., & Lam, K. (2002). A combined genetic algorithm and nonlinear least squares method for material characterization using elastic waves. *Computer Methods in Applied Mechanics and Engineering, 191*(17–18), 1909–1921.

Liu, G., Nguyen-Thoi, T., Nguyen-Xuan, H., & Lam, K. (2009). A node-based smoothed finite element method (ns-fem) for upper bound solutions to solid mechanics problems. *Computers & Structures, 87*(1–2), 14–26.

Luo, H., Ci, S., Wu, D., & Tang, H. (2010). Adaptive wireless multimedia communications with context-awareness using ontology-based models. In *Proceedings of IEEE Global Communications Conference*.

Lynch, J. P., Wang, Y., Loh, K. J., Yi, J.-H., & Yun, C.-B. (2006). Performance monitoring of the geumdang bridge using a dense network of high-resolution wireless sensors. *Smart Materials and Structures, 15*(6), 1561.

Mäkinen, O. (2015). Streaming at the edge: Local service concepts utilizing mobile edge computing. In *Proceedings of The 9th International Conference on Next Generation Mobile Applications, Services and Technologies*.

Melhem, H. & Cheng, Y. (2003). Prediction of remaining service life of bridge decks using machine learning. *Journal of Computing in Civil Engineering, 17*(1), 1–9.

Mohan, R., Purohit, Y., & Liang, Y. (2012). Deformation behavior of nanoscale material systems with applications to tensile, flexural and crack propagation. *Journal of Computational and Theoretical Nanoscience, 9*(5), 649–661.

Monaghan, J. J. (1992). Smoothed particle hydrodynamics. *Annual Review of Astronomy and Astrophysics, 30*, 543–574.

Morcous, G. (2006). Performance prediction of bridge deck systems using markov chains. *Journal of Performance of Constructed Facilities, 20*(2), 146–155.

Naess, A. & Rúyset, J. (2000). Extensions of turkstra's rule and their application to combination of dependent load effects. *Structural Safety, 22*(2), 129–143.
Nick, W., Asamene, K., Bullock, G., Esterline, A., & Sundaresan, M. (2015). A study of machine learning techniques for detecting and classifying structural damage. *International Journal of Machine Learning and Computing, 5*(4), 313–318.
Nunna, S., et al. (2015, April). Enabling real-time context-aware collaboration through 5g and mobile edge computing. In *Proceedings of 12th International Conference on Information Technology - New Generations (ITNG)*, Las Vegas, NV.
Pakzad, S. (2008). *Statistical approach to structural monitoring using scalable wireless sensor networks*. Berkeley: University of California.
Petyt, M. (2015). *Introduction to finite element vibration analysis* (2nd ed.). Cambridge: Cambridge University Press.
Quek, S. & Liu, G. (2003). *Finite element method: A practical course*. London: Butterworth-Heinemann.
Robelin, C.-A. & Madanat, S. M. (2007). History-dependent bridge deck maintenance and replacement optimization with markov decision processes. *Journal of Infrastructure Systems, 13*(3), 195–201.
Roshandeh, A. M., Poormirzaee, R., & Ansari, F. S. (2014). Systematic data management for real-time bridge health monitoring using layered big data and cloud computing. *International Journal of Innovation and Scientific Research, 2*(1), 29–39.
Sak, H., Senior, A. W., & Beaufays, F. (2014, September). Long short-term memory recurrent neural network architectures for large scale acoustic modeling. In *Proceedings of Interspeech*, Singapore (pp. 338–342).
Salakhutdinov, R. & Hinton, G. E. (2009). Deep boltzmann machines. In *Proceedings of 12th International Conference on Artificial Intelligence and Statistics*, Clearwater Beach, FL, USA (pp. 1–8).
Samaan, N. & Singh, C. (2002). A new method for composite system annualized reliability indices based on genetic algorithms. In *2002 IEEE Power Engineering Society Summer Meeting*, Chicago, IL, USA (pp. 850–855).
Simiu, E. & Scanlan, R. H. (1986). *Wind effects on structures: An introduction to wind engineering* (2nd ed.). New York: Wiley-Interscience.
Sofge, D. A. (1994, 29 November–2 December). Structural health monitoring using neural network based vibrational system identification. In *Proceedings of the Australia and New Zealand Conference on Intelligent Information Systems*, Brisbane, Queensland, Australia.
Sohn, H., Czarnecki, J. A., & Farrar, C. R. (2000). Structural health monitoring using statistical process control. *Journal of Structural Engineering, 126*(11), 1356–1363.
Thoft-Christensen, P., & Baker, M. J. (1982). *Structural reliability theory and its applications*. Berlin, Heidelberg: Springer.
Torres-Toledano, J. G., & Sucar, L. E. (1998). Bayesian networks for reliability analysis of complex systems. In *IBERAMIA '98, Proceedings of the 6th Ibero-American Conference on AI: Progress in Artificial Intelligence*, London, UK (pp. 195–206).
Tran, L., Banerjee, D., Wang, J., Kumar, A. J., McKenzie, F., Li, Y., et al. (2013). High-dimensional mri data analysis using a large-scale manifold learning approach. *Machine Vision and Applications, 24*, 995–1014.
Wang, J., Liu, G., & Lin, P. (2002). Numerical analysis of biot's consolidation process by radial point interpolation method. *International Journal of Solids and Structures, 39*(6), 1557–1573.
Wang, H., Ullah, M. M., Klaser, A., Laptev, I., & Schmid, C. (2009). Evaluation of local spatio-temporal features for action recognition. In *University of Central Florida, USA*.
Wu, D., Ci, S., Luo, H., Wang, H., & Katsaggelos, A. (2009). A quality-driven decision engine for service-oriented live video transmission. *IEEE Wireless Communications, Special Issue on Service-Oriented Broadband Wireless Network Architecture, 16*(4), 48–54.
Wu, D., Ci, S., Luo, H., Ye, Y., & Wang, H. (2011). Video surveillance over wireless sensor and actuator networks using active cameras. *IEEE Transactions on Automatic Control, 56*(10), 2467–2472.

Wu, D., Chatzigeorgiou, D., Youcef-Toumi, K., Mekid, S., & Mansour, R. (2014). Channel-aware relay node placement in wireless sensor networks for pipeline inspection. *IEEE Transactions on Wireless Communications, 13*(7), 3510–3523.

Wu, D., Chatzigeorgiou, D., Youcef-Toumi, K., & Mansour, R. (2015a). Node localization in robotic sensor networks for pipeline inspection. *IEEE Transactions on Industrial Informatics, 12*(2), 809–819.

Wu, Z., Wang, X., Jiang, Y., Ye, H., & Xue, X. (2015b, October 26–30). Modeling spatial-temporal clues in a hybrid deep learning framework for video classification. In *MM '15 Proceedings of 23rd ACM International Conference on Multimedia*, Brisbane, Australia (pp. 461–470).

Ye, W., Heidemann, J., & Estrin, D. (2002). An energy-efficient mac protocol for wireless sensor networks. In *Proceedings of INFOCOM 2002* (pp. 1567–1576).

Ye, X., Ni, Y., Wong, K., & Ko, J. (2012). Statistical analysis of stress spectra for fatigue life assessment of steel bridges with structural health monitoring data. *Engineering Structures, 45*, 166–176.

Zhao, Z. & Chen, C. (2001). Concrete bridge deterioration diagnosis using fuzzy inference system. *Advances in Engineering Software, 32*(4), 317–325.

Zhou, J. T. & Yang, J. (2013). Prediction of bridge life based on svm pattern recognition. *Intelligent Automation and Soft Computing, 17*(7), 1009–1016.

Zhu, P., Ma, Z. J., Cao, Q., & French, C. E. (2011). Fatigue evaluation of transverse u-bar joint details for accelerated bridge construction. *Journal of Bridge Engineering, 17*(2), 191–200.

Part III
Applications in Medicine

Chapter 13
Nonlinear Dynamical Systems with Chaos and Big Data: A Case Study of Epileptic Seizure Prediction and Control

Ashfaque Shafique, Mohamed Sayeed, and Konstantinos Tsakalis

13.1 Introduction

Mathematical models have enabled us to understand the universe we live in and to a great extent manipulate behaviors of physical, chemical and biological systems. While the accuracy of a model in an application is debatable, the fact that a fairly accurate model helps predict the system's behavior is not. This was clearly demonstrated by Pierre-Simon Laplace with his concept of universal determinism (Simon et al. 1951). In his 1778 work, Laplace ingeniously explained the motions of the known celestial bodies in his time (sun, moon, and planets) with Newton's laws of motion and reduced the study of planets to a set of differential equations.

A century later, Henri Poincaré developed what is known as *state-space*—a set of system states defined over time. This helped study the evolution of a physical system over time by characterizing the system based on laws of physics with the application of differential equations (Poincaré 1992). It is during this work that Poincaré observed the phenomenon of *sensitivity to initial conditions*. While studying the classical *three-body problem* involving the earth, the sun, and the moon, he found that they have orbits that are non-periodic, yet their periods are neither stable nor unstable. In other words, the periods are bounded within a region in their state-space, neither escaping to infinity nor settling to a fixed-point. However, as Poincaré pointed out, knowledge of the initial conditions of the states do not allow for their long-term predictability. Thus breaking away from Laplace's work of universal determinism. This point in time is regarded as the birth of Chaos Theory, although it would take another century for this field of science to become mainstream.

A. Shafique • M. Sayeed (✉) • K. Tsakalis
Arizona State University, 650 E. Tyler Mall, Tempe, AZ 85281, USA
e-mail: ashfaque@asu.edu; msayeed@asu.edu; tsakalis@asu.edu

S. Srinivasan (ed.), *Guide to Big Data Applications*, Studies in Big Data 26,
DOI 10.1007/978-3-319-53817-4_13

A nonlinear dynamical system is said to be chaotic when it is sensitive to differences in initial conditions, is topologically mixing and has dense periodic orbits. The high sensitivity to initial conditions is often referred to as *the butterfly effect* in pop culture. These minuscule differences in initial conditions may appear due to noise or machine accuracy in computation and can yield widely diverging outcomes for such nonlinear chaotic dynamical systems. Long-term prediction thus becomes impossible in general. Chaotic systems need not be complex in nature, they can be completely deterministic, implying that their future states are completely predictable with no stochasticity involved. Therefore, the study of mathematics that deals with deterministic systems with chaotic behavior is known as Chaos Theory and was first formalized by Edward Lorenz in 1963 (Lorenz 1963). In Lorenz's own words he summed up chaos as:

> Chaos: When the present determines the future, but the approximate present does not approximately determine the future.

Ever since, Lorenz's discovery of chaotic phenomena in weather modeling, the field of chaos theory has exploded into many aspects of science and engineering. Chaos theory is used in cryptography by computer scientists, creating population models in biology, studying turbulence in fluid mechanics. In economics, it is utilized to predict stock market behavior but has had mixed results due to the tremendous complexity of such systems. Feedback, where humans anticipate and react to changes in the market by buying or selling stocks, exacerbates this complexity in stock market behavior. In astronomy, chaos has been used to describe the motion of many planetary bodies and in particular to better predict asteroid paths and whether or not they may come in contact with Earth. And, in more recent years it has been applied in the prediction and/or control of human brain dynamics. While this list of applications is by no means comprehensive, it does provide us with a notion of how ubiquitous chaos is in our modern lives.

One such interest in modeling brain dynamics through chaos theory is in the search of a remedy for epilepsy. Epilepsy is a chronic, noncommunicable medical condition that results in seizures affecting a wide range of mental and physical functions in humans. It is the fourth most common neurological disorder after stroke and Alzheimer's, and affects at least 50 million people worldwide and an estimated 2.4 million people are diagnosed with epilepsy each year (World Health Organization 2016). Currently, antiepileptic drugs (AEDs) are the principal form of chronic epilepsy treatment. However, in addition to the lack of efficacy for complete seizure control, there also is a substantial morbidity associated with the use of AEDs in patients, especially when polypharmacy is required. The goal of epilepsy management is to make the patient completely seizure free, with or without minimal side effects from the anti-epileptic treatment. While surgical removal of the seizure focus is an important and effective therapeutic intervention for some of the people with uncontrollable epilepsy, because of multiple foci, seizure foci located within non-resectable areas of the brain and possible post-operative complications, resective surgery is unlikely to ever replace chronic treatment as the primary mode of epilepsy management in the large majority of patients with epilepsy.

There are an estimated 0.4 million Americans whose epilepsy cannot be treated with AEDs or surgery (Citizens for Research in Epilepsy 2016). Currently, there are only two FDA-approved devices for treatment. One is a neurostimulator called RNS System developed by NeuroPace to reduce the frequency of seizures and is an implantable device (Food and Drug Administration 2015). The other is a vagus nerve stimulator developed by Cyberonics; this device is implanted in the chest to prevent seizures and has been FDA approved since the 90s. Both these devices claim to capture a patient's unique seizure patterns and apply appropriate electrical stimulation. The RNS worked only in a subset of patients and could provide only a 50% or more reduction in the rate of seizures 2 years post implant. None of these treatments have provided a cure for the disease and even their effectiveness is limited (Miller 2014). Hence, there is a great need to develop a wearable device that can make this a treatable disease. Quite a number of research efforts, over the past two decades, have been carried out in combining control engineering and physiological functioning of the brain to develop theoretical and computational models combined with experiments on animals for epileptic seizure prediction and intervention (Tsakalis and Iasemidis 2004). One of the key impediments to the advances in this field comes from the fact that recording, analyzing and modeling brain activity is a data intensive task. Until recently our resources were largely limited by the computational tools available to us. Recent advances in Big Data analytics are driving seemingly impossible tasks from the past, towards fruition today.

The dawn of the internet era and a decade of technological advancement in computational hardware and software along with new developments in modeling and simulations in science and engineering has ushered in an exciting and challenging problem of creating, mining and handling large data sets. Now, every discipline, from the arts and social sciences, to science and engineering, is seeing an explosion of data (National Science Foundation 2016). In particular two recent flagship initiatives standout—the *European Human Brain Project* (Human Brain Project 2016) and the United States' *The Brain Initiative project* (The White House 2014). These are mega size research projects focused on accelerating progress towards a multilevel understanding of the human brain, better diagnosis and treatment of brain diseases, and brain-inspired Information and Communications Technologies (ICT) (Human Brain Project 2016). The BRAIN 2025 Report states, theory, modeling, and statistics will be essential to understanding the brain. The computational validation of theory and models requires an easy to use, high-performance rapid prototyping platform for complex data analysis, control, and visualization (National Institute of Health 2014). Thus, the task of solving the brain dysfunction, epilepsy, is being reformulated as a big data problem under the banner of these large scale projects. As part of our current work, we have developed a framework titled *HPCmatlab* (based on *Matlab*) which enables us to perform big data analytics on a high-performance computational platform that enables multi-scale modeling, simulation, data analyses and intuitive visualization of not only the specific problem of epilepsy management but any other field of research involving large data sets.

This book chapter focuses on the process of developing nonlinear systems analysis methods, particularly by utilizing the Lyapunov exponent, in order to treat epilepsy with the help of a big data analytics tool HPCmatlab. In the following sections, we discuss in more detail—the background of epilepsy as a disease; past and current work done in the field, information about HPCmatlab and parallel computational tools; Nonlinear systems, chaos theory, and Lyapunov exponents; a case study of epilepsy management utilizing Lyapunov exponents. Finally, we conclude the chapter with some remarks about possible future directions to be taken in this topic.

13.2 Background

This section dives a little deeper into the topics introduced previously. First, we start with a historical treatment of epilepsy and move into the current practices of its management. The engineering approach to curing epilepsy is not a new one, however, advances in computational technology is allowing for better, more accurate treatment methods. These include development of efficient algorithms, data capture and storage techniques and analysis platforms for the challenges resulting from Big Data. The discussion is then followed by one on massively parallel computing within the context of the computational tool—HPCmatlab and Big Data. These topics may seem disjoint at first and quite correctly may be so, however, from our work we show that great strides can be made when supercomputing tools are utilized in cybernetics problems such as the treatment of epilepsy from an engineering standpoint.

13.2.1 Epilepsy

The World Health Organization depicts that epilepsy affects 50 million people worldwide and annual new cases are between 30 and 50 per 100,000 people in the general population (World Health Organization 2016), making it the fourth most common neurological disorder after migraine, stroke, and Alzheimer's (Hirtz et al. 2007). Although Epilepsy occurs in all age groups, the highest incidence rates are among children and the elderly (Forsgren 1990). The known factors that cause epileptogenesis in humans are traumatic brain injuries, central nervous system infections, brain tumors and genetic abnormalities amongst many others (Annegers et al. 1996). Currently antiepileptic drugs (AEDs) are the principal form of chronic epilepsy management; however, they are ineffective on 30% of the patients (Dodson and Brodie 2008). Thus there are nearly 15 million people worldwide who need a more efficient means of treatment.

Epilepsy manifests itself in patients as epileptic seizures. Seizures in epileptic patients occur due to synchronous neural firing in the cerebral cortex. It has been found that these paroxysmal electrical discharges may begin in a small neighborhood of the brain, in which case they are called partial seizures or focal seizures with single or multiple foci. Or, they may begin simultaneously in both cerebral hemispheres in which case they are classified as primary generalized seizures (Engel 2013). After the onset of a seizure, partial seizures may remain localized and cause relatively mild cognitive, psychic, sensory, motor relapses or may spread to other regions of the brain and cause symptoms similar to generalized seizures. Generalized seizures at their onset bring about altered states of consciousness and can also have a variety of motor symptoms, ranging from a minimal loss of motor action or brief localized body jerks to heavy convulsive actions known as tonic-clonic activity. Immediately after the seizure, the patient may lose complete consciousness and wake up a few minutes later, never realizing what they went through.

The ancient Greeks thought of epilepsy as a disease that was "given by the Gods" to the patient. The Latin word *Epilepsia* is a translation of the Greek term *Epilambanein* which means "to take hold of, or seize" (Temkin 1994). This was primarily due to the fact that seizures would come and go, in a seemingly unpredictable fashion. For some patients, it happens hundreds of times a day and for others, once every few years. The definition of epilepsy is to have two or more unprovoked seizures (World Health Organization 2016). This elusive property of seizures is what makes them difficult to treat. The long-standing clinical practice of temporal or frontal lobectomy is still the most successful means of treatment; however, not everyone is a candidate for such a procedure, especially if their focus sites are in critical brain function areas. Patients with seizures who cannot be treated with AEDs and are not candidates for lobotomies are termed to have refractory epilepsy. This leaves little choice but to look for viable alternative means of treatment with stimulations such as Deep Brain Stimulation (DBS), Vagus Nerve Stimulation (VNS), Transcranial magnetic stimulation (TMS) etc.

In their study, Tassinari et al. (2003) and Boroojerdi et al. (2000) have reported on how TMS had affected seizure rates to decrease. Also, Lulic et al. (2009) have shown promising results for VNS in patients with refractory epilepsy. Similarly, DBS, principally of thalamic structures, has been reported to reduce seizure frequency in humans (Hodaie et al. 2002). In a focal model of epilepsy, such stimulations have been shown to result in seizure elimination in 48% and improvement in seizures in 43% of patients (Good et al. 2009). However, there is still plenty of room for research in the control of this "divine" disease. One such area for study is the stimulation parameters for seizure control. As Kuncel and Grill (2004) have shown that there is a total of 12,964 possible combinations of stimulation parameters. Of these, certain frequencies and intensities have been identified as "safe", to be used in neurostimulators, such as those developed by NeuroPace and Cyberonics.

An endeavor to control the occurrence of seizures must begin with identifying them accurately and in a timely fashion. Ever since its discovery in the 1920s by Hans Berger (Niedermeyer and da Silva 2005), the Electroencephalogram (EEG) has been utilized to diagnose conditions of the brain including seizures (Thiel et al. 2013). Primarily, a trained electroencephalographer would notice changes in the EEG and be able to diagnose that a seizure is occurring. Also in cases of partial seizures, they would mark the focal electrodes. However, this method has a major limitation in the fact that a timely intervention either by the use of drugs or electrical stimuli via DBS cannot be made effective since the seizure is identified only when a visible change occurs in the EEG, not prior to seizure initiation. To this end, research has progressed greatly in the last few decades, aided by mathematical tools commonly used in the engineering principles of signal processing to help detect and even in some cases predict seizures.

In the late 1980s, Iasemidis and colleagues have reported on such mathematical methods that utilize nonlinear systems theory to assess the state of the brain (Iasemidis et al. 1988). In their work they analyzed EEG data from epileptic patients to compute the Short Term Maximum Lyapunov Exponent (STL_{max}). It was shown that as the brain transitions from *interictal* (between seizures) to *preictal* (immediately before seizure) to *ictal* (during a seizure) and then to *postictal* (after a seizure) and back to interictal states, the nonlinear dynamics of the brain changes from a chaotic to a more ordered and back to a chaotic state. Lasemidis et al. (2000) shows how the choice of different parameters in the STL_{max} algorithm can be tuned to reveal the change of the state of the brain from chaos to order in the order of minutes or even hours. These detection/prediction algorithms implied that finally, it would be possible to think of a real time intervention strategy minutes before a seizure occurs.

The problem of control of epileptic seizures has been approached by different researchers in two major ways. *Open-loop control* is where a stimulation is applied periodically or in some cases the patient initiates the stimulator whenever they get the "aura" of an impending seizure. *Closed-loop control* techniques involve monitoring of EEG data and using them in a feedback loop to provide stimulus if and when necessary. Chakravarthy et al. (2008) have shown that the control of synchrony in a neural mass model of the brain can be achieved via simple feedback control utilizing a Proportional-Integral (PI) controller; such a design methodology for PI controllers using an optimization procedure is described in Shafique and Tsakalis (2012). Following results from these simulations, it was shown that epilepsy is when the internal feedback mechanism of a pathological brain is disrupted and that it can be restored via an additional feedback path through the use of electrical stimulation. Likewise, Kalitzin et al. (2010) have reported that application of a periodic burst of stimulus can help predict seizures better and be used to control them; in a sense this can be thought of as an observability problem, where the chance of predicting the seizures is increased through minute stimulations. To achieve closed loop feedback control, sensory information about the system's states is necessary. One of the most prominent ways to achieve this is by the use of observers, a.k.a filters. Aram et al. (2013) uses a neural mass model based on that by Jansen and Rit (1995) and implemented an Unscented Kalman Filter to estimate model parameters to be used eventually for control. Even with the advent of these new tools, control of epileptic seizures still remains an open ended question and the search for the "holy grail", the best prediction and control scheme, continues.

13.2.2 Parallel Computing

Recent decades have seen rapid technological advancements in both computing hardware and software. However, the past dependance on hardware to improve performance of applications has stalled, since the physical limit of the number of transistors that can be packed on a single processor chip is reached. And, the current trend towards multi-core and many-core processors (also called accelerators such as GPUs and Intel Xeon Phi co-processors), will continue to progress rapidly. Interestingly, these single computer systems can now offer enormous computational muscle, in the teraflops to tens of teraflops (FLOPS-floating point operations per second), with the help of accelerators that they can outperform a parallel supercomputer only a couple of decades earlier. Current generation of top ranked supercomputers are usually in the tens to hundreds of petaflops peak theoretical performance (TOP 500 2016). These supercomputers, are also referred to as high performance computing (HPC) systems or clusters. These clusters, small or big, are built using commodity processors—CPUs and GPGPUs (general purpose GPUs), memory and storage devices and networked together using sophisticated usually proprietary interconnects with high bandwidth and low latency.

These HPC systems are complex and require advanced knowledge of both the hardware and programming models. Parallel programming for such systems usually involves two main paradigms—distributed memory and shared memory. Distributed memory programming requires explicit message passing using the message passing interface (MPI) standard application programming interface (API) (MPI Forum 2016) for internode communication. The shared memory programming model uses threads for intranode local memory accesses (within a node of a large cluster or your multicore/many-core computer) using POSIX threads (pthreads) (IEEE 2004) or OpenMP (Open Multi-Processing) standards API (OpenMP 2016), and the OpenCL (Open Computing Language) standard (OpenCL 2016) for heterogeneous architecture platforms i.e., nodes with accelerators. The shared memory parallel programming model using OpenMP, a directive-based approach is the simplest to program. However, large shared memory systems are very expensive to build and are not scalable. Hence, most of the systems, whether they are small, medium or large petaflops machines, are commodity distributed memory systems. The de-facto programming model of choice for such distributed memory systems is explicit message passing for communication between processes using MPI. The MPI standard defines a language independent API for both point-to-point, collective, one-sided communication and parallel I/O between processes (MPI Forum 2016). Over the period of more than two decades MPI has evolved and offers a truly high performance, scalable and portable solution for distributed memory parallel programming. Parallel application development on such platforms is traditionally performed using low level programming languages such as Fortran, C and C++ with the above mentioned APIs.

Given the challenges with application development, data storage and distribution, scientific and other communities have now evolved to using flagship applications, centralized data repositories, and compute/analytics platforms. Many approaches that allow ease of programming usually involve domain specific languages (DSL). Matlab (or it's clone Octave) is a popular DSL for matrix programming or numerical computing. While, Matlab is a commercial software package from Mathworks it's use is widespread in both academia and industry. Being an interpreted language, performance is usually sacrificed for ease of programmability. In particular, for big data applications, it may not even be an ideal platform for developing applications except for algorithm testing and rapid prototyping. The HPCmatlab platform used in this research study addresses some of the challenges and shortcomings.

13.2.3 Challenges

Traditionally scientific computing has been at the forefront of data driven knowledge discovery, both in terms of the amount of data generated from modeling and simulations and the data aggregated from observations for analysis. Numerous computational science and engineering examples exist that are all noticeably big data applications. These are commonly referred to as e-science applications. These applications have the characteristics of big data—the 3V's or 4V's namely Volume, Velocity, Variety and/or Value. Although, e-science has developed and applied techniques for data capture, storage, searching, sharing, analysis, and visualization, it does not fully address the current or future challenges of Big Data (Chen and Zhang 2014). Given the amount of data or information generated is growing exponentially and the state of the art analysis tools are inadequate for realistic problems (to solve in real time) or their relatively slow evolution. Other important challenges with Big Data analysis include inconsistent and incomplete data sets, data security, scalability and timelines (Kouzes et al. 2009). Usually, given the diversity of tools and information gathering techniques, appropriate data preprocessing techniques including data cleaning, data integration, data transformation and date reduction, need to be applied to remove noise and correct inconsistencies (Han et al. 2011).

One of the major drawbacks of current system architectures is the persistent bottleneck from slower I/O (Input/Output) units at multiple levels (a) network communication in case of distributed computing (b) storage—both network attached storage (NAS) or local hard drive and (c) memory hierarchy, compared to the speed at which floating point operations can be performed. Usually. this is aggravated for big data applications as they are I/O intensive. Some of these are discussed and addressed in the context of the present application of EEG data for epileptic seizure prediction and control. Moreover, these challenges become somewhat tractable with parallel computing.

13.2.4 Current State of Art

Most certainly Big Data has become the buzz word because of the power of social networks and the overall potential being touted of e-commerce based on data mining. Unfortunately, the current techniques and technologies to capture, curate, analyze and visualize Big Data are varied and far from meeting the needs of different disciplines. While the techniques employed actually rely on advances already made in statistics, signal processing, optimization, artificial neural networks, deep machine learning and visualization etc., the tools being developed and employed to address Big Data issues are recent. These tools, depending on the application are broadly classified under, batch processing (or offline analysis e.g., Apache/Map-Reduce, Dryad, Apache-Mahout etc.,), Stream processing (or online/real time analysis e.g., Storm, Splunk, S4 etc.,) and, interactive analysis—Dremel (Google) and Drill (Apache) (Chen and Zhang 2014). Briefly, Map-Reduce is based on the divide and conquer strategy with Apache's Hadoop implementation providing the Hadoop kernel, Map/Reduce and Hadoop distributed file system (HDFS) infrastructure. Applications that are suited to the Map-Reduce paradigm are the only suitable candidates that can use it productively. Only one particular scientific domain namely bioinformatics is well suited and seem to be exploiting it.

The Dryad tool (currently unavailable) uses data flow graph processing for implementing parallel and distributed programs on clusters. A Dryad application runs as a computational directed graph, with vertices corresponding to computational programs and edges forming the communication channels. A report comparing it's performance and use for scientific applications highlights several limitations are presented in Salsa Group (2010). Apache Mahout provides machine learning techniques for large-scale and intelligent data analysis applications on top of Hadoop platform via the Map/reduce framework. The real-time stream processing big data tools Storm, Splunk, S4 etc., mentioned above promise scalable, fault-tolerant and easy to use platforms. These tools, to the best knowledge of the authors, have not been used extensively for any real scientific applications. Similarly, the interactive tools, Dremel (now called BIGQUERY) and Drill offer scalability and support for different query languages, data formats and data sources. They offer large-scale ad hoc querying of data with capability to process petabytes of data and trillions of records in seconds (Chen and Zhang 2014).

While these tools and platforms certainly take away some of the complexity and challenges with big data, they have many limitations and should be carefully chosen for particular application needs. Nevertheless, HPCmatlab provides the needed functionality for HPC platforms with a very flexible and robust environment for Big Data analysis.

13.3 Nonlinear Dynamical Systems with Chaos

A *dynamical system* is represented mathematically as a set of differential equations. The function that describes a dynamical system and establishes the time dependence of its states is generally represented in the form

$$\dot{x} = f(x, u, t) \tag{13.1}$$

where, x, u and t represent the state, input and time variables respectively. If the system is *time-invariant* then t is no longer a variable and is dropped from the function. In this case the system is said to be *autonomous*. Sometimes an output equation will be added and the whole description is said to be a *state-space representation* of the dynamical system. Due to the nature of differential equations (difference equations in discrete time systems), dynamical systems exhibit *memory*; that is, the value of the current states would depend on its past values. For instance, a swinging pendulum is a system where the current states, position and velocity of the pendulum, relies on knowledge of its past states. For most real-world systems, the differential equations that describe their behavior are nonlinear in nature. As an example, Eq. (13.2) describes the motion of a pendulum with force input at its pivot.

$$\ddot{\theta} + \frac{c}{mL^2}\dot{\theta} + \frac{g}{L}sin(\theta) = \frac{T_{max}}{mL^2}T \tag{13.2}$$

Here, θ (output) is the angle between the pendulum rod and the vertical and $\dot{\theta}$, $\ddot{\theta}$ are the corresponding higher order derivatives (angular velocity and acceleration). c is the coefficient of friction, m is the mass of the pendulum, assumed to be centered at the end of the rod of length L. T is the applied torque and is normalized to the maximum applicable value, T_{max} (dependent on the maximum torque of a motor). Typically, for the pendulum, the states are considered to be the angle θ and its angular velocity $\dot{\theta}$. As can be seen in this case, the non-linearity in the model appears from the presence of the sine term on the system output θ; however non-linearity can appear in any other mathematical form for other systems. The validity of the model is also in question; we can have models that represent a system "closely" but never perfectly. This is one of the fundamental limitations of modeling real-world systems. The science of modeling is a juggling act between time and money spent in creating an "accurate" model versus the benefits of the increased accuracy. Albert Einstein's statement aptly captures it:

> As far as the laws of mathematics refer to reality, they are not certain, and as far as they are certain, they do not refer to reality.

The *state space* (for discrete-time systems) or *phase space* (for continuous time systems) describes the set of values that the states in a system can take. For instance, in the case of the pendulum, if the two states were represented on the two axes of a coordinate system, then a ring centered about $[0,0]$ would describe the phase-space plot of the system. At no point in time can the pendulum have values for states outside the ring. Certain classes of systems exhibit a behavior in which the phase-space is such that a large set of initial conditions will lead to the state-space trajectory converging to a point or an area; the region of initial conditions that do so is called a *basin of attraction*. Regions of phase-spaces with such behavior are called *attractors*. The four types of attractors are described in brief:

- Fixed-point: A fixed point attractor is one that comes out of a system whose eigenvalues are in the left-half-plane (stable systems). Typically systems that lose energy have fixed-point attractors; e.g. a pendulum with friction will always converge to the bottom position ($\theta = 0, \dot{\theta} = 0$), if there is no force input. The phase-space will show trajectories converging to the fixed-point from all local initial conditions (angular position and velocity). Intuitively—no matter where we start the pendulum off, after a certain amount of time, the bob will come to rest at the vertical downwards position.
- Limit Cycle: A limit cycle appears from systems whose states display periodic behavior. The trajectory takes the shape of a ring. A good example is the oscillator circuit used to generate tuning frequencies in a radio.
- Limit Torus: A limit torus, like a limit cycle has a periodic trajectory. However, in this case the system exhibits more than one natural frequency and any two of these frequencies being an irrational ratio makes for the trajectory to take the shape of a torus in the phase-space. An example would be an oscillator with two sinusoids where the frequencies form an irrational fraction.
- Strange attractor: Strange attractors are "strange" simply because the trajectories neither converge entirely into a fixed-point, nor do they escape like an unstable system, nor form a periodic orbit of a concentric nature and the trajectories are not quite on the same plane. The trajectories are locally bounded but never overlap. Nearby trajectories would seem to escape from each other, yet the distance cannot grow beyond a specific value. The dimension of these attractors are also *fractal* in nature due the trajectories not being on the same plane. A strange attractor is the cornerstone of a chaotic system's phase-space. The Lorenz attractor (Fig. 13.3) is an example of such an attractor.

David Ruelle and Takens are credited to have introduced the term *strange attractors* after Ruelle studied the Lorenz attractor in depth (Ruelle and Takens 1971). It is worth mentioning that not all dynamical systems would have attractors in their phase space. Having briefly introduced the concept of phase-space and attractors, we now move on to its relevance to chaotic systems. In the following, state-space and phase-space are used interchangeably with the assumption that the reader understands that state-space is meant for systems described by discrete time systems and phase-space for continuous ones.

Edward Lorenz's discovery of chaotic behavior came as a culmination of two centuries of research in systems theory and mathematics. He found that using the same equations, the computer solution with rounding in three digits versus six, produced solutions that were entirely different from each other (Oestreicher 2007). This is known as sensitive dependence to initial conditions and is one of the building blocks of chaos theory. In his 1972 paper titled "Predictability: Does the Flap of a Butterfly's Wings in Brazil Set Off a Tornado in Texas?" (Edward 1972), Lorenz popularized the notion of the *Butterfly effect*. That the insignificant effect of the flapping of a butterfly's wings can lead to a tornado in another part of the world, all because a slight variation in the initial conditions was brought about, was the gist of the publication. However, a formal definition of chaos would not be put forward until much later when Robert Devaney laid out the three essential properties for a system to be chaotic (Hasselblatt and Katok 2003):

- Sensitivity to initial conditions.
- Topological mixing.
- Dense periodic orbits.

Figure 13.1 shows the effect of sensitivity to initial conditions. The output was generated from the differential equations of a Lorenz system, which is a nonlinear dynamical system comprised of three states. The initial conditions for states $[x, z]$

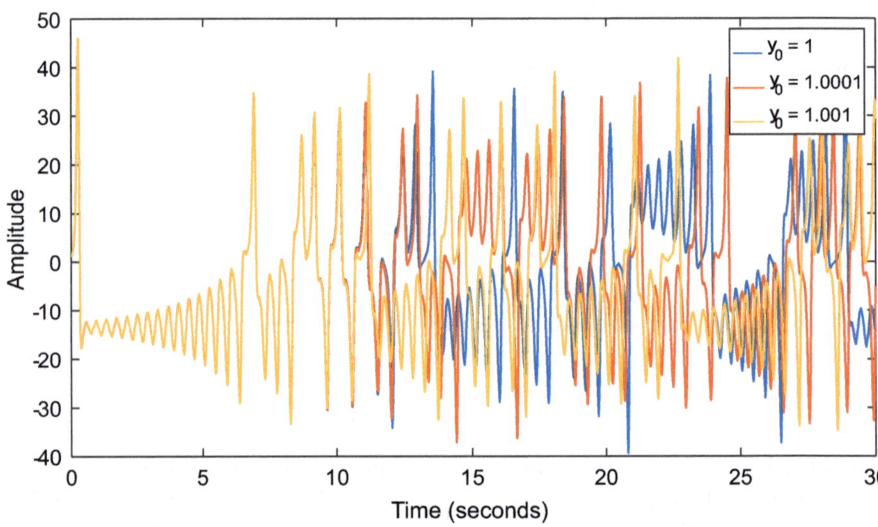

Fig. 13.1 Plot of the y variable from the Lorenz equation. The initial conditions, y_0, were chosen as 1, 1.0001 and 1.001. As can be seen, even the slightest variation in initial conditions can make the output of y vary greatly over a 50 s evolution of time; even though they start off being similar near the beginning of the simulation. No noise was added to this data to emphasize that the sensitivity is indeed to initial conditions and not any external influence. The values for the parameters were $\sigma = 45.92$, $\rho = 16$ and $\beta = 4$, while the initial conditions for states other than y were fixed to $x_0 = 0$ and $z_0 = 1.05$

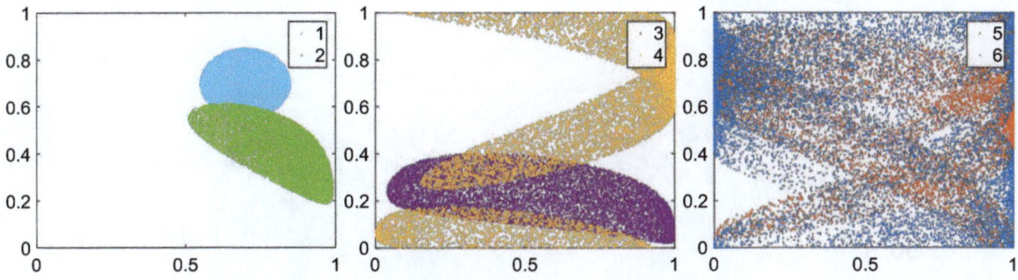

Fig. 13.2 Six iterations of a set of states $[x, y]$ passed through the logistic map. (**a**) the *blue plot* (legend 1) shows the first iterate (initial condition), which essentially forms a circle. Plots in (**a**)–(**c**), show the first to the sixth iteration of the circular initial conditions. It can be seen that *mixing* occurs as we progress in iterations. The sixth iteration shows that the points are almost completely scattered in the phase space. Had we progressed further in iterations, the mixing would have been homogeneous and irreversible. The logistic map has equation $x_{k+1} = 4x_k(1 - x_k)$. In order to expand the state-space of the logistic map into two dimensions, a second state, y, was created as $y_{k+1} = x_k + y_k$, if $x_k + y_k < 1$ and $y_{k+1} = x_k + y_k - 1$ otherwise. The y variable being depicted modulo one at each step makes the points fold over within the unit square, otherwise the points may have escaped the region. This modulo operation implies that in (**b**) and (**c**) the points near the top and bottom edge are in fact closer to each other than they appear; this is because the operation creates a cylinder parallel to the x axis

were kept the same while that for y was varied between $[1, 1.0001, 1.001]$. As can be seen from the figure, in a short time the state y evolved with a rather significant difference.

Topological mixing, implies that the system will evolve over time such that an *open set* in its state-space will eventually overlap with any other given region in the state-space. Mixing is a non-reversible process. Turbulence in fluids is an example of a chaotic system; the analogy allows us to imagine how two fluids can interact within a volume. As an example, Fig. 13.2 shows how a set of points bounded in $[0, 1]$ evolves over six iterations of the *Logistic Map*, another example of a chaotic system. As is evident from the images, the process of mixing smears the blue circle from the first iteration into almost a blur in the closed region. after just six iterations.

Finally, dense periodic orbits refer to the state-space trajectory of a system. The informal meaning or intuition of dense periodic orbits can be attained by thinking of trajectories in the state-space that can be arbitrarily close to each other. Figure 13.3 shows the trajectory of the Lorenz attractor in three-dimensional space. The solutions to the equation were computed numerically using MATLAB and only 50 s worth of data are presented. This is why the "dense" nature of the orbits is not apparent, however longer simulations would have produced an output whereby the trajectories would seem to overlap each other; trajectories never overlap in a chaotic attractor. More data would not imply that the attractor would necessarily grow (or shrink) in size, only that the trajectories would show more density. It is perchance that the shape of the Lorenz attractor looks similar to that of a "butterfly", finding another coincidence to the coining of the term "The Butterfly Effect".

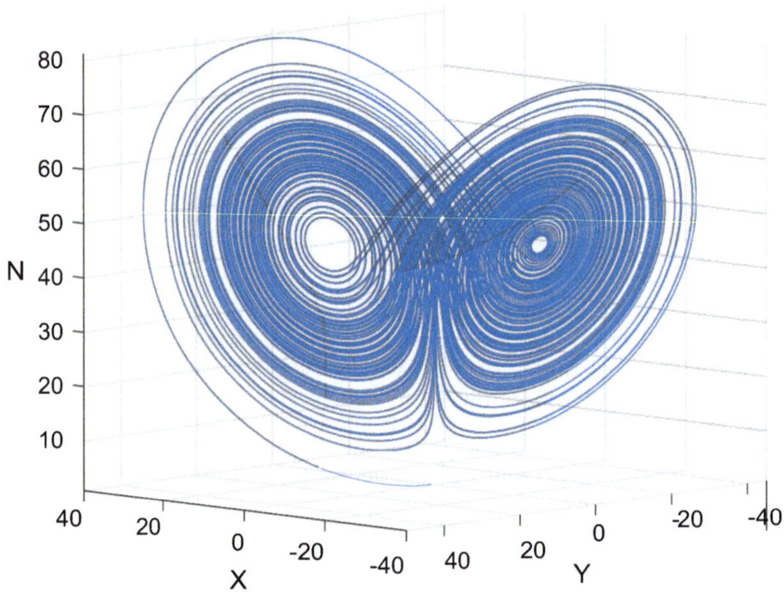

Fig. 13.3 A diagram of the Lorenz attractor. This plot was generated via a 3-D plot in MATLAB. The Lorenz equations were solved using MATLAB *ode45* tool with a step-size of 0.001 s. The simulation was run for 50 s. It is worth noting that the picture could be made to look *dense* as a chaotic attractor should be by simply allowing for much more time evolution of the trajectory. For this simulation the values for the parameters were the same as used in Fig. 13.1

So far, we have given conditions and their examples of how chaos comes about. We now look at two systems that show chaotic behavior. Equation (13.3) describes the logistic map in two dimensions. The logistic map, is an example of how chaos can arise from a simple system, i.e. complex systems are not a necessity for chaotic phenomena. It must also be noted that the logistic map, a discrete-system that is represented by difference equations, can display chaoticity with a single state variable(x_k); y_k is artificially created in this example for the purpose of visualization in Fig. 13.2 and is not considered a true second state of the system.

$$x_{k+1} = 4x_k(1 - x_k), \tag{13.3}$$

$$y_{k+1} = \begin{cases} x_k + y_k, & \text{if} \quad x_k + y_k < 1 \\ x_k + y_k - 1, & \textit{otherwise} \end{cases} \tag{13.4}$$

On the other hand we have Eq. (13.5) which describes the Lorenz system. According to the Poincarè-Bendixson theorem, unlike discrete systems, continuous systems must have at least three dimensions in their phase-space to produce chaotic behavior. The lorenz system is a result of Edward Lorenz's work on weather prediction and forms the basis of a mathematical model that describes atmospheric convection.

By inspection, it is apparent that this system is also deterministic; that is, knowing the exact initial conditions, we can tell the output after a certain time. This stands as an example that deterministic systems too can become chaotic under certain conditions. For the Lorenz system, those conditions are the values for the parameters σ, ρ and β (refer to Fig. 13.1 for a set of values for these parameters).

$$\begin{aligned} \dot{x} &= \sigma(y - x) \\ \dot{y} &= x(\rho - z) - y \\ \dot{z} &= xy - \beta z \end{aligned} \tag{13.5}$$

A description of chaotic system behavior is incomplete without comparing it to random or stochastic systems. It is imperative to understand the difference between chaotic systems and stochastic ones and be able to separate the two from given data. While a detailed discussion is irrelevant in our work, it suffices to say that in order to differentiate between a chaotic output and a stochastic output we need to start with a test state in the trajectory of the two systems and find a nearest neighbor to their respective test states in a nearby trajectory and measure the difference in the two states after a few discrete time evolutions. A chaotic system will have a difference which increases exponentially over time; the time evolution of the differences in a stochastic system, however, will be randomly distributed.

This has been but a concise introduction to the topic of systems theory and chaos. Dynamical systems, nonlinear dynamical systems and chaos are by themselves topics of their own, creating enormous interest in the scientific community. A more detailed description of these topics is out of the scope of this chapter and would lose focus from our subject matter—the treatment of epilepsy and the role of chaos as applied to its solution based on principles of mathematics and engineering. For a more detailed study of nonlinear systems and chaos, the avid reader is referred to an excellent book chapter in Socolar (2006).

13.4 Lyapunov Exponents

For dynamical systems, the Lyapunov exponent is a quantity that characterizes the separation rate of two nearest neighbor trajectories in the phase-space of the system. Nearest neighbor trajectories are found based on a distance metric, euclidean distance for example. Mathematically, the Lyapunov exponent can be described by the following linearized approximation

$$\delta L(t) \approx e^{\lambda t} \delta L_0 \tag{13.6}$$

where, δL_0 is the initial separation of the nearest neighbors and $\delta L(t)$ is their distance after time evolution t. In Eq. (13.6), λ is the lyapunov exponent along the direction being considered. In a dynamical system, there will be as many lyapunov exponents as there are dimensions in its phase-space. The lyapunov exponent in each case represents the separation rate in the directional orientation of the initial separation vector δL_0.

Of this spectrum of lyapunov exponents, one in particular is of utmost interest in the study of chaotic systems—the maximum lyapunov exponent (L_{max}). Its importance is due to the fact that L_{max} being positive is an indication of a system being chaotic. However, for a chaotic attractor to be present, the overall dynamics must be dissipative; i.e. the system must be globally stable and the sum of all the lyapunov exponents must be negative (Rosenstein et al. 1993). The larger L_{max} is, the more chaotic the system. A system being more chaotic simply implies that the rate of separation is higher. This has been characterized as the lyapunov time for chaotic systems and is strictly dependent on its dynamics. Electrical circuits that are chaotic have very short lyapunov times (milliseconds) versus the solar system which has lyapunov time in the order of millions of years.

Given a system model with n equations, computing the spectrum of lyapunov exponents is achieved by solving all the equations for a set of nearby initial conditions and allowing them to expand through the equations. The growth of the vectors defined by the initial conditions is measured and at every time evolution a Gram-Schmidt Reorthonormalization procedure is performed to ensure that the vectors maintain proper phase-space orientation and do not all lean towards the direction of the most rapid growth (Wolf et al. 1985). The rate of the growth of these vectors can then be used to compute all the lyapunov exponents. However, there are cases when a mathematical model of a system is not available, only experimental data. In those cases, particularly to estimate the chaoticity of a system, only L_{max} is required and can be estimated using the method described in the following sections.

13.4.1 Numerical Computation of Lyapunov Exponents

It has been shown in Iasemidis et al. (1988) that EEG in humans and animals are not random stochastic signals as they were thought of in the past. Rather, they can be described as electrical activity generated by a chaotic oscillator that is part of the brain's mechanism. Figure 13.4 is a graphical representation in three dimensions of an attractor generated from EEG data of an epileptic animal. This data is proof that the epileptic attractor is strange in nature. As such, the chaoticity of a strange attractor can be quantified by means of its Maximum Lyapunov Exponent (L_{max}). In the absence of a dynamical model (state equations), computing the Lyapunov exponent means relying on Takens embedding theorem which allows a single observed variable to be expanded into a higher dimensional state-space. The Lyapunov exponent is then computed as a mean logarithmic deviation of the trajectories in the higher dimensional space over time. The following describes a few notable algorithm's that can achieve just that.

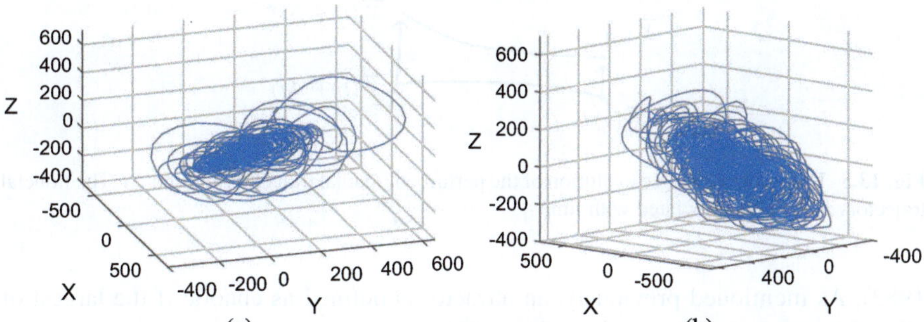

Fig. 13.4 Strange attractors from an epileptic animal EEG. (**a**) shows the attractor during an interictal epoch of 10 s whereas (**b**) shows the same for an ictal epoch. These attractors were created using 10 s of EEG data from the same first channel, expanding it into seven dimensions using Takens embedding theorem methodology and then compressing into three dimensions, using orthogonal projection, for visual representation. The Trajectories may seem like they cut each other in 3D but that will not be the case in higher dimensions of embedding. The attractor at seizure seems to occupy more volume indicating that the trajectories separate from each other more. This intuition will be corroborated with data presented later. The axis limits in both images are the same for *X*, *Y* and *Z*

13.4.1.1 The Wolf Algorithm

The first step in computing the Lyapunov exponent is to create a delayed vector of observed values from a time series. Takens (1981) has shown that this delayed signal contains within them all the state variables of the system. In the case of the EEG, this method can be used to reconstruct the multidimensional state space of the brain's electrical activity from each EEG channel. For example, if $x(t)$ is a $n \times 1$ dimensional vector of duration T recorded from an EEG channel and sampled every T_s seconds, then $\bar{X}_i(t)$ is the $n \times p$ dimensional reconstructed signal such that

$$\bar{X}_i(t) = [x(t_i), x(t_i + \tau), \ldots, x(t_i + (p-1) * \tau)] \tag{13.7}$$

where, τ is the delay between successive components of $\bar{X}_i(t)$ and it rarely, if ever, is the same as T_s. If a phase space plot of the $n \times p$ dimensional vector $\bar{X}_i(t)$ were to be created, then it would look like that of a strange chaotic attractor (Iasemidis et al. 1988). For reference, such an attractor is shown in Fig. 13.4. The complexity of this attractor is measured by its dimension D and Iasemidis *et al* have shown that for a sinusoidal attractor, the value of $D = 1$ and for that of a chaotic attractor, such as those found in EEGs of epileptic patients, to be within $D = [2.5, 2.7]$.

A description of how D can be estimated from time series data via its state space correlation dimension ν is given by Grassberger et al. (1991). The measure of chaoticity of these attractors can be defined via either their Kolmogorov Entropy (Grassberger et al. 1991) or their Lyapunov exponents (Grassberger and Procaccia

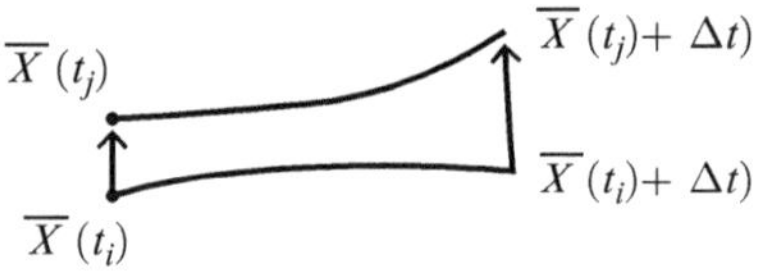

Fig. 13.5 Diagram of a single evolution of the perturbed fiducial trajectory in time Δt. The fiducial trajectory is the one associated with time t_i

1983). As mentioned previously, an attractor is defined as chaotic if the largest of all its Lyapunov exponents (L_{max}) is positive.

The method for choosing p, the embedding dimension of the state space of the signal $x(t)$, was proposed by Takens (1981) to be $p \geq (2 * D + 1)$. Although the dimension of an attractor can be fractal, that of the embedded signal, p, cannot. It is worthwhile to mention that the brain is a nonstationary system and as such never reaches steady state; so its value for D is never constant. This is why the time window of $T = 10\,\text{s}$ is chosen so as to better satisfy the assumption of stationarity for the signal. While the embedding dimension should be changed from epoch to epoch, the value of $p = 7$ is kept fixed for pre-ictal, post-ictal and interictal stages. The justification is that the existence of irrelevant information in dimensions higher than 7 might not influence the estimated dynamical measure by a great degree and also the reconstruction of the state space with high p suffers more from the short length of moving windows that are used to handle non-stationary data (Iasemidis et al. 1988).

Originally, Wolf had proposed an algorithm to estimate L_{max} from stationary data (Wolf et al. 1985), however, later, Iasemidis et al. modified this algorithm to compute what is known as the average short-term maximum Lyapuonv exponent (STL_{max}) for non-stationary EEG data on short time windows (Lasemidis et al. 2000). STL_{max} can be calculated as follows:

$$STL_{max} = \frac{1}{N_a \Delta t} \sum_{i=1}^{N_a} log_2 \frac{|\delta \bar{X}_{i,j}(\Delta t)|}{|\delta \bar{X}_{i,j}(0)|} \tag{13.8}$$

where, $\delta \bar{X}_{i,j}(0) = \bar{X}(t_i) - \bar{X}(t_j)$ is the displacement vector at time t_i, i.e. a perturbation of the fiducial orbit at t_i, and $\delta \bar{X}_{i,j}(\Delta t) = \bar{X}(t_i + \Delta t) - \bar{X}(t_j + \Delta t)$ is the evolution of this perturbation after time Δt (Fig. 13.5). In other words, Δt is the time over which $\delta \bar{X}_{i,j}(0)$ is allowed to evolve in the state space. When the evolution time, Δt, is given in seconds, STL_{max} has units in bits/sec. N_a is the number of local Lyapunov exponents that are estimated within a duration T of the data segment. This gives us the following relation between T, the length of a segment of data, and Δt the evolve time:

$$T = (N - 1)\Delta t \approx N_a \Delta t (p - 1)\tau \tag{13.9}$$

13.4.1.2 The Rosenstein and Kantz Algorithm

The principal flaw in Wolf's algorithm was in the reselection process wherein after every evolve time, Δt, a new candidate for the nearest neighbor to the fiducial trajectory is selected. This process lends itself to increased error in the presence of even the slightest amount of noise in the data; rendering the Wolf method useless to any real data. In the 1990s two other groups formulated new methods to estimate the Lyapunov exponent in order to circumvent this error (Rosenstein et al. 1993; Kantz 1994).

In the Rosenstein et al. (1993) method, Takens embedding theorem is applied to reconstruct the state-space in higher dimensions. Then an initial fiducial trajectory is chosen along with its nearest neighbor trajectory. The average divergence between two points, one on the fiducial and another on its nearest neighbor trajectory, $L(t)$, at time t, is computed using the following equation,

$$L(t) = Ce^{L_{max}t} \tag{13.10}$$

where, C is a constant that normalizes the initial separation. In order to compute L_{max} numerically, the trajectories are allowed to evolve over the entire finite data set of time T. Then the log distance versus evolution time is plotted and the gradient of the initial part of the log distance curve is evaluated as the L_{max} for one trajectory pair. Similarly, for the rest of the data set, the process is repeated by moving forward, a number of samples at a time, in the phase-space and computing an estimate for L_{max}. The average of all such L_{max} values is the final result from the Rosenstein algorithm.

The Kantz (1994) algorithm, also starts from computing the expanded phase-space trajectory. However, unlike the Rosenstein algorithm, instead of selecting a single nearest neighbor to track, the Kantz algorithm chooses a radius around the fiducial and computes the average of how all the trajectories within the radius separates from the fiducial. The log distance vs evolve time is plotted and like the Rosenstein algorithm, the gradient at the beginning of the plot is the value for L_{max}. This uses an exponentially greater number of trajectories and thus is more robust to perturbations in the signal. It must be noted that the first few points in the evolution is discarded since the maximal Lyapunov exponent at that time has not overtaken the other stable Lyapunov exponents.

Figure 13.6, shows how these two methods are employed in computing L_{max} for a Lorenz attractor. The data was sampled at 100 Hz. The log distance curve in both cases was allowed to evolve for 500 samples. The straight line has a gradient that is the estimate of L_{max}. The Kantz algorithm correctly estimates the value for L_{max} at 1.4877. The Kantz algorithm, for its higher reliability in estimating L_{max}, is what we utilize in our work.

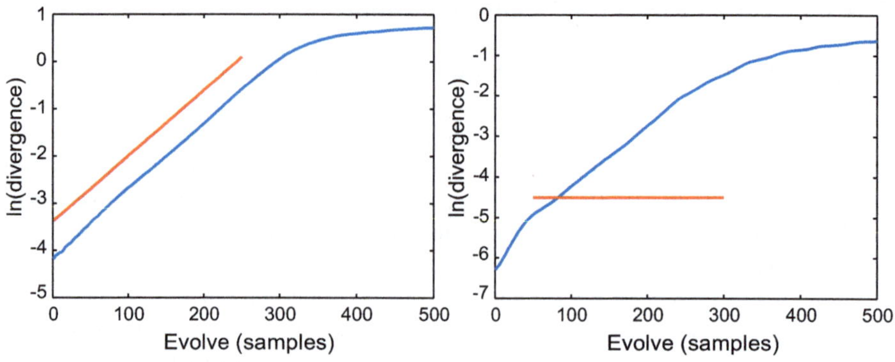

Fig. 13.6 Time vs log distance curves of the (**a**) Rosenstein and (**b**) Kantz algorithm applied to data from a Lorenz attractor sampled at 100 Hz. The straight lines show the fit that is used and its gradient is the value for L_{max}. The value for L_{max} from (**a**) and (**b**) respectively are 1.3870 and 1.4877. 1.5 is the theoretical value; thus Kantz algorithm works better in this case

13.5 Rapid Prototyping HPCmatlab Platform

Matlab is an ideal platform for prototyping, design explorations, and algorithm development, for modeling and simulations. However, to retain its appeal of ease of programming, Mathworks offers limited parallel computing capability through its Parallel Computing Toolbox (PCT) to enable intranode simple loop-based parallelism, and a separate Distributed Computing Server (DCS) for internode distributed computing using MPI. Both of these require a separate license and the DCS license is rarely purchased even at large institutions as it is prohibitively expensive. In addition, the performance of DCS is unsatisfactory and provides limited functionality in terms of point-to-point and collective communication calls only. The user is thus limited to using only these functions (Guo et al. 2016).

To address the limitations of Matlab without the DCS license and challenges associated with big data applications, the HPCmatlab framework enables shared memory and distributed memory parallelism using POSIX threads and MPI. In addition, it supports parallel I/O using MPI I/O and Adaptable IO System (ADIOS) (Liu et al. 2014). As noted earlier parallel I/O is extremely important for Big Data applications and currently not supported in Matlab. Our novel approach is based on MEX (Matlab EXecutable Engine) API (Mathworks 2016). It enables Matlab programs to interact with functions or programs written in other languages such as Fortran, C and C++. We have tried to retain the semantics of the POSIX, MPI and OpenCL API standards in HPCmatlab. So any code translation or transformation from prototype development to a programming language such as Fortran or C is almost trivial. The HPCmatlab framework currently offers several advantages including performance compared to Mathworks offering (for details see Guo et al. 2016).

The ease of use of Matlab is a key factor that has enabled its success. However, HPCmatlab emphasizes the use of standard APIs providing greater functionality and control in the hands of the user, rather than defining another new API like most tools including Matlab, in order to simplify and ease the burden of parallel programming. It serves multiple purposes—(a) put the control in the hands of the application developer and (b) the effort put in learning the standards for parallel programming will payoff in the long run (as you do not have to learn a new API for another programming package/tool) and (c) enable parallel programming use become main stream as Matlab has a large user base.

13.5.1 Bigdata and Matlab

The big data trend is likely to continue in the near future with increasing digitization of information. The Mathworks Inc. realizes this opportunity and outlines key features of its product offering at this URL (The Mathworks 2016). A set of tools i.e., functions are available, namely, *memmapfile* to map a file, or a portion of a file, to a MATLAB variable in memory, *matfile* to access MATLAB variables directly from MAT-files on disk, using MATLAB indexing commands, without loading the full variables into memory and, *datastore* function to access data that doesn't fit into memory. All these functions allow one to perform in-core (memory) or out-of-core (disk) computations and are limited to a single multicore processor. It offers limited parallel computing capabilities through its PCT and DCS toolboxes. It supports accelerators, mainly NVIIDA GPU's using the CUDA toolkit, which again is not a standard. In addition, there is limited GPU computing capability, only a few GPU enabled functions, few multithreaded intrinsic functions, distributed arrays using PCT or DCS to work with large datasets that do not fit in a single computer's memory. Thus, very limited parallel computing capability is enabled via these tools to simplify the ease of use and other reasons.

However, it provides builtin functions to the popular Map-Reduce paradigm (Dean and Ghemawat 2010), using Hadoop platform (Apache Software Foundation 2016), which does not provide good performance. It offers many algorithms for analysis in its Statistics, Machine Learning and Neural Network Toolbox. On the visualization front limited functionality is available via the Image Processing Toolbox using *blockproc* function in PCT. Currently no large volume rendering of datasets or real time visualization is possible. Thus, although a ubiquitous numerical computing platform, support for large scale parallel computing with big data is currently lacking in Matlab.

13.5.2 Parallel Computing and HPCmatlab

The HPCmatlab platform seeks to address some of the above limitations to enable large number of Matlab users to carry out big data computations, analysis and visualization in an already familiar environment of Matlab (or Octave). As mentioned earlier, the two main parallel programming models and HPC hardware architectures are either shared or distributed memory with addition of the accelerator hardware. Message passing programs reflect the underlying architectures distributed address space. In MPI, programmers must explicitly partition the data and use message passing to communicate between processors that own data and those that require data. Although intuitive, distributed memory programming can be difficult and an error prone process for complex and adaptive applications.

The main features of the framework $v1.0$ are shown above in Fig. 13.7. It provides all basic MPI communication routines for performing point to point, collective and one-sided communication. A Pthreads based interface is also provided for shared memory parallelism or a hybrid MPI-Pthreads approach to carryout large scale computations. The Pthreads module can be used to parallelize Matlab functions after translating them to C language program. At the minimum it provides same functionality as Matlab's *parfor* command to parallelize *for* loop iterations but can also be used to exploit task based functional parallelism (if available in an application). For many scientific applications I/O is a potential source of bottleneck. The framework provides parallel I/O capability using Adaptable IO System (ADIOS).

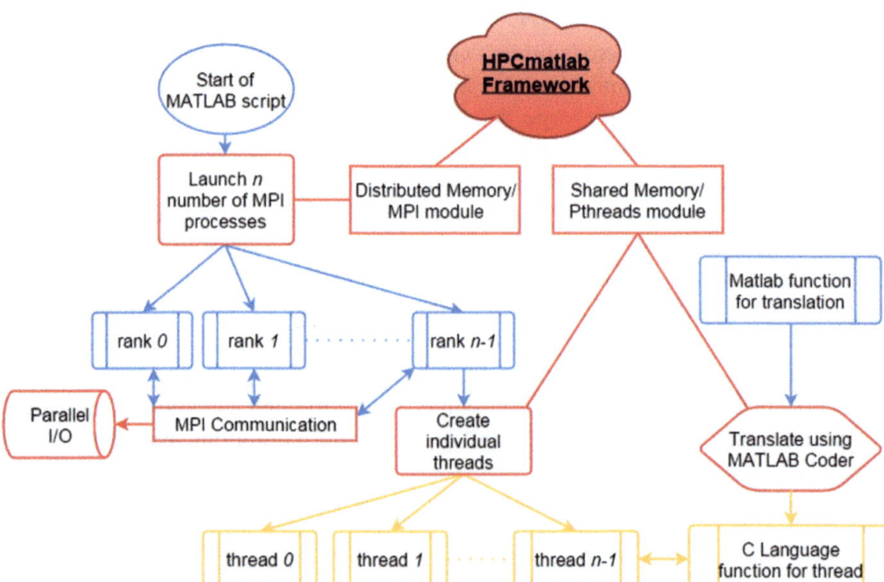

Fig. 13.7 Shows current features available in HPCmatlab's *v1.0* computational framework

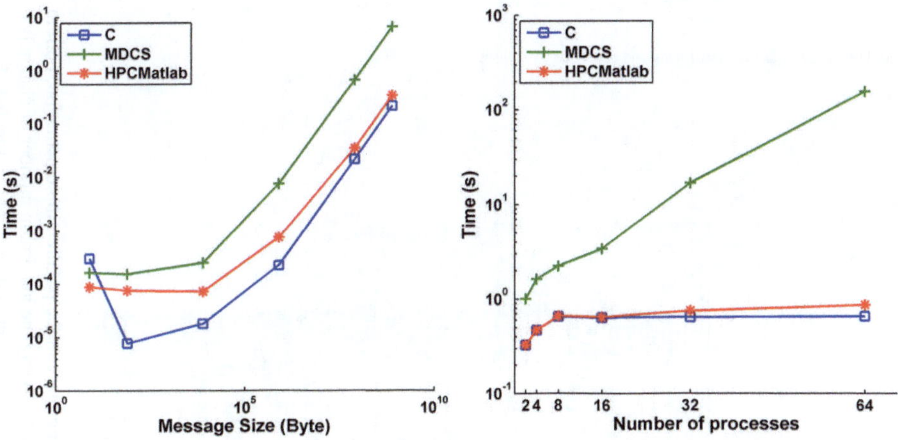

Fig. 13.8 Performance comparison of MPI point to point (*left*) and collective communication (*right*) in native C language program, HPCmatlab and, Matlab DCS (reproduced from Guo et al. 2016)

ADIOS is an easy-to-use package for fast, scalable and portable parallel IO with a simple, transparent, and flexible mechanism to define how to read, write or process data in simulations (Liu et al. 2014). A similar tool called *HPCoctave* based on free GNU open source Octave (Watson 2014) provides similar functionality. For details about HPCmatlab framework ($v1.0$) implementation, specific use, and performance results see Guo et al. (2016).

In general, the performance of parallel applications depends on the communication overhead. Briefly, for completeness, the performance of our framework is shown in Fig. 13.8. It is very close to a C based MPI program for both point-to-point and collective communication operations. The performance of Matlab DCS is relatively poor and does not show good scalability.

13.5.2.1 Parallel Computation of Lyapunov Exponents

Extensive offline investigations are required to optimize the model parameters and decide the level and location of electrical stimulation. Our choice of Kantz algorithm for computation of Lyapunov exponents of the chaotic dynamical system was presented earlier. The flowchart in Fig. 13.9 shows the algorithm and it's parallel implementation using Matlab's *parfor* and MPI in HPCmatlab. These computations are naturally parallel and lend themselves easily to rapid prototyping for this application. In the implementation, Lyapunov exponents are calculated over windows of 10 s of data. Note, a single EEG recording can be as long as 178 h (1 week). These segments overlap with an offset of 2 s between two consecutive segments. The computations over 10 channels of EEG data (iteration counter i) are parallelized using *parfor*, a loop based parallelism mechanism provided in Matlab. But this

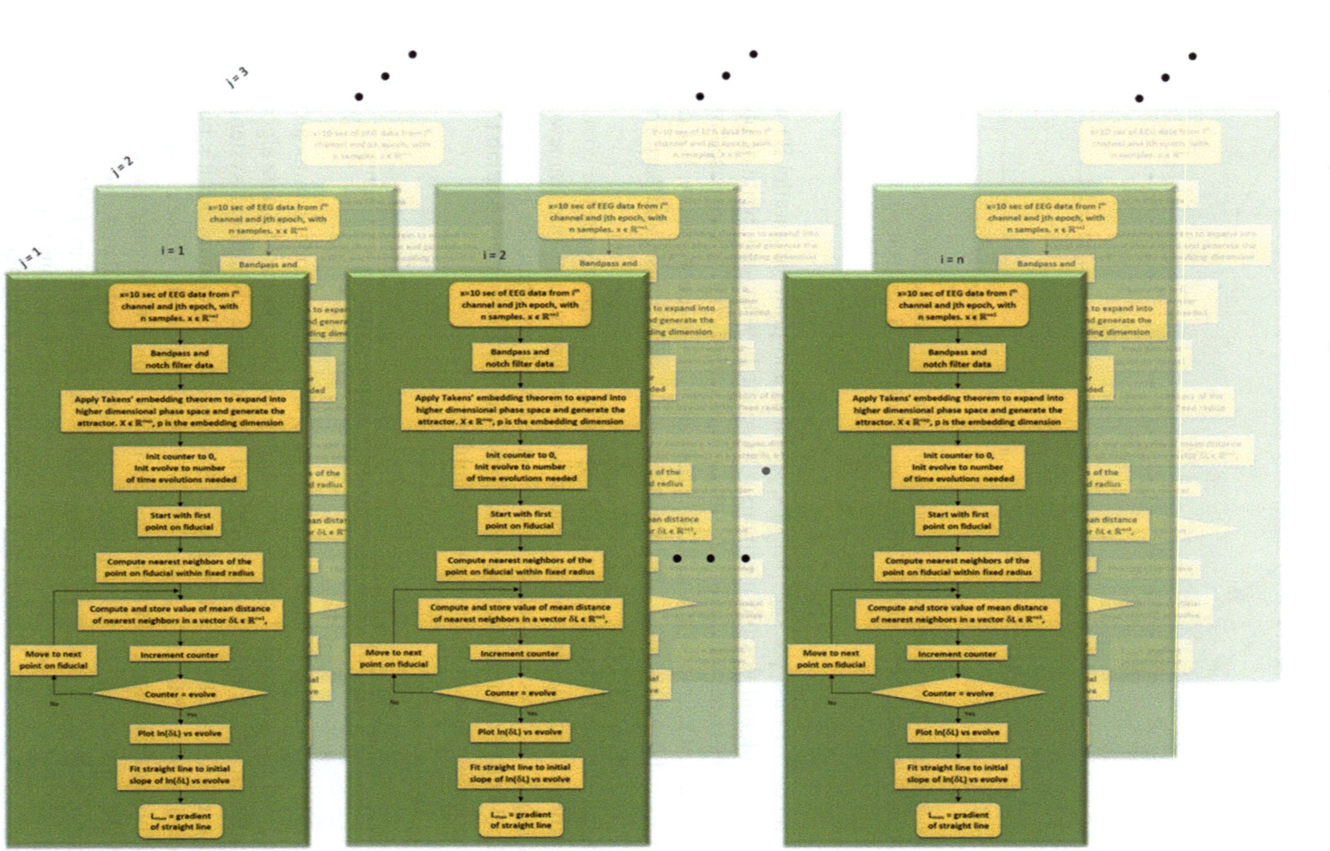

Fig. 13.9 Flowchart showing the Kantz algorithm. The repetitions in *i* and *j* are to emphasize the process taking place in the parallel implementation, both in online and offline execution of the code. *i* is the iterate for the number of channels parallelized using *parfor* with *n* being the maximum number of channels. *j* is the iterate of epochs, parallelized using HPCmatlab with MPI. For offline computation, the value is limited by the amount of data stored on the disk. For the online case, however, it can go on as long as data is being collected

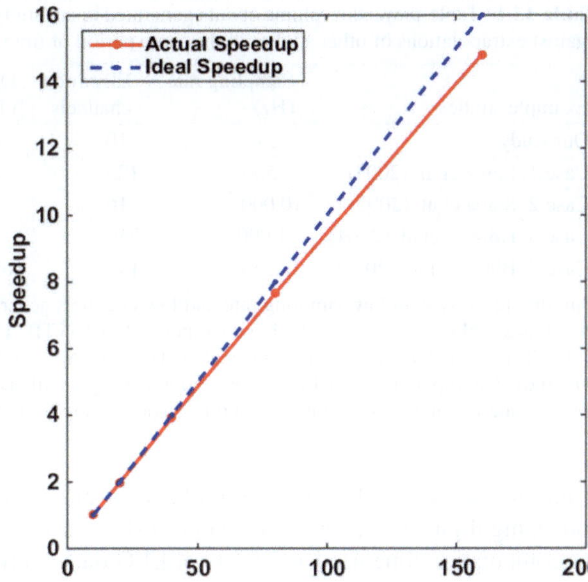

Fig. 13.10 Speedup plot showing scalability of parallel seizure prediction and control code used for the computation of Lyapunov exponents. The actual speedup value should be multiplied by a factor of 10 to account for the use of 10 *parfor* threads

limits the computation to just one node with multiple cores in a cluster. This, loop level coarse-grained parallelization over channels was used as the first step towards parallelization. It was found inadequate for the large number of offline computations needed for parameter optimizations. So, HPCmatlab was used to divide the EEG data as work segments (j iterate) or epoch among MPI processes. Each MPI process calculates the Lyapunov exponent for its part of data segments (10 s) in the EEG data file and at the end, result from all processes are gathered in one single array on the root process. This has resulted in a hybrid programming model where data segments are distributed on different nodes via MPI and parfor is used within a node to parallelize the computation on ten different channels. Figure 13.10 shows very good scaling performance of HPCmatlab for this application. The application scaling performance and MPI point-to-point (Fig. 13.8) benchmarking results were obtained on the Gordon cluster at San Diego Supercomputing Center (SDSC) using Matlab version 2014b and HPCmatlab $v1.0$.

13.5.2.2 Epilepsy as a Big Data Problem

Undoubtedly, epilepsy management is a big data problem as several studies already refer to it as such (Ben-Menachem 2016; Devinsky et al. 2016; The Neurology Lounge 2016). The recommended protocol for clinical management of epilepsy requires prolonged inpatient monitoring of brain activity using EEG (ambulatory or video telemetry) to accurately access each patient's day-to-day event pattern (The Neurology Lounge 2016). In addition, new devices allow monitoring of clinical and subclinical seizure activity at home, resulting in large volumes of data which

Table 13.1 Table shows the volume of data generated in our study over a 1 week period compared against extrapolations of other studies in the same period of time in column 4

Example studies	Sampling rate (Hz)	Electrode channels	Data volume (TB)	Time taken (s)
Our study	512	10	0.18	0.0927
Case 1, Lantz et al. (2003)	500	125	2.2	1.132
Case 2, Staba et al. (2002)	10,000	16	5.63	2.898
Case 3, Holmes et al. (2004)	1,000	256	9.01	4.6369
Case 4, Blanco et al. (2011)	32,556	144	165.04	84.9139

Our simple study with low sampling rate and low channels generate 180 GB of data compared to the biggest volume in case 4, which is extrapolated to 165 TB. The study truncated their volume to 12 TB to keep data processing manageable. The fifth column shows how much time it takes in our study to compute L_{max} for 10 s of data and how long it will take in the case of the other studies (extrapolated) using a 6 core Intel Xeon processor running 6 Matlab parfor workers

is interrelated with other disease markers for personalized treatments. Table 13.1 above, highlights some interesting studies showcasing the volume of EEG data and the computation time needed for 10 s of EEG data analysis on a desktop computer.

13.6 Case Study: Epileptic Seizure Prediction and Control

As described in Sect. 13.2.1, epilepsy is a neurological disorder characterized by seizures which are recurrent perturbations of normal brain function. Of the 2.2 million troops returning from Iraq and Afghanistan, 100,000 are estimated to develop post-traumatic epilepsy (PTE) (Citizens for Research in Epilepsy 2016). The CDC estimates 5.1 million adults and children combined, in the US, have been diagnosed of epilepsy as of 2016 (1.6% of total US population) (Centers for Disease Control and Prevention 2016). The total indirect and direct cost of epilepsy in the United States is estimated to be $15.5 billion yearly. This estimate is based on a reported cost of $12.5 billion in 1995 converted to 2004 dollar value using Bureau of Labor Statistics data. Approximately 60% of new onset epilepsy cases respond to existing antiepileptic drugs (AEDs) but 30% are pharmaco-resistant, having seizures that cannot be fully controlled with available medical therapy or without unacceptable side effects (Dodson and Brodie 2008). Thus, there are at least 15 million people world-wide for whom the development of more effective epilepsy treatment paradigms would be greatly beneficial.

The use of electrical stimulation presents itself as an attractive alternative for the treatment of epilepsy but the development of effective stimulation strategies and the mechanisms of operation are still not well understood. In addition to the extremely high complexity of brain operation and the variety of seizure types, there are only limited modes of data collection that can be used, both because of physical constraints and because of ethical regulations. The net result is that we are currently lacking a solid and widely acceptable model of seizure development, including seizure precursors or predicting mechanisms. As a few examples of the variety of models we cite Howbert et al. (2014), showing how using spectral power features from pre-ictal and inter-ictal data in a logistic regression classifier can improve chances of seizure prediction in dogs with focal epilepsy (Mormann et al. 2003), where phase synchronization between different regions of the brain using intracranial EEG recordings from 18 patients, shows a characteristic decrease from a few minutes to several hours before a seizure. On the other hand, using a maximum "Short-Time Lyapunov" exponent metric, Iasemidis et al. (2004) have observed an increasing entrainment in minutes to hours before a seizure in human EEG data. In Kalitzin et al. (2010) and Suffczynski et al. (2008), the authors showed that prediction and detection of seizures can be facilitated by using low levels of stimulation and examining the properties of the identified input-output dynamical system. In a completely different approach, Kramer et al. (2011) describes a new portable method of seizure detection using motion sensors, that can prove very useful in clinical practice; 91% of seizures were detected within a median period of 17 s for tonic-clonic seizures.

One conclusion drawn from the collection of these studies is that the modeling of epilepsy can potentially involve a variety of metrics computed from real-time signals. The volume of the data can increase very rapidly when we consider the number of electrodes used (10 in a simple rodent study but beginning from 32 all the way up to 256 in most modern systems involving human subject studies) and the frequency of sampling (from 240 Hz of the earlier EEG sampling rates to tens of kHz that may appear in some relatively modern studies). And, while the intent and the paradigm followed by each investigator is to use one metric as a seizure predictor or detector, the possibility that the computational algorithm may require some form of tuning for different cases clearly demonstrates the need for extremely high computing capabilities aka, HPC. Finally, a great variety in the data can be found from simultaneous acquisition of EEG data with ambient data, muscle motion data and video data that, at the very least, can relate the observed electrical signals to the clinical behavior of the subject. The variety of metrics, the volume of data and its varying types is what has made the treatment of epilepsy a big data problem in recent years. In our work, we have focused on one type of data, i.e. EEG only.

From our work, we have seen that the use of Kantz algorithm enhances the results for L_{max} computation not only in simulated chaotic data but also in real EEG data collected from animals. Figure 13.11 shows a comparison of the three different methods applied to the same set of EEG data. A seizure as marked visually by inspecting EEG is seen to occur at about 13 min. The plot shows L_{max} over time, where L_{max}, for each channel, was computed for 10 s of data moving forward in time with a 2 s overlap. It is interesting to observe that during the seizure, while the Kantz algorithm shows a distinct peak, in the L_{max} time profile, that from the Wolf shows none and from the Rosenstein algorithm shows a drop but it is not consistent over many other seizures. Also, the Rosenstein algorithm seems to be more susceptible to noise as is evidenced by the differing value of one channels compared to the rest; this channel had more noise in it in the EEG data. At this point it would be prudent

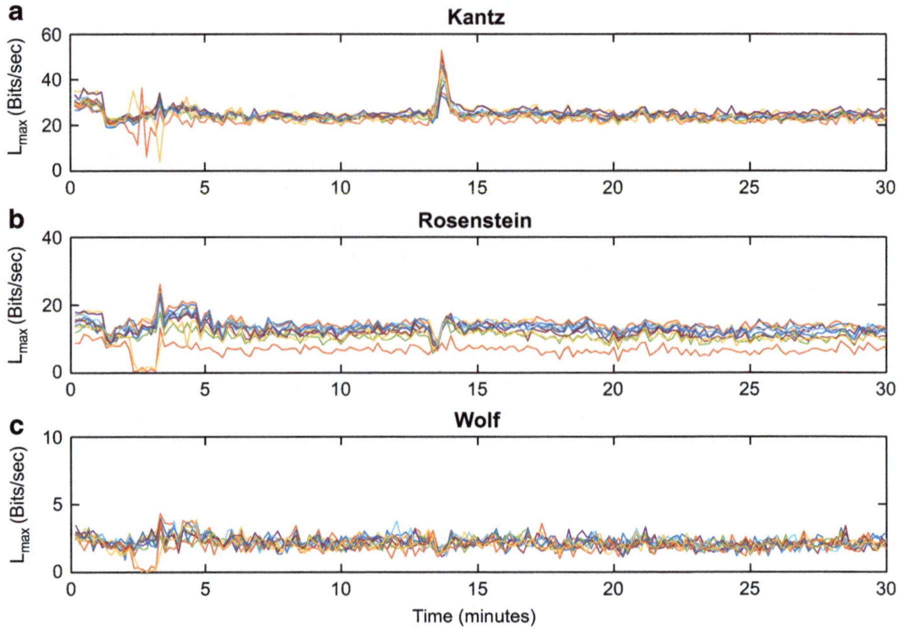

Fig. 13.11 Comparison of L_{max} vs time, where L_{max} is estimated from (**a**) Kantz, (**b**) Rosenstein, (**c**) Wolf algorithm. Clearly the one computed from Kantz shows a distinct shape during the seizure at time 13 min. The blue vertical line marks the beginning of the seizure

to mention that the intuition gained about the shape of the epileptic attractor during interictal and ictal epochs (refer to Fig. 13.4) does support the idea that during the seizure, L_{max} would have a higher value since the trajectories seem to separate from each other more in Fig. 13.4b than in Fig. 13.4a. Another interesting observation is that the L_{max} profile drops from its mean level of roughly 30 bits/sec to a lower value almost 5 min prior to the seizure start in the case of the Kantz algorithm; this in turn can become useful as a seizure precursor measure and be used to stimulate in order to attempt a seizure abortion. The caveat to using the Kantz algorithm in such a setting is that it requires almost 200 times more computation due to using a neighborhood of nearby trajectories compared to the one trajectory used in the Wolf and Rosenstein algorithms. Without the use of super-computing platforms offline data analysis would take days to weeks and online computations would be infeasible in the absence of accelerators.

The consideration of seizure suppression or seizure reduction by means of electrical stimulation only adds to our challenge. The more "traditional" line of work looks at the injection of a stimulus at the epileptic focus. Here, it is important to model the effect of the different type of stimuli on the observed brain behavior. For example, Beverlin and Netoff (2013) shows that it is possible to either extend or truncate the tonic or clonic phases of the seizure by changing the frequency of stimuli. Mina et al. (2013) shows that deep-brain stimulation (DBS) strongly reduced the sustained epileptic activity of the focal cortical dysplasia (FCD) for low-frequency (LFS, < 2 Hz) and high-frequency stimulation (HFS, > 70 Hz) while intermediate-frequency stimulation (IFS, around 50 Hz) had no effect. Our own studies and preliminary results (further discussed below) also indicate that one needs to consider multiple stimulation points and durations to account for varying effectiveness of the stimulation. The development and maintenance of such "control efficacy" maps adds new dimension to the seizure control problem.

The difficulties in designing and implementing effective seizure-suppressing controllers motivated a new look at the problem; one that integrates our collective experience in physiology, nonlinear signal processing and adaptive feedback systems. The brain is not an unstructured, random collection of neurons guided by statistics. Thus, there is no need for brain models to be so. Our approach was to consider "functional models" of the epileptic brain, i.e. networks of elements that have feedback internal structure, perform high level operations, and produce a seizure-like behavior when they fail. In our work with such theoretical models we have shown that "seizure" suppression can be well achieved by employing a *feedback decoupling* control strategy (Tsakalis et al. 2005). The implementation of such theoretical models required only weak knowledge of their detailed structure and was guided by two principles: (a) synchronization between elements increases as the coupling between them gets stronger (also observed in experimental EEG data from epileptic rats and humans), and (b) the pathology of hyper-synchronization in the network lies primarily in its elements' internal feedback connections. These principles were introduced to explain the transitions from normal to epileptic behavior as well as the triggering of epileptic seizures caused by external stimuli (e.g., strobe lights).

Based on our prior work on seizure prediction we have found that measures of spatial synchronization of dynamics, in particular the behavior of Lyapunov exponents of critical brain sites, provide a good characterization of the transition to seizures and can be utilized in the reliable issuance of warnings on impending seizures. Results from our most recent project shows that the maximum Lyapunov exponents show pronounced and persistent drops prior to seizures (refer to Figs. 13.11 and 13.12a). This observation is consistent in epileptic subjects although it is not deterministically followed by a seizure at all times. As such, and in view of our postulated models in Chakravarthy et al. (2008), it can be interpreted as an increased seizure susceptibility state of the brain. This may not be entirely useful as a prediction measure, especially since in epileptic subjects such periods do arise frequently, compared to healthy subjects where they do not occur at all, but it creates the possibility of being used as an appropriate signal for feedback control and stimulation.

While seizures can be predicted with good sensitivity and specificity (Iasemidis et al. 2004), the question remains if we can intervene effectively to change the brain dynamics and prevent a seizure from occurring. Results from our work shows that the applied electrical stimulation dynamically disentrains the brain sites. This constitutes evidence that electrical stimulation can actually change the spatio-temporal dynamics of the brain in the desired direction (Fig. 13.12b), and hence

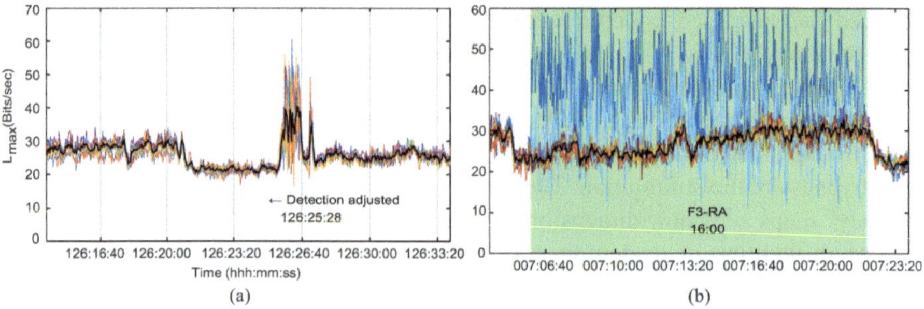

Fig. 13.12 Plots of maximum Lyapunov exponents computed using Kantz algorithm on all 10 channels of a rodent; the *thick black line* is the mean of the L_{max} from the channels. The L_{max} was computed using 10 s of data, sampled at 512 Hz, every 2 s moving in time. (**a**) Shows a seizure event at time 126 h 25 min and 26 s from the beginning of recording. The *red vertical line* marks the beginning of the seizure. As can be seen the Lyapunov drops a little a few minutes before the seizure, then at the seizure, rises to a higher value than its mean. (**b**) Shows the effect of stimulation on F3-RA electrodes was started based on a drop of mean Lyapunov after a certain threshold. As can be seen stimulation gradually pulled the Lyapunov exponent back up to average levels before the drop occurred. Two channels show L_{max} profiles that seem spurious, these are the electrodes that were stimulated and at that moment were in open circuit. Refer to Sect. 13.7 for explanation and location of the electrodes

has been used as actuation in a control scheme for epileptic seizure prevention in our most recent work. This work entails an expansion to our proof-of-concepts for the development of an efficacious controller for the epileptic brain using adaptive spatially distributed control. The developed controller (online) is to be validated in vivo. Our main thrust includes multivariable stimulation with guidance from focus localization techniques and multivariable modeling that we have recently developed, and the use of pulse-train-modulated signals for implementation of a realistic adaptive stimulation.

During the course of the ongoing study we have shown that impulsive electrical stimulation does desynchronize the rat's epileptic brain dynamics, provided that the stimulus duration and the electrode location is chosen wisely, i.e., based on a prior "control efficacy" experiment (Fig. 13.14). So our choice of electrical stimulation as a control input appears to be a viable candidate. For a realistic and feasible implementation of the desired effect, the input to the system is a biphasic control signal with its pulse-width or pulse-number modulated to accomplish the desynchornization of the brain dynamics. Figure 13.13 shows the efficacy of applying stimulation to a particular epileptic rat over a 10 week period. In this study, the animal was labeled Rat 13 and was allowed 4 weeks of rest after Status Epilepticus (SE) was induced so that the seizure frequency stabilizes. This was followed by 5 weeks of baseline recording and on the sixth week stimulation was applied to a set of electrodes every time a "seizure warning" was generated by the computer system when it detected a drop of L_{max} beyond a threshold. The case of 5 min stimulation had an adverse effect and seemed to have increased both seizure length and average seizure rate per week. However, weeks 7 through 9 were

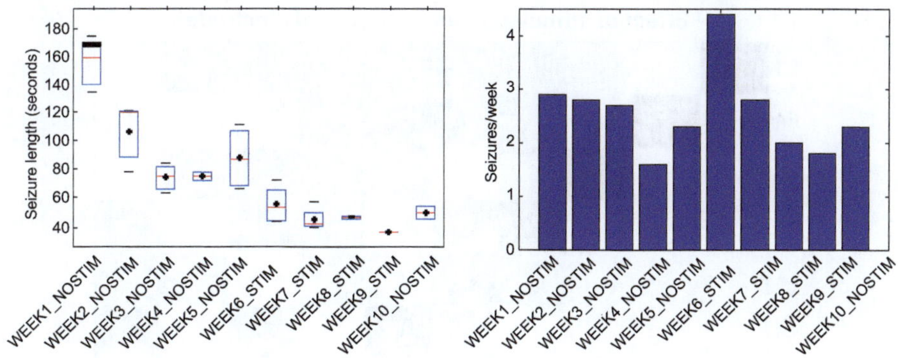

Fig. 13.13 (**a**) Box and whisker plot of seizure lengths for Rat 13 over the course 10 weeks preceded by 4 weeks of rest after status epilepticus induction. The *black dots* indicate the mean of each box. (**b**) Average number of seizures per day for the same rat over the same 10 week period of experimentation. Here week essentially imply a recording file which in general is never exactly 168 h. The first 5 weeks were baseline recording, thus no stimulation was provided. Week 10 was another case of no stimulation but this was recorded 5 weeks after the end of the week 9 file. No stimulation was provided during those 5 weeks and it can be seen that the seizure lengths and the seizure rate per week both had started to increase gradually. Refer to Sect. 13.7 for explanation and location of the electrodes

followed by stimulations of 10 min or more and as Fig. 13.13a, b shows, both the seizure lengths and the average seizure frequency started coming down. Following week 9, the rat was given a period of stimulation cessation in order to see if any plasticity effects of the stimulation would wear out. Sure enough after 4 weeks, the tenth week of recording showed that the seizure length and frequency both went up slightly. While we do recognize that results from only one animal is not statistically relevant, it does shine light on our conjecture that controlling seizures will require proper tuning of stimulation length and possibly more parameters in the stimulation signal.

Another potential difficulty in the design of a seizure control system arises from the multivariable nature of this stimulation as well as its relation to the energy of the stimulus. The following figure (Fig. 13.14) shows a characterization of the stimulus efficacy (as a binary decision over a week-long set of experiments), when the stimulation is applied across a given electrode pair. The top triangular part shows the result after a 2 min stimulation to each electrode pair while the lower triangular part shows same for 15 min. It is quite apparent that stimulation of arbitrary electrodes may not bring the desired effect. The same is true if the duration (i.e., energy) of the stimulus is not high enough as given by the stimulation current, pulse width, frequency etc.

A partial direction towards the selection of an appropriate pair of electrodes for stimulation may be obtained by analyzing the EEG data for focus localization, using the GPDC method of Baccalá and Sameshima (2001), as refined by

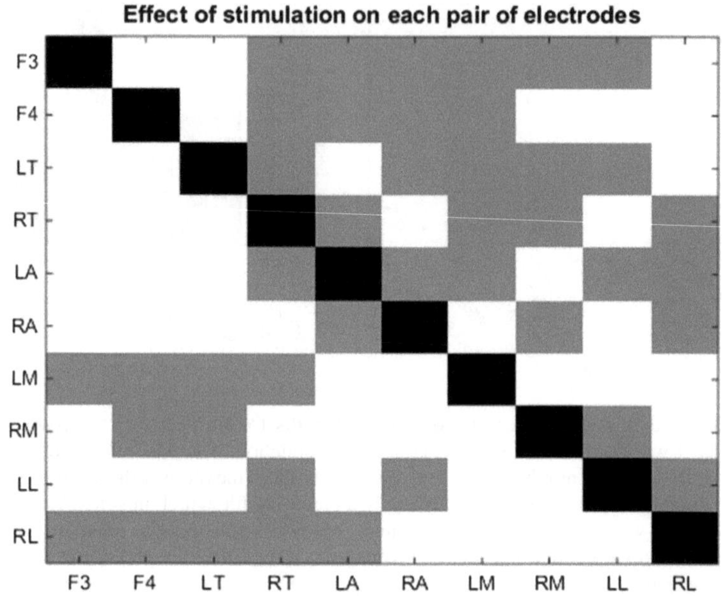

Fig. 13.14 Plot of effect of stimulation on electrode pairs ("control efficacy"). Upper triangle shows effect of 2 min stimulation on each pair. Lower triangle shows same for 15 min. A box color of *grey* indicates no effect whereas white indicates Lyapunov increase due to stimulation. It can be seen that 2 min stimulation has little effect in bringing the Lyapunov up whereas 15 min stimulation is quite capable of doing so. Refer to Sect. 13.7 for explanation and location of the electrodes

Vlachos et al. (2013). This method can provide an indication of the brain site that is the most likely candidate for a focus, and hence a good first choice for stimulation. However, the drawback of this method is that the results are (so far) based on the analysis of averaged short-term data. Long term effects and possible plasticity have been observed in the work of Iasemidis et al. (2009), and therefore further analysis of the data is necessary to provide quantifiable relationships. Another direction that we are currently investigating is what follows from the result shown in Fig. 13.12b, where stimulation of a particular site with a fixed amount of time will start to produce disentrainment as is validated by the L_{max} profile being pulled back to its mean level over time; the mean level in this case is the level before which it dropped sharply. These experiments are long, in the order of weeks, thus analysis of results and producing meaningful interpretations take quite a while. Added to the difficulty is when an animal will cease due to health issues. In such cases the experiments have to be restarted with a new animal. Mortality rates for animals are typically between 40–60%. Another miscreant to the mortality rate is headcap breakage, whereby during a seizure a rat would violently hit the cage walls and the recording headcap will break off; in such cases euathanizing the animal is the only remaining option. In the following sections we describe the experimental setup utilized to gather data and provide both offline and online stimulation as well as how the animals are made epileptic.

13.7 Future Research Opportunities and Conclusions

As the ability to predict leads to the possibility of control, research in control of seizures is expected to flourish in the near future, much to the benefit of the epileptic patients. Investigations in stimulation and control of the brain have attracted the attention of the academic community, and medical device companies have started off designing and implementing intervention devices for various neurodegenerative diseases (e.g. stimulators for Parkinsonian patients) in addition to the existing ones for cardiovascular applications (e.g. pacemakers, defibrillators). For epilepsy, there is currently an explosion of interest in academic centers and medical industry, with clinical trials underway to test potential prediction and intervention methodology and devices for FDA approval.

Electromagnetic stimulation and/or administration of anti-epileptic drugs at the beginning of the preictal period, to disrupt the observed entrainment of normal brain with the epileptogenic focus, may result in a significant reduction of the number and severity of epileptic seizures. Our underlying hypothesis is that an epileptic seizure will be prevented if an external intervention successfully resets the brain prior to the seizure's occurrence. Preliminary results from our experiments have shown that both the length of seizures and the rate can be lowered by such means. However, it is very important to investigate the parameters that lead to maximum efficacy and minimum side effects of such an intervention.

We have shown that a successful and robust controller should be correcting the pathological part of the system, that is where the coupling between brain sites increases excessively, a situation the existing internal feedback in the brain cannot compensate for. To this end, we have shown how L_{max} computed using the tuned Kantz algorithm can be treated as a synchronization measure (output) of interest. A possible future direction in this work will entitle generating dynamical system models from the applied stimulus (input) to the Lyapunov exponents. The inputs, as always, will be time modulated with a biphasic impulse train since this is the type of signal that gave us better decoupling results so far. A parameterization of these signals in terms of their duration, frequency, and location should also be considered in order to eventually develop a comprehensive input-output model from the average stimulus power to the output of interest. It is worth mentioning that the measure of synchronization can include, but not be limited to Lyapunov exponents computed from EEG and other biomarkers such as heart rate, pulse oxygen levels, body temperature, etc. Using such disparate data as potential markers, undoubtedly presents significant challenges and complexity resulting in all issues related to big data, namely, volume, velocity, variety and value being addressed. Tools such as HPCmatlab suited for HPC platforms ameliorate and enable such problems to be tractable. Future compute platforms will be heterogeneous and comprise accelerator devices such as GPU's and FPGA's. Standard API's like MPI, POSIX threads, and OpenCL provide a uniform underlying approach to distributed memory, shared memory and accelerated computing. Overall, the scientific community has embraced these approaches and the trend will continue in the near future towards new uses of accelerated and reconfigurable computing with completely new or modified algorithms tuned to these platforms.

The method exploited in our work involving the reliable computation of Lyapunov exponents can be utilized in generating a highly accurate automated seizure detection system. A reliable seizure detection mechanism will aid medical practitioners and patients with epilepsy. The established practice is to admit patients into the hospital, once identified with epilepsy, and record long-term EEG data. Trained technicians and doctors then have to sift through hours of patient data in order to identify the time of seizures to make their diagnosis. Typically, the number of epileptic patients admitted, at a time, in a hospital can become overwhelming owing to the fact that epilepsy is such a common neurological disorder. Technicians and doctors simply cannot keep up with the volumes of EEG generated in such short periods of time. Studies show that about 30% of Intensive Care Unit (ICU) patients have undiagnosed seizures due to this labor-intensive process that is prone to human error (Claassen et al. 2004). Untimely detection of seizures increase the morbidity of patients and can lead to mortality in certain cases. It also means that the patients have to stay in the hospital longer, thus adding to cost. With some fine-tuning the envisioned automated seizure detection mechanism will help alleviate much of that burden equipped with its high true positive and low false positive detection rates.

Based on the above, we expect that the envisioned active real-time seizure detection and feedback control techniques will mature into a technology for intervention and control of the transition of the brain towards epileptic seizures and would result in a novel and effective treatment of epilepsy. The "heavy machinery" of computing Lyapunov exponents is a significant overhead to be paid, as compared to existing linear measures, in order to improve the reliability of seizure detection and prevention. Computational power is definitely a great concern, especially if the technology is to be applied in a portable fashion so that patients can go on with their day-to-day lives without interference. Considerations must be taken into account of how the devices can perform big data operations while at the same time being small enough to be portable. The ultimate goal is to provide a seizure-free epileptic brain capable to function "normally", with minimum time-wise and power-wise intervention and side effects with the help of these advancements. We envision that this technology will eventually enable a long anticipated new mode of treatment for other brain dynamical disorders too, with neuromodulation, anti-epileptic drugs and electromagnetic stimuli as its actuators.

Acknowledgements The authors work was supported by the National Science Foundation, under grant ECCS-1102390. This research was partially supported through computational resources provided by the Extreme Science and Engineering Discovery Environment (XSEDE), which is supported by National Science Foundation grant number ACI-1053575.

Appendix 1: Electrical Stimulation and Experiment Setup

Deep Brain Stimulation (DBS) protocols in several "animal models" of epilepsy have shown some effectiveness in controlling epileptic seizures with high frequency stimulation targeting the subthalamic nucleus, anterior thalamic nucleus, caudal superior colliculus, substantia nigra, and hippocampus (Vercueil et al. 1998; Lado et al. 2003). All these investigators used stimulation parameters in the following ranges: frequency from 50–230 Hz, bipolar constant current pulses (30–1000 μs) at current intensities from 0.1 to 2 mA. In contrast, low frequency (between 1 and 30 Hz) stimulation resulted in increase of seizure susceptibility or synchronization of EEG. The electrical stimulation we used in our experiments had a pulse width of 300 μs at a current intensity ranging from 100–750 μA and pulse-train duration of 10–20 min. Considering possible induced tissue damage by electrical stimulus, the maximum current intensity of 750 μA falls under the safe allowable charge density limit of 30 $\mu C/cm^2$ as reported by Kuncel and Grill (2004) for deep brain stimulation; given the size of the stimulating electrode and pulse parameters chosen. While these specific values of the stimulation parameters are being utilized in our work, their optimization can become a project on its own right and is not considered at this point.

The stimulation is applied between pairs of electrodes (amygdalar, hippocampal, thalamic, frontal) according to the localization analysis results, whenever a seizure warning is issued by our seizure warning algorithm or in the case of the offline stimulation in an open-loop manner for a fixed duration on various pairs of electrodes. A stimulation switching circuitry was developed in-house for the purpose of stimulating two sites at will. This equipment consisted of an Arduino Mega and printed circuit board containing electronic switches. Figure 13.15 shows a sketch of hardware setup used for all experiments.

The Intan RHD2000 development board and its 16 channel amplifier was setup as the EEG acquisition machine. The rats used in our study have 10 EEG channels, a ground and a reference channel. We collect EEG data from 10 electrodes located in different parts of the rat brain (see Fig. 13.16). The EEG channels go through a switch board into the Intan EEG data acquisition system. The Intan board has an amplifier that conditions the signal so that its 16 bit ADC has sufficient resolution in the EEG waveforms which are typically in the 100s of μV range. Once digitized, the data is collected in an FPGA buffer until a MATLAB program polls it from the buffer over USB. The MATLAB code operates every 2 s and it brings in 2 s worth of data that are sampled from the ADC at 512 Hz from all channels. On MATLAB, either the offline fixed stimulation code or the seizure warning algorithm decide on when and how to stimulate and those parameters are sent over emulated Serial Port (USB Virtual COM port) to an Arduino MEGA. The MEGA then commands the stimulator to provide stimulation on its output port. The amplitudes are fixed using analog knobs and are not programmable. Another function of the Arduino is to command the switch board so that it disconnects a chosen pair of electrodes from the rat to the EEG board and connects them to the stimulator so that the stimulation

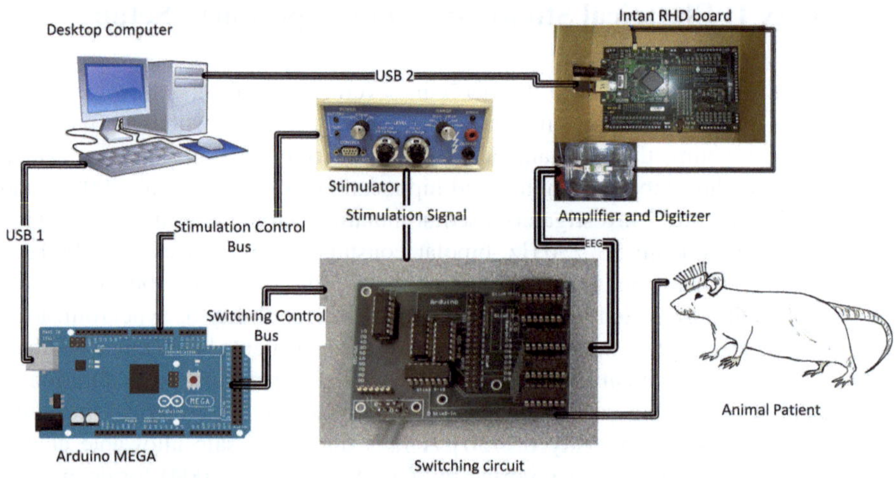

Fig. 13.15 Block diagram showing the experimental setup used. The Switching circuit is controlled by the Arduino Due which in turn gets commands from the MATLAB program running on the computer. The switching circuit enables us to stimulate any pair of electrode at will. The stimulation signal is generated from the A-M systems stimulator

signal can pass through to the rat brain. Once stimulation needs to be switched off, the Arduino commands the stimulator to switch off and reconnects the EEG channels to the rat electrodes.

Appendix 2: Preparation of Animals

The animals used in this study were male Spraque Dawley rats, weighing between 200–225 g, from Harlan Laboratories. All animal experimentation used in the study were performed in the Laboratory For Translational Epilepsy Research at Barrow Neurological Institute (BNI) upon approval by the Institutional Animal Care and Use Committee (IACUC). The protocol for inducing chronic epilepsy was described previously by Walton and Treiman (1988). This procedure generates generalized convulsive status epilepticus (SE). Status epilepticus was induced by intraperitoneal (IP) injection of lithium chloride (3 mmol/kg) followed by subcutaneous (SC) injection of pilocarpine (30 mg/kg) 20–24 h later. Following injection of pilocarpine, the EEG of each rat were monitored visually for clinical signs of SE noted behaviorally by the presence of a Racine level 5 seizure (rearing with loss of balance, Racine 1972). At EEG Stage V (approximately 4 h after pilocarpine injection) SE was stopped using a standard cocktail of diazepam 10 mg/kg, and Phenobarbital 25 mg/kg, both IP. The rats were then kept under visual observation for 72 h within which all measures were taken to stop them from deceasing. In the event that none of the methods to keep them alive worked, the animals were euthanized.

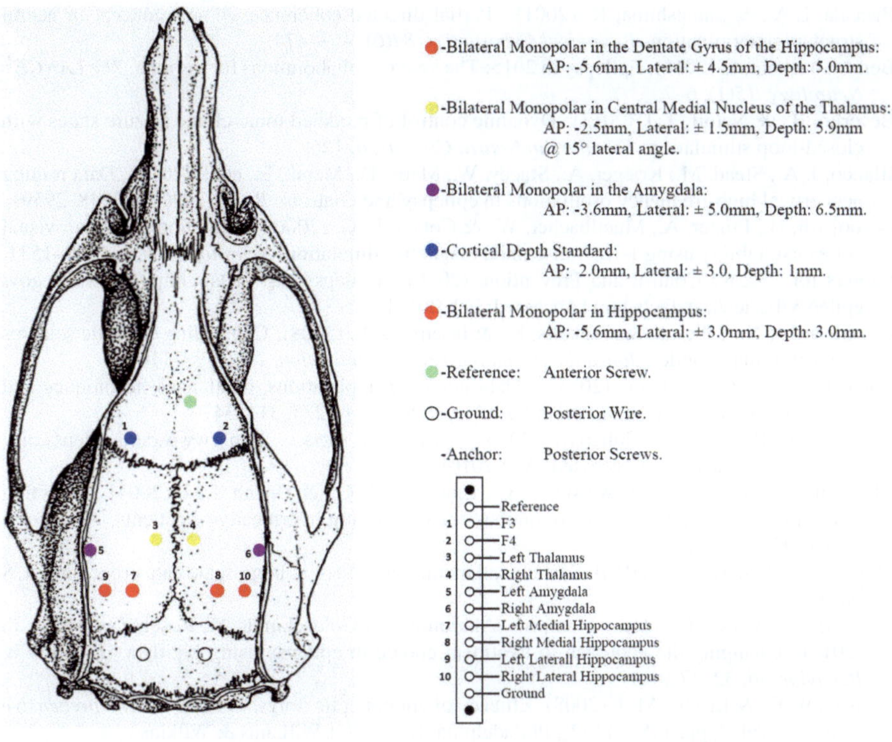

Fig. 13.16 Diagram of surgical placement of electrodes in the rat's brain (top view)

After SE was successfully induced in the animals, they were allowed 5 weeks for the seizure frequency to stabilize. Following this 5 week period, the animals were taken into surgery and an electrode array, as shown in Fig. 13.16, were implanted into their brain. Not including the reference and ground connections, each rat had 10 electrodes implanted. After surgery, each animal was allowed a week before being connected to an EEG machine. The referential voltages from each of the 10 electrodes mentioned was then recorded using an EEG machine (Intan RHD2000 development board).

References

Annegers, J. F., Rocca, W. A., & Hauser, W. A. (1996). Causes of epilepsy: Contributions of the Rochester epidemiology project. *Mayo Clinic Proceedings, 71*, 570–575; Elsevier.

Apache Software Foundation. (2016). Hadoop. https://hadoop.apache.org/ [Accessed: Aug 2016].

Aram, P., Postoyan, R., & Cook, M. (2013). Patient-specific neural mass modeling-stochastic and deterministic methods. *Recent advances in predicting and preventing epileptic seizures* (p. 63). River Edge: World Scientific.

Baccalá, L.A., & Sameshima, K. (2001). Partial directed coherence: A new concept in neural structure determination. *Biological Cybernetics, 84*(6), 463–474.

Ben-Menachem, E. (2016). Epilepsy in 2015: The year of collaborations for big data. *The LANCET Neurology, 15*(1), 6–7.

Beverlin, B., & Netoff, T. I. (2013). Dynamic control of modeled tonic-clonic seizure states with closed-loop stimulation. *Frontiers in Neural Circuits, 6*, 126.

Blanco, J. A., Stead, M., Krieger, A., Stacey, W., Maus, D., Marsh, E., et al. (2011). Data mining neocortical high-frequency oscillations in epilepsy and controls. *Brain, 134*(10), 2948–2959.

Boroojerdi, B., Prager, A., Muellbacher, W., & Cohen, L. G. (2000). Reduction of human visual cortex excitability using 1-Hz transcranial magnetic stimulation. *Neurology, 54*(7), 1529–1531.

Centers for Disease Control and Prevention. (2016). Epilepsy fast facts. http://www.cdc.gov/epilepsy/basics/fast-facts.htm [Accessed: Feb 2016].

Chakravarthy, N., Sabesan, S., Tsakalis, K., & Iasemidis, L. (2008). Controlling epileptic seizures in a neural mass model. *Journal of Combinatorial Optimization, 17*(1), 98–116.

Chen, C. P., & Zhang, C.-Y. (2014). Data-intensive applications, challenges, techniques and technologies: A survey on big data. *Information Sciences, 275*, 314–347.

Citizens for Research in Epilepsy. (2016). Epilepsy facts. http://www.cureepilepsy.org/aboutepilepsy/facts.asp [Accessed: Aug 2016].

Claassen, J., Mayer, S. A., Kowalski, R. G., Emerson, R. G., & Hirsch, L. J. (2004). Detection of electrographic seizures with continuous eeg monitoring in critically ill patients. *Neurology, 62*(10), 1743–1748.

Dean, J., & Ghemawat, S. (2010). System and method for efficient large-scale data processing. US Patent 7,650,331.

Devinsky, O., Dilley, C., Ozery-Flato, M., Aharonov, R., Goldschmidt, Y., Rosen-Zvi, M., et al. (2016). Changing the approach to treatment choice in epilepsy using big data. *Epilepsy & Behavior, 56*, 32–37.

Dodson, W. E., & Brodie, M. J. (2008). Efficacy of antiepileptic drugs. *Epilepsy: A comprehensive textbook* (Vol. 2, pp. 1185–1192), Philadelphia: Lippincott Williams & Wilkins .

Edward, L. (1972). Predictability: Does the flap of a butterfly's wings in Brazil set off a tornado in Texas? Washington, DC: American Association for the Advancement of Science.

Engel, J. (2013). *Seizures and epilepsy* (Vol. 83). Oxford: Oxford University Press.

Food and Drug Administration. (2015). Rns system. http://www.fda.gov/MedicalDevices/ProductsandMedicalProcedures/DeviceApprovalsandClearances/Recently-ApprovedDevices/ucm376685.htm [Accessed: Aug 2016].

Forsgren, L. (1990). Prospective incidence study and clinical characterization of seizures in newly referred adults. *Epilepsia, 31*(3), 292–301.

Good, L. B., Sabesan, S., Marsh, S. T., Tsakalis, K., Treiman, D., & Iasemidis, L. (2009). Control of synchronization of brain dynamics leads to control of epileptic seizures in rodents. *International Journal of Neural Systems, 19*(03), 173–196. PMID: 19575507.

Grassberger, P., & Procaccia, I. (1983). Characterization of strange attractors. *Physical Review Letters, 50*(5), 346–349.

Grassberger, P., Schreiber, T., & Schaffrath, C. (1991). Nonlinear time sequence analysis. *International Journal of Bifurcation and Chaos, 01*(03), 521–547.

Guo, X., Dave, M., & Mohamed, S. (2016). HPCmatlab: A framework for fast prototyping of parallel applications in Matlab. *Procedia Computer Science, 80*, 1461–1472.

Han, J., Kamber, M., & Pei, J. (2011). *Data mining: Concepts and techniques* (3rd ed.). Burlington: Morgan Kaufmann.

Hasselblatt, B., & Katok, A. (2003). *A first course in dynamics: With a panorama of recent developments*. Cambridge: Cambridge University Press.

Hirtz, D., Thurman, D. J., Gwinn-Hardy, K., Mohamed, M., Chaudhuri, A. R., & Zalutsky, R. (2007). How common are the "common" neurologic disorders? *Neurology, 68*(5), 326–337.

Hodaie, M., Wennberg, R. A., Dostrovsky, J. O., & Lozano, A. M. (2002). Chronic anterior thalamus stimulation for intractable epilepsy. *Epilepsia, 43*(6), 603–608.

Holmes, M. D., Brown, M., & Tucker, D. M. (2004). Are "generalized" seizures truly generalized? evidence of localized mesial frontal and frontopolar discharges in absence. *Epilepsia, 45*(12), 1568–1579.
Howbert, J. J., Patterson, E. E., Stead, S. M., Brinkmann, B., Vasoli, V., Crepeau, D., et al. (2014). Forecasting seizures in dogs with naturally occurring epilepsy. *PLoS ONE, 9*(1), e81920.
Human Brain Project. (2016). Human brain project. https://www.humanbrainproject.eu/ [Accessed: Aug 2016].
Iasemidis, L., Sabesan, S., Good, L., Tsakalis, K., & Treiman, D. (2009). Closed-loop control of epileptic seizures via deep brain stimulation in a rodent model of chronic epilepsy. In *World Congress on Medical Physics and Biomedical Engineering, September 7–12, 2009, Munich, Germany* (pp. 592–595). New York: Springer.
Iasemidis, L., Shiau, D.-S., Sackellares, J., Pardalos, P., & Prasad, A. (2004). Dynamical resetting of the human brain at epileptic seizures: Application of nonlinear dynamics and global optimization techniques. *IEEE Transactions on Biomedical Engineering, 51*(3), 493–506.
Iasemidis, L., Zaveri, H., Sackellares, J., Williams, W., & Hood, T. (1988). Nonlinear dynamics of electrocorticographic data. *Journal of Clinical Neurophysiology, 5*, 339.
IEEE (2004). Ieee posix standard. http://www.unix.org/version3/ieee_std.html [Accessed: Aug 2016].
Jansen, B. H., & Rit, V. (1995). Electroencephalogram and visual evoked potential generation in a mathematical model of coupled cortical columns. *Biological Cybernetics, 73*(4), 357–366.
Kalitzin, S. N., Velis, D. N., & da Silva, F. H. L. (2010). Stimulation-based anticipation and control of state transitions in the epileptic brain. *Epilepsy & Behavior, 17*(3), 310–323.
Kantz, H. (1994). A robust method to estimate the maximal Lyapunov exponent of a time series. *Physics Letters A, 185*(1), 77–87.
Kouzes, R. T., Anderson, G. A., Elbert, S. T., Gorton, I., & Gracio, D. K. (2009). The changing paradigm of data-intensive computing. *Computer, 42*(1), 26–34.
Kramer, U., Kipervasser, S., Shlitner, A., & Kuzniecky, R. (2011). A novel portable seizure detection alarm system: Preliminary results. *Journal of Clinical Neurophysiology, 28*(1), 36–38.
Kuncel, A. M., & Grill, W. M. (2004). Selection of stimulus parameters for deep brain stimulation. *Clinical Neurophysiology, 115*(11), 2431–2441.
Lado, F. A., Velíšek, L., & Moshé, S. L. (2003). The effect of electrical stimulation of the subthalamic nucleus on seizures is frequency dependent. *Epilepsia, 44*(2), 157–164.
Lantz, G., Spinelli, L., Seeck, M., de Peralta Menendez, R. G., Sottas, C. C., & Michel, C. M. (2003). Propagation of interictal epileptiform activity can lead to erroneous source localizations: a 128-channel eeg mapping study. *Journal of Clinical Neurophysiology, 20*(5), 311–319.
Lasemidis, L. D., Principe, J. C., & Sackellares, J. C. (2000). Measurement and quantification of spatiotemporal dynamics of human epileptic seizures. *Nonlinear biomedical signal processing: Dynamic analysis and modeling* (Vol. 2, pp. 294–318). New York: Wiley
Liu, Q., Logan, J., Tian, Y., Abbasi, H., Podhorszki, N., Choi, J. Y., et al. (2014). Hello adios: the challenges and lessons of developing leadership class i/o frameworks. *Concurrency and Computation: Practice and Experience, 26*(7), 1453–1473.
Lorenz, E. N. (1963). Deterministic nonperiodic flow. *Journal of the Atmospheric Sciences, 20*(2), 130–141.
Lulic, D., Ahmadian, A., Baaj, A. A., Benbadis, S. R., & Vale, F. L. (2009). Vagus nerve stimulation. *Neurosurgical Focus, 27*(3), E5.
Mathworks (2016). Mex library API. http://www.mathworks.com/help/matlab/mex-library.html [Accessed: Aug 2016].
Miller, K. (2014). Seizures, in theory: Computational neuroscience and epilepsy. http://biomedicalcomputationreview.org/content/seizures-theory-computational-neuroscience-and-epilepsy, [Accessed: Aug 2016].
Mina, F., Benquet, P., Pasnicu, A., Biraben, A., & Wendling, F. (2013). Modulation of epileptic activity by deep brain stimulation: a model-based study of frequency-dependent effects. *Frontiers in Computational Neuroscience, 7*, 94.

Mormann, F., Kreuz, T., Andrzejak, R. G., David, P., Lehnertz, K., & Elger, C. E. (2003). Epileptic seizures are preceded by a decrease in synchronization. *Epilepsy Research, 53*(3), 173–185.
MPI Forum. (2016). Mpi forum. http://www.mpi-forum.org/ [Accessed: Aug 2016].
National Institute of Health. (2014). Brain initiative. http://braininitiative.nih.gov/pdf/BRAIN2025_508C.pdf [Accessed: Aug 2016].
National Science Foundation. (2016). Empowering the nation through discovery and innovation. http://www.nsf.gov/news/strategicplan/nsfstrategicplan_2011_2016.pdf [Accessed: Aug 2016].
Niedermeyer, E., & da Silva, F. H. L. (2005). *Electroencephalography: Basic principles, clinical applications, and related fields*. Philadelphia: Wolters Kluwer Health.
Oestreicher, C. (2007). A history of chaos theory. *Dialogues in Clinical Neuroscience, 9*(3), 279–289.
OpenCL. (2016). The open standard for parallel programming of heterogeneous systems. https://www.khronos.org/opencl/ [Accessed: Aug 2016].
OpenMP. (2016). The OpenMP API specification for parallel programming. http://openmp.org/wp/ [Accessed: Aug 2016].
Poincaré, H. (1992). *New methods of celestial mechanics* (Vol. 13). New York: Springer.
Racine, R. J. (1972). Modification of seizure activity by electrical stimulation: II. Motor seizure. *Electroencephalography and Clinical Neurophysiology, 32*(3), 281–294.
Rosenstein, M. T., Collins, J. J., & De Luca, C. J. (1993). A practical method for calculating largest Lyapunov exponents from small data sets. *Physica D: Nonlinear Phenomena, 65*(1), 117–134.
Ruelle, D., & Takens, F. (1971). On the nature of turbulence. *Communications in Mathematical Physics, 20*(3), 167–192.
Salsa Group. (2010). Applicability of DryadLINQ to scientific applications. Pervasive Technology Institute, Indiana University http://salsaweb.ads.iu.edu/salsa/ [Accessed: Aug 2016].
Shafique, A. B., & Tsakalis, K. (2012). Discrete-time PID controller tuning using frequency loop-shaping. In *Advances in PID Control* (Vol. 2, pp. 613–618).
Simon, P., de Laplace, M., Truscott, F. W., & Emory, F. L. (1951). *A philosophical essay on probabilities* (Vol. 166). New York: Dover Publications.
Socolar, J. E. S. (2006). *Nonlinear dynamical systems* (pp. 115–140). Boston: Springer.
Staba, R. J., Wilson, C. L., Bragin, A., Fried, I., & Engel, J. (2002). Quantitative analysis of high-frequency oscillations (80–500 Hz) recorded in human epileptic hippocampus and entorhinal cortex. *Journal of Neurophysiology, 88*(4), 1743–1752.
Suffczynski, P., Kalitzin, S., da Silva, F. L., Parra, J., Velis, D., & Wendling, F. (2008). Active paradigms of seizure anticipation: Computer model evidence for necessity of stimulation. *Physical Review E, 78*(5), 051917.
Takens, F. (1981). Dynamical systems and turbulence (detecting strange attractors in fluid turbulence). *Lecture notes in mathematics*. New York: Springer.
Tassinari, C. A., Cincotta, M., Zaccara, G., & Michelucci, R. (2003). Transcranial magnetic stimulation and epilepsy. *Clinical Neurophysiology, 114*(5), 777–798.
Temkin, O. (1994). *The falling sickness: a history of epilepsy from the Greeks to the beginnings of modern neurology*. Baltimore: Johns Hopkins University Press.
The Mathworks. (2016). Best practices for a matlab to c workflow using real-time workshop. http://www.mathworks.com/company/newsletters/articles/best-practices-for-a-matlab-to-c-workflow-using-real-time-workshop.html?requestedDomain=www.mathworks.com.
The Neurology Lounge. (2016). 12 fascinating advances in epilepsy: big data to pacemakers. https://theneurologylounge.com/2015/12/28/12-fascinating-advances-in-epilepsy-big-data-to-pacemakers/ [Accessed: Nov 2016].
The White House. (2014). Brain initiative. https://www.whitehouse.gov/share/brain-initiative [Accessed: Aug 2016].
Thiel, M., Schelter, B., Mader, M., & Mader, W. (2013). Signal processing of the EEG: Approaches tailored to epilepsy. In R. Tetzlaff, C. E. Elgar, & K. Lehnertz (Eds.), *Recent advances in preventing and predicting epileptic seizures* (pp. 119–131). Singapore: World Scientific.
TOP 500. (2016). Top 500. https://www.top500.org/lists/2016/06/ [Accessed: Aug 2016].

Tsakalis, K., Chakravarthy, N., & Iasemidis, L. (2005). Control of epileptic seizures: Models of chaotic oscillator networks. In *Proceedings of the 44th IEEE Conference on Decision and Control* (pp. 2975–2981).

Tsakalis, K., & Iasemidis, L. (2004). Prediction and control of epileptic seizures. In *International Conference and Summer School Complexity in Science and Society European Advanced Studies Conference V, Patras and Ancient Olympia, Greece* (pp. 14–26).

Vercueil, L., Benazzouz, A., Deransart, C., Bressand, K., Marescaux, C., Depaulis, A., et al. (1998). High-frequency stimulation of the sub-thalamic nucleus suppresses absence seizures in the rat: Comparison with neurotoxic lesions. *Epilepsy Research, 31*(1), 39–46.

Vlachos, I., Krishnan, B., Sirven, J., Noe, K., Drazkowski, J., & Iasemidis, L. (2013). Frequency-based connectivity analysis of interictal iEEG to localize the epileptogenic focus. In *2013 29th Southern Biomedical Engineering Conference*. New York: Institute of Electrical & Electronics Engineers (IEEE).

Walton, N. Y., & Treiman, D. M. (1988). Response of status epilepticus induced by lithium and pilocarpine to treatment with diazepam. *Experimental Neurology, 101*(2), 267–275.

Watson, J. W. (2014). Octave. https://www.gnu.org/software/octave/ [Accessed: Aug 2016].

Wolf, A., Swift, J. B., Swinney, H. L., & Vastano, J. A. (1985). Determining Lyapunov exponents from a time series. *Physica D: Nonlinear Phenomena, 16*(3), 285–317.

World Health Organization. (2016). Epilepsy fact sheet No. 999. http://www.who.int/mediacentre/factsheets/fs999/en/ [Accessed: Aug 2016].

Chapter 14
Big Data to Big Knowledge for Next Generation Medicine: A Data Science Roadmap

Tavpritesh Sethi

14.1 Introduction

Living systems are inherently complex. This complexity plays out as health and disease states over the lifetime of an organism. Deciphering health has been one of the grand endeavors of humanity since times immemorial. However, it is only in the recent decades that a disruptive transformation of healthcare and its delivery seems imminent. Like other disciplines, this transformation is being fueled by exponentially growing Big-data. This has sparked a widespread move for transitioning to Next-generation medicine which aims at being *Preventive, Predictive, Personalized, Participatory*, i.e., P4 (Auffray et al. 2009) and *Precise* (Collins and Varmus 2015). However, this also requires a major upgrade of our scientific methods and approach. Our current model of medical discovery has evolved over the past 500 years and relies upon testing pre-stated hypotheses through careful experimental design. This approach of "*hypothesis-driven medical discovery*" has been instrumental in advancing medicine and has consistently led to breakthroughs in newer treatments, vaccines, and other interventions to promote health over the past 500 years. However, for the first time, medicine is at a threshold where the rate of data-generation has overtaken the rate of hypothesis generation by clinicians and medical researchers. It has been estimated that biomedical Big-data will reach 25,000 petabytes by 2020 (Sun 2013) largely attributable to digitization of health-records and the pervasive genomic revolution. Therefore, a new paradigm of "*data-driven medical discovery*" has emerged and is expected to revolutionize

T. Sethi (✉)
Department of Computational Biology, Indraprastha Institute of Information Technology, New Delhi, India

Department of Pediatrics, All India Institute of Medical Sciences, New Delhi, India
e-mail: tavpriteshsethi@iiitd.ac.in

S. Srinivasan (ed.), *Guide to Big Data Applications*, Studies in Big Data 26,
DOI 10.1007/978-3-319-53817-4_14

the next hundred years of medicine (Kohane et al. 2012). The biggest challenge in this direction will be to incorporate the complex adaptive properties of human physiology into Big-data technologies.

Genomics has been the poster child of the Big-data movement in medicine as the cost of sequencing has been falling faster than the limits imposed by Moore's law (NHGRI 2016). However, the genomics revolution has also taught us a sobering lesson in science. The scientific community realized that Big-data is a *necessary condition*, but not a *sufficient condition* for translation to bedside, community or policy. Even before the advent of genomics era, there were glaring gaps in our understanding of biology and it was hoped that Big-data would fill these gaps. On the contrary, what followed was quite the opposite. The more Big-data we generated, more we realized our lack of understanding of the complexity of living systems. Following conventional statistical approaches, the predictive power of genomics was found to be low for common diseases and traits. This is the well-known problem of missing heritability, which arises partly due complex (and often unpredictable) biological interactions and partly due to limitations of currently available statistical techniques. For example, Type II Diabetes, a complex disease, the number is estimated to be around 10% (Ali 2013). This is because the genetic code in DNA, which was thought to be a major health determinant is now known to be just one of the layers of the multiscale influences. These layers include the *Exposome* consisting of sum-total of environmental exposures (Miller 2014), the *Microbiome* consisting of resident micro-organisms (Turnbaugh et al. 2007) and even the heath influences spreading upon *Social-networks* (Christakis 2007), in addition to other "omics" operating in an individual such as transcriptomics, proteomics, and metabolomics (Fig. 14.1a). These layers can be thought as "Russian doll" hierarchies with DNA (genome) as the blueprint for dictating the RNA (transcriptome) which further gets translated into proteins (proteome) and metabolites (metabolome). This is a simplified picture of the hierarchical organization with feedback loops and more "omics" layers added each day. Therefore, with these characteristics, the current approaches of Big-data analytics are not sufficient by themselves to tackle the challenges of Next-generation medicine. The rich complexity, multiscale nature and interactions between these scales necessitates a Data-science roadmap incorporating the distinguishing features of Biomedical Big-data as presented in Table 14.1.

The aim of this chapter is to propose a Data-science roadmap for knowledge-generation from Big-data through a combination of modern machine learning (data-driven) and statistical-inference (hypothesis-driven) approaches. The roadmap is not a comprehensive one, and could be one of the many possible approaches that could be geared towards the final objective of delivering better clinical (*Personalized, Predictive*) and community (*Preventive, Predictive, Participatory*) care through Big-data.

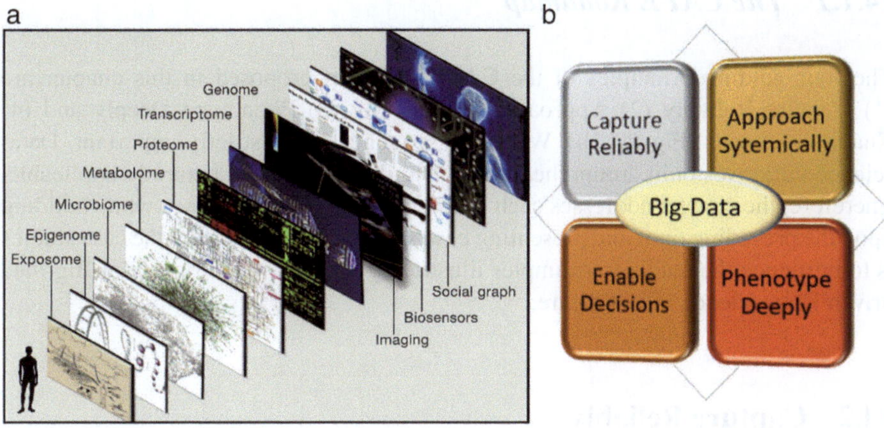

Fig. 14.1 Eric Topol's vision is to synthesize various layers of information ranging from environmental exposures (exposome) to the individual's genetic blueprint (genome) to enable the Next-generation medicine (**a**). The CAPE roadmap (**b**) proposed in this chapter serves as a Data-science blueprint for executing such a vision. With the explosion of data across genetic, cellular, organismal and supra-organismal layers, there is an immediate need for such roadmaps and CAPE is one of these directions discussed in this chapter. (Permission to use the image authored by Eric. J Topol obtained from the publisher, Elsevier under License Number 3967190064575, License date Oct 13, 2016)

Table 14.1 Challenges in biomedical big-data that make it unique

	Unique challenges in healthcare big-data (D) and generative physiology (P)	Corresponding challenge in conventional big-data	Non-exhaustive list of possible solutions
1.	Heterogeneity (P, D)	Variety	Large scale observational studies (Hripcsak et al. 2016), Patient Aggregation (Longhurst et al. 2014), Stratified Medicine (Athey and Imbens 2016; Sethi et al. 2011)
2.	Messy (D)	Veracity	Imputation (Longford 2001), Natural Language Processing (Doan et al. 2014)
3.	Bedside Potential (D)	Value	Machine Learning (Rothman et al. 2013; Dinov et al. 2016)
4.	Inter-connected (P)	Not defined as a component	Omics (Topol 2014), Graphs and Networks (Barabási et al. 2011)
5.	Dynamically Adaptive (P)	Not defined as a component	Complex Adaptive Systems (Kottke et al. 2016; Coveney et al. 2016)
6.	Multiscale Integration (P)	Not defined as a component	Multiscale Modeling (Walpole et al. 2013)
7.	Data Privacy & Open Data	Not defined as a component	Cybersecurity, Citizen-science (Follett and Strezov 2015)

14.1.1 The CAPE Roadmap

The four guiding principles of the CAPE roadmap proposed in this chapter are (1) **C**apture Reliably (2) **A**pproach Systemically (3) **P**henotype Deeply and (4) **E**nable Decisions (Fig. 14.1b). While the sequence is not strictly important, Data-science initiatives built around healthcare Big-data will find it naturally applicable. Therefore, the chapter addresses each of these principles in sequence while building upon the preceding ones and presenting case studies. The purpose of the case studies is to introduce the reader to examples illustrating the real-world impact of Big-data driven Data-science in healthcare.

14.2 Capture Reliably

Biomedical data are notoriously messy. Unless data quality is ensured, the maxim of "garbage in- garbage out" may constrain the scientific and clinical utility of Big-data in healthcare. Very often the measurements are missing, incorrect, or corrupted. In addition to measurement errors in structured data, biomedical data is also messy because a large fraction of it is unstructured. It is estimated that up to 80% of healthcare data resides in the form of text notes, images and scans (Sun 2013), yet contains highly valuable information with respect to patients' health and disease. Hence reliable capture of big-data is one of the most crucial step in successful execution of healthcare applications. Ensuring reliable capture not only involves ensuring the fidelity of existing systems such as Electronic Health Records (EHRs) and Electronic Medical Records (EMRs) but also in newer technologies such as mHealth. While challenges of the former (such as interoperability of EMRs) need to be addressed at a healthcare policy level, the latter offers a different set of scientific challenges. mHealth leverages mobile devices such as mobile phones and tablet computers for recording health parameters through sensors and wireless technologies. This has led to the recent surge in health-tracking and wellness monitoring (Steinhubl et al. 2015). If captured reliably, these could reveal key factors in regulation of health and disease thus enabling *Precision* and *P4 medicine*.

The scientific potential of this approach is already being evaluated and has sparked interest in personal "individualomes" (Snyder 2014) i.e., longitudinal tracking of multiple variables reflecting an individual's health. Further, consensus guidelines on collection and analysis of temporally tracked data have started emerging in order to establish its scientific validity (Wijndaele et al. 2015). However, in the current scenario, most wellness tracking and home monitoring devices are not certified to be research-grade. Important features such as sampling rates and filters are often not specified and vendors typically do not seek approvals from regulatory agencies such as the Food and Drug Administration (FDA).

Therefore, from the standpoint of reliable capture, the key open challenges in healthcare are (i) creating data standards, and (ii) developing tools that can recover data from noisy and/or missing measurements.

14.2.1 Biomedical Data Quality and Standards

Mining of Electronic Health Records (EHRs) is one of the most promising directions for Precision medicine. Defining medical ontologies such as International Classification of Diseases (ICD) has been a key enabling feature in this direction. Despite this enforcement of ontologies, the lack of interoperability has led to siloing of Big-data. Typically, in any Biomedical Big-data project, data harmonization and dealing with messy data alone take up to 80% of the time and effort thus leading to initiatives such as those being implemented for cancer (Rolland et al. 2015) and HIV (Chandler et al. 2015). These approaches are expected to evolve as more Individualome data such as genomes and environmental exposures become available. It is anticipated that harmonization of such multidimensional data sources would require expertise from diverse domains and hence and would be possible to be achieved only through community efforts. Community efforts to create open-source frameworks such as the R Language for Statistical Programming, Python, Hadoop and Spark are already revolutionizing data-science. In a similar fashion, a major role of the open-source movement for sharing of codes and API's for reliable. Secure and inter-operable capture of healthcare data is anticipated (Wilbanks and Topol 2016; Topol 2015). In addition, reliable capture must also ensure implementation of better data-security standards and mechanisms for protecting the privacy of individuals.

14.2.2 Data Sparsity

Sparsity is a double-edged sword in data-science. While sparsity is important for removing redundancy, efficient storage and transmission of data (e.g. Telemedicine), it should not be induced at the cost of completeness of data. Many times, biomedical data suffer from both the problems at the same time. While some variables might be redundantly represented, others might not have acceptable fidelity. The latter kind of sparsity in observations is more often observed and may result from factors which are technical (such as missing data) or human (such as inaccurate recording, corrupted data, textual errors, etc.). Data-science approaches to deal with the latter problem include variable removal, data imputation and data reconstruction. However, data-scientists are advised to inspect the possible generative mechanisms of such missing data through statistical tests and assumptions must be tested. For example, a naïve assumption may be about data *Missing Completely At Random* (MCAR). Simple statistical computations to check whether the missing values are distributed uniformly across all samples and groups of data must be applied before imputing the data. As an illustrative example, this assumption may be invalidated if data missingness was coupled with the generative process. This leads to a non-random pattern in the missingness of data. One of the simplest and most common form of non-random missingness in healthcare data is the *monotone pattern*. A monotone pattern of missingness arises in situations such as patient drop-out or death in longitudinal data (Fig. 14.2). More complex patterns such as stratification of missing data by sub-groups arise when a sub-set of patients is more prone to drop-out or yield erroneous measurements. While imputation with a central tendency (mean/median/mode) may suffice in the MCAR scenario, missing data with structure might need more sophisticated techniques such as k-nearest neighbors, Robust Least Squares Estimation with Principal Components (RLSP), Bayesian Principal Component Analysis and multiple imputation (Hayati et al. 2015). Sophisticated algorithms for testing the assumptions and performing imputations are available in most of the standard statistical software such as R, Stata, SPSS etc.

14.2.3 Feature Selection

While on one hand, data-imputation aims to recover missing data, on the other hand many datasets suffer the problem of redundancy in data. In such datasets, it is desirable to induce a sparsity of variables and preserve only the ones that may be of relevance to the data-science question. This procedure of selecting relevant features is most useful in the presence of multi-collinearity (some features being a linear combination of others) or even nonlinear relationships in the data. The presence of linear and nonlinear correlations between variables makes it a mathematically underdetermined system. Further, Occam's razor suggests that most parsimonious models should be selected and adding variables often leads to over-fitted models. In many such Big-data problems, traditional methods of statistics for dimensionality reduction such as Principal Component Analysis (PCA) and Multi-dimensional Scaling (MDS) are sufficient to reduce the dimensionality using linear transformations to an orthonormal basis. However, in other Big-data situations such as genomics, where the number of features can run into millions per patient, specific statistical and machine learning approaches are often required. Briefly, such feature selection methods can be classified into (i) Filter (ii) Wrapper and (iii) Embedded approaches and the interested reader is referred to (Wang et al. 2016) for

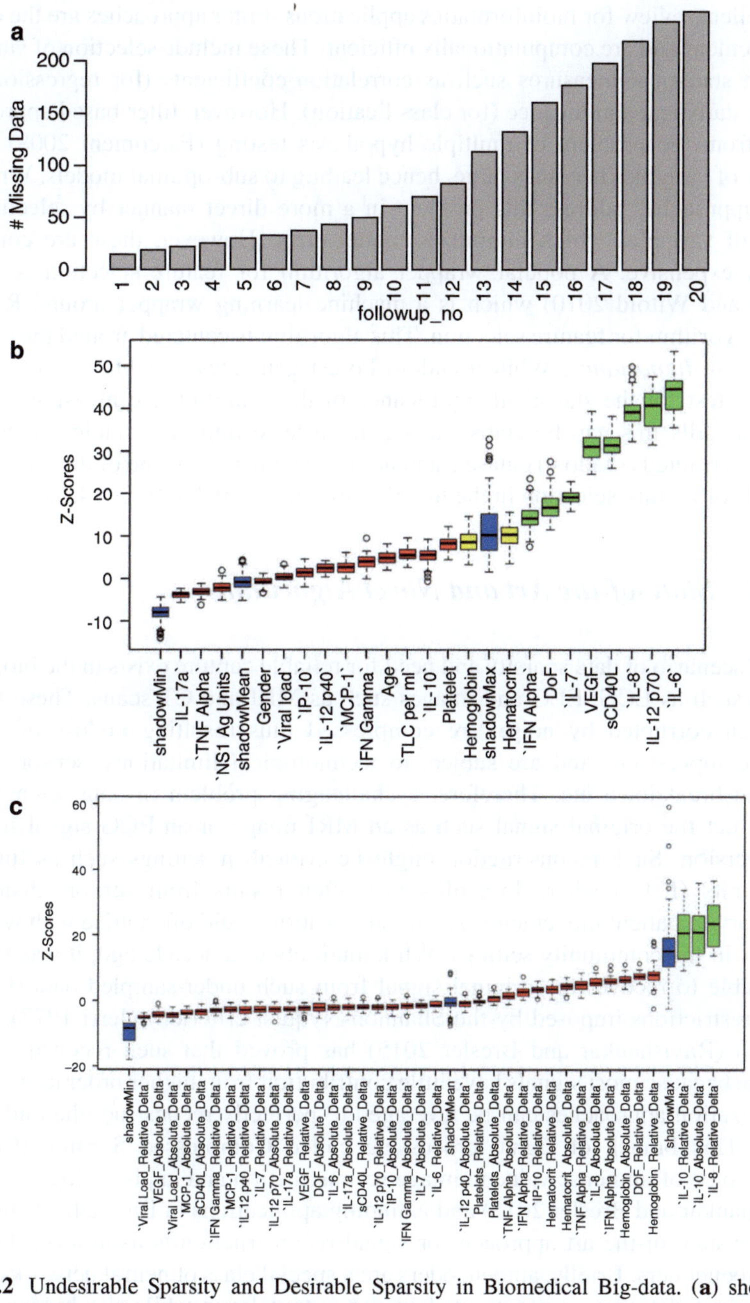

Fig. 14.2 Undesirable Sparsity and Desirable Sparsity in Biomedical Big-data. (**a**) shows the monotone pattern of missing data with patients dropping out over the length of the study, hence invalidating the statistical assumption of Missing Completely at Random (MCAR). (**b, c**) Deliberate induction of sparsity of variables in the data (feature selection) by using machine learning algorithms such as Boruta. The example shows a run of variable selection carried out by the author for selecting variables important for predicting Dengue severity (**b**) and recovery (**c**) from Severe Dengue illness (Singla et al. 2016)

an excellent review for bioinformatics applications. Filter approaches are the easiest to implement and are computationally efficient. These include selection of variables through statistical measures such as correlation-coefficients (for regression) and tests of statistical significance (for classification). However, filter-based approaches suffer from the problem of multiple hypothesis testing (Farcomeni 2008) as the number of variables becomes large, hence leading to sub-optimal models. Wrapper-based approaches address this problem in a more direct manner by selecting the subset of variables which minimizes model error. However, these are computationally expensive. A popular wrapper algorithm for feature selection is Boruta (Miron and Witold 2010) which is a machine learning wrapper around Random Forest algorithm for feature selection. This algorithm is centered around the concept of *Variable Importance*. While Random Forest generates variable importances, it does not test for the statistical significance of these importance measures. Boruta algorithm fills this gap by statistical significance testing of variable importances against permuted (shadow) datasets and has been shown to be one of the most robust methods of feature selection in the toolkit currently available (Fig. 14.2b, c).

14.2.4 State-of-the Art and Novel Algorithms

A third scenario of data sparsity and need for reliable capture exists in the biological signals such as ECG, EEG and images such as MRI and CT scans. These signals are often corrupted by noise, are compressed thus resulting in loss of fidelity (lossy compression) and are subject to technological limitations, sensor failure, machine breakdown etc. Therefore, a challenging problem in data science is to reconstruct the original signal such as an MRI image or an ECG signal from its lossy version. Such reconstruction might be critical in settings such as Intensive Care Units (ICUs) where loss of signal often results from sensors dislodging secondary to patient movements. A similar situation could often arise with wellness trackers in the community settings. While until about a decade ago, it was thought impossible to recover the original signal from such under-sampled data (because of the restrictions imposed by the Shannon-Nyquist criterion) (Jerri 1977), recent research (Ravishankar and Bresler 2015) has proved that such reconstruction is possible because most signals have little contribution from higher order terms. Thus, regularization approaches can be leveraged to perfectly reconstruct the underlying signal. This branch of signal processing called Compressed Sensing (CS) has proven to be of immense value in image reconstruction and in physiological signals (Ravishankar and Bresler 2015) and is finding applications in Biomedical Big-data. Another state-of-the-art approach for signal reconstruction is using deep learning and autoencoders. Briefly, autoencoders are a special class of neural networks where the original data are re-constructed in the output layer while the hidden layers learn the features and representations of the input signals. A full discussion on autoencoders is beyond the scope of this chapter and the interested reader is referred to (Goodfellow et al. 2016).

14.2.5 *Physiological Precision and Stratified Medicine*

The final case for reliable capture of data stresses upon the physiological heterogeneity of health and disease. The original classification of diseases was based mostly upon signs, symptoms and pathological features of a disease. However, it is being realized that many different mechanisms lead to similar pathophysiological outcomes and hence manifestations of diseases. This is especially valid for chronic multifactorial disorders such as Asthma, Diabetes, Obesity and Cardiovascular disorders which have common underlying themes such as inflammation (Arron et al. 2015). With the advent of Big-data and data-science technologies, it is now possible to address this heterogeneity and imprecise disease classification. This understanding was instrumental in for the proposition of the *Precision Medicine Initiative* (Collins et al. 2015). Development of state-of-the-art computational mathematical techniques of clustering multidimensional data (Hinks et al. 2015) are being used for characterization of individuals on millions of features and provide a precision to diagnosis. Evidently, active development of *unsupervised clustering* methods is one of the most important advances in this direction. *Data-driven aggregation* of patients based upon multivariate similarity measures derived from data has led to the approach is known as *Stratified Medicine* (Fig. 14.3a). These algorithms of patient stratification and aggregation attempt to define pure-subclasses by minimizing the ratio of within-group to between-group variability. In most medical conditions stratified so far (e.g. breast cancer), the discovered clusters are expected to have differential disease evolution or response to therapy (Nielson et al. 2015).

On the mathematical side, most clustering algorithms rely upon the notion of a *dissimilarity measure*. This may range from simple *distance metrics* such as *Euclidean* (2-norm), *Manhattan* (taxicab norm) and *Mahalanobis* metrics to

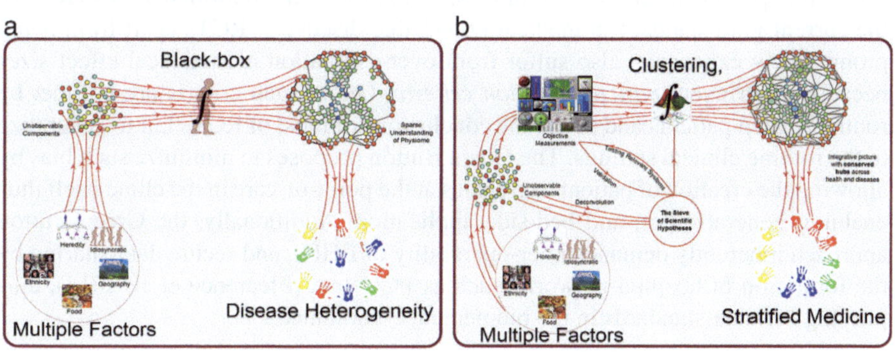

Fig. 14.3 Stratified Medicine. Human physiology integrates the diverse layers mentioned in Fig. 14.1 to produce complex health and diseased phenotypes. (**a**) Our understanding of this integration and the physiological network is sparse. (**b**) Application of data-science to recapitulate physiological integration may create a more complete understanding of physiology and stratification of diversity. This stratification is helping doctors in tailoring of therapies to sub-groups of patients (see text)

dissimilarities obtained through machine learning algorithms such as unsupervised Random Forests (Breiman 2001). For the healthcare data-scientist, the choice of distance metric may prove to be critical. Often, this choice is dictated by the type of the variables (numerical, categorical or mixed) and the presence of noise in the data. Following the choice of a metric and the application of a clustering algorithm, cluster quality inspection and visualization algorithms such as Multidimensional Scaling (MDS), Partitioning around Medoids (PAM) further help the data-scientist in making an informed choice on the strata that may exist in a particular disease. In addition to standard clustering algorithms, newer methods and approaches have focused on clustering complex data of arbitrary shapes and include multivariate density based clustering (Ester et al. 1996), Self-organizing maps (Kohonen 1982), Message passing between the data-points (Frey and Dueck 2007) and Topological Data Analysis (Nielson et al. 2015).

14.2.6 The Green Button: A Case Study on Capture Reliably Principle of the CAPE Roadmap

Tailoring of medical decisions, practices and products are the goals of precision medicine and personalized medicine. As discussed in the preceding section, stratified medicine is a key step towards this goal. Green Button (Longhurst et al. 2014) is an example of stratification applied to Big-data that resides in the Electronic Health Records (EHRs). Since every patient is unique, this proposition aims at creating patient cohorts "on the fly" by matching characteristics across millions of patient records present in the EHRs. This would enable the new paradigm of practice-based evidence combined with the current paradigm of *evidence-based practice.* Such an approach would also complement *Randomized Controlled Trials* (RCTs), the current gold standard in medical knowledge discovery. RCTs apart from being prohibitively expensive, also suffer from over-estimation of statistical effect sizes because of stringent *patient selection criteria.* Often, such criteria are not met by routinely seen patients and hence the conclusions of most of RCTs fail to generalize to the routine clinical settings. The Green Button proposes to minimize such bias by allowing the creation of patient aggregates at the point-of-care in the clinic itself thus enabling generalization and bed-side application. Additionally, the Green Button approach inherently demands inter-operability of EHRs, and secure data-sharing by the formation of hospital-networks such as PCORnet (Fleurence et al. 2014), thus pushing for data-standards in the biomedical community.

14.3 Approach Systemically

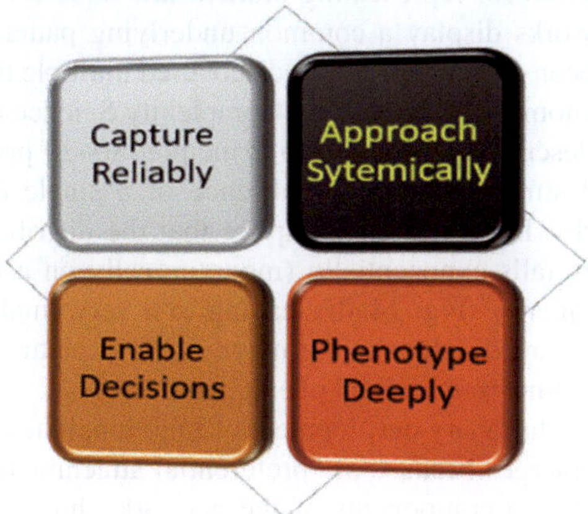

Dense inter-connectivity, regulatory phenomena, and continuous adaptations distinguish biological systems from mechanical and physical systems. Therefore, reductionist approaches, while immensely successful in physical systems (e.g. automobiles and space-crafts), have met with limited success in biology and medicine. Hence, holistic approaches to biomedical data with interdisciplinary application of mathematics, computer science and clinical medicine are required. This understanding led to the birth of *Systems Biology* and to the recent fields of *Networks Medicine* and *Systems Medicine*. The umbrella term of *Systems Medicine* is proposed to be the next step for healthcare advancement and has been defined by (Auffray et al. 2009) as, "*the implementation of Systems Biology approaches in medical concepts, research, and practice. This involves iterative and reciprocal feedback between clinical investigations and practice with computational, statistical, and mathematical multiscale analysis and modeling of pathogenetic mechanisms, disease progression and remission, disease spread and cure, treatment responses and adverse events, as well as disease prevention both at the epidemiological and individual patient level. As an outcome Systems Medicine aims at a measurable improvement of patient health through systems- based approaches and practice*".

Therefore, this section reviews the most common approaches that are being applied to achieve a holistic and data-driven systems medicine.

14.3.1 Networks Medicine

In the Networks Medicine paradigm, the states of health, disease, and recovery can be thought of as networks of complex interactions and this approach has recently gained much popularity in biomedical Data-science (Barabási et al. 2011). A network is data structure with variables represented as 'nodes' and connections between objects represented as 'edges'. Therefore, the network representation is not only an excellent tool for visualizing complex biological data but also serves as a mathematical model for representing multivariate data. It has been found that most biological networks display a common underlying pattern called scale free behavior which has been discovered and re-discovered multiple times and in various contexts such as Economics, Statistics and Complexity Science (Newman 2005). It has been variously described as the Pareto Principle, 80–20 principle and Power-law distribution and simply reflects the absence of a single characteristic scale in biological networks. Intuitively, this implies that the distribution of number of connections of nodes falls exponentially (monotonically on a log-log scale) as a function of node-frequency (Fig. 14.4b) leading to a very small number of nodes sharing most of the share of edges in the network. This picture is consistent with many of the known natural and societal phenomena.

In the context of evolutionary development of function, it has been proposed that scale free behavior emerges because of "preferential attachment" of new functions to already well-developed components of the network, thus implying a "rich get richer" scenario. Being a quantifiable property, this strategy has been exploited to target the key components of a network, reveal communities, interactions and the spread of information, and interventions that may disrupt this communication. This strategy has found recent uses in effective understanding of drug development (Rodriguez-Esteban 2016) and for understanding community health dynamics (Salvi et al. 2015).

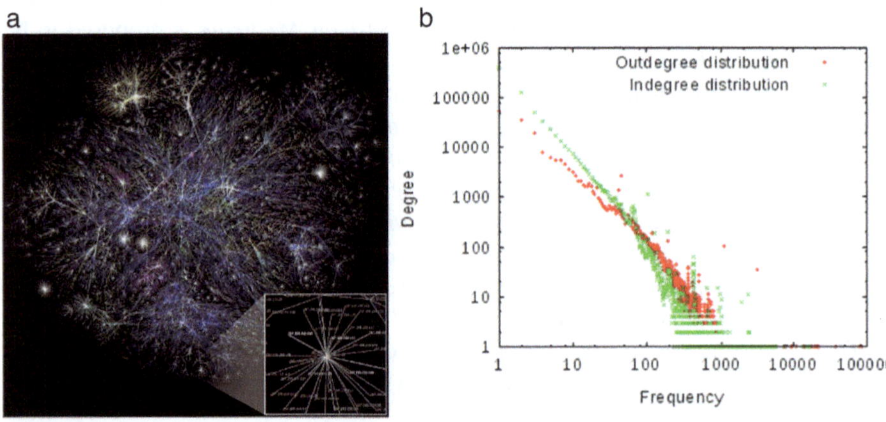

Fig. 14.4 Scale free property of complex networks. Most complex networks including those encountered in healthcare and biology display the presence of hubs at each scale of the network. These hubs are important in understanding not only technological networks such as (**a**) The Internet but also biological networks. (**b**) Illustration of the power law distribution of degree-connectivity that defines a scale free network

14.3.2 *Information Theory for Biology*

In mid-nineteenth century, Ludwig Eduard Boltzman revolutionized the study of physical systems by abstracting away the microscopic properties of a system into a macroscopic quantity called Thermodynamic Entropy,

$$\mathrm{S} = \mathrm{k.logW}$$

Almost a century later, Claude Shannon proposed the theory of information (Shannon 1948) to calculate the information content of any system and proposed the famous equation,

$$\mathrm{H}\ (\mathrm{X}) = -(1) \sum\nolimits_{i=1}^{n} pi\ \mathrm{logpi}$$

The two concepts and equations share a deep theoretical relationship through the Landauer Principle (Landauer 1961), that connects thermodynamic entropy with change in information content of a system. Since physiological and cellular functioning essentially involves information transfer through mediators such as neural tracts or chemical messengers, the concept of entropy and information has found applications designing Big-data applications in biology at a fundamental level (Rhee et al. 2012). For example, quantification of entropy and complexity of a heart-beat patterns has profound applications in critical care settings with a lower complexity of inter-beat intervals shown to be a risk factor for a higher five-year mortality and a poor prognosis after critical events such as an ischemic heart attack (Mäkikallio et al. 1999).

14.3.3 *Agent Based Models*

Agent Based Models (ABMs) are a class of computational models that are particularly suited for holistic modeling of a system in the presence of interactions and rules. These models are based upon the theory of planned behavior and allow autonomous entities (called agents) to interact in space-time, hence allowing collective dynamics of the system to emerge. ABMs allow learning from data in a fashion like societal interactions i.e., peer-to-peer interaction where each agent can be thought of as an individual with a set of rules to enable decisions while interacting. The set of rules are updated every time an interaction occurs in accordance to the perceived gain or loss to the individual. Combined with the principles of behavioral psychology (such as reward and punishment) this leads to a powerful tool in the arsenal of data-science known as Reinforcement Learning, which has been applied to healthcare at clinical (e.g. optimal planning to prevent sepsis in Intensive Care Units), (Tsoukalas et al. 2015) as well as community levels (Fig. 14.5).

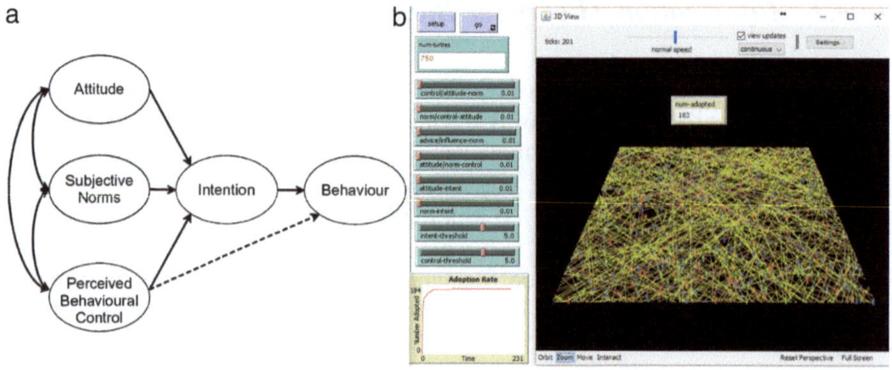

Fig. 14.5 Agent Based Modeling for Behavioral adoption of critical health practices in the State of Bihar (**a**) (Aaron Schecter). This example simulation run (**b**) shows the conversion rate in a community as a function of advisors and influencers in the local community. Such models are expected to be extremely useful for modifying community behavior towards adoption of better health practices in the community

14.3.4 Prevalence of Symptoms on a Single Indian Healthcare Day on a Nationwide Scale (POSEIDON): A Case Study on Approach Systemically Principal of the CAPE Roadmap

This case study (Salvi et al. 2016) illustrates application of Networks Analysis upon a unique patient data resource of 2,04,912 patients collected on a single day across India by Chest Research Foundation, Pune, India. The purpose of the study was to get a snapshot of "what ails India". India is amongst the countries with highest disease burden in the world as per its Annual Report, 2010 and an epidemiologic transition from infectious to life-style disorders has been documented. The methodology adopted for the POSEIDON study was to conduct a one-day, point prevalence study, across India using an ICD-10 compliant questionnaire. In addition to the standard statistical approaches, ta Networks based approach to understand the global structure of the "Symptome" of India was carried out by the author of this chapter. Data were divided by decades of age to dissect the change in network structure across the different age groups of the Indian population. Edges were derived based upon significance achieved by the Fisher's exact test, a standard test applied to test for associations. Since a weighted network analysis gives better information about the community structure, the negative log of p-value was used as weights for pairwise associations of symptoms in each age group. Mapequation algorithm (Rosvall and Bergstrom 2008) was then used to detect community structure (modularity) in the network and the dynamics of the modules were represented as alluvial diagram (Rosvall and Bergstrom 2008). As can be seen in the association network (Fig. 14.6) each symptom/disease is represented

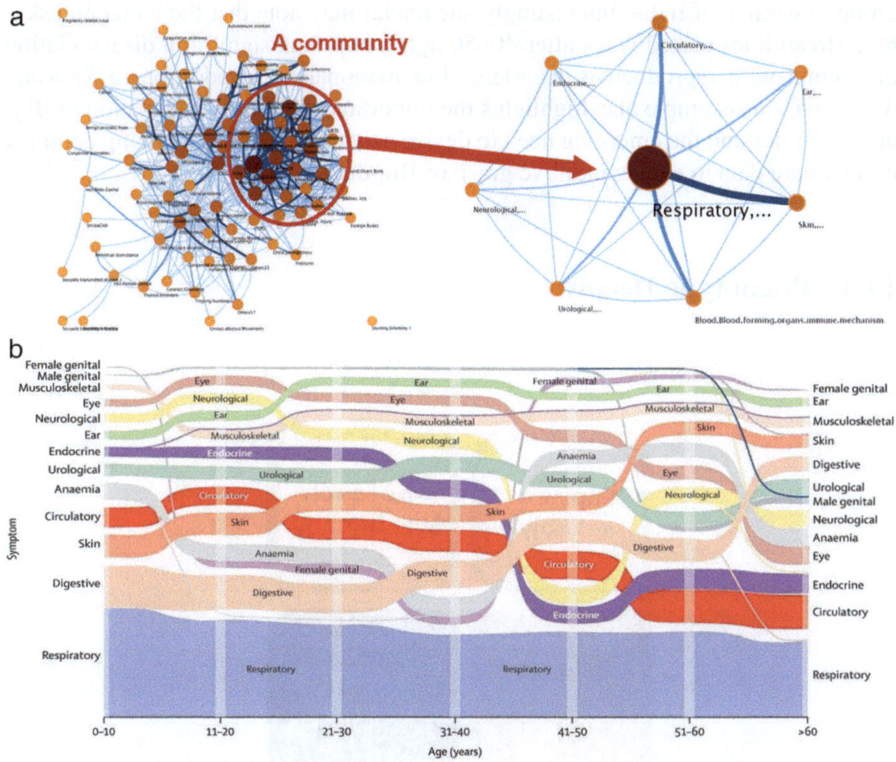

Fig. 14.6 Association Networks and Alluvial Mapping for quantitative insights. In the POSEIDON case study (**a**) Diseases (nodes) formed communities which merged together (*right*). (**b**) These communities were found to change across the age groups of the population in the visualization in the form of an alluvial mapping of 2,04,912 Indian OPD patients (the POSEIDON study, Salvi et al. 2015)

as a node and each pair of symptoms/diseases was tested for association. Force-directed and Kamada-Kawai algorithms of graph layout were then used to visually inspect the structure of these networks which were found to be strikingly different between young and elderly age groups (Fig. 14.6a). Further, the dynamic nature of these community patterns was represented through an alluvial mapping that showed the change in communities represented as flow of streamlines. It was seen that the respiratory group of comorbidities (blue, thick streamline in the figure) was most prevalent across all age groups. Most interestingly, the *Circulatory* (red) streamline was seen to merge with *Endocrine* (purple) comorbidities later in life. This was confirmed to be due to *Diabetes* being most common comorbid endocrine disorder in this age. Hence, a data-driven merger of nodes represented the co-morbid association of cardiovascular diseases and diabetes. Similarly, a merger of *Female Genitalia* (violet) with *Anemia* (sky-blue) was seen in the reproductive age group, anemia due to menstrual disorders being hugely prevalent in this age group in women of India. Interestingly, the reader may note that the violet and sky-blue streamlines parted ways after 40–50 age group thus signifying disassociation of anemia with reproductive disorders after menopause. In addition to *Systemic Approach*, this example also highlights the importance of data-visualization in Big-data analytics and the emerging need to devise new methods of visualizing complex multivariate data to enable intuitive grasp of Big-data.

14.4 Phenotype Deeply

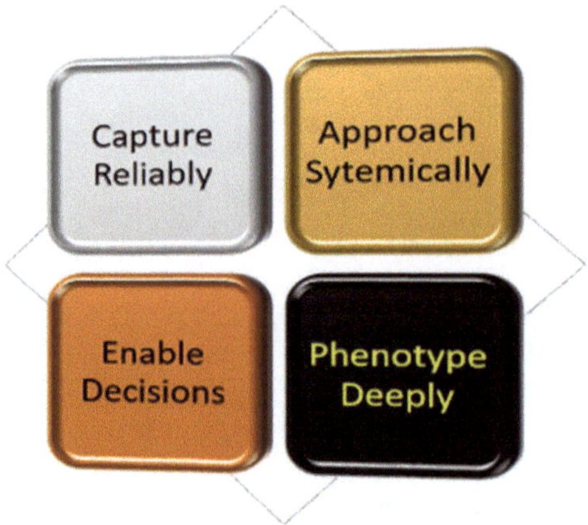

While "Approaching Systemically" addresses the breadth and scope of Big-data, there is an equally important need for deep phenotyping of healthy individuals and patients. Hence, the next CAPE principle, *Phenotyping Deeply* aims to discover meaningful "patterns within noise" and to exploit these for understanding the healthy and disease states. In contrast to gross changes in summary statistics (such as average measurements), pre-disease states are often characterized by subtle changes in dynamical behavior of the system (such as change in variability, fractal behavior, long-range correlations etc.). As an example, the human body is made up of about 30 trillion cells consisting roughly of about 200 cell-types arranged into tissues that perform orchestrated physiological functions. Despite sharing the same genetic code, there is enormous functional and phenotypic variability in these cells. Further each cell-type population (tissue) has considerable amount of heterogeneity within the tissue itself. Most cellular level experiments ignore this variation which is often summarized into average properties (e.g. gene expression) of the cell-population. It was only the advent of single cell sequencing technology that has now shed light into enormous diversity and distinct functional sub-types even within a cellular population. In a manner reminiscent of scale-free networks, similar heterogeneity (noisiness) of function exists at the phenotypic and physiological levels, heart rate variability being a prominent example.

14.4.1 Principal Axes of Variation

Since it is impossible to measure the entire physiology of an individual with the available technologies, a natural question that a biomedical data-scientist faces is "where to start with deep phenotyping of an individual?" The answer may lie in a combination of expert driven and data-driven approaches. Leveraging the key network players may be combined with expert-knowledge of human physiology and the disease in question. In most scientific approaches to Big-data, it is prudent to form scientific hypothesis which may be tested through Big-data analytics. One of such scientific hypothesis consistently validated is the existence of physiological axes that form the core of many complex disorders (Ghiassian et al. 2016). These axes (also called endophenotypes, endotypes or shared intermediate patho-phenotypes) can be thought of as major relay stations for the development of a multitude of diseases including complex diseases such as diabetes and cardiovascular diseases (Ghiassian et al. 2016). At the cellular level, such key axes of health-regulation are found to be (i) inflammation, (ii) fibrosis, and (iii) thrombosis. Similarly, at the physiological level, systems which are known to integrate regulatory influences and maintain homeostasis are expected to be strong candidates for deep-phenotyping approaches. Since Autonomic Nervous System (ANS) is a natural choice for being an integrator of physiological networks, its quantification through Heart Rate Variability may be one of the key factors in untangling the complexity of diseases as discussed before and in the following case study.

14.4.2 Heart Rate Variability: A Case Study on Phenotype Deeply Principle of the CAPE Roadmap

A large majority of the automatic and unconscious regulation of human physiological functions happens through the Autonomic Nervous System (ANS). The nerve supply from ANS controls some of the most vital physiological processes including heart rate, breathing rate, blood pressure, digestion, sweating etc. through rich nerve supply bundled into two opposing components that dynamically balance out each other. These components are the sympathetic component ('fight or flight') response and the parasympathetic component ("rest and digest") respectively. Thus, the assessment of sympathetic and parasympathetic components can yield insights into the delicate dynamical balance of this physiological axis. Interestingly, this axis has been shown to be perturbed early in the presence of a variety of diseases and these perturbations can be measured non-invasively through heart beat intervals (Task Force for Heart rate variability 1996). The beating of the heart is not a perfectly periodic phenomenon as it constantly adapts to the changing demands of the body. Therefore, the heart rate exhibits complex behavior even at rest and this variation in the rhythm of the heart is known as heart rate variability (HRV). An illustration of inter-beat intervals time series obtained from an ECG is depicted in Fig. 14.7. One of the most significant uses of heart rate variability is in the prediction of a devastating and often fatal blood stream infection (sepsis) in the newborn children admitted to an Intensive Care Unit (ICU). In the study conducted by (Fairchild et al. 2013), the heart rhythm showed lower complexity (as evidenced by a fall in entropy) by up to 72 h before clinical recognition of sepsis. A pervasive application of this technique can therefore allow a window of early recognition in which newborn babies could be treated for sepsis. This was further validated in a clinical trial to test for the clinical value of this test and the results supported the predictive value of these features in decreasing the deaths due to sepsis by about 5% (Fairchild et al. 2013).

The lack of popularity of deep phenotyping stems from the highly mathematical nature of such patterns. However, it is anticipated that interdisciplinary application of mathematics and computer science shall be the driving force for next-generation medicine. Thus, adequate mathematical training and comfort with computational tools cannot be over-emphasized for the development of Big-data and Data-science for medicine. The mathematical features of interest in HRV are defined under three broad types, i.e. *Time Domain*, *Frequency Domain* and *Nonlinear Features* (Task Force for Heart rate variability 1996).

(a) *Time domain analysis* involves calculating the summary statistics and include,

i. *Mean* of the *Normal* RR intervals given by:

$$\bar{RR} = \frac{1}{N} \sum_{i=1}^{N} RR_i$$

ii. *SDNN*: *SDNN* is the standard deviation of the NN time series calculated as:

$$SDNN = \sqrt{\frac{1}{N-1} \sum_{i=1}^{N} \left(RR_i - \bar{RR}\right)^2}$$

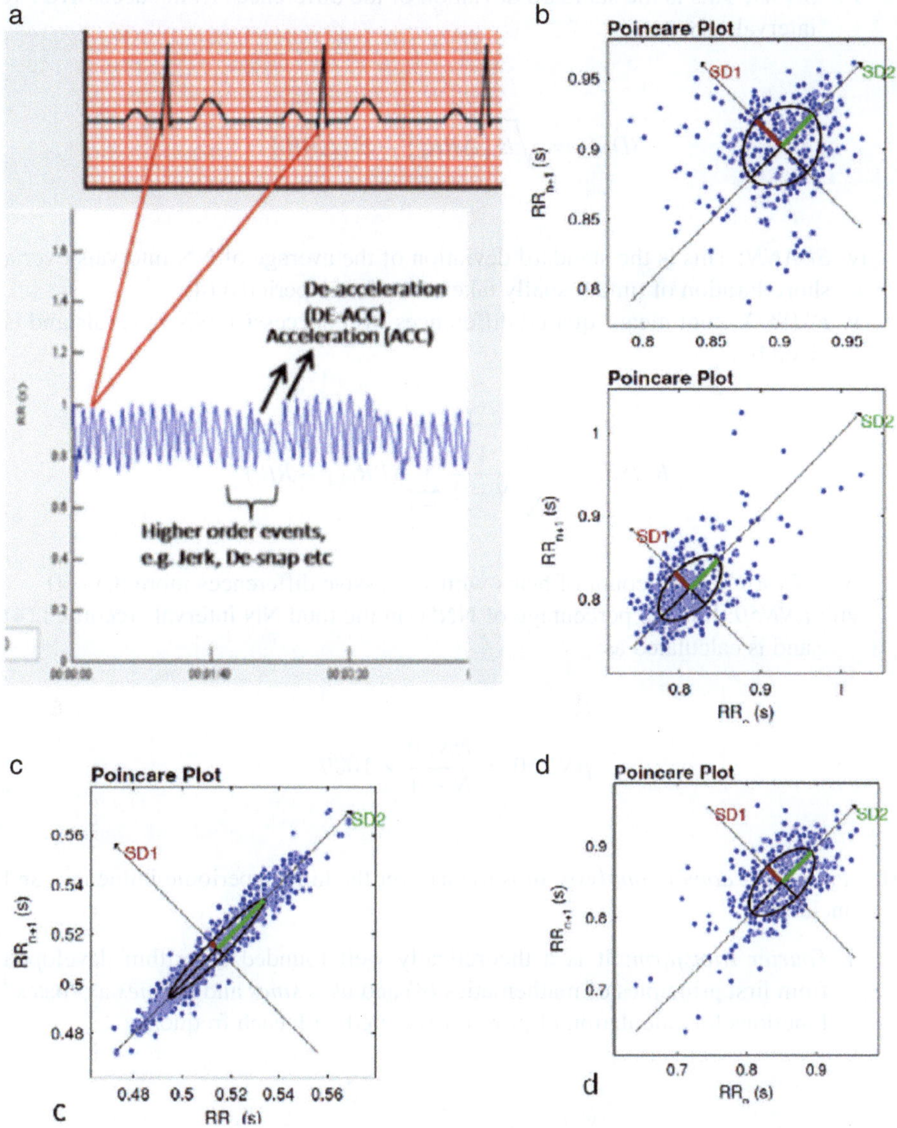

Fig. 14.7 Heart Rate Variability. (**a**) Heart rate time series is derived from peak-to-peak time difference in R waves from normal QRS complexes in an ECG. The heart accelerates when these intervals get shorter and vice versa. A few such events are marked. Even within healthy individuals (**b–d**) there is a considerable heterogeneity of physiological patterns that needs to be deciphered through methods of data-science such as pattern mining

iii. *SDSD*. This is the standard deviation of the differences from successive RR intervals given as:

$$SDSD = \sqrt{E\{\Delta RR_i^2\} - E\{\Delta RR_i\}^2}$$

iv. *SDANN*: This is the standard deviation of the average of NN intervals over a short duration of time (usually taken over 5 min periods) (4).

v. *RMSSD*: Root mean squared differences over successive NN intervals and is given by:

$$RMSSD = \sqrt{\frac{1}{N-1}\sum_{i=1}^{N-1}(RR_{i+1} - RR_i)^2}$$

vi. *NN50*: It is the count of beats with successive differences more than 50

vii. *pNN50*: It is the percentage of NN50 in the total NN intervals recorded (4) and is calculated as:

$$pNN50 = \frac{NN50}{N-1} \times 100\%$$

(b) *Frequency domain analysis* aims to discover the hidden periodic influences and includes

i. *Fourier transform*: It is a theoretically well founded algorithm developed from first principles of mathematics (8) and uses *sines* and *cosines* as "basis" functions for calculation of power associated with each frequency

$$R_n \equiv \sum_{t=0}^{N-1} RR_t e^{2\pi itn/N}, n = 0, \ldots, N-1$$

where, $f_n \equiv \frac{n}{N\Delta}, n = \frac{-N}{2}, \ldots, \frac{N}{2}$

The power spectrum is defined by the following equations:

$$P_0 = \frac{1}{N^2} |RR_0|^2, \textit{ at zero frequency}$$

$$P_n = \frac{1}{N^2} \left[|RR_n|^2 + |RR_{N-n}{}^2| \right], at\ n = 1, 2, \ldots, \left(\frac{N}{2} - 1 \right)$$

$$P_c = \frac{1}{N^2} \left[\left| RR_{N/2} \right|^2 \right], \textit{at Nyquist critical frequency}$$

The application of Fourier transform in HRV has given many insights of physiological relevance. It was found first from the Fourier spectrum that heart rate had at least two distinct *modes* of oscillation, the low frequency (LF) mode centered between 0.04–0.15 Hz and the high frequency (HF) mode centered between 0.15 Hz–0.40 Hz and was found from experiments in animals and humans that HF component arose from parasympathetic control of the heart. This has been found to be associated with the breathing frequency, hence also known as *Respiratory Sinus Arrhythmia* (RSA).

(c) *Nonlinear analyses of HRV* for complexity quantification include:

i. *Poincare plot*: In this method, *time-delayed embedding* of a signal is accomplished by reconstructing the phase space by using the lagged values of the signals. Poincare plots analysis is one of the simplest and a popular methods of phase space reconstruction of cardiac inter-beat (RR) interval series where *RR(n)* is plotted against *RR(n + d)*.
ii. *Fractal analysis using Detrended Fluctuation Analysis*: Fractal structures are self-similar structures which exhibit the property of long range correlations. *Detrended Fluctuation Analysis* (DFA) is a robust method of fractal analysis and its use for physiological signals and is described in the (Task Force for Heart rate variability 1996)
iii. *Entropies*: As discussed earlier Entropy methods such as *Shannon Entropy*, *Approximate entropy* and *Sample Entropy* (analyze the *complexity* or *irregularity* of the time series. Mathematically, these entropy measures represent the conditional probability that a data of length N having multiple patterns of length m, within a *tolerance* range r, will also have repeated patterns of length $m + 1$. For example, sample entropy is defined as:

$$SampEn\,(m, r, N) = -\log\left(\frac{C\,(m+1, r)}{C\,(m, r)}\right)$$

14.5 Enable Decisions

Until recently, medical knowledge discovery has solely relied upon the rigor of statistical approaches like hypothesis testing. With the availability of Big-data, complementary paradigms of *Machine Learning* and *Artificial Intelligence* have emerged. These approaches place more emphasis upon predictive modeling rather than hypothesis testing. Therefore, in contrast to traditional statistics, the data-drive paradigm takes a more flexible approach in the beginning, relaxing the assumptions about data such as parametric distributions. This relaxation of assumptions is countered by rigorous testing of predictive power of the models thus learnt through cross-validation and testing sets. The most popular approaches to machine learning include *Support Vector Machines* (SVM), Random Forests (RF), Shallow and Deep Neural Networks. Although many doubts are raised about flaws and spurious results of many such approaches, machine learning approaches have consistently outperformed statistical modeling in many clinical situations because of better handling of nonlinearity in these data (Song et al. 2004). Although statistical approaches are more rigorous, these have often led to fishing for significance without clinical translation. A meta-analysis of published findings shows that fishing for p-values has led to a proliferation of scientific studies which are either false or non-reproducible (Ioannidis 2005). In this situation, machine learning approaches can bring sanity by emphasizing predictive value rather than mere statistical support for the proposed hypotheses. Therefore, these approaches have a definite potential if applied in a robust manner. The key algorithms that are particularly important from the standpoint of the CAPE approach are discussed below.

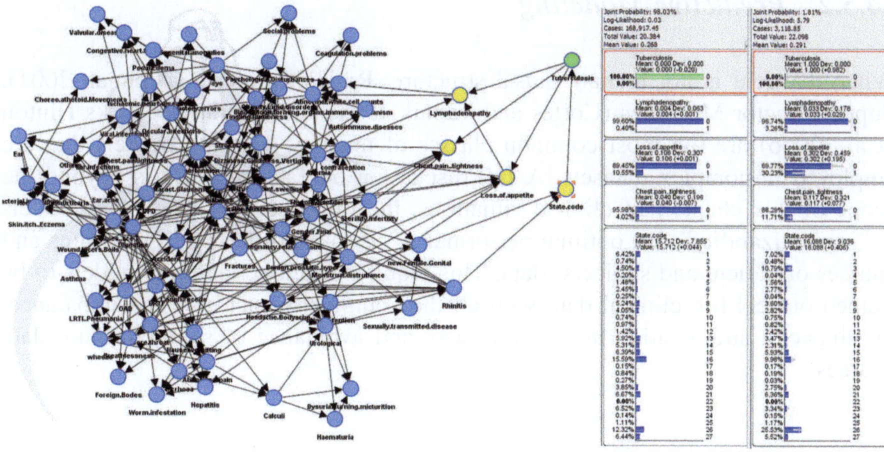

Fig. 14.8 Going Beyond Association networks to Enable Decisions through Causal Networks. A causal model on the POSEIDON data is shown and can help data-scientist s and clinicians in making informed decisions. This example shows the probability of tuberculosis in an Indian patient presenting in the OPD with Lymph Node enlargement, in the absence of Anemia. Notice that a particular Indian state (code 26) is associated with a higher probability of tuberculosis than other states if everything else is held constant

14.5.1 Causal Modeling Through Bayesian Networks

Bayesian Networks (BN) are Probabilistic Graphical Models that extend the argument on "Approach Systemically" and enable decisions. Unlike association networks where the edges might merely represent indirect or spurious relationships, Bayesian Networks discover direct causal relationships through conditional probabilities and repeated application of the Bayes Rule. Thus, BNs are one of the most advanced analytical tools for enabling causal decisions (Pearl 2010). Having adjusted for the possibility of spurious relationships, BNs are typically sparser than association networks and hence combining feature reduction, systemic approach and decision making, all in a single algorithm. Moreover, these reduce the possibility of false relationships by fitting a joint multivariate model upon the data in contrast to finding pairwise associations as done for association networks. Further, these models allow statistical inference to be conducted over the nodes of interest, thus enabling actionable decisions and policy thus making these one of the tools of choice for community health models. An example of a Bayesian Network constructed upon the POSEIDON data described earlier is shown in Fig. 14.8. Notice that this network allows the data-scientist to take actionable decisions based on quantitative inferences in contrast to pattern-mining for associations.

14.5.2 Predictive Modeling

When the goal is not to find causal structure, Random Forests (Breiman 2001), Support Vector Machines (Cortes and Vapnik 1995) and Neural Networks Hinton et al. (2006) are the most common classes of machine learning models that are employed in complex datasets. A full discussion on each of these is beyond the scope of this chapter. In clinical situations, the litmus test of predictive models is "generalizability", i.e. optimal performance against a different clinical sites and sources of patient and subjects' data. This requires machine learning models to be trained on real-life clinical data with all the complexity such as class-imbalance, missingness, stratification and nonlinearity and avoidance of synthetic pure data sources.

14.5.3 Reproducibility of Data Science for Biomedicine

Reproducibility is one of the biggest challenges of Biomedical Data-science. Hence there are global efforts to create standards for predictive modeling and machine learning to enable reproducibility. One of such standards is *Transparent reporting of a multivariable prediction model for individual prognosis or diagnosis* (TRIPOD) (Collins et al. 2015) formulated under the *Enhancing the QUAlity and Transparency Of health Research* (EQUATOR) consortium. TRIPOD provides guidelines and a checklist of twenty-two criteria that must be met by a well conducted predictive modeling study for medicine. From the data-science perspective, the key criteria are to clearly specify (i) data missingness, (ii) handling of predictors, (iii) type of model (iv) validation strategy, (v) model performance and model comparison (vi) model recalibration, (vii) limitations of the study. It is expected that such initiatives will improve the generalizability of predictive models and would help these to be adopted across laboratories and clinics.

14.5.4 SAFE-ICU Initiative: A Full Spectrum Case Study in Biomedical Data-Science

Intensive Care Units are one of the biggest source of biomedical Big-data. However, the application of Data-science to Biomedical Big-data is relatively nascent. At All India Institute of Medical Sciences, New Delhi, India an end-to-end initiative based upon the CAPE principles for Pediatric Intensive Care Units has been launched. The overarching goal of this initiative is to create a *Sepsis Advanced Forecasting Engine for ICUs (SAFE-ICU)* for preventing mortality due to killer conditions such as sepsis. Delay in recognition of sepsis in the ICUs can be devastating with a mortality as high as 50% in developing countries like India. The CAPE principles for SAFE-ICU have been built from the ground-up with in-house design and customization of pipelines for reliable capture and analysis of Big-data (Fig. 14.9). Multi-parameter

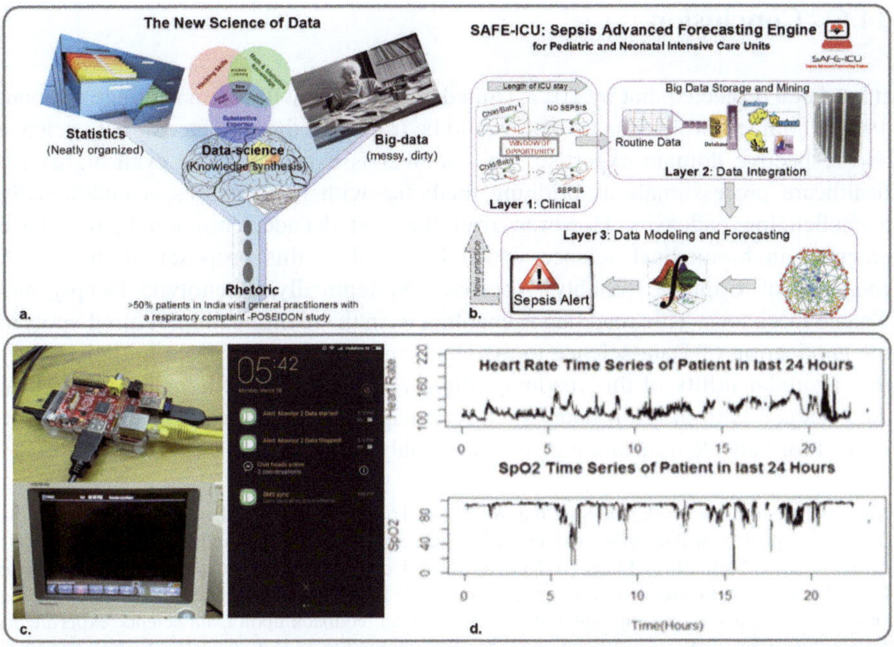

Fig. 14.9 (**a**) While traditional statistics likes to deal with neat data (analogous to library catalogue), Big-data often starts with messy data (analogous to Einstein's messy desk). The emerging role of Data-science is analogous to Einstein's brain that synthesizes these two into new knowledge. Full Spectrum demonstration of the CAPE principles through SAFE-ICU (**b**). In-house pipelines for warehousing Big-data from Pediatric ICU. Lean prototyping of the pipelines was initially carried out using Raspberry Pi (**c**) and finally deployed on a server. Reliable capture was ensured by deploying alert mechanisms (**c**). The reliably captured data were then structured using text mining (**d**) followed by exploratory data analysis of the multivariate time series (**d**). These time series include clinically relevant features such as Heart Rate, Oxygen Saturations, Respiratory Rates, Pulse Rate, Blood Pressure, End Tidal CO_2 etc. These data are integrated with laboratory investigations and treatment charts entered by clinicians and have led to the creation of a unique pediatric intensive care Big-data resource

monitoring data are being warehoused using open-source platforms such as *Spark*, *Python* and *R* and are documented for reproducibility using markdown documentation. Text-mining upon unstructured files such as treatment notes has been carried out and structured data is being warehoused alongside the multiparameter monitoring data for building graphical models. A unique Pediatric ICU resource of over 50,000 h of continuous multivariate monitoring data followed by deep-phenotyping using mathematical and computational analyses has been generated and models are being developed and tested with the aim improving delivery of care and saving lives.

14.6 Conclusion

It has been projected that by 2017, United States alone will face a shortage of about 190,000 trained professionals who would be trained at the interface of data-science and respective domain expertise. This number is expected to be even higher for healthcare professionals as bridging medicine with mathematics is undoubtedly a challenging endeavor. However, over the past decade, common themes have emerged in biomedical science which have led to this proposal of the *CAPE roadmap* of **C**apture Reliably, **A**pproach Systemically, **P**henotype Deeply and **E**nable Decisions. This roadmap is enabling us in the identification of blind-spots in the application of Data-science to medicine and other data-science initiatives may find a similar utility of this roadmap. Finally, the need of the hour for biomedical data-science is to develop many such roadmaps and adopt principles that may be critical for bedside translation of Big-data analytics.

Acknowledgements I acknowledge the Wellcome Trust/DBT India Alliance for supporting the SAFE-ICU project at All India Institute of Medical Sciences (AIIMS) and Indraprastha Institute of Information Technology Delhi (IIIT-Delhi), New Delhi, India. I also express deep gratitude to Dr. Rakesh Lodha, Professor-in-charge of the Pediatric Intensive Care Unit at AIIMS for providing an immersive clinical environment and constant clinical feedback upon Data-science experiments for creating a SAFE-ICU. I also acknowledge the mentorship of Prof. Charles Auffray and Prof. Samir K. Brahmachari and analytics support provided by Mr. Aditya Nagori.

References

Ali, O. (2013). Genetics of type 2 diabetes. *World Journal of Diabetes, 4*(4), 114–123. doi:10.4239/wjd.v4.i4.114.

Arron, J. R., Townsend, M. J., Keir, M. E., Yaspan, B. L., & Chan, A. C. (2015). Stratified medicine in inflammatory disorders: From theory to practice. *Clinical Immunology, 161*(1), 11–22. doi:10.1016/j.clim.2015.04.006.

Athey, S., & Imbens, G. (2016). Recursive partitioning for heterogeneous causal effects: Table 1. *Proceedings of the National Academy of Sciences of the United States of America, 113*(27), 7353–7360. doi:10.1073/pnas.1510489113.

Auffray, C., Chen, Z., & Hood, L. (2009). Systems medicine: The future of medical genomics and healthcare. *Genome Medicine, 1*(1), 2. doi:10.1186/gm2.

Barabási, A., Gulbahce, N., & Loscalzo, J. (2011). Network medicine: A network-based approach to human disease. *Nature Reviews Genetics, 12*(1), 56–68. doi:10.1038/nrg2918.

Breiman, L. (2001). Random forests. *Machine Learning, 45*(1), 5–32. doi:10.1023/A:1010933404324.

Chandler, R. K., Kahana, S. Y., Fletcher, B., Jones, D., Finger, M. S., Aklin, W. M., et al. (2015). Data collection and harmonization in HIV research: The seek, test, treat, and retain initiative at the National Institute on Drug Abuse. *American Journal of Public Health, 105*(12), 2416–2422. doi:10.2105/ajph.2015.302788.

Christakis, N. A., Fowler, J. H. (2007). The spread of obesity in a large social network over 32 years. *New England Journal of Medicine, 357*(4), 370–379.

Collins, F. S., & Varmus, H. (2015). A new initiative on precision medicine. *New England Journal of Medicine, 372*(9), 793–795. doi:10.1056/nejmp1500523.

Collins, G. S., Reitsma, J. B., Altman, D. G., & Moons, K. G. (2015). Transparent reporting of a multivariable prediction model for individual prognosis or diagnosis (TRIPOD): the TRIPOD statement. *BMJ, 350*, g7594. doi:10.1136/bmj.g7594.

Cortes, C., & Vapnik, V. (1995). Support-vector networks. *Machine Learning, 20*(3), 273–297. doi:10.1007/BF00994018.

Coveney, P. V., Dougherty, E. R., & Highfield, R. R. (2016). Big data need big theory too. *Philosophical Transactions Series A, Mathematical, Physical, and Engineering Sciences, 374*(2080). http://rsta.royalsocietypublishing.org/content/374/2080/20160153.long.

Dinov, I. D., Heavner, B., Tang, M., Glusman, G., Chard, K., Darcy, M., Madduri, R., Pa, J., Spino, C., Kesselman, C., Foster, I., Deutsch, E. W., Price, N. D., Van Horn, J. D., Ames, J., Clark, K., Hood, L., Hampstead, B. M., Dauer, W., & Toga, A. W. (2016). Predictive big data analytics: A study of Parkinson's disease using large, complex, heterogeneous, incongruent, multi-source and incomplete observations. *PLoS One, 11*(8), e0157077. doi:10.1371/journal.pone.0157077.

Doan, S., Conway, M., Phuong, T. M., & Ohno-Machado, L. (2014). Natural language processing in biomedicine: A unified system architecture overview. *Methods in Molecular Biology Clinical Bioinformatics, 1168*, 275–294. doi:10.1007/978-1-4939-0847-9_16.

Ester, M; Kriegel, H-P; Sander, J; Xu, X (1996). A density-based algorithm for discovering clusters in large spatial databases with noise. In E. Simoudis, J. Han, & U. M. Fayyad (Eds.), *Proceedings of the Second International Conference on Knowledge Discovery and Data Mining (KDD-96)* (pp. 226–231). California: AAAI Press. ISBN 1–57735–004-9.

Fairchild, K. D., Schelonka, R. L., Kaufman, D. A., Carlo, W. A., Kattwinkel, J., Porcelli, P. J., Navarrete, C. T., Bancalari, E., Aschner, J. L., Walker, M. W., Perez, J. A., Palmer, C., Lake, D. E., O'Shea, T. M., & Moorman, J. R. (2013). Septicemia mortality reduction in neonates in a heart rate characteristics monitoring trial. *Pediatric Research, 74*(5), 570–575. doi:10.1038/pr.2013.136.

Farcomeni, A. (2008). A review of modern multiple hypothesis testing, with particular attention to the false discovery proportion. *Statistical Methods in Medical Research, 17*(4), 347–388.

Fleurence, R. L., Curtis, L. H., Califf, R. M., Platt, R., Selby, J. V., & Brown, J. S. (2014). Launching PCORnet, a national patient-centered clinical research network. *Journal of the American Medical Informatics Association: JAMIA, 21*(4), 578–582. doi:10.1136/amiajnl-2014-002747.

Follett, R., & Strezov, V. (2015). An analysis of citizen science based research: Usage and publication patterns. *PLoS One, 10*(11), e0143687. doi:10.1371/journal.pone.0143687.

Frey, B. J., & Dueck, D. (2007). Clustering by passing messages between data points. *Science, 315*(5814), 972–976. doi:10.1126/science.1136800.

Ghiassian, S. D., Menche, J., Chasman, D. I., Giulianini, F., Wang, R., Ricchiuto, P., Aikawa, M., Iwata, H., Müller, C., Zeller, T., Sharma, A., Wild, P., Lackner, K., Singh, S., Ridker, P. M., Blankenberg, S., Barabási, A. L., & Loscalzo, J. (2016). Endophenotype network models: Common core of complex diseases. *Scientific Reports, 6*, 27414. doi:10.1038/srep27414.

Goodfellow, I., Bengio, Y., & Courville, A. (2016). *Deep Learning*. Cambridge, MA: MIT Press.

Hayati, R. P., Lee, K. J., & Simpson, J. A. (2015). The rise of multiple imputation: a review of the reporting and implementation of the method in medical research. *BMC Medical Research Methodology, 15*, 30. doi:10.1186/s12874-015-0022-1.

Hinks, T., Zhou, X., Staples, K., Dimitrov, B., Manta, A., Petrossian, T., et al. (2015). Multidimensional endotypes of asthma: Topological data analysis of cross-sectional clinical, pathological, and immunological data. *The Lancet, 385*(Suppl 1), S42. doi:10.1016/s0140-6736(15)60357-9.

Hinton, G. E., Osindero, S., & Teh, Y. (2006). "A fast learning algorithm for deep belief nets" (PDF). *Neural Computation, 18*(7), 1527–1554. doi:10.1162/neco.2006.18.7.1527.

Hripcsak, G., Ryan, P. B., Duke, J. D., Shah, N. H., Park, R. W., Huser, V., et al. (2016). Characterizing treatment pathways at scale using the OHDSI network. *Proceedings of the National Academy of Sciences of the United States of America, 113*(27), 7329–7336. doi:10.1073/pnas.1510502113.

Ioannidis, J. P. (2005). Why most published research findings are false. *PLoS Medicine, 2*(8), e124.

Jerri, A. (1977). The Shannon sampling theorem—Its various extensions and applications: A tutorial review. *Proceedings of the IEEE, 65*(11), 1565–1596. doi:10.1109/proc.1977.10771.

Kohane, I. S., Drazen, J. M., & Campion, E. W. (2012). A Glimpse of the next 100 years in medicine. *New England Journal of Medicine, 367*(26), 2538–2539. doi:10.1056/nejme1213371.

Kohonen, T. (1982). Self-organized formation of topologically correct feature maps. *Biological Cybernetics, 43*(1), 59–69. doi:10.1007/bf00337288.

Kottke, T. E., Huebsch, J. A., Mcginnis, P., Nichols, J. M., Parker, E. D., Tillema, J. O., & Maciosek, M. V. (2016). Using principles of complex adaptive systems to implement secondary prevention of coronary heart disease in primary care. *The Permanente Journal, 20*(2), 17–24. doi:10.7812/TPP/15-100.

Landauer, R. (1961). "Irreversibility and heat generation in the computing process" (PDF). *IBM Journal of Research and Development, 5*(3), 183–191. doi:10.1147/rd.53.0183.

Longford, N. (2001). Multilevel analysis with messy data. *Statistical Methods in Medical Research, 10*(6), 429–444. doi:10.1191/096228001682157643.

Longhurst, C. A., Harrington, R. A., & Shah, N. H. (2014). A 'green button' for using aggregate patient data at the point of care. *Health Affairs, 33*(7), 1229–1235. doi:10.1377/hlthaff.2014.0099.

Mäkikallio, T. H., Høiber, S., Køber, L., Torp-Pedersen, C., Peng, C. K., Goldberger, A. L., & Huikuri, H. V. (1999). Fractal analysis of heart rate dynamics as a predictor of mortality in patients with depressed left ventricular function after acute myocardial infarction. TRACE Investigators. TRAndolapril Cardiac Evaluation. *The American Journal of Cardiology, 83*(6), 836–839.

Miller, G. W. (2014). *Exposome: A Primer*. Waltham: Elsevier Academic Press.

Miron, B. K., & Witold, R. R. (2010). Feature selection with the Boruta Package. *Journal of Statistical Software, 36*(11), 1–13. http://www.jstatsoft.org/v36/i11/.

Newman, M. (2005). Power laws, Pareto distributions and Zipf's law. *Contemporary Physics, 46*(5), 323–351. doi:10.1080/00107510500052444.

NHGRI. (2016). *The cost of sequencing a human genome – national human*. Retrieved October 22, 2016, from https://www.genome.gov/27565109/the-cost-of-sequencing-a-human-genome/

Nielson, J. L., Paquette, J., Liu, A. W., Guandique, C. F., Tovar, C. A., Inoue, T., et al. (2015). Topological data analysis for discovery in preclinical spinal cord injury and traumatic brain injury. *Nature Communications, 6*, 8581. doi:10.1038/ncomms9581.

Pearl, J. (2010). An introduction to causal inference. *International Journal of Biostatistics, 6*(2), 7. doi:10.2202/1557–4679.1203.

Ravishankar, S., & Bresler, Y. (2015). Efficient blind compressed sensing using sparsifying transforms with convergence guarantees and application to magnetic resonance imaging. *SIAM Journal on Imaging Sciences, 8*(4), 2519–2557. doi:10.1137/141002293.

Rhee, A., Cheong, R., & Levchenko, A. (2012). The application of information theory to biochemical signaling systems. *Physical Biology, 9*(4), 045011. doi:10.1088/1478-3975/9/4/045011.

Rodriguez-Esteban, R. (2016). A drug-centric view of drug development: How drugs spread from disease to disease. *PLoS Computational Biology, 12*(4), e1004852. doi:10.1371/journal.pcbi.1004852.

Rolland, B., Reid, S., Stelling, D., Warnick, G., Thornquist, M., Feng, Z., & Potter, J. D. (2015). Toward rigorous data harmonization in cancer epidemiology research: One approach. *American Journal of Epidemiology, 182*(12), kwv133. doi:10.1093/aje/kwv133.

Rosvall, M., & Bergstrom, C. T. (2008). Maps of random walks on complex networks reveal community structure. *Proceedings of the National Academy of Sciences of the United States of America, 105*(4), 1118–1123. doi:10.1073/pnas.0706851105.

Rothman, M. J., Rothman, S. I., & Beals, J. (2013). Development and validation of a continuous measure of patient condition using the Electronic Medical Record. *Journal of Biomedical Informatics, 46*(5), 837–848. doi:10.1016/j.jbi.2013.06.011.

Salvi, S., Apte, K., Madas, S., Barne, M., Chhowala, S., Sethi, T., Aggarwal, K., Agrawal, A., & Gogtay, J. (2016). Symptoms and medical conditions in 204 912 patients visiting primary health-care practitioners in India: a 1-day point prevalence study (the POSEIDON study). *The Lancet Global Health, 3*(12), e776–e784. doi:10.1016/S2214-109X(15)00152-7.

Sethi, T. P., Prasher, B., & Mukerji, M. (2011). Ayurgenomics: A new way of threading molecular variability for stratified medicine. *ACS Chemical Biology, 6*(9), 875–880. doi:10.1021/cb2003016.

Shannon, C. E. (1948). A mathematical theory of communication. *Bell System Technical Journal, 27*(3), 379–423. doi:10.1002/j.1538-7305.1948.tb01338.x.

Singla, M., Kar, M., Sethi, T., Kabra, S. K., Lodha, R., Chandele, A., & Medigeshi, G. R. (2016). Immune response to dengue virus infection in pediatric patients in New Delhi, India–Association of Viremia, inflammatory mediators and monocytes with disease severity. *PLoS Neglected Tropical Diseases, 10*(3), e0004497. doi:10.1371/journal.pntd.0004497. Erratum in: PLoS Neglected Tropical Diseases. 2016 Apr;10(4):e0004642.

Snyder, M. (2014). IPOP and its role in participatory medicine. *Genome Medicine, 6*(1), 6. doi:10.1186/gm512.

Song, X., Mitnitski, A., Cox, J., & Rockwood, K. (2004). Comparison of machine learning techniques with classical statistical models in predicting health outcomes. *Studies in Health Technology and Informatics, 107*(Pt 1), 736–740.

Steinhubl, S. R., Muse, E. D., & Topol, E. J. (2015). The emerging field of mobile health. *Science Translational Medicine, 7*(283), 283rv3. doi:10.1126/scitranslmed.aaa3487.

Sun, J. (2013). *Big data analytics for healthcare – SIAM: Society for Industrial and Applied Mathematics*. Retrieved October 22, 2016, from https://www.siam.org/meetings/sdm13/sun.pdf

Task Force for Heart rate variability. (1996). Standards of measurement, physiological interpretation, and clinical use. Task Force of the European Society of Cardiology and the North American Society of Pacing and Electrophysiology. *European Heart Journal, 17*(3), 354–381.

Topol, E. J. (2014). Individualized medicine from prewomb to tomb. *Cell, 157*(1), 241–253. doi:10.1016/j.cell.2014.02.012.

Topol, E. J. (2015). The big medical data miss: Challenges in establishing an open medical resource. *Nature Reviews Genetics, 16*(5), 253–254. doi:10.1038/nrg3943.

Tsoukalas, A., Albertson, T., & Tagkopoulos, I. (2015). From data to optimal decision making: a data-driven, probabilistic machine learning approach to decision support for patients with sepsis. *JMIR Medical Informatics, 3*(1), e11. doi:10.2196/medinform.3445.

Turnbaugh, P. J., Ley, R. E., Hamady, M., Fraser-Liggett, C. M., Knight, R., & Gordon, J. I. (2007). The Human Microbiome Project. *Nature, 449*(7164), 804–810. doi:10.1038/nature06244.

Walpole, J., Papin, J. A., & Peirce, S. M. (2013). Multiscale computational models of complex biological systems. *Annual Review of Biomedical Engineering, 15*, 137–154. doi:10.1146/annurev-bioeng-071811-150104.

Wang, L., Wang, Y., & Chang, Q. (2016). Feature selection methods for big data bioinformatics: A survey from the search perspective. *Methods, 111*, 21–31. doi:10.1016/j.ymeth.2016.08.014.

Wijndaele, K., Westgate, K., Stephens, S. K., Blair, S. N., Bull, F. C., Chastin, S. F., Dunstan, D. W., Ekelund, U., Esliger, D. W., Freedson, P. S., Granat, M. H., Matthews, C. E., Owen, N., Rowlands, A. V., Sherar, L. B., Tremblay, M. S., Troiano, R. P., Brage, S., & Healy, G. N. (2015). Utilization and harmonization of adult accelerometry data: Review and expert consensus. *Medicine and Science in Sports and Exercise, 47*(10), 2129–2139. doi:10.1249/MSS.0000000000000661.

Wilbanks, J. T., & Topol, E. J. (2016). Stop the privatization of health data. *Nature, 535*(7612), 345–348. doi:10.1038/535345a.

Chapter 15
Time-Based Comorbidity in Patients Diagnosed with Tobacco Use Disorder

Pankush Kalgotra, Ramesh Sharda, Bhargav Molaka, and Samsheel Kathuri

15.1 Introduction

Every fifth adult in the United States of America uses tobacco products (Control and Prevention 2012). Tobacco use is one of the leading causes of preventable deaths in the world (Rutten et al. 2008). According to a report by US Department of Health and Human Services and CDC in 2014, nearly half a million patients die annually due to the disorders caused by the use of tobacco in USA (US Department of Health and Human Services 2014).

The primary addictive substance in the tobacco products is nicotine. The tobacco products are addictive in nature and there is no quick fix solution to control its use (Rigotti 2002). The dependence on tobacco causes tobacco use disorder (TUD) and several other diseases. Tobacco use disorder is highly related to many health disorders (Merikangas et al. 1998). It is related to anxiety disorders such as post-traumatic stress disorder, substance use disorder and obsessive–compulsive disorder (Morissette et al. 2007). It is also related to coronary heart disease, chronic obstructive pulmonary disease (COPD), lung cancer, and tuberculosis (Critchley and Capewell 2003; Lin et al. 2008).

The presence of other conditions in addition to an important disease (index disease) is known as comorbidity (Feinstein 1970). The distinction in the conceptualization of comorbidity is made in different ways by Valderas and colleagues (Valderas et al. 2009). In one conceptualization, the co-occurrence of the diseases is considered. However, in another conceptualization, the chronology or the sequence of disease development is considered.

P. Kalgotra (✉) • R. Sharda • B. Molaka • S. Kathuri
Oklahoma State University, Stillwater, OK 74074, USA
e-mail: pankush@okstate.edu; ramesh.sharda@okstate.edu; bhargav.molaka@okstate.edu; samsheel.kathuri@okstate.edu

S. Srinivasan (ed.), *Guide to Big Data Applications*, Studies in Big Data 26,
DOI 10.1007/978-3-319-53817-4_15

Past studies have largely focused on finding comorbidities in TUD patients based on their co-occurrences. However, the research on finding the comorbidities developed over time in TUD patients is rare. In this study, we adopt the second conceptualization briefed by Valderas and colleagues, and discover time-based comorbidities in TUD patients. Discovering comorbidities over time can provide additional understanding on how different diseases develop sequentially in TUD patients and help physicians take preemptive actions to prevent the future diseases.

We use Electronic Medical Records (EMRs) of the TUD patients to find time-based comorbidities. During a hospital visit, a patient can develop multiple diseases. Thus, when multiple diseases are diagnosed in a patient over multiple hospital visits, this makes the data multidimensional and time-variant (MDTV). Due to MDTV nature of the Electronic Medical Records, the traditional data mining techniques might not be useful for the analysis. The novel temporal machine learning methods are required (Gubbi et al. 2013). The methods should be able to synchronize multidimensional data events and generate meaningful patterns. In this paper, we adapt traditional sequence analysis to find time-based comorbidities. SA is an efficient way to analyze time-ordered data (Agrawal and Srikant 1995; Srikant and Agrawal 1996). It can find sequential patterns in the time-ordered transactional dataset.

First, we describe our approach to prepare multidimensional time variant events in a form to be able to perform sequence analysis (SA). Then we adapt the sequence analysis to find time-based comorbidities in the TUD patients.

In the next section, we describe the dataset used to study the comorbidities in TUD patients followed by the preparation of data. We also present the step-by-step process to find comorbidities over time. In the following section, we describe the results. Finally, we conclude by discussing the results.

15.2 Method

15.2.1 Data Description and Preparation

We obtained dataset from the Cerner Corporation, a major Electronic Medical Records (EMR) provider. The database is housed in the Center for Health Systems Innovation (CHSI) at Oklahoma State University. The database contains EMRs on the visits of more than 50 million unique patients across US hospitals (1999–2014). In this paper, we focused on EMRs of the patients diagnosed with the Tobacco Use Disorder (TUD). The step-by-step process followed to prepare and analyze the data is presented in Fig. 15.1.

First of all, information about the patients, hospitals, types of visits and diseases developed by the patients were synchronized for further analyses. The diseases were recorded according to the International Classification of Diseases, Ninth Revision, Clinical Modification (ICD-9-CM). Next, the emergency and in-patients

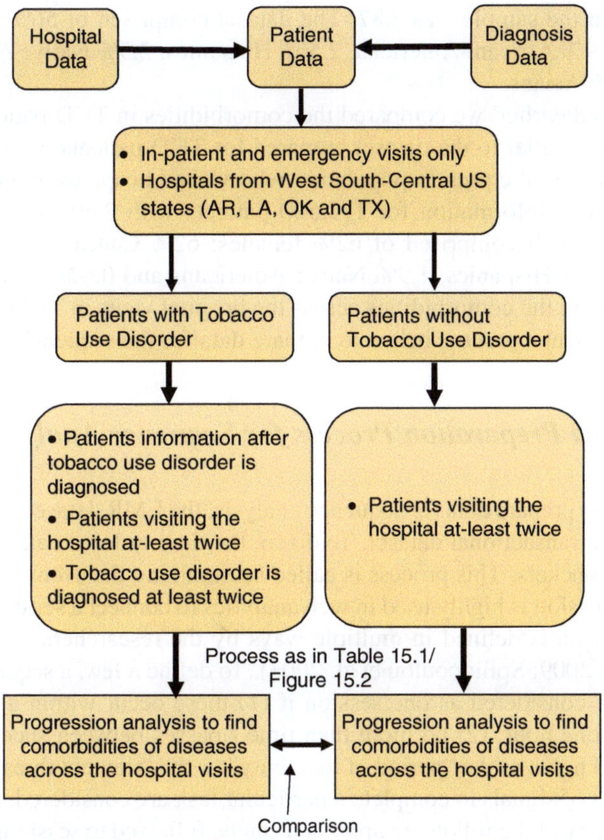

Fig. 15.1 Data preparation and analysis flowchart

were extracted for further analysis. To minimize the differences in comorbidities across regions, we only used hospital visits from the West South-Central region of United States (AR, LA, OK and TX).

To find the unique comorbidities in the Tobacco Use Disorder (TUD) patients, we compare the comorbidities in the patients who were never diagnosed with TUD (non-TUD). We first perform independent analyses on two types of patients (TUD and non-TUD) and then compare the results.

To prepare the dataset containing information about TUD patients, we considered their hospital visits starting from the time they were diagnosed with TUD. We also made sure that a patient visited the hospital at least twice and TUD was diagnosed at least in two visits. These steps were taken to reduce the sample bias as TUD may not be a prevalent disease in a patient. If TUD is diagnosed at least twice, the disease progressions are the indicative evidence of the TUD effect.

After cleaning and preparing the dataset of TUD patients, the final dataset contained information about 25,330 patients. The average number of hospital visits by a patient in the sample was 3.87. The dataset comprised of 56% females; 75% Caucasians, 19% African–American, 2.5 % Hispanics, 2.5% Native Americans and less than 0.2% Asians.

As discussed earlier, we compared the comorbidities in TUD patients with non-TUD patients. Similar to the dataset prepared for TUD patients, we extracted non-TUD patients from the pseudo-population who visited hospitals at least twice. The dataset contained information for 172,500 patients with 2.69 as average number of hospital visits. It comprised of 62% females; 62% Caucasians, 28% African-American, 6.4 % Hispanics, 1.9% Native Americans and 0.6% Asians. Finally, to find and compare the comorbidities across the hospital visits in TUD and non-TUD patient, we present a process below to prepare data for the sequence analysis.

15.2.2 Data Preparation Process for Sequence Analysis

As our aim is to prepare data for sequence analysis, the EMR dataset is required to be converted to a transactional dataset. To do so, the time-ordered dataset is separated into different buckets. This process is called sessionizing the events or signals. The concept of a session is highly used in web analytics to connect a series of user events online. A session is defined in multiple ways by the researchers in web analytics (Gayo-Avello 2009; Spiliopoulou et al. 2003). To define a few, a sequence of events or queries are considered as one session if (1) these occur within a specific time, for example, one hour; (2) no more than time t passes between successive events. This time t is known as the “period of inactivity” or the “timeout threshold”; and (3) the collections of signals to complete a particular task are considered. In a particular MDTV case, any of the following approach can be followed to sessionize the signals generated by the signals.

The steps followed to sessionize the EMRs and find the progressions in the diseases across hospital visits are presented in Table 15.1. The process is illustrated with an example in Table a–d (Fig. 15.2). To illustrate the process, we use a hypothetical dataset containing a patient (P) who visited the hospital three times (Visit No.) as presented in Table a (Fig. 15.2). On his first visit, the patient developed two diseases (A and B); on the second visit, two new unique diseases were diagnosed (C and D); and on the third visit, E and F were developed. Our aim is to find disease associations over hospital visits as in Fig. 15.2.

To create a sessionized dataset appropriate for creating paths or sequences of diseases as in Table d (Fig. 15.2), we first divided the dataset into three buckets, each having information about a visit as shown in Table b (Fig. 15.2). The left join of first visit dataset with the second visit dataset is performed, and its output is further left joined with the third visit dataset. In our hypothetical dataset, we have three visits. However, one can extend the analysis to any number of visits. The resultant table from multiple joins is presented in Table c (Fig. 15.2), in which each combination of the diseases is labeled by a different session. Finally, the dataset is converted as in Table d (Fig. 15.2) by appending different visits. This dataset is now appropriate for sequence analysis.

Table 15.1 Steps to prepare data for disease progressions

Step 1: For n patients (i=1 to n) with v visits and k diseases in each visit, the dataset contains P_i, $V_{ij\,(j=1\ to\ v)}$, $D_{ijk\,(k=1\ to\ m)}$
Step 2: A separate dataset for each visit is extracted; a total of v number of datasets are resulted: $(P_i, V_{ij\,(j=1)}, D_{ijk\,(k=1\ to\ m)})$, $(P_i, V_{ij\,(j=2)}, D_{ijk\,(k=1\ to\ m)})$ … $P_i, V_{ij\,(j=v)}, D_{ijk\,(k=1\ to\ m)}$
Step 3: The dataset with J=1 is left joined with J+1. The resultant dataset contains $(P_i, (V_{ij}, D_{ijk\,(k=1\ to\ m)})_{(j=1)}, (V_{ij}, D_{ijk\,(k=1\ to\ m)})_{(j=2)} \ldots (V_{ij}, D_{ijk\,(k=1\ to\ m)})_{(j=v)})$. *In each join,* $D_{jk} \neq D_{(j-l)k}$, *where l=1 to (j-1).*
Step 4: Each record is assigned a unique session. Total number of sessions per patient are $S=\prod_1^v C_i$, *where C is count of diseases in a visit i.*
Step 5: A union of P_i, V_j, D_K *and* S_S *are taken as* $(P_i, V_{ij\,(j=1)}, D_{ijk\,(k=1\ to\ m)}, S_{(1\ to\ s)})$ U $(P_i, V_{ij\,(j=2)}, D_{ijk\,(k=1\ to\ m)}, S_{(1\ to\ s)})$ U … $(P_i, V_{ij\,(j=v)}, D_{ijk\,(k=1\ to\ m)}, S_{(1\ to\ s)})$
Step 6: The Sequence Analysis is run to find the comorbidities over time.
Step 7: The number of comorbidities are adjusted by calculating the number of patients following the specific disease path from the result of Step 3.

The dataset prepared in Table d (Fig. 15.2) is the input for sequence analysis. The sessions in Table d (Fig. 15.2) are analogous to the transactions in the sequence mining method. The sequence analysis can be run to discover common sequence patterns. As the same disease can be repeated over time, the adjustment in the number of pair-wise combinations is made to find the comorbidities over hospital visits.

As the process involves joining of multiple large tables, it becomes difficult for the traditional data mining tools to analyze the data. Therefore, we used Teradata Aster, a Big Data platform. To run the analysis, the nPath function of Teradata Aster was used. This function is able to create paths with a session/transaction sorted by the time.

15.3 Results

The process described above is used to find comorbidities developed over time. We analyze maximum of three hospital visits. We first find common comorbidities in TUD patients and then compare with the non-TUD patients. The results of the sequence analysis to find time-based comorbidities are presented below in four subsections.

1. Top 20 diseases in TUD and non-TUD patients as in Figs.15.3 and 15.4 respectively.
2. Top 20 diseases in TUD patients and corresponding prevalence in non-TUD patients as in Fig. 15.5.
3. Top 15 comorbidities in TUD patients across two visits as in Fig. 15.6 and corresponding prevalence in non-TUD patients as listed in Table 15.2.

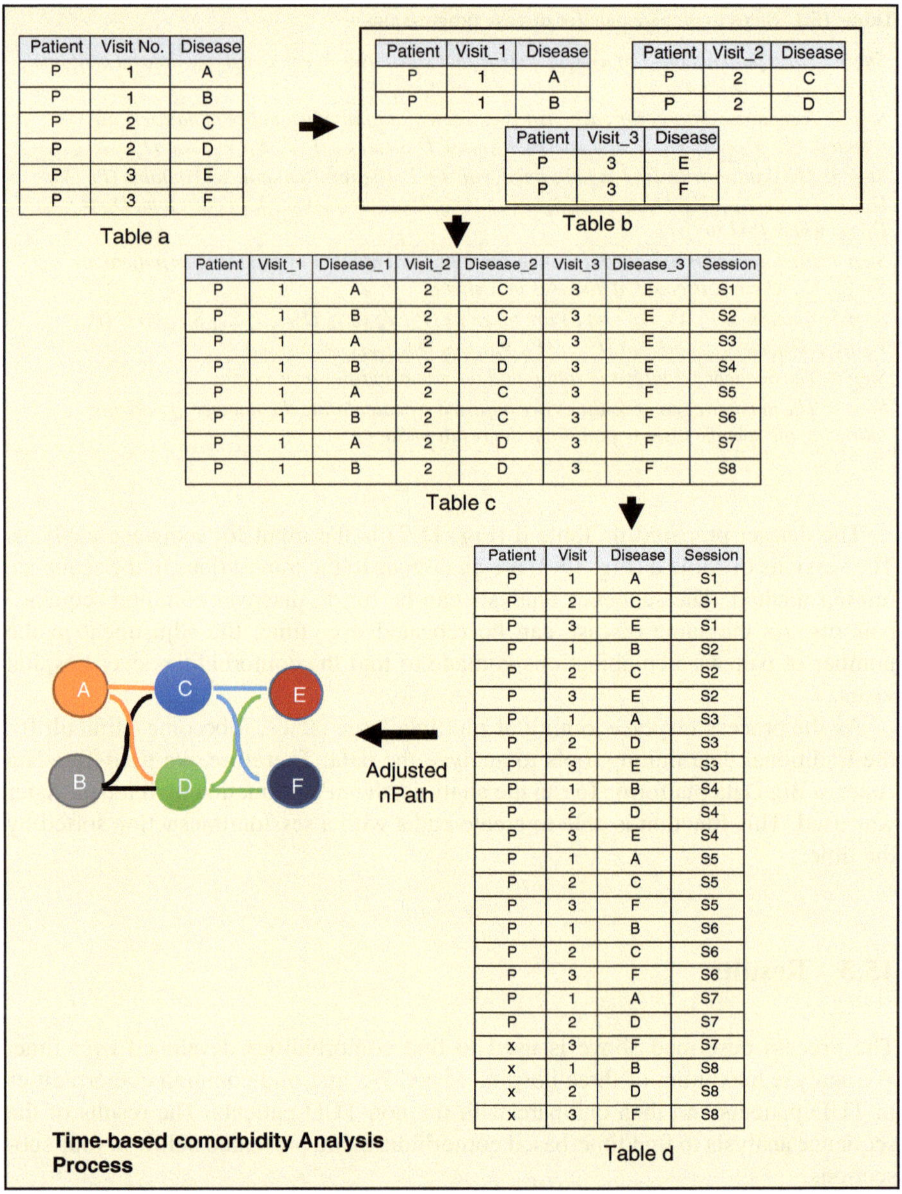

Fig. 15.2 Chronological comorbidities

4. Given comorbidities across two visits, top 5 comorbidities during third visit and corresponding prevalence in non-TUD patients as listed in Table 15.3 and Fig. 15.6.

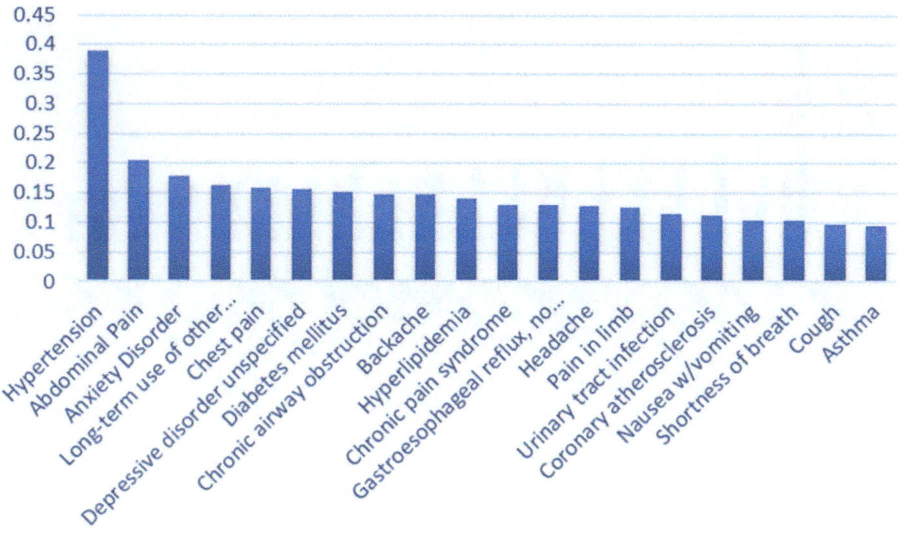

Fig. 15.3 Twenty most common disorders in TUD patients

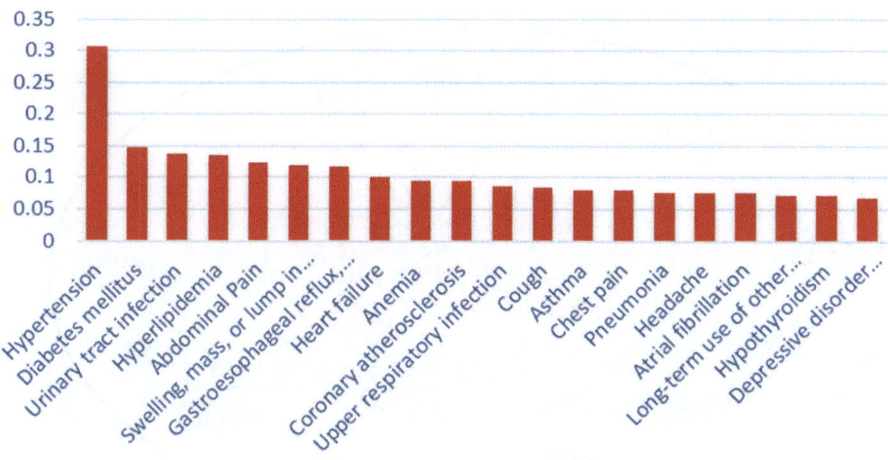

Fig. 15.4 Twenty most common disorders in non-TUD patients

15.3.1 Top 20 Diseases in TUD and Non-TUD Patients

The most prevalent diseases in TUD patients are Hypertension, Abdominal pain, Anxiety state, Long term use of opiate analgesic and Chest pain. Other chronic disorders are also common such as diabetes mellitus, chronic obstructive pulmonary disease (COPD), urinary tract infection and coronary atherosclerosis. The list of top twenty diseases based on the International Classification of Diseases, Ninth

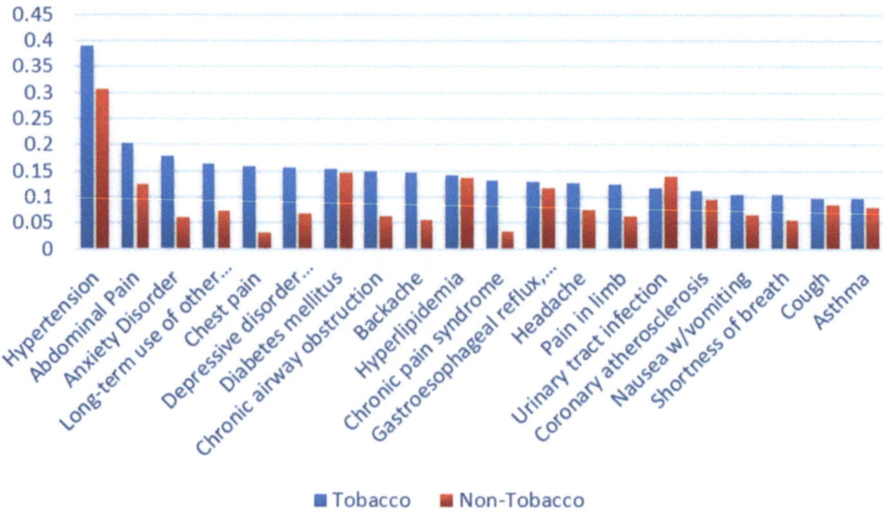

Fig. 15.5 Twenty most common disorders in TUD patients and corresponding prevalence in non-TUD patients

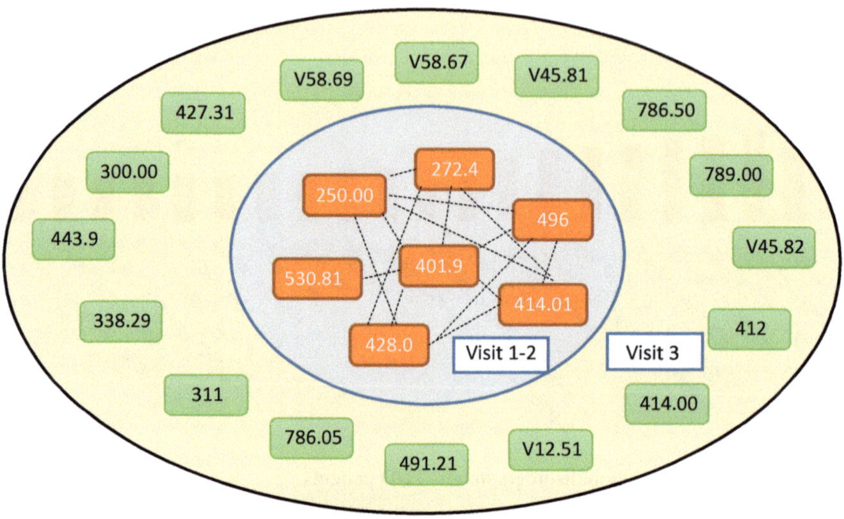

401.9- Hypertension
250.00-Diabetes mellitus
272.4-Hyperlipidemia
414.01-Coronary atherosclerosis
428.0-Congestive heart failure
496-Chronic airway obstruction
V58.67- Long-Term use of Insulin
V45.81- Aortocoronary bypass
V58.69- Long-term use of other medications
786.50-Chest Pain
789.00-Abdominal pain
412-Old myocardial infarction
414.00-Coronary atherosclerosis
786.05-Shortness of breath
311-Depressive disorder unspecified
338.29-Other chronic pain
443.9-Peripheral vascular disease
300.00-Anxiety disorder
427.31-Unspecified atrial fibrillation
530.81- Gastro-esophageal reflux disease without esophagitis
491.21-Obstructive chronic bronchitis with exacerbation
V12.51-Personal history of venous thrombosis and embolism
V45.82- Percutaneous transluminal coronary angioplasty

Fig. 15.6 Common comorbidities in the TUD patients over three hospital visits

Table 15.2 Common disease associations across two hospital visits

Comorbidity – Pair of ICD-9 codes (diseases) across first two hospital visits
Tob Patient Count – Number of tobacco use disorder patients having a comorbidity across first two hospital visits.
% of Tobacco Patients – Percent tobacco use disorder patients having a comorbidity across first two hospital visits.
Non-Tob Patient Count – Number of patients without tobacco use disorder having a comorbidity across first two hospital visits.
% of Tobacco Patients – Percent patients without tobacco use disorder having a comorbidity across first two hospital visits.

S. No.	Comorbidity	Tob Patient Count	% of Tobacco Patients	Non-Tob Patient Count	% of Non-tob Patients
1	401.9 ↔ 250.00	1959	7.73	12128	7.03
2	401.9 ↔ 272.4	1865	7.36	11645	6.75
3	401.9 ↔ 496	1542	6.09	3605	2.09
4	401.9 ↔ 414.01	1399	5.52	7488	4.34
5	401.9 ↔ 530.81	1169	4.62	7385	4.28
6	401.9 ↔ 428.0	908	3.58	7090	4.11
7	272.4 ↔ 250.00	839	3.31	6308	3.65
8	272.4 ↔ 414.01	788	3.11	4559	2.64
9	250.00 ↔ 496	641	2.53	1874	1.09
10	414.01 ↔ 496	639	2.52	1612	0.93
11	414.01 ↔ 250.00	612	2.42	4122	2.39
12	496 ↔ 428.0	501	1.98	2171	1.26
13	250.00 ↔ 428.0	493	1.95	4675	2.71
14	414.01 ↔ 428.0	487	1.92	3612	2.09
15	272.4 ↔ 428.0	458	1.81	3720	2.16

Revision, Clinical Modification (ICD-9-CM) are presented in Fig. 15.3. The proportion of patients with a particular disease can be observed.

The most prevalence diseases in non-TUD patients include hypertension, diabetes mellitus, urinary tract infraction, hyperlipidemia, abdominal pain and others. A list of twenty diseases with the proportion of patients can be observed in Fig. 15.4.

It is interesting to see that some diseases such as hypertension, diabetes, hyperlipidemia and others are not unique to TUD patients. Hence, these diseases cannot be related to the tobacco use.

15.3.2 Top 20 Diseases in TUD Patients and Corresponding Prevalence in Non-TUD Patients

To find the diseases related to tobacco use, we compared the prevalence of top twenty diseases in TUD patients, observed in previous section, with the non-TUD patients. In Fig. 15.5, the proportion of diseases in TUD patients is plotted against the non-TUD patients. Blue colored bars represent proportion of TUD patients and red colored bars represent proportion of non-TUD patients. We clearly see that some diseases are highly prevalent only in TUD patients and not in non-TUD patients such as abdominal pain, anxiety and depressive disorders, chest pain, backache, chronic pain syndrome and chronic airway obstruction. However, it is even more interesting to observe that diseases such as hypertension, diabetes mellitus, hyperlipidemia, gastroesophageal reflux and urinary tract infection are not unique to TUD patients. It indicates that these disorders might not be related to TUD.

Table 15.3 Comorbidities in tobacco and non-tobacco patients across three hospital visits

The top row of each sub-table describes the comorbidity across first two hospital visits. Top five associated diseases diagnosed in third visit are listed given the comorbidity across first two visits. Blue colored column in every table represents percentage of patients with tobacco use disorder. Red colored column in every table represents percentage of patients without tobacco use disorder.

ICD9 – Diagnosis Code

Tob – Percentage of TUD patients having third visit and diagnosed with a disease given first two diagnoses.

Ntob - Percentage of non-TUD patients having third visit and diagnosed with a disease given first two diagnoses.

Table4a. 401.9 ↔ 250.00

ICD9	TUD	NTUD
V58.69	12.67	7.56
V58.67	10.76	9.08
V45.81	9.90	9.31
786.50	9.05	6.08
789.00	8.10	5.70

Table4b. 401.9 ↔ 272.4

ICD9	TUD	NTUD
V45.81	15.16	13.91
V45.82	11.98	7.90
412	11.65	9.48
414.00	11.54	11.32
V12.51	10.44	3.24

Table4c. 401.9 ↔ 496

ICD9	TUD	NTUD
491.21	14.55	10.68
V45.81	12.97	10.10
786.05	11.15	8.93
311	10.18	8.01
338.29	10.06	2.93

Table4d. 401.9 ↔ 414.01

ICD9	TUD	NTUD
V45.82	19.29	15.19
V45.81	19.15	18.20
412	16.83	12.90
786.50	13.27	8.07
414.00	12.59	12.56

Table4e. 401.9 ↔ 530.81

ICD9	TUD	NTUD
V58.69	11.21	8.42
300.00	10.60	6.19
786.50	10.45	6.83
789.00	10.45	7.04
V45.81	9.52	8.64

Table4f. 401.9 ↔ 428.0

ICD9	TUD	NTUD
V45.81	16.36	14.04
491.21	15.43	4.28
786.05	13.57	7.40
427.31	12.64	26.15
414.00	11.90	11.18

Table4g. 250.00 ↔ 428.0

ICD9	TUD	NTUD
491.21	17.59	4.42
V45.81	15.96	15.02
V45.82	14.01	8.22
786.05	13.68	8.17
V58.67	13.03	11.07

Table4h. 496 ↔ 428.0

ICD9	TUD	NTUD
V45.81	18.53	14.08
427.31	17.25	26.34
491.21	15.97	11.04
V45.82	14.38	6.69
414.00	14.06	10.33

Table4i. 250.00 ↔ 272.4

ICD9	TUD	NTUD
V45.81	13.19	14.38
V58.69	12.73	7.78
V45.82	12.04	8.19
443.9	11.81	5.87
412	11.11	10.02

Table4j. 496 ↔ 250.00

ICD9	TUD	NTUD
V45.81	14.59	11.31
491.21	13.79	10.71
786.05	11.67	9.87
443.9	10.08	5.54
V45.82	9.81	5.78

Table4k. 272.4 ↔ 414.01

ICD9	TUD	NTUD
V45.82	25.44	18.26
V45.81	24.94	20.76
412	20.91	14.85
414.00	19.90	14.85
786.50	14.86	8.36

Table4l. 414.01 ↔ 250.00

ICD9	TUD	NTUD
V45.82	21.53	15.37
V45.81	20.96	19.64
412	15.01	13.32
414.00	14.73	12.98
443.9	12.46	8.14

Table4m. 414.01 ↔ 428.0

ICD9	TUD	NTUD
V45.81	21.00	18.00
V45.82	18.86	12.78
412	18.86	13.08
414.00	16.73	13.14
786.05	16.01	9.40

Table4n. 272.4 ↔ 428.0

ICD9	TUD	NTUD
V45.81	21.13	20.33
412	16.98	13.32
414.00	16.60	16.67
V45.82	15.85	9.72
491.21	13.96	4.36

Table4o. 414.01 ↔ 496

ICD9	TUD	NTUD
V45.82	21.90	13.12
V45.81	21.33	17.96
412	17.00	11.88
491.21	14.12	8.98
786.50	13.83	5.39

15.3.3 Top 15 Comorbidities in TUD Patients Across Two Hospital Visits (Second Iteration) and Corresponding Prevalence in Non-TUD Patients

Here, we describe fifteen most common comorbidities or pairs of diseases across first two hospital visits in TUD patients. In addition, we also compare the comorbidities in non-TUD patients. The percentages of patients having a specific comorbidity in both categories of patients are listed in Table 15.2. In addition, the same comorbidities can be seen in the inner circle of Fig. 15.6. Across first two hospital visits of TUD patients, the most common comorbidity is hypertension and diabetes mellitus. More than 7% TUD patients were diagnosed with hypertension and diabetes mellitus across first two visits. However, this comorbidity is common in non-TUD patients as well with 7% patients developing it. Similarly, there are other comorbidities listed in Table 15.2 where there is no difference in TUD and non-TUD patients such as hypertension and Gastro-esophageal reflux disease without esophagitis, hypertension and Congestive heart failure, diabetes mellitus and hyperlipidemia and several others.

However, there are comorbidities that are unique in TUD patients. The comorbidity such as hypertension and chronic airway obstruction is more common in TUD patients than non-TUD. Similarly, pairs of diabetes mellitus and chronic airway obstruction, diabetes mellitus and chronic airway obstruction, and coronary atherosclerosis and chronic airway obstruction are more prevalent in TUD than non-TUD. The prevalence of each pair-wise combination of diseases can be seen in Table 15.2. In addition, the inner circle of Fig. 15.6 shows the comorbidities represented by the connections between the diseases.

15.3.4 Comorbidities in TUD Patients Across Three Hospital Visits (Third Iteration) and Comparison with Non-TUD Patients

In our sample, about 51% TUD and 33% non-TUD patients visited hospitals at least three times. To find the comorbidities across the first three visits, we report the most prevalent diseases during third hospital visit given the second iteration comorbidities reported in the previous section and Table 15.2. Given the top comorbidities across first two visits as described in Table 15.2, we report the top five diseases during the third visit. Different sub-tables of Table 15.3 list different diseases during third hospital visit with the two-visit comorbidities. We report unique diseases observed during third visit and compare their prevalence in non-TUD patients.

Given the comorbidity including hypertension and diabetes mellitus (401.9 $\leftrightarrow$ 250.00) across first two visits, about 12% of patients who visited hospitals at least thrice get affected by the long-term use of medication (ICD-9 code: V58.69) and 10% patients are reported long-term use of insulin (V58.67). Similarly, other diseases developed during the third visit in presence of specific comorbidities can be interpreted from different sub-tables of Table 15.3.

Other unique diseases diagnosed in TUD patients during the third visit include old myocardial infarction, coronary atherosclerosis, shortness of breath, unspecified depressive disorder unspecified, chronic pain, peripheral vascular disease, anxiety disorder, unspecified atrial fibrillation, pneumonia, obstructive chronic bronchitis with exacerbation, personal history of venous thrombosis and embolism and percutaneous transluminal coronary angioplasty. These diseases can also be seen in the outer circle of Fig. 15.6.

The prevalence of each comorbidity is compared with the non-TUD patients in the red columns of Table 15.3. It is quite interesting to observe that some comorbidities are common in both TUD and non-TUD patients across first two visits but there are differences during the third visit. For instance, there is no difference in TUD and non-TUD patients with respect to the first reported comorbidity across two visits in Table 15.2 (401.9 $\leftrightarrow$ 250.00) i.e. hypertension and diabetes mellitus. However, during third visit, long-term use of medication has more impact on TUD patients than the non-TUD patients. Similarly, other comorbidities can be observed in different sub-tables of Table 15.3 where there are large differences between two types of patients.

15.4 Discussion and Concluding Remarks

We presented a method to find time-based comorbidities from the electronic medical records. Specifically, comorbidities in the patients diagnosed with Tobacco Use Disorder (TUD) across three visits were discovered. To assess the validity of our results, we compared the results with a sample of patients who never developed Tobacco Use Disorder. To find time-based comorbidities, the traditional sequence mining method was adapted. First, the pair-wise combination of diseases across two hospital visits were found. Second, their relationships with other diseases in the third hospital visits were analyzed. In a way, we discovered triangles or clusters containing three diseases across three hospital visits.

We found several interesting results that can be very helpful to the medical community. We found several comorbidities that are not unique in TUD patients across two visits. However, it was very interesting to observe that during third visit, such comorbidities lead to different types of comorbidities. This indicates that different comorbidities have different long-term effect on the patient's health condition. The information on chronological comorbidities can help physicians take preemptive actions to prevent future diseases.

We restricted our analysis to find comorbidities over time. However, it will be more interesting to see the impact of a specific comorbidity on health over time. For instance, it is possible that some specific time-based comorbidities can lead to death of a patient but not others. Classifying such cluster of diseases will be a part of our future research.

In our paper, we focused on TUD patients. However, this process can be applied to patients of any other primary diagnosis. Moreover, the application of our process is not limited to the healthcare problems. It is a generalized process and can be applied to any multidimensional and time-variant (MDTV) problem.

The contributions of this paper are two-fold. First, we contribute to the methodology literature by presenting a generalized process to prepare MDTV data to be able to perform sequence analysis. Second, we contribute to the problem of comorbidity literature by adapting the sequence analysis. We enhance the understanding about the time-based comorbidities in TUD patients.

Acknowledgments We are thankful to Cerner Corporation, which has shared the electronic medical records data through the Center for Health Systems Innovation to enable us to carry out this analysis.

References

Agrawal, R., & Srikant, R. (1995). *Proceedings of the Eleventh International Conference on Mining Sequential Patterns*. Paper presented at the data engineering.

Centers for Disease Control and Prevention. (2012). Current cigarette smoking among adults-United States, 2011. *Morbidity and Mortality Weekly Report, 61*(44), 889.

Critchley, J. A., & Capewell, S. (2003). Mortality risk reduction associated with smoking cessation in patients with coronary heart disease: a systematic review. *JAMA, 290*(1), 86–97.

Feinstein, A. R. (1970). The pre-therapeutic classification of co-morbidity in chronic disease. *Journal of Chronic Diseases, 23*(7), 455–468.

Gayo-Avello, D. (2009). A survey on session detection methods in query logs and a proposal for future evaluation. *Information Sciences, 179*(12), 1822–1843.

Gubbi, J., Buyya, R., Marusic, S., & Palaniswami, M. (2013). Internet of Things (IoT): A vision, architectural elements, and future directions. *Future Generation Computer Systems, 29*(7), 1645–1660.

Lin, H.-H., Murray, M., Cohen, T., Colijn, C., & Ezzati, M. (2008). Effects of smoking and solid-fuel use on COPD, lung cancer, and tuberculosis in China: a time-based, multiple risk factor, modelling study. *The Lancet, 372*(9648), 1473–1483.

Merikangas, K. R., Mehta, R. L., Molnar, B. E., Walters, E. E., Swendsen, J. D., Aguilar-Gaziola, S., et al. (1998). Comorbidity of substance use disorders with mood and anxiety disorders: results of the International Consortium in Psychiatric Epidemiology. *Addictive Behaviors, 23*(6), 893–907.

Morissette, S. B., Tull, M. T., Gulliver, S. B., Kamholz, B. W., & Zimering, R. T. (2007). Anxiety, anxiety disorders, tobacco use, and nicotine: a critical review of interrelationships. *Psychological Bulletin, 133*(2), 245.

Rigotti, N. A. (2002). Treatment of tobacco use and dependence. *New England Journal of Medicine, 346*(7), 506–512.

Rutten, L. J. F., Augustson, E. M., Moser, R. P., Beckjord, E. B., & Hesse, B. W. (2008). Smoking knowledge and behavior in the United States: sociodemographic, smoking status, and geographic patterns. *Nicotine and Tobacco Research, 10*(10), 1559–1570.

Spiliopoulou, M., Mobasher, B., Berendt, B., & Nakagawa, M. (2003). A framework for the evaluation of session reconstruction heuristics in web-usage analysis. *Informs Journal on Computing, 15*(2), 171–190.

Srikant, R., & Agrawal, R. (1996). *Mining sequential patterns: Generalizations and performance improvements*. Paper presented at the International Conference on Extending Database Technology.

US Department of Health and Human Services (2014). *The health consequences of smoking—50 years of progress: a report of the Surgeon General*, 17. Atlanta, GA: US Department of Health and Human Services, Centers for Disease Control and Prevention, National Center for Chronic Disease Prevention and Health Promotion, Office on Smoking and Health.

Valderas, J. M., Starfield, B., Sibbald, B., Salisbury, C., & Roland, M. (2009). Defining comorbidity: implications for understanding health and health services. *The Annals of Family Medicine, 7*(4), 357–363.

Chapter 16
The Impact of Big Data on the Physician

Elizabeth Le, Sowmya Iyer, Teja Patil, Ron Li, Jonathan H. Chen, Michael Wang, and Erica Sobel

The practice of medicine has historically been a very unilateral process. Physicians go through a great deal of training to possess near exclusive knowledge of human health and illness. They are trained to collect information from patients on a subjective level, combine this with objective data in the form of the physical exam, laboratory values, and imaging studies and then independently process this information into a diagnosis with a recommended course of treatment for the patient. Patients have been traditionally expected to accept and follow this recommendation, trusting that their physician has been adequately trained and is giving the best recommendation for their health without having any particular knowledge of the condition or treatment themselves. As the age of technology and now big data has evolved, this process and relationship is undergoing dramatic changes. While the advances are remarkable, the amount of information generated can be overwhelming and ultimately stifling to individual physicians. As such, a whole other realm of technology is being created to assist in analyzing this data, ensuring that physicians are utilizing it to the highest potential, and adhering to the most proven treatment regimens. As technology has evolved to support a deeper understanding of illness,

E. Le • S. Iyer (✉) • T. Patil
Veterans Affairs Palo Alto Healthcare System, Palo Alto, CA 94550, USA
e-mail: elizabeth.le@va.gov; sowmya.iyer@va.gov; teja.patil@va.gov

R. Li • J.H. Chen
Stanford University, Palo Alto, CA 94305, USA
e-mail: ronl@stanford.edu; ronc101@stanford.edu

M. Wang
University of California, San Francisco, CA 94720, USA
e-mail: Michael.wang4@ucsf.edu

E. Sobel
Kaiser Permanente Santa Clara Medical Center, Santa Clara, CA 95051, USA
e-mail: erica.m.sobel@kp.org

S. Srinivasan (ed.), *Guide to Big Data Applications*, Studies in Big Data 26,
DOI 10.1007/978-3-319-53817-4_16

so has a multitude of new types and ways to collect data. Data is being both passively and actively collected in every aspect of life, from specific biometric data including glucose readings and blood pressure to everyday data including tracking an individual's daily steps and calorie counts for each meal eaten. Increasingly, these once ordinary activities of daily life are being analyzed as components of health and living, and becoming a portion of the medical chart. Social engagement and interaction with the health system is also growing and changing in directions never before anticipated or experienced. Patients now have the opportunity to directly compare their doctors and hospitals. Information about medical conditions and healthcare is more readily available to consumers than ever before. Historically, one would have had to speak to a physician to learn about their condition and treatment. Now, anyone can simply "Google" their symptoms and be provided with a list of diagnoses and potential treatments. Patients seek to learn more about the therapies being recommended, as well as form communities of individuals with similar diagnoses to compare treatment plans and lend support. With increasing adoption of electronic health records (EHRs), and increasing innovation in areas of big data and healthcare, the ways that physicians interact with patients, approach diagnosis and treatment, and strive for improved performance at an individual and a health system level are evolving. This chapter will discuss many of the big data applications that practicing physicians and their patients encounter.

16.1 Part 1: The Patient-Physician Relationship

16.1.1 Defining Quality Care

Consumers in the healthcare economy strive to be cared for by the "best" doctors and hospitals, in particular as society increasingly moves towards individualized, consumer-centered healthcare. However, physicians and hospitals are not chosen based solely on the quality of care they provide. Geography and health insurance often play a big role in pairing patients with physicians. Beyond this, patients largely depend on word of mouth recommendations to infer quality of care (Tu and Lauer 2008). The result is a system in which patients choose healthcare providers based on subjective and non-standardized metrics. The challenge lies not only in quantifying quality, but then making that information transparent to the consumer. A data science solution can lend rigor and clarity to answering the question: With whom and where can I get the best care?

A general guideline for what can be considered a data science solution to assessing healthcare quality contains two major criteria. First, the methods must process large amounts of data from multiple sources, and second, the relevant results of the analysis must be presented in an accessible, interpretable and thus usable manner. This work is done in the belief that it can lead to better health decisions. However, quantifying healthcare quality is incredibly challenging as so many variables affect physician and hospital success. Factors to be considered include facets of patient experience, availability of appointments/access, treatment outcomes, and complication rates. Some measures are more easily quantifiable than others, and determining how heavily each factor should be weighed is quite subjective. Several tools have been recently developed to allow healthcare consumers to compare and contrast physicians and hospitals on key objective measures.

16.1.2 Choosing the Best Doctor

The Internet has given consumers an immense amount of new information regarding products and services. One may be tempted to cite online review sites such as Yelp.com as a big data solution for connecting patients with the best physicians. These online review forums report patient satisfaction through individual anecdotal experience in an open text format and a self reported star rating of overall experience. It is interesting to note that multiple studies have shown a strong positive correlation between a hospital's patient experience scores and strong adherence to clinical guidelines and better outcomes (Chaterjee 2015). Still, while patient satisfaction is an important measure of quality care and correlates well with other measures of success, patient satisfaction is only one metric. Additionally, not only is the data sourced through a single method, online review forums do not analyze nor present the data in an optimized manner.

Health Grades, Inc. (Denver, CO, USA), a data science company, compiles quality metrics on individual physicians as well as hospitals. The physician rating is comprised of demographic information about the individual (i.e. education, specialty, board certification, malpractice claims) and a separate patient satisfaction survey. While the survey portion has a free text Yelp-like review segment, it improves upon this model by standardizing the reviews through a series of questions. Patients rank the physician on a scale of 1–5 stars. Some questions include: level of trust in a physician's decisions, how well a physician listens and answers questions, ease of scheduling urgent appointments, and how likely the patient is to recommend this physician to others. Unlike Yelp, this scale provides more quantitative information about the patient experience in a standardized and thus comparable format. Despite this comparative data, a weakness of the Health Grades system is that it does not recommend any doctor over another and has no outcomes data. Therefore, two physicians with similar education and patient satisfaction scores will appear equal even if one physician has much higher complication rates than another. In addition, there is no single algorithm or composite score to compare physicians to one another, and the results are not customized to the patient. Instead, it is a platform through which demographic information about the physician and patient satisfaction review results are easily accessed.

A more personalized data science solution has been put forth by Grand Rounds, Inc. (San Francisco, CA, USA) which uses a multivariate algorithm to identify top physicians. The proprietary "Grand Rounds Quality Algorithm" uses clinical data points from 150 million patients in order to identify "quality verified" physicians (Freese 2016). 96% of practicing physicians in the United States (∼770,000) have been evaluated by the Grand Rounds Quality Algorithm (Grand Rounds n.d.). The algorithm scores variables such as physician's training, publications, affiliated institution, procedural volumes, complication rates and treatment outcomes. These and other non disclosed variables are combined to create a composite quality score. The algorithm takes into account the quality score as well as location, availability, insurance networks and expertise in specialty topics. Finally, patient characteristics derived from the patient medical record (i.e. languages spoken) are also included in order to then match the individual patient to a "quality verified" physician. As a result, the recommended cardiologist for one patient will not necessarily be the best for their family member or friend. The great benefit of this model is that it synthesizes a massive amount of information on a vast number of physicians, and then works to individualize the results for each patient.

The major criticism of the Grand Rounds model is that the proprietary algorithm is not transparent. It is not known what data sources the company uses nor how it extracts individual physician level outcomes data. The exclusion criteria used and how different variables are weighted in the algorithm is also unknown. As a result of this lack of transparency, the model has not been externally validated.

16.1.3 Choosing the Best Hospital

Compared to physician quality metrics, patients have more options when examining hospital quality using tools such as Health Grades, U.S. News World Report, and Leap Frog. As an example, Health Grades evaluates hospitals in three segments: clinical outcomes, patient experience and safety ratings. The clinical outcomes results are derived from inpatient data from the Medicare Provider Analysis and Review (MedPAR) Database, as well as all payer state registries (Healthgrades 2016a, n.d.). From these databases it presents either mortality rate or complication rate for 33 conditions and procedures including heart failure, sepsis, heart valve surgery, and knee replacement. Each of the 33 conditions has a prediction model which takes into account confounding variables such as patient age, gender and relevant comorbidities. Using logistic regression, Health Grades then compares the health care systems actual mortality/complication rate against their own predicted mortality/complication rate for that condition in that hospital. Finally, hospitals are presented to the patient as "below expected", "as expected" or "better than expected". The second segment reports patient satisfaction based on the Hospital Consumer Assessment of Healthcare Providers and Systems (HCAHPS) (Healthgrades 2016b, n.d.). HCAHPS data is available from the Centers of Medicare and Medicaid Services (CMS). It is a 32 question survey administered to recently discharged patients with questions focusing on factors including doctor and nurse communication, pain control, hospital cleanliness and noise levels, post discharge instructions, and whether they would recommend this hospital to others. The final segment, patient safety, is constructed from Medicare claims (Healthgrades 2016c, n.d.). Health Grades has chosen fourteen patient safety indicators (PSI) such as catheter related bloodstream infections, development of pressures ulcers and mortality rates in post surgical patients. Like the clinical outcomes segment, Health Grades calculates predicted complication rates for each hospital. The predicted complication rates are adjusted for complexity of cases using the Medicare Case Mix Index. So hospitals with more complicated cases are given higher predicted rates of PSI. The actual rates are then compared to the predicted and the hospital is ranked either "below expected", "as expected" or "better than expected" for each PSI. Health Grades presents its final results in a user-friendly format with simple graphics using star ratings and pie charts.

The major benefits of the Health Grades model are that the methodology and data sources are transparent and results are displayed in a user-friendly manner. However, there are also many limitations, both with the statistical analysis and with the final product. For instance, comparing a hospital's actual complication rate to a predicted complication rate is fraught with statistical challenges. Even though the predicted complication/mortality rates are corrected for age and comorbidities, it is very hard for a logistic regression to take into account all confounding variables. This is especially true because patient's comorbidities are not always accurately reflected in the billing data. Billing data is also significantly delayed, thus present day analysis is often carried out on data from several years ago. In addition, the results for each hospital are not generalizable to every condition or patient. Since patients can only see data regarding 33 conditions, this leaves out information on less common diseases or rare surgeries. In addition, data from some patients, such as those who are discharged to hospice or have metastatic cancer, are not included in the Health Grades outcomes analysis. Aside from issues of generalizability, there is a notable limitation of the final product application. While the Health Grades platform provides a report card for individual hospitals, it does not offer an easy way to compare multiple institutions.

Today patients have many more tools for making a data driven decisions in choosing their healthcare provider. However many challenges still exist both for determining physician and hospital quality, and for matching patients and healthcare providers in an optimal manner. In an ideal future state, treatment outcomes data at a physician, department and institutional level would be available and sourced directly from the electronic medical record as opposed to deduced from billing data. Additionally, treatment outcomes, patient satisfaction and safety metrics would be available at faster turnaround times. If available, the applications could extend beyond patient's using data to select physicians and allow providers themselves to have more real time feedback on their individual performance, which could ultimately influence practice patterns.

16.1.4 Sharing Information: Using Big Data to Expand the Patient History

The practice of medicine is undergoing a cultural shift in which the paternalism ingrained in the patient-physician relationship is no longer expected, and patients are increasingly and rightfully becoming partners in their medical care. At the same time, technology has advanced to create near limitless access to information and data generation, further supporting the patient bid for greater autonomy. From creating communities and promoting patient advocacy to increasing use of shared decision-making (Shay and Lafata 2015), patients are playing larger and larger roles in their health.

Despite this shift, one area in which patients still have surprisingly little input is with regards to their health data. Currently a patient's history, whether obtained directly from the patient or their care-taker, in verbal or written form, resides within an electronic health record (EHR) controlled by the physician or medical system. In the most common EHRs, patients may at most be able to view their health data but do not have the ability to add to or edit their record. As patients become more engaged and involved in their care, it follows that they should have more direct input and ownership over their health data. This increasing responsibility and management over personal health data has the potential to lead to increasing patient activation, a concept in which patients have the knowledge, skills, and confidence to become more effectual managers of their own health (Hibbard et al. 2004). Studies of patient activation, a cornerstone in the management of chronic disease (Clark 2003), have demonstrated improvement in healthy behaviors (Rask et al. 2009) and overall health outcomes (Dentzer 2013; Greene and Hibbard 2012). Giving patients ownership over their health data should therefore help promote activation for improved health outcomes.

16.1.5 What is mHealth?

What will this expansion and shared ownership in patient health data look like? Where will this data come from? How will it enter the electronic record? How will it be used? Alongside this change in the patient-physician partnership has been a rapid expansion in healthcare oriented devices and applications, or "apps," giving rise to the field of mobile health (mHealth). mHealth describes mobile technologies used in healthcare diagnostics, monitoring, and systems support, and currently encompasses monitoring for the collection of individual biometric and environmental data, personal emergency response systems (PERS), telemedicine, mobile medical equipment, Radio Frequency Identification (RFID) tracking, health and fitness software, mobile messaging, and electronic medical records. mHealth has developed and flourished under the jurisdiction of the medical establishment, with remote patient monitoring (RPM) of chronic conditions, most prominently using telehealth, capturing the largest portion of the mHealth market. As it stands, telehealth services are projected to cover 3.2 million patients in 2018, up from 250,000 in 2013 (Japsen 2013). The next frontier in mHealth is shifting from a focus on physician-collected data via RPM to patient-collected data using consumer targeted devices and applications.

16.1.6 mHealth from the Provider Side

Remote patient monitoring (RPM) has largely developed within integrated health-care systems and institutions with the goal of better managing chronic conditions. Two leading examples of RPM are with the Veterans Health Administration (VHA) Care Coordination/Home Telehealth (CCHT) program and Partners Healthcare in Boston, both of which have demonstrated substantial improvements in patient health care and utilization (Agboola et al. 2015 and Darkins et al. 2008).

Since 2003 the VHA, within the Office of Telehealth Services, has used the CCHT program to help veterans with chronic medical issues including diabetes, congestive heart failure, hypertension, posttraumatic stress disorder, chronic obstructive pulmonary disease, and other chronic conditions coordinate their care as well as receive remote patient monitoring. The home telehealth devices are configured to communicate health status including symptoms as well as capture biometric data. This data is transmitted and remotely monitored by a care coordinator working in conjunction with the patient's primary care provider. For the VHA, one of the primary goals of monitoring has been to reduce use of institutional medical care, particularly for veterans who may live remotely, and instead promote patient activation, self-management, and the earlier detection of complications. For patients with chronic illnesses enrolled in CCHT, monitoring has resulted in a 25% reduction in bed days of care, 19% reduction in hospital admissions, and high overall rates of patient satisfaction. By 2010, over 70,000 veterans have been enrolled in CCHT with plans for expansion to cover more.

Partners Healthcare (Partners Healthcare, Boston, MA, USA), a private integrated health system, has been using its own RPM system to improve outcomes in patients with heart failure. The Connected Cardiac Care Program (CCCP) provides home telemonitoring combined with nursing intervention, care coordination and education for patients with heart failure. Telemonitoring consists of the daily transmission of weight, heart rate, pulse and blood pressure by patients with the program demonstrating a significantly lower rate of hospitalization for up to 90 days, and decreased mortality within the first four months after enrollment at discharge. This model has thrived on strong institutional leadership as well as the overarching goal of activating and engaging patients in greater self-care through technology.

These examples demonstrate how data gathered from patients remotely can be used to drive real time clinical decisions and improve healthcare outcomes. Still, they represent a model of provider driven data collection. This data is reliant on large teams of care coordinators, often nursing based, for interpretation and use, and does not have the benefit of further analytics to distill and detect trends. The emerging frontier in health related data generation will come from the hands of patients and their caregivers, from data captured on consumer marketed devices including smartphones and wearable devices, and processed with sophisticated analytics to better support provider workflow.

16.1.7 mHealth from the Patient Side

In the U.S., smartphone ownership has increased from 35% of adults in 2011, to 64% of adults in 2014, with 62% of smartphone owners using their smartphone to look up information about a health condition within the past year (Smith 2015). 34% of adults have downloaded an app meant to support healthy living, and one in five have downloaded and regularly use an mHealth app (Witters and Agrawal 2014). Widespread ownership and use of smartphones is now evolving into ownership and use of wearable devices, including fitness trackers and smartwatches. A recent report by PriceWaterhouseCoopers has shown that over 20% of American adults now own a wearable device with the market share of these devices increasing each year (PwC 2014).

The rapid innovation in mobile consumer devices, including smartphones and wearables enabled to capture biometric and health-related data, and the increasing accessibility of these devices and apps has allowed consumers to capture their own health related data, or patient-generated health data (PGHD). PGHD encompasses health-related data including history, symptoms, biometric, and environmental data that is created, recorded, gathered, or inferred by or from patients or their care partners to help address a health concern (Shapiro et al. 2012). It is unique because patients determine with whom it will be shared. Use of consumer targeted mHealth devices and applications has allowed for greater capture of patient selected and controlled data, in contrast to provider-selected variables traditionally employed in remote patient monitoring.

With the generation of such large volumes of raw data, a new industry has emerged to determine how to best integrate and translate PGHD in a clinically useful manner into EHRs. One of the main pillars in emerging technologies is in the design and use of ecosystem-enabling platforms intended to bridge the divide between humans and technology (Gartner 2016). One particular platform-enabling technology as it relates to healthcare is seen in the Internet of Things (IoT). The IoT is an emerging concept that at its simplest is considered solutions generated from interconnected devices, and is used to describe "the range of new capabilities brought about by pervasive connectivity... involving situations where network connectivity and computing capability expand to objects, sensors, and everyday items that exchange data with little to no human involvement" (Metcalf et al. 2016). As it pertains to health, the IoT promotes a user-focused perspective in how data is managed and information exchanged, particularly between consumers and their devices, allowing consumers to engage in greater self-monitoring and management.

The IoT is an opportunity to passively transmit data from consumer devices, which can then be transformed into big data using sophisticated analytics, and ultimately translated into information and insights that can aid in improving individual health.

Further spurring innovation in PGHD is Stage 3 of the meaningful use electronic health record incentive program by the Office of the National Coordinator for Health Information Technology, which has placed a high priority on bringing PGHD into the EHR (U. S. Department of Health and Human Services 2013). Stage 3 specifies that EHRs should "provide 10% of patients with the ability to submit patient-generated health information to improve performance on high priority health conditions, and/or to improve patient engagement in care." With this mandate, several companies including eClinicalWorks (Westborough, MA, USA) and Apple, Inc. (Cupertino, CA, USA) have begun to build and deploy solutions, using platforms like the Internet of Things, to seamlessly integrate and translate PGHD captured by consumer devices into EHRs and create analytics to support clinical decision-making.

eClinicalWorks, a leading ambulatory EHR vendor, is integrating data from wearable devices into its Health & Online Wellness personal health record healow® with the goal of enhancing patient engagement in their health (Caouette 2015a). To date 45 million patients have access to their health records through healow®. According to a survey conducted online by Harris Poll and commissioned by eClinicalWorks, 78% of patients with wearable devices using them more than once a month feel that their physicians would benefit from access to the information collected (Caouette 2015b). Using its cloud platform, the Internet of Things, third party hardware platforms can collect, store, and analyze PGHD from home monitoring and wearable devices. These devices can include activity trackers, weight scales, glucometers, and blood pressure monitors. The healow® IoT contains analytics and dashboards designed to provide patients and physicians with high yield data culled from PGHD for more informed clinical decision-making. Though still early, this partnership has the promise to use big data to improve patient care using patient owned devices.

Although not a traditional corporation in the field of healthcare, Apple has been leveraging its existing technologies to collect, capture, and integrate PGHD. Apple has been rapidly evolving its HealthKit app, initially launched in 2014, into a platform to allow health data interoperability. It has partnered with numerous EHRs and over 900 health devices and apps and continues to expand its partnerships and potential applications. Working with large centers like Duke, Stanford, and most recently, in its largest patient integration of more than 800,000 patients, Cedars-Sinai Medical Center, Apple's HealthKit allows for the centralization and transmission of data from consumer devices to healthcare operating platforms (Higgins 2015). Given the potential for large amounts of unfiltered data to be transmitted, HealthKit incorporates flow sheets to keep PGHD separate from other data and allows providers to specify the frequency of data transmission. As there are no clear recommendations for how PGHD is to be incorporated or handled once within EHRs, healthcare systems like Duke have created modified consents so that patients understand that the PGHD data transmitted has no guarantee to be viewed or acted upon in a real time manner so as to avoid creating a false sense of security for patients (Leventhal 2015). Ultimately though, in contrast to the current state of health data in the EHR, patients retain control their PGHD, deciding to whom and what data is transmitted.

In one of the most promising published applications of mHealth and big data in clinical care, Apple's HealthKit has recently been used in a pilot to improve the management of pediatric diabetics in Stanford's outpatient clinics. Stanford partnered with Apple, a major glucose monitoring company Dexcom (San Diego, CA, USA), and its EHR vendor Epic (Verona, WI, USA) to automatically integrate patient glucose readings into its EHR and provide analytics to support provider workflow and clinical decision making (Kumar et al. 2016). Prior processes to input data from continuous glucose monitoring devices into the EHR have required either manual entry or custom interfaces, both of which limit widespread applicability and also impose a time delay variable, as data is only available to providers at clinic visits. Using the prior mentioned technologies, the study investigators were able to establish a passive data communication bridge to link the patient's glucometer with the EHR via the patient/parents smartphone. In this setting the patient wears an interstitial glucose sensor connected to a transmitter which sends blood glucose readings by Bluetooth connection to the Dexcon Share2app on the patient's Apple mobile device. The Dexcom Share2 app then passively shares glucose values, date, and time with the HealthKit app, which transmits the data passively to the Epic MyChart app. The MyChart patient portal is a part of the Epic EHR and uses the same database so that the PGHD glucose values are populated into a standard glucose flow sheet in the patient's chart. Importantly, this communication bridge results in the passive integration of the patient's data into the chart but can only occur when a provider places an order in the patient's electronic chart and the patient accepts the request within their MyChart app, placing ultimate control of the data with the patient/parent. Hesitations on the part of providers to accept integration of PGHD into the EHR have often centered on concerns over receiving large and potentially unmanageable volumes of data, which may lead to increasing liability and unrealistic patient expectations, as well as to what degree data is actionable (National eHealth Collaborative 2013). Providers have also expressed concern about the financial impact of using staff and physician time to review the data. The study investigators addressed these concerns by creating an analytic report to triage patients and identify actionable trends between office visits based on home glucose readings, supporting rather than hindering provider workflow. This report was not meant to replace real-time patient/parent monitoring and thus verbal and written notification was used to establish patient/parent expectations regarding only intermittent provider monitoring. The report was generated every two weeks to identify trends, such as episodic nocturnal hypoglycemia, rather than provide real time glucose monitoring. When viewed by the provider, these reports could also be shared and discussed with the patient/parent via MyChart to create an ongoing dialogue for care between visits. In this study, several actionable trends including nocturnal hypoglycemia in a toddler and incorrect carbohydrate counting in a teenager were identified leading to improvements in overall glycemic control between office visits. Additionally, participants were noted to express gratitude that actionable trends were brought to their attention and there was no report of frustration regarding lack of contact for specific hypo or hyperglycemic episodes, further highlighting the importance of early expectation management.

This pilot study, thus far the only study to demonstrate automatic integration of PGHD into the EHR, demonstrated that integration using widely available consumer technology is not only possible, but when combined with smart and intuitive analytics can improve provider workflow for reviewing data and communicating with patients leading to better care for patients. Notably, this workflow did not require any institution-level customization or a specific EHR vendor. Additionally, other companies like Microsoft Health and Google Fit are both in the midst of developing similar patient-generated data platforms making it likely that any mobile device, not specifically Apple devices, may be able to be configured to perform similarly.

16.1.8 Logistical Concerns

The enormous promise of consumer devices and integrated health data platforms to revolutionize health care is two-fold in the goal of the establishment of the patient as an owner and active user of their health care data, and in the implementation of big data to support physician workflow and clinical decision-making. Inherent in the development and adoption of any new technologies, particularly within health care, are the logistics of addressing accessibility, privacy, security, and liability.

16.1.9 Accessibility

Despite rapidly expanding ownership and usage of mobile devices, economic disparities and technological literacy both stand as potential barriers to accessibility. Use of new technologies is most often associated with younger and more financially stable demographic categories. This is particularly seen in the wearable devices market, where approximately half of consumers are between the ages of 18–34 and one-in-three have a household income of greater than $100,000 (Nielsen 2014). However, approximately two thirds of Americans now own a smartphone of some sort with 10% owning a smartphone without any other form of high-speed internet access at home, making them heavily dependent on their smartphone (Smith 2015). Those who tend to be most heavily dependent on a smartphone for online access include younger adults, those with low household incomes and levels of educational attainment, and non-whites. This implies that shifting technology toward smartphones may still be a reasonable strategy to cover diverse demographic groups. Interestingly, as they relate to health, consumers have indicated that they are not willing to pay much for wearable devices but would be willing to be paid to use them (PwC 2014). This may signal a more broad interest in wearable technologies for health independent of financial status, and may offer an opportunity for insurers, providers, and employers to step in and potentially level the playing field for patients. In fact, consumers have noted that they are more willing to try a wearable technology provided by their primary care doctor's office than they are for any other brand or category (PwC 2014).

16.1.10 Privacy and Security

With regards to privacy and security, health care providers must address HIPAA requirements and malpractice issues while developers must pay attention to standards for product liability including Federal Drug Administration (FDA) and Federal Trade Commission (FTC) rules to protect patients (Yang and Silverman 2014). In the United States, it is generally held that an individual's medical record is owned by the provider or institution that retains the record, not the individual patient the record describes. Patients are however still covered by privacy provisions included in the Health Insurance Portability and Accountability Act (HIPPA) of 1996, that ensures the confidential handling and security of protected health information (PHI) (U.S. Department of Health and Human Services 2013). With increasing use of mHealth technologies, Congress has expanded the use of HIPAA through the Health Information Technology for Economic and Clinical Health (HITECH) Act (U.S. Department of Health and Human Services 2009), which sets forth requirements for mandatory breach notifications. Despite this, HIPAA coverage as it pertains to mHealth technologies remains complex. When a patient's health data is in the possession of health providers, health plans, or other "covered entities", it is protected under HIPAA. When it is transmitted among individuals or organizations that are not covered entities under HIPAA, it is not protected (Fisher 2014). For example, if a patient checks their heart rate and the data is recorded on their mobile device, it is not covered by HIPAA. However, if those readings are sent to their physician for use in clinical care, the data becomes HIPAA protected. This and many other scenarios will need to be identified and clarified to ensure data from mHealth apps and devices incorporated into health care are appropriately protected.

Patients themselves are becoming more and more aware of privacy concerns and even within the context of traditional methods for health information transfer (fax, electronic transfer), more than 12% of patients noted withholding health information from a health care professional due to privacy concerns (Agaku et al. 2014). Inspection of 600 of the most commonly used and rated English-language mHealth apps showed that only 30.5% had a privacy policy, and that bulk of these policies required college-level literacy to understand and were ultimately irrelevant to the app itself instead focusing on the developer (Sunyaev et al. 2015). More stringent attention from developers with regards to mHealth data privacy will be needed to both protect data and ensure consumer confidence in their devices.

Given the nature of mHealth technologies, data from mobile devices (that are themselves easily lost or stolen), is transmitted to cloud-base platforms over wireless networks that themselves are prone to hacking and corruption. Even prior to the expansion of mHealth devices, examples of lost or stolen patient data have already populated new cycles. In 2012, Alaska's Medicaid program was fined \$1.7 million by HHS for possible HIPPA violations after a flash drive with patient health information was stolen from the vehicle of an employee (U.S. Department of Health and Human Services 2012). Interestingly, health hacking has also become increasing prevalent given the relative ease of hacking medical systems and devices, and the increasing worth of health care data. It is estimated that health care data is currently more lucrative than credit card information for fraudulent purposes (Humer and Finkle 2014). As more patients, providers, and healthcare organizations use mobile health technologies to augment and conduct patient care, more attention to security features and protocols will be needed to ensure privacy.

16.1.11 Regulation and Liability

The regulation of mHealth as it relates to medical licensure and liability is a complex issue without clear guidelines or answers. In the current state, mHealth devices and apps may be regulated in a piecemeal fashion by several agencies. Potential regulatory agencies may include the Federal Communications Commission, Food and Drug Administration (FDA), Federal Trade Commission (FTC), Office for the Civil Rights of the Department of Health and Human Services (HHS), and National Institute of Standards and Technology, each with a unique role. These agencies can help create standards for mHealth technology, authorize carriers for access and transition of information from devices connected to networks, ensure appropriate use of information by health providers, and regulate advertising related to the app use (Yang and Silverman 2014). The FDA has expressed it will focus on the subset of mobile apps that are intended to be used as an accessory to a regulated medical device or transform a mobile platform into a medical device using attachments, display screens, sensors, and other methods (U. S. Department of Health and Human Services 2015). In essence, as part of its risk-based approach to cover apps with the potential to cause the most harm, it will focus its attention to mHealth apps that transform consumer devices into medical devices but leave the large majority of other health apps unregulated.

Aside from unclear regulatory domains, jurisdiction, and liability also poses an issues for mHealth technology. When providers use health apps to communicate with other providers in different locations, issues regarding the cross-jurisdictional practice of medicine may arise. As most medical silencing requirements are state specific, there exist over 50 different sets of requirements. Telemedicine has led the way in helping to clarify cross-jurisdictional practice (State Telehealth Laws and Medicaid Program Policies: A Comprehensive Scan of the 50 States and District of Columbia 2016), but more clarity will be needed as data from mHealth devices is potentially transmitted and acted upon across state lines. Additionally, without a clear standard of care with respect to the use of mHealth technologies, coverage of malpractice laws come into question. Traditional malpractice liability is based on a physician-patient relationship with direct contact and care. Given mHealth can be used to capture and monitor health data using a patient's own device or application, it is unclear what a physician's liability would be if patient injury resulted from faulty or inaccurate information from the patient's device. As such malpractice in the setting of use of mobile health technologies remains an open question.

Advancements in mobile health technologies promise to make healthcare more accessible and to more effectively engage patients in their medical care, strengthening the patient-provider relationship. Applying a big data approach with thoughtful analytics to the massive amounts of data generated, captured, and transmitted on mobile devices can not only supply providers with additional information, but also present that information in such a way to enhance provider workflow. Given the rapid growth of mobile health technologies and their increasing integration into health records and use in clinical decision-making, more attention will be needed to clarify privacy, security, and regulatory concerns.

16.1.12 Patient Education and Partnering with Patients

The meeting of a physician and patient is the start of a therapeutic relationship, but much of the work that happens to improve health occurs beyond the appointment time. Whether a physician meets a patient in the clinic, hospital, nursing home, or any other care setting, for the physician's recommendations to be successfully implemented by the patient there has to be a trusting partnership. Physicians are the experts in medicine and patients are the experts in their own lives, so both parties must be engaged for meaningful changes to occur.

Much research has been done in the areas of patient engagement, activation, and the patient-physician partnership. Patient engagement includes both the patient activation to participate in care, and the resulting positive health behaviors from this motivation (Hibbard and Greene 2013). Patient engagement has been linked to participating in more preventative activities including health screenings and immunizations, adherence to a healthy diet, and regular physical exercise. Patients who are very activated are two or more times more likely than their less activated counterparts to prepare for doctor visits, look for health information, and know about the treatment of their conditions (Hibbard and Greene 2013; Fowles et al. 2009; Hibbard 2008). Patient engagement in the therapeutic process has been studied and encouraged in specific areas such as chronic condition management and adverse event reporting, but the degree to which patients are able to be engaged in their healthcare in the current age of social media and Google is unprecedented.

As electronic health records are creating accessible mobile platforms, all patients, not just those with chronic conditions, are able to engage in their healthcare on a basic level to schedule appointments, ask questions to medical staff, and initiate medication refills online. Despite the fact that not all patients have the technical knowledge, health awareness, and engagement to interact with their physicians in this manner between appointments, this is still becoming an increasingly common way that patients and their caregivers are getting engaged with their medical care.

Understanding the role that patient engagement plays in the physician-patient interaction and disease treatment and management sets the stage for understanding the new ways patients can now engage in their healthcare through online access to medical information. A 2013 study in the journal *Pediatrics* sums up what many physicians perceive in the current age of medical care, that "Dr Google is, for many Americans, a de facto second opinion" (Fox 2013). The Pew Research Center has studied health information online since 2000. The latest national survey in 2014 showed that 72% of adults search online for health issues, mostly for specific conditions and their treatments. Meanwhile, 26% say they have used online resources to learn about other people's health experiences in the past 12 months, and 16% have found others with the same health concerns online in the past year (Fox 2014). Caregivers and those with chronic health conditions are most likely to use the internet for health information. Health professionals are still the main source of health information in the US but online information, especially shared by peers, significantly supplements the information that clinicians provide (Fox 2014). Many survey respondents report that clinicians are their source of information for technical questions about disease management, but nonprofessionals are better sources of emotional support (Fox 2013).

16.1.13 Developing Online Health Communities Through Social Media: Creating Data that Fuels Research

PatientsLikeMe Inc. (Cambridge, MA, USA) and the rare disease community exemplify the use of social media to both connect people with similar health conditions, and provide real-time data feedback to healthcare providers, healthcare systems, pharmaceutical companies and insurance companies. This feedback is unique in that it collects large amounts of patient-level data to help in the development of new care plans for specific patient populations.

PatientsLikeMe allows users to input and aggregate details on symptoms, treatments, medications, and side effects of various illness, connect to support groups dedicated to challenging health conditions, and creates a platform that can potentially set the stage for clinical trials and medical product development (Sarasohn-Kahn 2008; Wicks et al. 2010). Patients who choose to enter their personal health data on the site are actively choosing to use their personal health information in a way that is different than the traditional method of encapsulating health information in the electronic health record, to be seen and used privately between the physician and patient. For example, those with Amyotrophic Lateral Sclerosis (ALS) may choose to share personal information about their symptoms, treatments, and outcomes with the ALS community within PatientsLikeMe. These community-level symptoms and treatments are aggregated and displayed so that users can discuss the data within the site's forums, messages, and comments sections. A study of PatientsLikeMe's data sampled 123 comments (2% of the total commentary posted) and noted that group members sought out answers to particular questions guided by this data, offered personal advice to those who could benefit, and made relationships based on similar concerns and issues. This study illustrated that individual patients who shared their personal health data benefitted by participating in conversations that may help with self-management of their disease. (Frost and Massagli 2008). In addition to individual patients benefitting from their usage of the site, PatientsLikeMe allows pharmaceutical companies to partner with patients to design clinical trials and research studies (PatientsLikeMe Services n.d.). Since 2007, PatientsLikeMe has achieved many milestones in patient-centered and patient-directed research. For example, in 2007 a study on excessive yawning in ALS patients was conducted in which the symptom was listed on the ALS page and each user had to rate the severity of the symptom experienced. This quickly created a method to evaluate the symptom in the context of the person's medications and disease course and ultimately data from the PatientsLikeMe users helped to identify that the excessive yawning was more likely a symptom of emotional lability associated with the disease state rather than a drug side effect, or a side-effect of respiratory issues with ALS. Soon after, it was found that the impact of the research extended beyond its clinical scope. While the potential physical pain of yawning in ALS patients may have been the impetus to study this issue, based on discussions on the site it came to light that people with ALS had lost friends due to the misinterpretation that the yawning represented lack of interest and was a sign of rudeness. With this study, patients/families and healthcare professionals are now better able to understand and be more sympathetic to this symptom, physicians can warn patients of this symptom, and researchers have a greater impetus to find a treatment for the emotional lability likely causing yawning (Wicks 2007). The example highlights the invaluable nature of user input in helping to guide research in a more patient-centered manner.

The rare disease community has also been a remarkable testament to the power of online communities for sharing healthcare data and furthering medical practice. When new findings are published as case reports in academic journals, the process relies on clinicians in various parts of the world to see those articles and recall the specific symptomatology at the time needed to make a diagnoses - a process that can take many years to diagnose individuals with the same rare diseases. With common online platforms like Facebook, Twitter, and blogs, parents of kids with rare diseases and people who have rare conditions are taking matters into their own hands to find and share more information. Patients and families are turning to social platforms to promote greater collaboration between patients, caregivers, and healthcare industry professionals.

Bertrand Might is a child with the first known case of NGLY-1 deficiency, a very rare illness that was not diagnosed until he was nearly 4 years old (Might and Wilsey 2014). His journey to diagnosis is extraordinary because of the hard work of his parents, a couple who epitomize a new perspective and the shift occurring in the work of clinical diagnosis. After years of moving from one genetic specialist to the next looking for an explanation for their son's condition, researchers at Duke finally gave the family an uncertain diagnosis about a rare new enzyme deficiency. Bertrand's parents blogged and documented his journey including evaluations, symptoms, visits to specialists, wrong diagnoses, and ultimately the gene mutations leading to his condition. A second patient with Bertrand's condition was discovered after a clinician came across this blog and realized that the two children likely shared the same enzyme deficiency. Yet another patient was identified when parents on another continent came across the blog after searching for similar symptoms and were motivated to have sequencing of the NGLY-1 mutation performed on their child. Now, 14 children from around the world have been diagnosed with this deficiency (Might and Wilsey 2014). Compared to how diseases have traditionally been discovered and disseminated through medical literature, this innovative new way of using social media to fuel recognition of symptoms and prompt genetic data analysis is quite rapid, often leading to a more timely diagnosis for those with extremely rare conditions. The implications of this phenomenon extend beyond the rapid diagnosis of rare diseases. Often rare-disease communities prompt pharmaceutical companies to consider researching and developing treatment options, support creating patient registries and push for clinical trials, and attend FDA meetings to advocate for the approval of new therapeutic options (Robinson 2016).

16.1.14 Translating Complex Medical Data into Patient-Friendly Formats

The value of data from online sources comes from the aggregation of opinions on a specific topic, such that the sum of different user input is more powerful than a single person's comment. A current example of a site making big data useable for patients and physicians alike is iodine.com.

16.1.15 Beyond the Package Insert: Iodine.com

Since 1968, the US Food and Drug Administration (FDA) has required that certain prescription medications contain package inserts that consist of usable consumer medication information (CMI). The suggested CMI includes the name and brand name of the medication, use, contraindications, how to take the medication, side effects, and gene years, the FDA's recommendations have supported changes to the package insert (Food and Drug Administration 2006).

Expanding upon this concept, Iodine.com (San Francisco, CA, USA) was founded in 2013 and currently offers free medication information for more than 1000+ drugs. This information is compiled from FDA drug side effect data and augmented with user input on drug side effects from Google Consumer Surveys (GCS) (Iodine 2014), creating a experience that has been called the "yelp of medicine" (Sifferlin 2014). Ultimately, these medication-related experiences become part of a growing database that can guide new insights into how drugs work in the real world. There are many other sources of drug information on the internet for consumers including drugs.com, medlineplus.gov, rxlist.com, webmd.com, and mayoclinic.org, but these sites lack the peer-to-peer recommendations that Iodine.com provides. Medication side effect profiles provided for physicians and that are listed in the drug package inserts are notoriously long, listing all side effects regardless of incidence frequency from less than a 1% chance of occurring to frequent side effects. This type of information is hard for consumers to interpret, and impossible for physicians to memorize, so often patients are prescribed medications with minimal guidance: only important contraindications such as "don't take with alcohol" or "take on an empty stomach" are communicated. Most physicians also do not have first-hand experience with these drugs, and so are unfamiliar with which side effects truly occur most often.

Iodine.com helps to fill the void that is common in traditional prescribing practices by adding personalization. The data for iodine.com is collected from a variety of sources including traditional medical research literature and pharmaceutical product labels, center for Medicare and Medicaid Services (CMS) data, non-research sources of data like insurance claims formularies, pharmacist reports, patient reported data from more than 100,000 Americans on Google Consumer Surveys, and social data sharing health experiences extracted from many other sites (Iodine Data n.d.).

Iodine users also complete online reviews of their medications, and this community-generated content contributes to the data on the site. The data from these sources covers 1000+ medications, and offers data subclassified by age, gender, and medical condition. This subcategorization allows users to see how similar populations experience and feel about different medications. This unique aspect, the user experience, is a valuable addition to the information provided by medical research literature primarily because medical research often has limited generalizability based on the study design and exclusion criteria. Results of medical studies can often only be extrapolated to populations similar to the study populations, which often exclude older adults and those with complex medical conditions limiting generalizability. These exclusions exist to make study results easier to interpret, but a consequence of such stringent exclusion criteria is that a large percentage of complex patients (precisely the ones for which clinicians most need guidance) do not have a lot of evidence-based management guidelines. If large amounts of data on patient experience can be captured from this complex population of older adults, this can help guide clinical decision making (Bushey 2015).

Despite the potential for significantly expanding the database of medication use and effects, there is a one major limitation of Iodine.com: validity. There is no mechanism for confirming the self-reported data within the system. The integrity of the reviews is potentially quite variable. In an attempt to combat this, the company manually reviews all of the user input. Iodine.com reports that the site rarely receives reviews that are not legitimate, but if detected, these reviews are removed. They also check data trends to see if the patterns of reviews are consistent with known population side effects (Bushey 2015).

16.1.16 Data Inspires the Advent of New Models of Medical Care Delivery

The life of a physician is typically a delicate balance between the time demands of seeing patients, extensive medical documentation of encounters, and billing and coding. Through all of this, certain interventions such as behavioral health discussions rarely occur due to low provider knowledge and confidence, insufficient support services, and little feedback from patients that behavior interventions are needed and are effective (Mann et al. 2014). There are several companies that have recognized the need to identify and engage high risk patients through technological solutions with the goal to yield behavioral change. Two such companies are Ginger.i.o and Omada health.

Ginger.io (San Francisco, CA, USA) offers a mobile application for patients with various mental health conditions to have a personalized health coach. It allows users to communicate through text or live video sessions with a licensed therapist specializing in anxiety and depression, and has 24/7 access to self care tools through its app. The company combines mobile technology with health coaches, licensed therapists, consulting psychiatrists, and medical providers to create an interface that is technology based, but with a human element. The app collects both active data through regular mood surveys, and passive data through mobile phone data on calling, texting, location, and movement. This passive data can generate a norm for the user's regular patterns. Once regular patterns are known, changes in communication and movement can help predict depression or changes in mental health (Feldman 2014). The active and passive data in addition to in-app activities to build skills around managing mood changes are synthesized into personalized reports that can be shared with the person's doctor and the Ginger.i.o care team. Accessible mental health care is challenging in the US for a variety of reasons-availability of providers, cost and cost sharing limitations by insurance companies, distance to care, and availability of appointments (Cunningham 2009). This type of mobile platform can help narrow this gap and provide timely and convenient behavioral care when needed through mobile devices (Ginger.io Evidence N.d.). Ginger.io combines its data with behavioral health research data from the National Institutes of Health and other sources to help provide insights from the aggregate data. One particularly interesting insight is that a lack of movement from a user could signal that a patient feels physically ill and irregular sleep patterns may precede an anxiety attack. (Kayyali et al. 2013)

Omada Health (San Francisco, CA, USA) is an innovative company that offers "digital therapeutics," which are evidence-based behavioral treatment programs that are cost-effective and potentially more accessible than traditional programs. Their 16-week program, *Prevent*, offers a targeted intervention to individuals who are at high risk for chronic illnesses such as diabetes and heart disease. Each participant is paired with a personal health coach and online peer group for regular feedback and support. This program includes weekly lessons about nutrition, exercise, and psychological barriers. In 2016, a study was published examining long term clinical outcomes of the Prevent Pilot and the effects of the program on body weight and Hemoglobin A1C, a marker of blood sugar control in diabetics. The 187 pre-diabetic participants who completed the four core lessons achieved an average of 5.0% and 4.8% weight loss at 16 weeks and 12 months, respectively and had some reduction in their A1C level at final measurement (Sepah 2015).

The type of behavioral change program developed by Omada health would require extensive staff and resources if replicated in person, rather than administered online. The benefits of these online interventions include increased access to care, convenience to patients by using mobile health delivery systems and avoiding travel time, and increased patient engagement in their healthcare. Potential limitations of these types of online behavioral intervention technology (BIT) programs are technological barriers, lack of engagement due to the design of the program, and issues with translating the program into actionable behavior change. Ideally, human support should complement BIT use, so that together all potential barriers are addressed and individuals can gain the maximum potential benefits from these behavioral change programs (Schueller et al. 2016).

16.2 Part II: Physician Uses for Big Data in Clinical Care

16.2.1 The Role of Big Data in Transforming How Clinicians Make Decisions

16.2.1.1 Imagine the Following Scenario

An elderly man, Mr. Williams, develops a fever, back pain, and nausea, and is admitted to the local hospital where he is diagnosed with a kidney infection. He tells his physician that he has several other medical conditions, such as heart disease and diabetes, and was hospitalized several times in the past year for different infections, although they were at another hospital where medical records are not available. The physician chooses to start him on ciprofloxacin, which is an antibiotic that she typically uses to treat kidney infections. She also faxes a form to the other hospital to request records, although the request would not be processed until the following day.

In the middle of the night, the physician receives a page from the nurse saying that Mr. Williams' heart rate is elevated. She walks over to his room to examine him, and notices that he is a bit anxious about being in the hospital. This was actually the fifth time she was woken up that night; the last four times were all for patients who were anxious and needed something to help them calm down. She takes a minute to scan Mr. Williams' chart and does not notice anything else that looked worrisome, so instructs the nurse to give him an anti-anxiety medication and goes back to sleep.

The following morning, Mr. Williams develops a high fever and becomes confused. The physician, now more concerned, looks through his chart and notices that his heart rate had continued to increase throughout the night. She diagnoses him with sepsis and immediately starts intravenous fluids and switches his antibiotic to piperacillin/tazobactam, which is the antibiotic that she usually uses if ciprofloxacin is not working. However, he continues to worsen throughout the day and is later transferred to the intensive care unit. That afternoon, the medical records from the other hospital are faxed over, which consist of 50 printed pages of progress notes, discharge summaries, medication lists, and lab reports. Although the physician's shift had ended two hours ago, she spends the time looking through the faxed charts because she is particularly worried about Mr. Williams (she had taken care of a patient with a kidney infection who ended up dying in the hospital just two weeks ago). After pouring through the pages of records, she finally notices a line of text in a lab report that is a critical piece of information: two months ago at the other hospital, Mr. Williams had developed another kidney infection and his urine at the time grew out a bacteria that was resistant to both ciprofloxacin and piperacillin/tazobactam. She remembers that meropenem would be the antibiotic of choice in this situation, but takes the extra time to look it up in her online reference since she has not used that medicine in over a year. At 5 pm, almost 24 h after Mr. Williams was admitted to the hospital, the physician starts him on the correct antibiotic. He eventually recovers, but was weakened because of the prolonged hospital stay and had to be discharged to a nursing home.

The above hypothetical scenario is not uncommon in modern hospitals. Although the resources needed to treat Mr. Williams' kidney infection were readily available and his physician did her best to care for him, the limitations in the process by which medical decisions are made in healthcare led to missed opportunities that may have improved the care that he received in the hospital. He could have received the correct antibiotic much earlier, and his sepsis could have been recognized and acted upon sooner. Clinical decisions are often made in situations of uncertainty with incomplete information and sometimes inconsistent levels of expertise. As demonstrated in the scenario with Mr. Williams, many decisions are heavily influenced by the experiences of the individual physician, which can lead to high variability in cost and quality of care (Institute of Medicine 2012). Existing clinical knowledge is often inconsistently applied, with compliance with evidence-based guidelines ranging from 20%–80%. Additionally, most clinical decisions including common diseases such as heart attacks, lack strong evidence from clinical research studies (Institute of Medicine 2012) with only $\sim$11% of clinical practice guideline recommendations supported by high quality evidence (Tricoci et al. 2009). This section will describe the potential for big data to address these gaps and transform how decisions are made in healthcare.

The methods by which physicians are currently trained to make medical decisions are not very different from those utilized one hundred years ago. Physicians undergo a long and arduous training process that heavily relies upon developing the individual's knowledge base and experience. The amount of data available for today's physicians to process, however, far exceeds the capacity of any individual and is increasing at a rapid rate. Advances in medical research have not only given us many more diagnostic and treatment options for long-standing diseases, but have also defined new diagnoses that add to the already expansive medical vocabulary. As our understanding of disease grows more granular, the management of patients is becoming increasingly complex. A physician in the intensive care unit will manage a range of 180 activities per patient per day (Institute of Medicine 2012). This rapidly accumulating knowledge base has allowed us to reimagine what medicine can accomplish, but also means that today's physicians are required to know and do more than ever before to deliver the standards of care that we now expect from modern healthcare systems. The cognitive tools that physicians are given, however, have not advanced at the same rate, thus creating a critical need for newer tools to help physicians make clinical decisions.

The promise of big data to address this need is driven by the increasing wealth of clinical data that is stored in the electronic medical record (EMR). Information such as patient histories, laboratory and imaging results, and medications are all stored electronically at the point of care. This critical mass of data allows for the development of clinical decision support computational tools that interact with the EMR to support physicians in making clinical decisions. The development of clinical decision support is an evolving field with different computational models that include earlier systems using probabilistic and rule-based models and newer data driven approaches that more effectively harness the power of big data.

16.2.2 Probabilistic Systems

The diagnosis of disease follows a Bayesian model. When approaching a new patient, a trained physician will come up with a list of potential diagnoses in order of likelihood, and adjust the probabilities of those diagnoses being present based on new information from the examination and diagnostic tests. A patient coming to the emergency room with chest pain could have anything from a heart attack to a muscle strain, but additional relevant information such as the character of the pain and the presence of elevated serum levels of cardiac enzymes will increase the posterior probability of the diagnosis being a heart attack and decrease that of the diagnosis being a muscle strain. Physicians use this mental model to diagnose diseases, but will often be affected by psychological biases. In the case of Mr. Williams, his physician mistakenly assigned a higher posterior probability to the diagnosis of anxiety as the cause of his elevated heart rate overnight, likely because she had just seen several other patients with anxiety. This phenomenon, known as the availability heuristic, is a common reason for misdiagnosis and physicians become increasingly susceptible to it when exhausted with data and tasks that exceed what can be processed by an individual person.

Probabilistic clinical decision support systems use computers to simulate the Bayesian model with which physicians are trained to think. For a presenting symptom, a computer can start by considering the prior probabilities of the differential diagnosis, which is typically based on the known prevalence of those conditions, and modify the posterior probabilities in a sequential approach based on new inputted data, such as symptoms and diagnostic test results (Shortliffe and Cimino 2014). These clinical decision support systems require a knowledge base of estimated conditional probabilities of a set of diseases for the pieces of data that are entered. VisualDx (Rochester, NY, USA) is an example of a commercially available probabilistic clinical decision support system that is designed to help users create a differential diagnosis at the point of care that is not affected by human error such as recall bias. A user can enter combinations of clinical data such as symptoms and laboratory results to generate lists of likely diagnoses that are ordered by their posterior probabilities.

16.2.3 Rule-Based Approaches

Rule-based clinical decision support systems rely on encoded concepts in a knowledge base that are derived from content experts to simulate recommendations that experts might provide. The knowledge base can include probabilistic relationships such as those between symptoms and diseases, and medications and side effects. For example, a commonly used rule-based clinical support system is the drug interaction alert system that is built into many modern EMRs. The system alerts the physician if a new medication ordered has an adverse interaction with an existing medication that the patient is taking. When the medication is ordered in the EMR, the action is then analyzed against a previously encoded knowledge base of all known medication interactions, which then triggers an alert. Knowledge bases can be created for specific diseases to create tools such as early disease detection systems. For example, many hospitals now use EMR embedded clinical decision support tools to detect severe sepsis, which is a life-threatening physiologic state that patients can develop when they have an infection. These alerts are powered by algorithms designed based on probabilistic rules derived from sepsis treatment guidelines (Rolnick et al. 2016). Such a clinical decision support system may have been able to help Mr. Williams' physician recognize earlier that he had sepsis based on clinical data in the EMR that the physician did not otherwise notice, or could have initiated an alert when the physician ordered the first antibiotic that his previous urine culture grew a resistant antibiotic.

16.2.4 Data Driven Approaches

Rule-based and probabilistic models, which comprise the majority of existing clinical decision support systems that are in use, rely on manually curated knowledge bases that are applied to clinical data in a top down approach. They act on, rather than use the data in the EMR to generate insights into clinical decisions. The paradigm of clinical decision support is now evolving towards a data driven approach, which employ machine learning techniques to mine EMR data for new knowledge that can guide clinical decisions. Rather than relying on a pre-formed knowledge base, a data driven approach seeks to "let the data speak for itself" by extracting patterns and insights from the data generated by everyday clinical practice. The most direct approach is to attempt discovery of new knowledge that can guide clinical decisions. This approach is particularly powerful for the majority of clinical decisions that do not have an existing evidence base in the form of clinical trials or guidelines, but for which many "natural experiments" occur in regular practice. Efforts are underway to develop systems that would allow physicians to generate real-time, personalized comparative effectiveness data for individual patients using aggregate EMR data (Longhurst et al. 2014). For example, there may have been thousands of other patients who share Mr. Williams' specific clinical history who were treated at the same hospital. His physician could query the EMR to find out how this cohort of patients responded to certain treatments in order to choose the most effective treatment for him. Reliable conclusions can be challenging due to confounding by indication, but can somewhat be mitigated by causal inference methods that risk adjust for different clinical factors (e.g., propensity score matching) as has been demonstrated through established methods in retrospective observational research.

Another example of an emerging data driven approach is known as collaborative filtering. Traversing the hierarchy of medical evidence, we first look to randomized controlled trials to guide our medical decision making, followed by observational studies, before accepting consensus expert opinion, or finally our own local expert (consultant) opinions and individual anecdotal experience. With only $\sim$11% of clinical practice guideline recommendations backed by high quality evidence and only about a quarter of real-world patients even fitting the profile of randomized controlled trial inclusion criteria, it should not be surprising that the majority of medical decisions we have to make on a daily basis require descending the entire hierarchy to individual experience and opinion. For a practicing clinician, the established norm is to consult with other individual local experts for advice. The advent of the EMR, however, enables a powerful new possibility where we can look to, not just the opinion, but the explicit actions of *thousands* of physicians taking care of similar patients. Right or wrong, these practice patterns in the collective community reflect the real world standard of care each individual is judged against. More so, these may reflect latent wisdom in the crowd, with clinical practices refined through years of hard won experience, but which never before had a fluid mechanism to disseminate through conventional social and publication channels. Such an approach can represent an entirely new way of generating and disseminating clinical practice knowledge and experience, owing heavily to methodology established in product recommender systems for companies such as Netflix and Amazon. Active research is underway to help discern how such approaches can separate the wisdom of the crowd from the tyranny of the mob, and the potential impacts of integrating such dynamic information into a physician's point of care decision making process (Chen et al. 2016). When the computer systems are trained to recognize established standards of care through readily available clinical data, they will be able to seamlessly anticipate clinical needs even without being asked. This will translate endpoint clinical big data into a reproducible and executable form of expertise and, deploying this right at the point-of-care, can close the loop of a continuously learning health system.

16.2.5 Challenges and Areas of Exploration

Significant challenges remain in the development and adoption of clinical decision support. Although the EMR stores clinical data electronically, much of the data is not in a format that is easily readable by computers. Inherently structured data such as medication lists and laboratory values are often the sources for data used in existing rule-based clinical decision support systems. The promise of data driven machine learning approaches to clinical decision support, however, require the use of data in the unstructured, free text narratives that comprise the majority of valuable, actionable EMR data generated by clinicians. The question of how to structure large amounts of clinical data into reliable variables that can then be used by computational tools remains one of the "grand challenges" of clinical decision support (Sittig et al. 2008). Natural language processing, which is a technique that has been used in other applied fields of computer science, is being explored as a way to translate clinician generated text into encoded data (Liao et al. 2015). How this data can then be organized into meaningful groups, or phenotypes, that can be analyzed is an area of active research. For example, EMR phenotypes can be used to create electronic cohorts of patients around specific disease states, either at the point of care to generate real time comparative effectiveness data for clinical decision support, or for the creation of electronic cohorts that can be used for clinical research (Xu et al. 2015).

The infrastructure for how clinical data is stored and shared is also in need of change to accommodate a data driven healthcare system. In our example with Mr. Williams, a critical piece of clinical data, the urine culture results, was actually located in another hospital and had to be faxed over as printed text to be read by the physician. Any clinical decision support system would be limited by the amount of clinical data available to analyze. Currently, electronic clinical data is stored throughout disparate EMR systems that are owned by individual healthcare delivery systems and often not shared electronically. Although the Health Information Technology and Clinical Health Act (HITECH), which was enacted as part of the American Recovery and Reinvestment Act of 2009, includes standards for EMR interoperability that envision a system where clinical data can be shared electronically among clinicians across the country in real time, issues such as privacy, misaligned financial incentives, and the lack of technological infrastructure continue to remain as barriers to adoption (Ball et al. 2016). Regional health information exchanges are beginning to have some success in allowing for data sharing among health systems, although the scope remains limited (Downing et al. 2016). Further, insights are needed to understand how to successfully scale implementation of clinical decision support systems into clinical enterprises. Issues such as physician workflow integration, system usability, and alignment with financial incentives of healthcare delivery systems all need to be considered. Nevertheless, in spite of these challenges, the convergence of the need for improved healthcare quality, an unprecedented amount of available clinical data, and the rapid development of powerful analytical tools is pushing the healthcare system to a tipping point and into an era of big data.

16.2.6 Using Big Data to Improve Treatment Options

One of the greatest promises of big data as it relates to medical care is in precision or personalized medicine. These terms are often used interchangeably, with precision medicine emerging more recently and persisting as the preferred term to describe the concept of taking individual variability into account in creating prevention and treatment plans (National Academies Press 2011). Precision medicine has existed for at least a century with the most prominent example seen in blood typing to more safely guide blood transfusions (Collins and Varmus 2015). Additionally, complete sequencing of the human genome at the beginning of this century has led to a wealth of data towards the better understanding of disease states, developmental variability, and human interaction with pathogens (Lander et al. 2001). More recently, precision medicine has come to define a framework to combine huge databases of patient health data with OMICS, primarily genomics but also proteomics, metabolomics, and so on, to facilitate clinical decision-making that is "predictive, personalized, preventive and participatory" (P4 Medicine) (Hood and Flores 2012). The overarching hope for precision medicine is to be able to select therapies for predictable and optimal responses, and identify potential side effects to therapies based on a patient's genetic makeup and individual characteristics.

In his 2015 State of the Union Address, President Obama announced details about the Precision Medicine Initiative (PMI), a $215 million research effort intended to be at the vanguard of precision medicine (The White House 2015). Funds have been invested into the National Institutes of Health (NIH), National Cancer Institute (NCI), Food and Drug Administration (FDA) and Office of the National Coordinator for Health Information Technology (ONC) to pioneer a new model of patient data-powered research intended to accelerate the pace of biomedical discoveries. The PMI's initial focus is on cancer treatment, but long-term goals emphasize preventive care and disease management.

The Precision Medicine Initiative encourages collaboration between the public and private sectors to accelerate biomedical discoveries using technology to analyze large health datasets alongside advances in genomics. This ambitious goal, while simple in framing, in practicality will require an immense amount of oversight and regulation alongside the actual research components to ensure the safety, privacy, and security of data used. As such, the NIH is tasked with building a voluntary national research cohort of over one million Americans to collect a broad collection of data including medical records, genetic profiles, microbes in and on the body, environmental and lifestyle data, patient-generated information, and personal device and sensor data. The ONC is specifically tasked with developing standards for interoperability, privacy, and secure data exchange across systems. Additional provisions within the PMI have been included to protect privacy and address other legal and technical issues. In sum, the Precision Medicine Initiative will help support and also make practical the transition into the era of precision medicine.

Currently, the best examples of precision medicine can be seen in the field of oncology, where patients increasingly undergo extensive molecular and genetic testing so that physicians can chose treatments best suited to improve survival and reduce side effects (Kummar et al. 2015). One particularly encouraging advancement in targeted oncology has been with the treatment of metastatic melanoma, which prior to 2011 was thought of as a rapidly fatal condition with a prognosis usually under one year (Jang and Atkins 2013). Studies of melanoma biology and immunology revealed that almost 50% of melanomas harbor mutations in BRAF, mainly at codon 600. Ipilimumab and vemurafenib, two BRAF Val600 selective inhibitors, demonstrated significant tumor response with improved progression-free survival. Despite this promising initial response, patients often suffered disease progression at a median of 5–7 months due to multiple resistance mechanisms within the tumors. It was then discovered that some patients with BRAF Val600 mutations were able to obtain more durable responses with the addition of certain immunotherapies like high-dose interleukin 2. While further studies are still needed to ascertain the extent of downstream mutations and optimal combination therapies for greater survival (Robert et al. 2015), targeted therapies like the BRAF V600 selective inhibitors remain the goal in increasing precision medicine in oncologic care.

While targeted therapies are being designed to treat tumors, testing for somatic germ line mutations is also being employed to assess risk and stratify management decisions for certain types of cancer. The most well known mutations, BRCA1 and BRCA2, can be tested in breast cancer patients to identify optimal surgical, radiotherapeutic, and drug choices for patients (Trainer et al. 2010). When tested in patients without cancer, identification of these mutations can significantly alter an individual's knowledge of their risk profile and affect downstream management of cancer screening and prevention (U.S. Preventative Services Task Force 2014). Studies are currently being undertaken to examine the potential benefits versus harms of BRCA genetic testing at the population level (Gabai-Kapara et al. 2014).

With respect to the management of chronic diseases, precision medicine has already yielded some concrete improvements in patient health. In 2014, direct and indirect mental health expenditures exceeded those of any organic health condition including cardiovascular disease and diabetes (Agency for Healthcare Research and Quality 2014). Treatment of mental health conditions, particularly refractory conditions, can be exceedingly challenging and is often based on trial and error with various medications. The field of pharmacogenomics has developed to identify genetic differences in the pharmacokinetic and pharmacodynamic profiles of individuals, and stratify their likely responses to different drugs. This can not only lead to more effective drug use but also mitigate adverse effects and potentially deliver cost savings to the health care system. The GeneSight Psychotropic test was developed to provide clinicians with a composite phenotype for each patient applied to the known pharmacology of certain psychiatric medications (Winner et al. 2013). A recent study showed that pharmacogenomic-guided treatment with GeneSight doubled the likelihood of response for patients with treatment resistant depression, and identified patients with severe gene-drug interactions enabling them to be switched to genetically preferable medication regimens. A later study showed that use of GeneSight in medication selection resulted in patient exposure to fewer medications with greater adherence and an overall decrease in annual prescription costs (Winner et al. 2015). These examples exemplify the improvements in individual clinical care that can come about as a result of the collaboration between informatics, research, and clinical medicine.

Harnessing big data can also transform the way physicians apply research to their daily practice. Clinical research has traditionally relied on time-consuming acquisition of data and human driven analysis to conduct studies. However, this exposes them to only a fraction of published data. Furthermore, clinically relevant research then is distributed through published journals that often take weeks to months at minimum to disseminate to clinicians. True changes in practice patterns take 17 years on average to then be fully implemented (Morris et al. 2011). Well designed EHR interfaces could one day intelligently match publications to clinical situations, significantly augmenting physicians' ability to apply new published evidence at the point of medical decision making. Meanwhile real-time aggregation of data at smaller levels such as a city or county can lead to timely public health interventions.

Despite the huge promise of big data in precision medicine, some of the largest issues to be addressed include variability and reliability of information (Panahiazar et al. 2014). Health records can potentially provide fragmentary information if the health record is not complete and there is not systematic quality control for data elements gathered to ensure data accuracy. Additionally, with such a large amount of information being gathered and inputted from various sources, issues with incongruent formatting and lack of interoperability exist. Several different strategies are being employed to address this, from platforms such as the Internet of Things to loop in data from devices to semantic web technologies to make information interpretable for search and query and integration. Data must additionally be standardized and processed to ensure high quality input.

The limits of what data may be captured, in particular, has potential for controversy. For example, healthcare organizations that wish to reduce healthcare costs may wish to see patients' credit card data to assess their risk for disease based on things such as travel history or food, alcohol, and tobacco purchases. However, further discussion is needed to determine where the boundaries of this big data "creep" may lie. Meanwhile, as the pure volume of data available exceeds what providers can process in a reasonable timeframe, physicians will increasingly rely on big data analytics to augment their clinical reasoning. Already, robust decision support systems are being developed with IBM's Watson to help clinicians match advanced molecular therapies with an individual's tumor (Parodi et al. 2016). Future providers and patients will need to determine the role of increasingly powerful clinical decision support without inappropriately making it a clinical decision maker.

Moving away from traditional fee-for-service payment models and historic medical disease management techniques is necessary to facilitate innovation and incorporation of big data into medical care. In many current health systems, payers and providers are incentivized differently and care is influenced by what will be covered by insurance and what will not, rather than all stakeholders focusing on high quality effective care (Kayyali et al. 2013). Political and financial pressures are already improving the landscape for standardized data. In some cases, this has taken the form of interoperability and led to immediate clinical benefit. Kaiser Permanente has fully integrated their electronic EHR HealthConnect, to allow information to cross over all health care settings, inpatient and outpatient, and across all facilities. The integrated system has shown improved outcomes in heart disease management and has created an estimated $1 billion savings from reduced office visits and redundant diagnostic tests (Kayyali et al. 2013). Meanwhile, the Health Observational Healthcare Sciences and Informatics collaborative has aggregated over 680 million records across Australia and many countries in Asia and has been working on the early identification of adverse drug reactions (Duke 2015). In other cases, larger institutions are intentionally curating more robust databases to drive advanced analytics. Both the VA and the NIH are creating databases with one million or more patients with a specific emphasis on combining large genomic datasets with clinical datasets (VA 2016, National Institutes of Health 2016). These two movements will ultimately create richer, standardized datasets to drive higher quality big data analytics.

This chapter is intended to be an overview of the applications of big data as they interface with a physician's practice. Currently, healthcare data is derived from a multitude of sources including those internal to a health system like electronic health records, computerized order entry systems, and data from devices and sensors. External sources including billing records from insurance companies, pharmacies, social media, and mobile and wearable consumer devices are becoming more and more prominent. The acquisition and aggregation of data from these new sources, ranging from standardized biometric data to highly variable patient captured data, adds both a wealth of potentially actionable health information as well as a layer of complexity related to systems interoperability, privacy and security concerns, and legal and regulatory challenges. Despite these challenges, promising examples of big data applications have emerged in both the public and private sector to transform the patient-physician relationship, expand knowledge of patient health factors, activate and engage patients in their health care, strengthen evidence-based physician decision-making, and accelerate research for more individualized patient care. Currently, much work is needed to create mechanisms to translate data and data-informed insights into useable data that can directly affect patient care and drive quality improvement in healthcare (Neff 2013). While the technical aspects of big data applications may be out of the scope of knowledge for many physicians, every aspect of medicine and health care will soon be influenced by the transformative potential of big data to achieve high quality, efficient, and effective patient-centered care.

References

Agaku, I. T., Adisa, A. O., Ayo-Yusuf, O. A., & Connolly, G. N. (2014). Concern about security and privacy, and perceived control over collection and use of health information are related to withholding of health information from healthcare providers. *Journal of the American Medical Informatics Association, 21*(2), 374–378. doi:10.1136/amiajnl-2013-002079.

Agboola, S., Jethwani, K., Khateeb, K., Moore, S., & Kvedar, J. (2015). Heart failure remote monitoring: Evidence from the retrospective evaluation of a real-world remote monitoring program. *Journal of Medical Internet Research, 17*(4), e101. doi:10.2196/jmir.4417.

Agency for Healthcare Research and Quality. (2014). *Total expenses and percent distribution for selected conditions by type of service*. Retrieved from https://meps.ahrq.gov/data_stats/tables_compendia_hh_interactive.jsp?_SERVICE=MEPSSocket0&_PROGRAM=MEPSPGM.TC.SAS&File=HCFY2014&Table=HCFY2014%5FCNDXP%5FC&_Debug=.

Ball, M., Weaver, C., Kim, G., & Kiel, J. (2016). Healthcare information management systems. *Journal of Information Systems Education, 14*. doi:10.1007/978-3-319-20765-0.

Bushey, R. (2015, March 5). *Meet thomas goetz, The co-founder of iodine*. Retrieved October 31, 2016, from http://www.dddmag.com/article/2015/03/meet-thomas-goetz-co-founder-iodine.

Caouette, H. (2015a). *eClinicalWorks announces integration between wearable devices and healow platform*. Retrieved from https://www.eclinicalworks.com/pr-eclinicalworks-announces-integration-between-wearable-devices-and-healow-platform/.

Caouette, H. (2015b). *healow Announces Internet of Things Integration*. Retrieved from https://www.eclinicalworks.com/pr-healow-announces-internet-of-things-integration-2/.

Chaterjee, P. (2015, October 09). *Delivering value by focusing on patient experience—AJMC*. Retrieved October 25, 2016, from http://www.ajmc.com/journals/issue/2015/2015-vol21-n910/Delivering-Value-by-Focusing-on-Patient-Experience.

Chen, J. H., Goldstein, M. K., Asch, S. M., Mackey, L., & Altman, R. B. (2016). Predicting inpatient clinical order patterns with probabilistic topic models vs. conventional order sets. *Journal of the American Medical Informatics Association*. doi:10.1093/jamia/ocw136 [Epub ahead of print].

Clark, N. M. (2003). Management of chronic disease by patients. *Annual Review of Public Health, 24*, 289–313. doi:10.1146/annurev.publhealth.24.100901.141021.

Collins, F. S., & Varmus, H. (2015). A new initiative on precision medicine. *New England Journal of Medicine, 372*(9), 793–795. doi:10.1056/NEJMp1500523.

Cunningham, P. J. (2009). Beyond parity: Primary care physicians' perspectives on access to mental health care. *Health Affairs, 28*(3), w490–w501.

Darkins, A., Ryan, P., Kobb, R., Foster, L., Edmonson, E., Wakefield, B., & Lancaster, A. E. (2008). Care coordination/home telehealth: The systematic implementation of health informatics, home telehealth, and disease management to support the care of veteran patients with chronic conditions. *Telemedicine and E-Health, 14*(10), 1118–1126. doi:10.1089/tmj.2008.0021.

Dentzer, S. (2013). Rx for the "Blockbuster Drug" of patient engagement. *Health Affairs, 32*(2), 202–202. doi:10.1377/hlthaff.2013.0037.

Downing, N. L., Adler-Milstein, J., Palma, J. P., Lane, S., Eisenberg, M., Sharp, C., et al. (2016). Health information exchange policies of 11 diverse health systems and the associated impact on volume of exchange. *Journal of the American Medical Informatics Association, 34*(13), 150–160. doi:10.1093/jamia/ocw063.

Duke, J. (2015, June 14). *An open science approach to medical evidence generation*. Retrieved October 31, 2016, from http://www.ohdsi.org/wp-content/uploads/2014/07/ARM-OHDSI_Duke.pdf.

Feldman, B. (2014, January 30). *Using big data and game play to improve mental fitness*. Retrieved September 20, 2016, from http://radar.oreilly.com/2014/01/using-big-data-and-game-play-to-improve-mental-fitness.html.

Fisher, M. (2014, June 12). *Apple's health app and healthkit and HIPAA*. Retrieved from http://www.hitechanswers.net/apples-health-app-healthkit-hipaa/.

Food and Drug Administration. (2006). *Guidance: Useful written consumer medication information (CMI)*. Rockville, MD: US Food and Drug Administration.

Fowles, J. B., Terry, P., Xi, M., Hibbard, J., Bloom, C. T., & Harvey, L. (2009). Measuring self-management of patients' and employees' health: Further validation of the Patient Activation Measure (PAM) based on its relation to employee characteristics. *Patient Education and Counseling, 77*(1), 116–122.

Fox, S. (2013). After Dr Google: Peer-to-peer health care. *Pediatrics, 131*(Suppl. 4), S224–S225.

Fox S. (2014). *The social life of health information*. Washington, DC: Pew Research Center's Internet and American Life Project. January 15, 2014, Available at http://www.pewresearch.org/fact-tank/2014/01/15/the-social-life-of-health-information/.

Freese, N. (2016, May 11). *It's high time for change—And health care data is paving the way*. Retrieved 25 October, 2016, from https://medium.com/@hdpalooza/its-high-time-for-change-and-health-care-data-is-paving-the-way-cac1e54ee088#.wtgembknu.

Frost, J., & Massagli, M. (2008). Social uses of personal health information within PatientsLikeMe, an online patient community: What can happen when patients have access to one another's data. *Journal of Medical Internet Research, 10*(3), e15.

Gabai-Kapara, E., Lahad, A., Kaufman, B., Friedman, E., Segev, S., Renbaum, P., et al. (2014). Population-based screening for breast and ovarian cancer risk due to BRCA1 and BRCA2. *Proceedings of the National Academy of Sciences, USA, 111*(39), 14205–14210. doi:10.1073/pnas.1415979111.

Gartner. (2016, August 16). *Gartner's 2016 hype cycle for emerging technologies identifies three key trends that organizations must track to gain competitive advantage*. Retrieved October 10, 2016 from http://www.gartner.com/newsroom/id/3412017.

Ginger.io Evidence (n.d.) Retrieved from https://ginger.io/evidence/.

Grand Rounds (n.d.). Retrieved October 25, 2016, from https://www.grandrounds.com.

Greene, J., & Hibbard, J. H. (2012). Why does patient activation matter? An examination of the relationships between patient activation and health-related outcomes. *Journal of General Internal Medicine, 27*(5), 520–526.

Healthgrades. (2016a). *Methodology: Mortality and complications outcomes*. Retrieved October 16, 2016, from https://www.healthgrades.com/quality/methodology-mortality-and-complications-outcomes.

Healthgrades. (2016b, April 05). *Patient experience measures data source*. Retrieved October 25, 2016, from https://www.healthgrades.com/ratings-and-awards/data-source-patient-experience.

Healthgrades (2016c). *Patient safety ratings and patient safety excellence award 2016 methodology*. Retrieved October 25, 2016, from https://www.healthgrades.com/quality/2016-patient-safety-methodology.

Healthgrades (n.d.). *Healthgrades find a doctor doctor reviews hospital ... (n.d.)*. Retrieved 25 October, 2016, from http://www.healthgrades.com/.

Hibbard, J. H. (2008). Using systematic measurement to target consumer activation strategies. *Medical Care Research and Review, 66*, 9s–27s.

Hibbard, J. H., & Greene, J. (2013). What the evidence shows about patient activation: Better health outcomes and care experiences; fewer data on costs. *Health Affairs, 32*(2), 207–214.

Hibbard, J. H., Stockard, J., Mahoney, E. R., & Tusler, M. (2004). Development of the patient activation measure (PAM): Conceptualizing and measuring activation in patients and consumers. *Health Services Research, 39*(4 Pt. 1), 1005–1026. doi:10.1111/j.1475-6773.2004.00269.x.

Higgins, T. (2015, April 26). Apple's healthkit linked to patients at big los angeles hospital. Bloomberg.com. Retrieved from http://www.bloomberg.com/news/articles/2015-04-26/apple-s-healthkit-linked-to-patients-at-big-los-angeles-hospital.

Hood, L., & Flores, M. (2012). A personal view on systems medicine and the emergence of proactive P4 medicine: Predictive, preventive, personalized and participatory. *New Biotechnology, 29*(6), 613–624. doi:10.1016/j.nbt.2012.03.004.

Humer, C., & Finkle, J. (2014, September 24). Your medical record is worth more to hackers than your credit card. *Reuters*. Retrieved from http://www.reuters.com/article/us-cybersecurity-hospitals-idUSKCN0HJ21I20140924.

Institute of Medicine. (2012). *Best care at lower cost*. Washington, DC: The National Academies Press. doi:10.17226/13444.

Iodine. (2014, September 23). Retrieved August 26, 2016, from http://www.prnewswire.com/news-releases/iodine-launches-service-that-transforms-100000-americans-real-life-experience-into-unique-data-driven-tools-about-medications-for-consumers-276858401.html.

Iodine Data. n.d.. Retrieved from http://www.iodine.com/data.

Jang, S., & Atkins, M. B. (2013). Which drug, and when, for patients with BRAF-mutant melanoma? *The Lancet Oncology, 14*(2), e60–e69. doi:10.1016/S1470-2045(12)70539-9.

Japsen, B. (2013, December 22). *ObamaCare, doctor shortage to Spur $2 billion telehealth market*. Retrieved October 10, 2016, from http://www.forbes.com/sites/brucejapsen/2013/12/22/obamacare-doctor-shortage-to-spur-2-billion-telehealth-market/.

Kayyali, B., Knott, D., & Van Kuiken, S. (2013). *The big-data revolution in US Health Care: Accelerating value and innovation* (pp. 1–13). New York City, NY: Mc Kinsey & Company.

Kumar, R. B., Goren, N. D., Stark, D. E., Wall, D. P., & Longhurst, C. A. (2016). Automated integration of continuous glucose monitor data in the electronic health record using consumer technology. *Journal of the American Medical Informatics Association, 23*(3), 532–537. doi:10.1093/jamia/ocv206.

Kummar, S., Williams, P. M., Lih, C.-J., Polley, E. C., Chen, A. P., Rubinstein, L. V., et al. (2015). Application of molecular profiling in clinical trials for advanced metastatic cancers. *Journal of the National Cancer Institute, 107*(4), djv003. doi:10.1093/jnci/djv003.

Lander, E. S., Linton, L. M., Birren, B., Nusbaum, C., Zody, M. C., Baldwin, J., et al. (2001). Initial sequencing and analysis of the human genome. *Nature, 409*(6822), 860–921. doi:10.1038/35057062.

Leventhal, R. (2015, April 21). *How duke is using HealthKit to get patient-generated data into the EHR*. Retrieved October 5, 2016, from http://www.healthcare-informatics.com/article/how-duke-using-healthkit-get-patient-data-ehr.

Liao, K. P., Cai, T., Savova, G. K., Murphy, S. N., Karlson, E. W., Ananthakrishnan, A. N., et al. (2015). Development of phenotype algorithms using electronic medical records and incorporating natural language processing. *BMJ, 350*(apr24_11), h1885. doi:10.1136/bmj.h1885.

Longhurst, C. A., Harrington, R. A., & Shah, N. H. (2014). A "green button" for using aggregate patient data at the point of care. *Health Affairs, 33*(7), 1229–1235. doi:10.1377/hlthaff.2014.0099.

Mann, D. M., Quintiliani, L. M., Reddy, S., Kitos, N. R., & Weng, M. (2014). Dietary approaches to stop hypertension: Lessons learned from a case study on the development of an mHealth behavior change system. *JMIR mHealth and uHealth, 2*(4), e41.

Metcalf, D., Milliard, S., Gomez, M., & Schwartz, M. (2016, October 3). *Wearables and the internet of things for health*. Retrieved from http://pulse.embs.org/september-2016/wearables-internet-of-things-iot-health/?trendmd-shared=1.

Might, M., & Wilsey, M. (2014). The shifting model in clinical diagnostics: How next-generation sequencing and families are altering the way rare diseases are discovered, studied, and treated. *Genetics in Medicine, 16*(10), 736–737.

Morris, Z., Wooding, S., & Grant, J. (2011, December). *The answer is 17 years, what is the question: Understanding time lags in translational research*. Retrieved October 30, 2016, from https://www.ncbi.nlm.nih.gov/pmc/articles/PMC3241518/.

National Academies Press. (2011). *Toward precision medicine: Building a knowledge network for biomedical research and a new taxonomy of disease*. Washington, DC: National Academies Press. Retrieved from http://www.nap.edu/catalog/13284.

National eHealth Collaborative. (2013, December). *Patient-generated health information technical expert panel final report*. Retrieved October 21, 2016, from https://www.healthit.gov/sites/default/files/pghi_tep_finalreport121713.pdf.

National Institutes of Health (2016). *About the Precision Medicine Initiative Cohort Program*. Retrieved October 30, 2016, from https://www.nih.gov/precision-medicine-initiative-cohort-program.

Neff, G. (2013). Why big data won't cure us. *Big Data, 1*(3), 117–123.

Nielsen. (2014, April 16). *Hacking health: How consumers use smartphones and wearable tech to track their health*. Retrieved October 3, 2016, from http://www.nielsen.com/us/en/insights/news/2014/hacking-health-how-consumers-use-smartphones-and-wearable-tech-to-track-their-health.html.

Panahiazar, M., Taslimitehrani, V., Jadhav, A., & Pathak, J. (2014). Empowering personalized medicine with big data and semantic web technology: Promises, challenges, and use cases. *Proceedings: IEEE International Conference on Big Data, 2014*, 790–795. doi:10.1109/BigData.2014.7004307.

Parodi, S., Riccardi, G., Castagnino, N., Tortolina, L., Maffei, M., Zoppoli, G., et al. (2016). Systems medicine in oncology: Signaling network modeling and new-generation decision-support systems. In U. Schmitz & O. Wolkenhauer (Eds.), *Systems medicine* (pp. 181–219). New York: Springer.

PatientsLikeMe Services(n.d.) Retrieved from http://news.patientslikeme.com/services.

PwC. (2014, October 21). *Wearable technology future is ripe for growth—Most notably among millennials, says PwC US*. Retrieved October 5, 2016, from http://www.pwc.com/us/en/press-releases/2014/wearable-technology-future.html.

Rask, K. J., Ziemer, D. C., Kohler, S. A., Hawley, J. N., Arinde, F. J., & Barnes, C. S. (2009). Patient activation is associated with healthy behaviors and ease in managing diabetes in an indigent population. *The Diabetes Educator, 35*(4), 622–630. doi:10.1177/0145721709335004.

Robert, C., Karaszewska, B., Schachter, J., Rutkowski, P., Mackiewicz, A., Stroiakovski, D., et al. (2015). Improved overall survival in melanoma with combined dabrafenib and trametinib. *New England Journal of Medicine, 372*(1), 30–39. doi:10.1056/NEJMoa1412690.

Robinson, R. (2016, February). *Patients and patient organizations power rare disease therapies—Pharmavoice*. Retrieved October 1, 2016, from http://www.pharmavoice.com/article/2016-02-rare-disease-therapies/.

Rolnick, J., Downing, N. L., Shepard, J., Chu, W., Tam, J., Wessels, A., et al. (2016). Validation of test performance and clinical time zero for an electronic health record embedded severe sepsis alert. *Applied Clinical Informatics, 7*(2), 560–572. doi:10.4338/ACI-2015-11-RA-0159.

Sarasohn-Kahn, J. (2008). *The Wisdom of Patients: Health Care Meets Online Social Media*. Retrieved from http://www.chcf.org/topics/chronicdisease/index.cfm?itemID=133631.

Schueller, S. M., Tomasino, K. N., Lattie, E. G., & Mohr, D. C. (2016). Human support for behavioral intervention technologies for mental health: The efficiency model. *Management, 21*, 22.

Sepah, S. C., Jiang, L., & Peters, A. L. (2015). Long-term outcomes of a web-based diabetes prevention program: 2-year results of a single-arm longitudinal study. *Journal of medical Internet research, 17*(4), e92.

Shapiro, M., Johnston, D., Wald, J., & Mon, D. (2012). *Patient-generated health data. White Paper: Prepared for Office of Policy and Planning*. Retrieved from http://healthitgov.ahrqdev.org/sites/default/files/rti_pghd_whitepaper_april_2012.pdf.

Shay, L. A., & Lafata, J. E. (2015). Where is the evidence? A systematic review of shared decision making and patient outcomes. *Medical Decision Making: An International Journal of the Society for Medical Decision Making, 35*(1), 114–131. doi:10.1177/0272989X14551638.

Shortliffe, E., & Cimino, J. (2014). *Biomedical informatics: Computer applications in healthcare and medicine* (4th ed.). London: Springer.

Sifferlin, A. (2014, September 24). *Look up your meds on this massive new drug database*. Retrieved November 1, 2016, from http://time.com/3425527/iodine-prescription-drug-reviews/.

Sittig, D. F., Wright, A., Osheroff, J. A., Middleton, B., Teich, J. M., Ash, J. S., et al. (2008). Grand challenges in clinical decision support. *Journal of Biomedical Informatics, 41*(2), 387–392. doi:10.1016/j.jbi.2007.09.00.

Smith, A. (2015, April 1). *U.S. smartphone use in 2015*. Retrieved from http://www.pewinternet.org/2015/04/01/us-smartphone-use-in-2015/.

Sunyaev, A., Dehling, T., Taylor, P. L., & Mandl, K. D. (2015). Availability and quality of mobile health app privacy policies. *Journal of the American Medical Informatics Association, 22*(e1), e28–e33. doi:10.1136/amiajnl-2013-002605.

The White House. (2015, January 30). *FACT SHEET: President Obama's Precision Medicine Initiative*. Retrieved October 20, 2016, from https://www.whitehouse.gov/the-press-office/2015/01/30/fact-sheet-president-obama-s-precision-medicine-initiative.

Trainer, A. H., Lewis, C. R., Tucker, K., Meiser, B., Friedlander, M., & Ward, R. L. (2010). The role of BRCA mutation testing in determining breast cancer therapy. *Nature Reviews Clinical Oncology, 7*(12), 708–717. doi:10.1038/nrclinonc.2010.175.

Tricoci, P., Allen, J. M., Kramer, J. M., Califf, R. M., & Smith, Jr. S. C. (2009). Scientific evidence underlying the ACC/AHA clinical practice guidelines. *Journal of the American Medical Association, 301*(8), 831–841.

Tu, H. T., & Lauer, J. (2008, December). *Word of mouth and physician referrals still drive health care provider choice*. Retrieved November 1, 2016, from http://www.hschange.com/CONTENT/1028/.

U.S. Department of Health and Human Services. (2013, January). *Realizing the full potential of health information technology to improve healthcare for americans: The path forward*. Retrieved October 4, 2016, from https://www.healthit.gov/sites/default/files/hitpc_stage3_rfc_final.pdf.

U.S. Department of Health and Human Services. (2015, February 9). *Mobile medical applications: Guidance for industry and food and drug administration staff*. Retrieved from http://www.fda.gov/downloads/MedicalDevices/DeviceRegulationandGuidance/GuidanceDocuments/UCM263366.pdf.

U.S. Department of Health and Human Services. (2009, October 28). *HITECH Act enforcement interim final rule*. Retrieved from http://www.hhs.gov/hipaa/for-professionals/special-topics/HITECH-act-enforcement-interim-final-rule/index.html.

U.S. Department of Health and Human Services. (2012, June 26). *Alaska DHSS settles HIPAA security case for $1,700,000*. Retrieved October 18, 2016, from www.hhs.gov/hipaa/for-professionals/compliance-enforcement/examples/alaska-DHSS/index.html.

U.S. Department of Health and Human Services. (2013, March 26). *HIPAA administrative simplification*. Retrieved from https://www.hhs.gov/sites/default/files/ocr/privacy/hipaa/administrative/combined/hipaa-simplification-201303.pdf.

U.S. Preventative Services Task Force. (2014, December 24). *BRCA-related cancer: risk assessment, genetic counseling, and genetic testing*. Retrieved from https://www.uspreventiveservicestaskforce.org/Page/Document/RecommendationStatementFinal/brca-related-cancer-risk-assessment-genetic-counseling-and-genetic-testing#citation27.

US Department of Veterans Affairs Office of Research and Development. (n.d.) *Million veterans program*. Retrieved October 30, 2016, from http://www.research.va.gov/mvp/veterans.cfm.

Wicks, P. (2007, July 7). *Excessive yawning in ALS/MND*. Retrieved October 1, 2016, from http://blog.patientslikeme.com/2007/07/07/excessive-yawning-in-alsmnd/.

Winner, J. G., Carhart, J. M., Altar, A., Allen, J. D., & Dechairo, B. M. (2013). A prospective, randomized, double-blind study assessing the clinical impact of integrated pharmacogenomic testing for major depressive disorder. *Discovery Medicine, 16*(89), 219–227.

Winner, J. G., Carhart, J. M., Altar, C. A., Goldfarb, S., Allen, J. D., Lavezzari, G., et al. (2015). Combinatorial pharmacogenomic guidance for psychiatric medications reduces overall pharmacy costs in a 1 year prospective evaluation. *Current Medical Research and Opinion, 31*(9), 1633–1643. doi:10.1185/03007995.2015.1063483.

Witters, D., & Agrawal, S. (2014, November 3). *How mobile technology can improve employees' well-being*. Retrieved October 10, 2016, from http://www.gallup.com/businessjournal/179111/mobile-technology-improve-employees.aspx.

Xu, J., Rasmussen, L. V, Shaw, P. L., Jiang, G., Kiefer, R. C., Mo, H., et al. (2015). Review and evaluation of electronic health records-driven phenotype algorithm authoring tools for clinical and translational. Journal of the American Medical Informatics Association, 22 (6), 1251–1260. doi:10.1093/jamia/ocv070

Yang, Y. T., & Silverman, R. D. (2014). Mobile health applications: The patchwork of legal and liability issues suggests strategies to improve oversight. *Health Affairs, 33*(2), 222–227. doi:10.1377/hlthaff.2013.0958.

Part IV
Applications in Business

Chapter 17
The Potential of Big Data in Banking

Rimvydas Skyrius, Gintarė Giriūnienė, Igor Katin, Michail Kazimianec, and Raimundas Žilinskas

17.1 Introduction

The emergence of the notion of Big Data (BD) has created substantial value expectations for organizations with huge collections of data. Banks, as all the other organizations, recognize the value potential contained in big data. Despite the fact that the term "Big Data" had emerged relatively recently, the concept of big data or vast collections of digital data has been around since the early days of computing; one might just remember all the comparisons of capacity of some new digital storage media with the contents of Library of Congress. The advances in IT and growing media capacity gradually ensured that larger and larger collections of data could be processed relatively easily and with low costs. This can be named the supply side of drivers of interest in BD (Bholat 2015), while the demand side is driven by the economic entities needs to enhance productivity and profits by using valuable insights from data analysis.

According to Turner et al. (2013), for financial services companies with no physical products to manufacture, data—the source of information—is one of arguably their most important assets, and the approach of Big Data Analytics (BDA) is especially promising and differentiating. As advances in business intelligence technologies spawned data mining in the last decade of the twentieth century, the following wave of big data analytical methods and applications found industries like banking, with billions of transactions recorded in the data bases, ripe and ready for advanced analytics. The current issue is to which extent these data assets may be leveraged to produce value and gain competitive advantage.

R. Skyrius (✉) • G. Giriūnienė • I. Katin • M. Kazimianec • R. Žilinskas
Vilnius University, Vilnius, LT-10222, Lithuania
e-mail: rimvydas.skyrius@ef.vu.lt; gintare.giriuniene@ef.vu.lt; igor@getweb.lt; kmichailas@gmail.com; raimundas.zilinskas@ef.vu.lt

S. Srinivasan (ed.), *Guide to Big Data Applications*, Studies in Big Data 26,
DOI 10.1007/978-3-319-53817-4_17

As Hopperman and Bennett (2014) indicate, banks, like any other organization, require big data handling solutions that are cost-effective, easy to manage, and provide high business value. However, due to the industry specifics, information activities in banks have features that reflect their requirements for precision, data protection and risk management.

This chapter is based mostly on literature research and professional experiences in dealing with information activities and challenges in banking. The statements and conclusions presented in this chapter are of inductive nature; however, several directions of possible empirical research have emerged in the course of writing this chapter.

The *structure* of the chapter is laid out according to the chosen logic of presenting the materials, and is organized as follows: the *first* subchapter discusses informing activities in banking business, including prevailing types of information systems, organizational culture in banks, and analytical information needs. The *second* subchapter relates Big Data prospects to banking activities, naming risks of doing and not doing Big Data analysis. It also discusses data collection and quality issues, potential directions of information discovery, encompassing possible risks, customer management and operations management. In the end of second subchapter there's a specific section assigned to Big Data issues in banking supervision, which covers several industry-wide Big Data-related issues, as opposed to approaches used by separate banks. The *third* subchapter discusses in more detail certain Big Data approaches, methods and instruments, as they are or may be used in banking. The *fourth* subchapter presents managerial implications regarding the use of Big Data in banking, concentrating on people issues and traits of intelligence culture. The chapter ends with conclusions and suggestions for research directions.

17.2 Informing Activities in Banking

17.2.1 The Nature of Bank Information Activities

Banking activities create an information environment that is rich with transaction data, and there is a long-term experience of handling this data with precision and caution. Over time, computer technologies and information systems have become the principal platforms of banking operations. Banking, as a largely data- and information-driven business, has substantial experience in managing transaction records: while in other types of business an entity's behavior did not leave much data on record in pre-digital times, in banking everything or nearly everything performed by the customers or personnel had to be recorded by definition. The powerful transaction processing systems are commonplace, and, as banking transactions have largely become commoditized, these systems cannot adequately serve as the basis for differentiation and subsequent competitive advantage. This incurs a need for analytical applications as one of the possible sources of such advantage, and this, at least in part, explains a current wave of interest in BD from the banking community.

As BD in its most cited general definitions relates to extraction of important insights and valuable information from vast and often chaotic data, in banking a significant part of the raw data is well-organized, cleaned and checked for its provenance. Following the same definitions of BD, it should carry one or more of the three key features—*volume*, *velocity* and *variety*. The volumes of recorded structured data in banking are traditionally high, while data velocity and the potential speed of reaction are offset by risk avoidance behaviors. Regarding the data variety, although unstructured data is collected through surveys, interviews and other channels, the dominating mode of sense extraction revolves around structured data sets implemented as relational databases (Bholat 2015). The multitude of definitions of Big Data is rather confusing, but for practical purposes the following set of questions, narrowing down the path of condensed potential value, could be useful:

What constitutes BD for a certain situation or activity?
What part of it is accessible?
What part of it is useful?
What part of it is processable?
What part of produced results is comprehensible?
What part of results is valuable?

A commonly encountered problem for banks is that the data very often sits in large, disparate legacy systems. Making data science tools work with legacy platforms and databases sitting in silos is a huge challenge. However, a feature of analytics in big companies is coexistence strategy: combining the best of legacy databases and data warehouses with tools of new analytic environments.

According to Cloudera white paper (Cloudera 2015), four main factors have driven the need for financial services companies to collect, store and analyze massive volumes of data:

1. Commoditization and digitization of financial products and services. Customers could work online with most of banks services.
2. Increased activity. The ease and affordability of executing financial transactions via online vehicles has led to ever-increasing activity and expansion into new markets.
3. New data sources. The digital revolution has led to new sources of data that are complex to ingest, such as data from derivative trading platforms, social media, blogs and other news feeds. This information, if combined with individual financial transactions and history, can help to paint a holistic picture of individuals, families, organizations, and markets.
4. Increased regulations. In recent years, federal stress tests have increased the demand for predictability and integrated solutions for capital asset management.

A similar, although not exactly the same list of factors influencing the use of Big Data approaches in banking, was presented in (Patwardhan 2016):

- Banking is becoming commoditized; a good information system does not provide competitive advantage, but good intelligence and analytics have more potential here (an assumption, based on examples from other industries);

- Cost of storing and analyzing data is falling;
- Alternative solutions (cloud etc) not always viable because of security requirements;
- The most acute applications (fraud and money laundering detection, sanctions and blacklists, monitoring of key financial indicators, enhanced account management) require top-grade analytics.

According to Deutsche Bank white paper (Deutsche Bank 2014), the drivers of Big Data technology in the financial industry are:

1. Explosive data growth
   Devices used in banking transactions grow in numbers and variety. The range of types of data to be analyzed is growing as well; many systems (e.g., insurance claims systems) store textual information that later can be combined with structured data on transactions.
2. Regulations
   Information from disparate data sets may require integration to understand certain events, manage risk or comply with regulations. The time window for liquidity estimation has shrunk to nearly real-time or at least same day, following Basel requirements. The scale of electronic trading also places significant strain on data capture and analytics.
3. Fraud detection and security
   A number of banks (e.g., Morgan Chase) have started to identify risky individual behavior and possible rogue traders. Also, more data is used for analysis, whereas earlier just limited samples have been in use.
4. Customer insight and marketing analytics
   The 360° customer view is potentially helpful where the services have been competing with the offerings by new entrants—PayPal, Amazon, Google etc. These institutions have managed to snatch a portion of services that traditionally belong to banking sector, so a more complete understanding of customers' interests and preferences is vital in ensuring customer loyalty, and external data from social platforms and the like may play an important role.

Deloitte, a consultancy, Deloitte (2016) presents an almost similar set of drivers for Big Data analytics: regulatory reform; operational efficiency; and customer profitability. Summarizing, the drivers for BD analytics in banking can be grouped as presented in Table 17.1.

As in any other industry, over time banking has experienced its share of mergers and acquisitions. For information activities, this has meant that well-tuned and reliable core systems had to be joined in some way to act as one system, although the systems could be vastly different. A final solution, often being some kind of compromise, may undermine the stability and clarity of information infrastructure and standards, further complication the intent to utilize Big Data approaches.

Table 17.1 Drivers of Big Data approaches in banking

External	Regulations	Better overall transparency, growing velocity for liquidity estimation
	Market forces	360-degree customer view, competitive intelligence
	External risks	Insights require top-grade analytics
	Growth of external data	Potentially useful sources of social information
Internal	Operational efficiency	Standard commoditized services to be run in a near-optimal mode
	Internal risks	Demand for predictability, detection of rogue traders
	Growth of internal data	Growth of the numbers and variety of devices used in banking transactions

17.2.1.1 The Corporate Culture of Banks

Banks possess a specific type of corporate culture that exerts influence on all information activities, including BD analysis. This culture is conditioned by the specifics of the banking activities—banks are conservative institutions acting in a rather competitive environment, where risk management has priority over opportunities, and control over creativity, including opening the systems to integrate their data with external content. The use of external data by banks will be discussed further in this chapter.

The practical actions in implementing BD solutions, suggested by PriceWaterhouseCoopers study (PwC 2013) are largely aimed at cultural issues: concentration on a specific business problem or a class of problems; development of a policy for data creation, access and sharing; coordination of skills and efforts; business insight-driven approaches. According to Wilcock (2013), the pinnacle of analytical thinking is to have a widespread "insight-driven culture". Other sources in (Wilcock 2013) define insight-driven as "real-time use of complex analytics to keep a picture of the changing business context, adjusting business models and driving operational decisions". Analytical thinking requires ongoing training, both formal and informal, to refine inquiring skills to detect problems and discover their primary root causes, as well as to keep up with the latest trends. Rewarding positive behaviors and discovered insights from the beginning should be promoted; non-crucial mistakes should not be penalized (Wells 2008). Healthy analytical culture, which is an important part of corporate culture, is created when BI programs focus on people before technology, and can be described by terms like enthusiasm, belief, confidence, support and competence. Such culture can hardly be created to a plan, but it can be shaped from existing culture.

17.2.2 Analytical Information Needs in Banks

For banks, the need for analytical activities is probably more expressed than in other types of business, largely due to the reliability requirements, fierce competition and needs to survive in unstable economic environment. This requires a 360-degree view not only in dealing with customers, products and operations, but in all other activity aspects as well. In bank management there are leaders whose positions are largely defined by data and analytics, e.g. chief risk officers.

In retrospect, banks have been quite proficient at performing analytics at product or transaction level; as structured internal data dominates, it is easily queried and analyzed by standard tools (Fuller 2015). While the pool of source data somewhat limits the scope of analysis, the strict provenance requirements positively affect data quality. While banks have historically been good at running analytics at a product level, such as credit cards, or mortgages, very few have done so holistically, looking across inter-connected customer relationships that could offer a business opportunity—say, when an individual customer works for, supplies or purchases from a company that is also a client of the bank. The current advent of Big Data approaches has had less effect on banking and financial services in general, as compared to other data-rich sectors, such as information and communications industry (Bholat 2015). However, judging by the recent rise of interest in Big Data technologies from the side of financial institutions, the situation seems to be changing.

Banks also should not be easy to research on analytical issues because of the often strategical importance and low publicity of analytical applications, despite the fact that their operations are commoditized, and core systems do not serve as a primary source for strategic advantage. The experience of earlier innovations in the field, e.g. introduction of ATM terminals or early e-banking systems, has shown that early adopters do create competitive advantage for a short time. Eventually, other market participants catch up, and innovation becomes a commonplace commodity that is virtually compulsory to conduct business. The issue of sustainable competitive advantage is no less important in the case of Big Data in banking—on one hand, there is a seemingly vast potential of discovery of important trends and relations that can be turned into competitive action. On the other hand, there are serious limitations stemming out from regulated industry, commoditized core activities, responses to same competitive threats, to name a few, and such limitations reduce the space for innovation.

The principal areas of analytical information needs in banks, as indicated by PwC (2013) are:

- Risk management, encompassing fraud detection, financial crime prevention, customer due diligence, compliance and other risk-related issues;
- Customer management, including development of a single view of a customer, flexible segmentation, satisfaction analysis;

– Operations management including product development, launch and tailoring, predictive models and forecasts in trading, support for internal intelligence culture.

There are many quite enthusiastic voices on the potential value of BD in banking, presenting the business opportunities to be discovered as a proven fact. While we support the point that there is certain (and often rather substantial) insight potential to be discovered, we advise that the discovery of such potential is a tricky and daunting task, based on a profound understanding of own information needs, organizational culture, discovery skills and last but not least—advanced information technology to support analytical activities and refining answers to important business questions.

17.3 Big Data and Banking Activities

17.3.1 Tradeoffs of Big Data Analytics

Banks seek for more powerful and flexible technologies to provide insights into their operations, customers or risk management. The banking sector has been doing this for decades, if not centuries; on the other hand, banks as a largely information-based business have to consider important innovations in information technologies and information management to deal with competitive threats or sources of competitive advantage. The current period, marked by the advent of big data, draws attention to the new sources of data and information, as well as technologies and approaches to effectively utilize them. While the number of banks with real implementations of big data projects is rather small (Hopperman and Bennett 2014), initial expectations are giving way to more structurized approaches, defining possible areas of application and benefits, and there's growing clarity of what can be achieved with BD approach. The Forbes report (Marr 2015) shows that, since 2011, substantially more companies are treating their data as a strategic corporate asset. The initial excitement about the possibilities presented by big data is shifting more towards strategic approach, defining which data initiatives will have the biggest and most immediate impact.

By using data science to collect and analyze Big Data, banks can improve, or reinvent, nearly every aspect of banking. It is worth noting that out of three principal areas of value creation, as indicated earlier—risk management, customer management, and operations management, the latter two areas are present in every other business, while risk management has a place of special importance in banking activities, and will be discussed in more detail in this chapter.

Risks of not doing Big data analysis:

- Failure to detect important changes starting to happen;
- Substandard informing impairs sustainable competitive advantage;

- Rising costs of non-compliance;
- No single view on the customer leading to customer frustration and churn;
- Risk of losing market to newcomers in the banking business, e.g. Facebook and PayPal entering the payments market; peer-to-peer financing; crowd funding.

Risks of doing Big data analysis:

- Big Data approach is novel and is more like blazing the trail than going a well-known path; it might require substantial investment with hard-to-prove returns;
- Risk, driven by expectations of discovery, to engage into many analytical directions at once and waste attention and effort; the danger of becoming stuck in "analysis paralysis" (Heskett 2012);
- The findings might be unrelated to business questions or be false positives (discovered relations and rules that are statistically valid yet are wrong or make no sense at all); Ziff-Davis study (2016) points to, albeit rhetorically, hundreds of business questions presented to data scientists that might lead to tens of discoveries and few, if any, improvements; put differently, there certainly is hidden stuff in data; not all of it is meaningful, not all of meaningful stuff is useful, and the useful stuff is difficult to discover;
- Sophistication of technology does not match the intelligence sophistication in the organization; without proper intelligence culture, acceptance is likely to encounter obstacles; the results of Big Data analysis will have limited trust if different groups of people engage in analysis and use of results.

Some sources (e.g., Kaisler et al. 2013) put Big Data just beyond the reach of today's technology, implying that it is a moving target, and its proper utilization is never to be achieved largely due to the permanent growth of the 3 V's: Volume, Velocity, and Variety. On the other hand, some doubts may arise whether the proper or complete utilization of BD is really a feasible goal. If the BD analysis provides information that is proved valuable to current or future practice, this result is probably far more important.

17.3.2 Data Collection and Integration

Traditionally, data collection in banks has not presented any significant problems—as a transaction-based business, banks collect records largely using own resources; the provenance of the collected data is rather clear. Additional issues with data collection and quality emerge with factors that disturb the standard data collection activities—mergers, global activities, multitude of service channels. One of the key challenges in BD is integration of data variety, emerging from multitude of sources and formats (Hashem et al. 2015). Some sources (Davenport and Dyche 2013) say that unstructured data in financial services is sparse—most of the records come in a structured format: transaction records, website clicks, responses to messages and offers, to name a few.

The study performed by IBM Institute for Business Value (IBM Institute for Business Value 2012) about the sources of big data used by financial organizations reveals the fact that most of early big data efforts relate to usage of internal data, such as transactions, logs, and event data. This suggests that banks act very pragmatically in big data adoption activities, and prefer focusing on their own well-structured data that has been collected for years, but has never been analyzed. The survey also shows that banks are less enterprising in terms of less structured data. Thus banks report about collection and analysis of social media data, images, emails, sensors data, and even audio that is produced in abundance in retail banks' call centers, but are still behind their peers in other industries.

Kaisler et al. (2013) name the challenges of internal data collection:

- How do we decide which data is irrelevant versus selecting the most relevant data?
- How do we ensure that all data of given type is reliable and accurate? Or, maybe just approximately accurate?
- How much data is enough to make an estimate or prediction of the specific probability and accuracy of a given event?
- How do we assess the "value" of data in decision making? Is more necessarily better?

To the above, we may add an issue of required data granularity—level of detail, at which data is collected and stored. Jacobs (2009) suggests that as BD in banks contains mostly temporal data (with time stamps), its volume swiftly increases. And temporal dimension is important, for it usually has to be ordered—for instance, to examine time series. So, another challenge emerges regarding whether all data should be time stamped.

Kaisler et al. (2013) provide other challenges or tradeoffs:

- Data input and output processes—access and processing, e.g., joining tables of related data incurs huge performance costs;
- Data volume growth in number of records, versus data variety expansion in adding new attributes and other data fragments;
- Structured versus unstructured data—problems of translation between structured and limited data tables, and unlimited data sources of unstructured data;
- Data ownership issues, especially regarding social data; e.g., who owns a person's Facebook data—a person or Facebook;
- Security—amassed personal data may be rather sensitive and make its objects vulnerable;
- Distributed data and distributed processing—communication between nodes may significantly increase performance degradation.

A special subject of discussion is the value potential created by information from external sources. According to (Hopperman and Bennett 2014), internal and traditional external data feeds will not satisfy the analytical requirements: multi-channel data, social platforms, location information and other available sources of potential interest will contribute to the competitive advantage. Another consultancy,

Cap Gemini (Cap Gemini Consulting 2014) indicates that less than half of banks analyze customers' external data available from alternative channels—social media, online behavior and others. When appending third-party data from external sources (Daruvala 2013), the risk discrimination models experienced a substantial improvement in their predictive power.

The importance of external data that is collected beyond the boundaries of bank operations is still a controversial issue, although some sources (e.g., PwC 2013) state: "While financial services companies are, for the most part, not yet buying third-party Big Data information, we recommend moving in that direction—especially for capabilities such as sentiment analysis." Bholat (2015) provides several examples of using heterogeneous external data to have a deeper understanding of banking industry and financial markets, and stresses that there are certain technical and legal challenges in blending different data collections. Technical challenges mostly amount to format, coding and naming differences, while legal challenges typically emerge when the use of external data is restricted only to purposes explicitly expressed in legislation.

Inability to overcome data integration and consolidation issues across various data silos has been a great challenge for enterprise organizations for years, and banks are not an exception, especially in cases of banks' acquisitions and mergers that result in a number of new internal data sources as a subject for integration. Integration of big data raises even more complex problems. Thus, data type variety imposes requirements for new infrastructure components like Hadoop, NoSQL, real-time streaming, visualization, etc. However, it is in these very technologies that financial organizations are behind their peers in other industries in the most cases, as is stated by IBM Institute study.

17.3.3 Data Quality Challenges

The quality of raw data has been an issue with every generation of information technologies and systems, and with the growing importance of analytical insights, the role of data quality has nowhere to go but to grow further. As the most common guidelines for data quality have been more or less agreed upon (e.g., Wand and Wang 1996; Kahn et al. 2002; Dyche 2004; Hoorn and van Wijngaarden 2010), in the context of Big Data it is worth looking whether there are significant differences or novel criteria that apply to Big Data, as opposed to data in general. The literature analysis did not reveal any significant differences between requirements for Big Data and those for data in general. According to Cai and Zhu (2015), big data quality faces the following challenges:

- The diversity of data sources, data types and data structures increases the difficulty of data integration
- Due to data volumes, its quality cannot be estimated within a reasonable amount of time

- Due to fast data updates, data timeliness is rather short-lived
- There are no unified and approved international data quality standards

A likely specific challenge for banking data is *data provenance* (Ram and Liu 2008). Data provenance covers the history of data—its origin, mode of record, events that have happened to data in its lifecycle. Many of generic data quality features, such as truthfulness, completeness or currency, are related to provenance. In the context of banking activities, provenance requirements are largely covered by regulations on banking activities and internal rules of banking entities, and are expected to be far more stringent than in other types of business. The research on data provenance is an interesting and developing area with specific conflicting criteria—e.g., a separate information system is required to manage provenance information, and with growing volumes of recorded transactions, the related provenance information further boosts the volumes to be recorded and managed.

The quality of analysis results is a separate important issue which, while subject to the generic requirements for data quality, aims at the completeness of the context in which the results are presented. A professional study by SurfWatch Labs (2014) presents several quality requirements that make analysis results actionable:

- *Comprehensive*—need to be joined into a model that provides the fullest picture and is easily distilled into useful intelligence.
- *Accurate*—any potential data fault or bias should be disclosed and processed accordingly.
- *Relevant*—applicable, easily integrated and flexible.
- *Timely*—reflects fresh trends, be it threats or opportunities.
- *Tailored*—data should be presented towards a specific purpose

Certainly the quality of source data is one of the most important factors for the quality of analysis results, influencing accuracy, relevance and timeliness in the above list. In addition, other higher-level issues like comprehensiveness and trust add up—the technology path from source data to analysis results may have deficiencies of its own that distort the results. The shortcomings of the analysis process may include incomplete data sets, incorrect rules, erratic analytical assumptions, to name a few. To our opinion, the production of analysis results satisfies complex information needs and is attributed to advanced informing, and it cannot be easily automated for the above risks that require human guidance and supervision.

17.3.4 Discovery and Detection

The goal of Big Data analysis is expected discovery of important information, possibly with adequate context, including important dimensions of what are the drivers behind discovered important information and how do they behave. In many cases this refers to asking the right questions. According to Marr (2015), companies are not asking for more data, but rather the *right* data to help solve specific problems and address certain issues.

Many sources present Big Data analytics as being data-driven, and another important issue related to Big Data comes up—how data-driven and question-driven approaches relate. This point has been argued in Skyrius et al. (2016): data-driven approaches see the availability of data and analytical functions as the primary driving factors for producing insights, whereas question-driven (or issue-driven, or problem-driven) approaches stress the primary role of well-pointed questions aimed at certain insights.

The expectations of discovery of important results in Big Data in banking may be grouped around the three directions, as stated earlier: risk management, customer and product management, and internal operations management. A report on the use of Big Data by PriceWaterhouseCoopers consultancy (PwC 2013) provides a structure of key expected benefits in the three areas:

1. Risk management and regulatory reporting:
   - Detection of enterprise risk across risk dimensions;
   - Regulatory reporting with reinforced agility for changing regulations.
2. Customer data monetization:
   - Customer centricity: development of a single view of the customer;
   - Customer risk analysis: analysis of behavior profiles, spending habits and cultural segmentation;
   - Customer retention: analysis of internal customer logs and activity in social media to detect dissatisfaction early.
3. Transactions and operations:
   - New products and services: use of social media to effectively define marketing strategies;
   - Algorithmic trading and analytics: large volumes of historical data feed predictive models and forecasts; analytics performed on complex securities using related data from different sources;
   - Organizational intelligence: use of employee collaboration analytics; BD culture of innovation empowers employees.

All three above areas are intertwined, and it is not uncommon that some interesting facts from one area find explanations in related areas. This is why a cross-functional analytical climate is one of the prime preconditions for the value of intelligence and analytics, and a specific type of intelligence culture has to be developed to boost the value of analytics. For banking activities, this is quite important to realize because of often-conservative and restrained corporate culture prevailing in banks.

17.3.4.1 Risks

Banking, as an activity with increased risk sensitivity, has to deal with huge variety of risks from all aspects of activity—internal and external, from customers, competitors or partners, market forces or economic policies, etc. Bank policy manuals contain hundreds of named risk sources. Risk management is a high-priority focus for banks, compared to customer data analytics or internal operations intelligence. Proactive and reactive approaches have to be combined with proper definition and redefinition of the monitoring scope. The advent of Big Data approaches, on one hand, creates conditions to radically improve awareness and detection of risks. On the other hand, there is a constant tradeoff between tried-and-tested methods of monitoring and discovery, and required agility to avoid losing the focus on constantly changing environment.

The holistic approach to risk detection and awareness is no simple task: the multitude of monitored entities from single customer credit risk to global economics, data sources with different formats and provenance, multiple regulations and rules, number of involved participants and stakeholders. Clearly, Big Data analytics are no magic pill for all risk management information needs, but industry cases show that well-pointed business questions and adequate analytical approaches lead to success stories. In an example, presented by Toos Daruvala from McKinsey consultancy (Daruvala 2013), a large US bank had their Ginni coefficient (ability to discriminate between good and bad risks) in the 40–45% range. After developing a 360-degree view of the customer from the internal cross-silo data, and integrating it with external third-party data, they improved the Ginni up to 75%. Another example, provided by Daruvala (2013), describes a bank with scarce data on customers that decided to use data of the local telecom company. It appeared that the telecom data on paying behavior is a great predictive indicator for credit behavior with the bank. CapGemini (Cap Gemini Consulting 2014) has pointed to one more possible source of risk management benefit—FICO expansion (FICO is a credit score model used in the US) by integrating it with demographic, financial, employment, and behavioral data. In this way, use of additional data sources, including CRM systems or social platforms, might increase awareness by revealing important facts like a gambling problem or an expensive purchase. In anti-money laundering activities, advanced statistical methods coupled to text mining of unstructured data may reveal hidden links between money movement and account location, detecting patterns that attract attention and invite a closer look.

In the context of risk detection, the advanced informing in banking in the form of early warning is of prime importance. The regular assessment of the financial stability and identification of early warning indicators informing about the risks in the banking system is one of the most important challenges for the banks administration, supervision of banks, and deposit insurance. The efficiency and stability of the banking system ensure a rational distribution of capital resources in the economy, and regulators therefore aim to prevent banking system crises and their associated adverse feedback effects on the whole economy.

It is important to note that the criminal sources of risk—money launderers, fraudsters and others—often are rather advanced users of technologies, and their skills develop surprisingly fast, so the above mentioned analysis tools and methods have to maintain required agility to catch up with advanced sources of risk. As well, in contrary to the customer management which centers on customers and aims to achieve a "360-degree view", risk management has numerous axis units, some of whom are rather vague and keep changing. This draws significant agility requirements for risk analytics.

17.3.4.2 Customers

The management of customers, services and products does not differ that much from other business activities, as compared to risk management that is specific to banking activities. There are, however, some features specific to banking.

A large proportion of the current Big Data projects in banking revolve around customers—driving sales, boosting retention, improving service, and identifying needs, so the right offers can be served up at the right time. The strict nature of banking records allows monitoring so-called customer journeys through assorted contact points like e-banking websites, tellers at physical branches, ATMs, contact centers (Davenport and Dyche 2013). The monitoring of such journeys supports better understanding and prediction of customer behavior in issues like attrition or accepting offers for additional services. By detecting many more patterns in the data than was previously possible, it is easier to detect fraud. Analysts can trace behaviors that allow retail banks to sell additional products by detecting important changes in lifestyle that might warrant the offering of new savings accounts and mortgages (The Banker Editorial 2013). The "next best offer", as well as cross-selling and upselling, may be targeted at micro-segments by combining past buying behavior, demographics, sentiment analysis and CRM data.

A white paper by Evry Innovation Lab (n.d.) reveals five fundamental use cases for acquiring, developing and retaining customers that we briefly observe below.

Sentiment analytics is used when examining customers' feedbacks gathered through social media to understand what customers think of company's products and services as well as to identify key customers for raising the word-of-mouth marketing. Tools for performing sentiment analytics are capable to process vast amounts of social media data and logs. Thus Barclays derived actionable insights from real-time social media analytics after the launch of the new mobile banking app. The app did not allow young app users under 18 to perform money transfers. After negative comments were caught by sentiment analysis tools, Barclays improved the app quickly and added access for 16- and 17-year-olds.

Customer 360-degree view provides much better understanding of the customer as a whole and provides new marketing opportunities. IBM Institute for Business Value (IBM n.d.) underlines several important tasks regarding Customer 360.

Customer profiling allows sending out personalized marketing messages. Even a very small personalization activity can improve customer engagement, security and loyalty. HDFC Bank increased the number of credit card activations as well as reduced cost per acquisition of each customer by personalized messages to every of the customer lifecycle segments that the bank had identified.

Analysis of the product use by the customer creates *understanding of the product engagement* and helps to position and sell products better. Bank of Austria takes actions for a product renewal when they recognize a specific customer behavior related with the product cancelation.

Determining the customer churn is one of the most important tasks of every organization with customer-centric strategy, as the cost of keeping the customer is lower than the cost of acquiring a new customer. With the use of predictive models, Tatra Bank showed well in decreasing customer churn from credit card holders. The bank established highly personalized retention campaigns for their customer segments.

Customer segmentation is known as finding a grouping of customers, where customers of the same group have common characteristics or behaviors. Understanding these groups is crucial for revealing customer needs. With the use of new Big Data processing technologies, the segmentation, being already well-adopted by many organizations, can be performed faster and with better quality level. Segmentation is used in *establishing marketing and loyalty programs* as well as in *optimizing pricing strategy*. For example, The Royal Bank of Canada (RBC) creates loyalty programs based on customer's card usage habits and then offers customer-oriented products and gifts. Fifth Third Bank uses analytics-based product pricing engine to help acquire new customers. Using analytics the bank runs scenarios on how varying price points influence customer acquisition and deposit levels. Thus the bank makes price predictions when interest rates will rise in the future, and creates scenarios how to attract customers when rates are changed.

By analyzing market basket and finding patterns between the products, organizations can create *next best offer*. For banks to survive in the very competitive environment, they need to offer tailored products or bundles of products and services. Usually customers have several banks with different products in each bank. Using bundles, based on preferences, banks can shift customer utilization and revenue significantly.

Big Data helps by taking a holistic view of the entire customer *journey* and experiences on each *channel*. This can be used to find patterns of usage that lead to better sales, or channels resulting in higher costs. From this knowledge, banks can optimize their communication with the customer and *provide relevant content*. Bank of China has created an online banking platform to examine data from various channels and to provide the right content at the right channel. The online platform integrates customer facing systems such as branch, phone, mobile and web services for Bank of China's 100 million customers.

While banks have historically been good at running analytics at a product level, such as credit cards, or mortgages, very few have done so holistically, looking across inter-connected customer relationships that could offer a business opportunity—say when an individual customer works for, supplies or purchases from a company that is also a client of the bank. The evolving field of data science facilitates this seamless view, provided that the data silos for different functional areas are opened to facilitate customer-centered data integration. The importance of cross-silo data integration as a prime concern for financial institutions is supported by Brian Yurcan in Banks, Big Data ... (SAS Institute Inc. and Bank Systems & Technology 2012).

17.3.4.3 Operations

Although the majority of analytical effort in banking deals with outside world, be it risk sources or consumer relations, the use of Big Data approaches may be as well directed inwards for help in managing operations—streamlining operations, maintaining efficiency and monitoring internal risks, to name a few. Monitoring of internal risks deserves separate attention, as the recent example of Wells Fargo bank has shown (McCoy 2016). The inadequate monitoring practices, coupled to sales incentives, have led to mass opening of deposit and credit card accounts unauthorized by the customers. This has resulted in huge restitution fees and fines for the bank.

The discoveries in the area of operations help innovating new business models, products and services by utilizing data obtained from actual use of products and services to improve their development or to create innovative after-sales service offerings. Big Data analytics may influence organizational structure by consolidating analytical competences and thus reinforcing intelligence culture component. For repeating actions of mass nature, use of Big Data analytics may lead to replacing or supporting human decision making with automated algorithms to unearth insights that would otherwise remain hidden (automatic warnings and alerts—e.g., flagging risk sources for additional attention).

17.3.5 Big Data Approaches in Banking Supervision

One of the key areas for risk management in banking business crosses the boundaries of individual banks and deals with banking supervision, executed by central banking authorities and deposit insurance funds. The regular assessment of financial stability and identification of early warning indicators is one of the most important challenges for the banks administration, supervision of banks, and deposit insurance. The efficiency and stability of the banking system ensure a rational distribution of capital resources in the economy. Regulators therefore aim to prevent banking system crises and their associated adverse feedback effects on the whole economy. Very important is the continuous and forward-looking assessment of a stability of the banking system which is used to set up early-warning indicators, values, and their potential impact on both regional banking and international financial markets.

The analytical information needs in banking supervision create approaches that differ from those in separate banks. As research shows, the principal features of

Big Data—volume, velocity and variety—are not much evident in the information environment of central banks and banking supervision (Bholat 2015). Data *volumes* are moderate, mostly structured and with strictly predefined formats because their primary sources are financial statements from the banks and aggregate data from statistics institutions. Because of relaxed urgency of collecting and processing this data, the *velocity* is not high. Although correspondence, surveys and interviews are recorded and collected, formal processing of this data uses predominantly structured sources in the form of relational tables, reducing *variety*. A few other features of the information environment in banking supervision include deductive approaches, extensive use of macroeconomic modeling, specific models for monitoring and testing—stress tests and the like.

Various resources—institutions, people, competencies, methods and technologies—have been developed to strengthen banking supervision in the form of monitoring and early warning systems. The early warning systems, enabling to alert on upcoming instability and to absorb them, are important not only to banking, but to the entire economy. The stability of the banking system is based on the macroeconomic, financial and structural factors. The market price of the assets, the key factors of the business cycle and monetary market are the principal indicators for the early warning system.

The banking sector can be very vulnerable by the financial market liberalisation and the competitive nature of banks, when banks take unmeasured new risks that often violate the balance of assets and liabilities. The lack of banking supervision and regulatory policy in these circumstances may also deepen these issues, because the banks are not enough limited to take the new risks. In addition, weaknesses in accounting, inadequate disclosure of information about the state of banks, lack of the adequate legal systems contribute to deepening the problem, because it allows to hide the extent of the banking problems. Supervision authorities or deposit insurance institutions sometimes fail to quickly identify problems of banks in order to take a prompt corrective action. Therefore, problems of all or part of the banking sector can grow and can affect the whole economy.

For banks, banking supervisory authorities, as well as deposit insurance institutions it is very important to understand and evaluate the key indicators to estimate the state of banks, its trends and to forecast the direction of a possible behaviour in the future, assessing not only the internal data of a bank, but a large quantity of various types of environmental data as well. The management of banks and supervisory authorities must regularly receive information on the status and trends in the banking business, suggestions for early action to be taken to improve the situation in order to avoid reaching the critical level. One of the most important roles goes to the early warning systems (EWS). These systems assist users to assess their relevant internal and environmental changes in an early phase and, on account of their origin and the potential impact on, set out the relevant steps to avoid the adverse impact. The weaknesses of such system or inappropriate assessment of its results may have a negative impact not only on a specific bank, but on the banking sector or even the economy as a whole.

EWS, like one of the supervision instruments, is closely linked to big data systems. The EWS based on the big data technologies is ensuring a maximum reliable and accurate information about potential threats or upcoming contingencies, that ensures the right assessment of the risk and its possible impact, and the finding of methods for the risk mitigation. EWS makes it possible to process and evaluate the warning signals, qualitatively assess the risk, make the reasonable decisions, take the prompt corrective actions and manage the situations (O'Brien 2010).

The information activities of EWS are best described by the following features (Žilinskas and Skyrius 2009):

- The activities require predominantly external data;
- Moderate volumes of raw data; their structure is predetermined, and there might be no structure at all (free form data);
- Strongly expressed influence of macroeconomic context on both analysis and decision making.

The banking sector worldwide is regulated by different sets of rules, although the logic of these rules is quite common. According to Basel Core Principles for Effective Banking Supervision (Bank for International Settlements 2012), supervising authorities use a variety of tools, including, but not limited to:

(a) Analysis of financial statements and accounts;
(b) Business model analysis;
(c) Horizontal peer reviews;
(d) Review of the outcome of stress tests undertaken by the bank; and
(e) Analysis of corporate governance, including risk management and internal control systems.

European Banking Authority (2014) publishes the results of European Union-wide stress tests that cover up to 12,000 data points for each EU bank, amounting to over one million data points covering banks' composition of capital, risk weighted assets (RWAs), profit and loss, exposures to sovereigns, credit risk and securitization. The Federal Reserve System in the USA provides the banking supervision examiners with CAMELS rating system covering six components of a bank's condition: capital adequacy, asset quality, management, earnings, liquidity and sensitivity to market risk.

Over the last two decades, the world banking system has experienced a substantial share of instability and global crises. The recent experiences somewhat contradict with the resources globally assigned for banking supervision. The rules for supervision are based on long-time experience and cover the most important aspects of banking activities. However, several questions arise when dealing with the recent experiences in global banking. The monitoring approaches are still far from being perfect, and are uninsured against missing important discoveries. As well, Big Data approaches seem to have potential to deal with deficiencies in monitoring activities, but they are not of much help if not properly coupled to human expertise.

The financial crisis of 2007–2008 rather painfully pointed out the need to introduce more powerful instruments of disclosure by financial companies of their data to supervision authorities. Bholat (2015) has presented several steps taken by the Bank of England on improving analysis of supervisory data: reporting of large exposures on a counterparty-by-counterparty basis on Common Reporting (COREP) templates; security-by-security reporting of insurers' assets; and the reporting of transactional derivatives data according to European Market Infrastructure Regulation (EMIR). These steps lead to using much more granular data, where Big Data approaches are starting to show importance. The resulting growth of data volumes might generate a need for a qualitatively new approach for analyzing the economic and financial system (Varian 2014).

17.4 Big Data Analysis Methods and Instruments in Banking

The volume and variety of Big Data potentially creates many ways to analyze it. However, over time, a set of most appropriate methods for analyzing Big Data has crystallized. For structured data, tools of multidimensional analysis (pivot tables, OLAP in various incarnations) and data mining approaches (rule detection, cluster analysis, machine learning etc.) prevail, while for unstructured data—free text, graphics, audio and video—different approaches are required. Tools like text mining, sentiment analysis, audio and speech analytics, video analytics, social media analytics are designed specifically for unstructured data (Gandoli and Haider 2015). Many definitions of Big Data state the inadequacy of traditional processing methods to deal with it either because of problems in algorithm scaling or need for different approaches (Kaisler et al. 2013; Deutsche Bank 2014). The recent emergence of non-relational databases like NoSQL or MongoDB has provided assistance in dealing with heterogeneous data collections.

The uneven sophistication of analytical solutions, as well as their uneven expected business value, has prompted researchers and industry analysts to relate sophistication to business value. In some sources (e.g., Everest Group Report 2014), the sophistication dimension reflects the suggested maturity of solutions and covers four stages—reporting, descriptive analytics, predictive analytics, and prescriptive analytics. While most analytical solutions for today lie in the area of reporting and descriptive analytics, with business value that is considered to be below average, it is projected that emerging areas of predictive and prescriptive analytics will have a considerably higher business impact.

The staging model raises several issues. The very separation of various informing modes as distinct stages with different business value probably aids in understanding the dynamics of development of analytical methods, but it neglects the ways these methods work together, supporting different aspects of informing. As well, there should be differences in reliability of methods—reporting and descriptive analytics may be simple to use and understand, but they can assist in detecting important issues, and their reliability leaves little doubt, while the reliability of advanced methods is often complicated to verify, mostly because of their complexity. The attempts to "softwarize" complex analytical functions like predictive and prescriptive analytics are another issue under discussion. The implied automation of complex information functions like insight-building and decision support leaves doubts whether in non-standard situations such solution would be feasible; there are many critical voices over the IT capability to automatically reason, present meaningful queries or interpret results. The more advanced predictive and prescriptive analytics use sophisticated methods, use or generate rules where the presence of human analysts is recognized by most sources, especially in practitioners' comments.

Several examples of attempts to automate the fulfillment of complex information needs are a set of technologies called Complex Event Processing, or CEP (Luckham and Frasca 1998; Hasan et al. 2012), Meaning-Based Computing, or MBC (Huwe 2011), optimizing real-time business intelligence (Walker 2009). If we look closer at, for example, the case of CEP, to the author's opinion, several important problems come up that might impair its adoption (Skyrius 2015):

- The combinations of signals are related to the business rules or knowledge structures in expert systems. Known combinations are based on available experience; unknown combinations are hard to evaluate, and a "black swan" problem arises—how to recognize a new important phenomenon.
- Does the performance of a system intended to recognize important signal combinations depend on the repetition ratio of the combinations?
- How the multiple and often overlapping signal combinations are going to be interpreted? Is it the task of people or software?
- If the number of events and signals is rather high, on what principles the volume of information presented to the users will be reduced? Is it going to be grouped, aggregated, or distilled into some initial sense?

17.4.1 Big Data Processing Methods: Data Mining, Text Mining, Machine Learning, Visualization

The principal goal of Big Data processing is refining and extraction of important information, or discovery hidden meanings, leading to important information. Gandomi and Haider (2015) present a collection of Big Data methods and techniques both for structured and unstructured data. For structured data, the techniques are categorized into two groups, based primarily on statistical methods: regression techniques and machine learning techniques. Although historically the banks have been dealing mostly with structured data collections, a recent rising interest in using unstructured data sources attracts more attention to appropriate methods and techniques. For unstructured data, different classes of data require different techniques:

1. *Text analytics*:

   – *Information extraction*, capturing structured data in unstructured text by *entity recognition* and *relation extraction*; recognized types of entities are "loaded" with semantic relationships between them;

- *Text summarization*, producing summaries of documents and other text collections by using *extractive summarization*, based on text unit (word) statistics, and *abstractive summarization*, using NLP (Natural Language Processing) and extracting semantics from the text;
- *Question answering* (e.g., Apple Siri, IBM Watson) that uses complex NLP techniques to extract question sense and formulate or select best possible answers;
- *Sentiment analysis* that evaluates polarity of a sentiment in a document or statement; sentiment analytics are typically associated with public sources and free-text format, although they as well may use internal data of past consumer feedback and written messages.

2. *Audio analytics*:
   - *LVCSR* (Large Vocabulary Continuous Speech Recognition) systems that use automatic speech recognition, also known as speech-to-text, and predefined dictionaries;
   - *Phonetics-based systems* that decode sounds or phonemes and formulate sequences of decoded phonemes.
3. *Video analytics*: although still in its immature phase, they are more and more used in surveillance, monitoring customer flow, studying buying behaviors; both analysis of video content and use of meta information about the video content are used.
4. *Social media analytics*: uses repositories of various content—social networks, blogs, microblogs, social news and syndication, media sharing, wikis and other; the subject of analysis is content (mostly user-created) as well as structure (relations between participants of a virtual platform); the analysis of structure may produce *social graphs* that reflect only connections, and *activity graphs* that reflect the intensity of exchange.

In can be noted that for all types of unstructured data analysis, eventually some kind of structured data is produced, and this makes possible to apply known algorithms for processing and discovery.

Another important Big Data processing aspect deals with a relation "theory-empirics". For banks, until recent times the dominant approach to analysis has been of deductive nature, starting from a general theory or assumption, and using empirical data to test it. With data volumes growing, an alternative approach—induction—starts from data and then attempts to discover general dependencies that eventually could be elevated to the status of theory, or universal rules at least. In practice, the two exploratory approaches are often mixed, and a third form called abduction has emerged (Bholat 2015). In this case, discovered patterns in the data define the path for the best explanation without looking for generalized theoretical claims. The widely used term "data-driven" has its roots in induction approaches; however, as stated before, many situations that require Big Data analysis are of a "question-driven" kind, or have a business question emerging in the first place.

17.4.1.1 Visualization

As one of the essential features of BD technologies is reduction of data volumes to comprehensible sets of aggregated information, visualization is one of the key tools in presenting these sets, managing attention and extending analytical capabilities. An offshoot of the Big Data emergence is advanced data visualization with ability to present huge, complex data sets in ways that can be read by non-experts. The goal of visualization can be defined as to join together business questions or assumptions, analysis algorithms and consistent visual representation to effectively manage and stimulate user attention. As complex discovery algorithms push the discovery logic away from the users, visualization aids the user perception of analysis process and results (Fayad et al. 2002). As well as data analysis itself, visualization can serve exploratory needs (without a hypothesis, assumption or clear question) or confirmatory needs (having a certain question or assumption in mind). It is unclear whether banking activities draw some specific requirements to visualization; the literature analysis did not reveal such specifics, and the existing cases of visualization application in banking show the use of the same tools used elsewhere.

A certain business situation might require combining different methods, tools and approaches. For example, anti-money laundering activities might use the following methods:

- Text Analytics: The capability to extract data from text files in an automated fashion can unlock a massive amount of data that can be used for transaction monitoring.
- Web analytics and Web-crawling: These tools can systematically scan the Web to review shipment and custom details and compare them against corresponding documentation.
- Unit price analysis: This statistic-driven approach uses publicly available data and algorithms to detect if unit prices exceed or fall far below global and regional established thresholds.
- Unit weight analysis: This technique involves searching for instances where money launderers are attempting to transfer value by overstating or understating the quantity of goods shipped relative to payments.
- Network (relationship) analysis of trade partners and ports: Enterprise analytics software tools can identify hidden relationships in data between trade partners and ports, and between other participants in the trade life cycle. They can also identify potential shell companies or outlier activity.
- International trade and country profiling analysis: An analysis of publicly available data may establish profiles of the types of goods that specific countries import and export, flagging outliers that might indicate trade-based money laundering activity.

17.4.2 Analytical Platforms

According to a study by Ziff-Davis (2016), the important attributes of a modern robust analytics platform contain:

- *Agility*—providing the users with required tools to perform independent analytics without the assistance of IT function
- *Intelligence*—built-in automation to support users' guidance towards required results
- *Mobility*—modern analysis tools have to support contemporary mobile devices in a consistent and transparent way
- *Trust*—with the multitude of data sources of uneven provenance, governance of data sources is expected to provide an adequate level of confidence and data quality

Apache Hadoop is one of the widest-known software platforms that support distributed applications for large data volumes. Although, as said before, banks traditionally base their information activities mostly on internal data, the rising interest for making use of external data creates an important issue for banks to organize data from different heterogeneous sources. Big data tools and methods help to integrate big amounts of data—structured and unstructured, using different data sources together. The core of the solution for this issue is to use big data hub or so called Hadoop cluster, which includes Apache Hadoop, HDFS file system and software to organize cluster and arm users with different tools for convenient use of data. Introduction of an enterprise data hub build on Apache Hadoop at the core of information architecture promotes the centralization of all data, in all formats, available to all business users. It mains full fidelity and security on lower capital expenditure compared to traditional data management technologies.

The enterprise data hub serves as a flexible repository to land all of an organization's unknown-value data, whether for compliance purposes, for advancement of core business processes like customer segmentation and investment modeling, or for more sophisticated applications such as real-time anomaly detection. It speeds up business intelligence reporting and analytics to deliver markedly better throughput on key service-level agreements.

One of the most important tasks, which could be solved applying big data hub to bank infrastructure, is the consolidation of many different data types, from many different sources, into a single, central, active repository. Due to Hadoop architecture, this repository is horizontally scalable and could change its size by adding or removing machines to cluster and these machines don't have special requirements. Other key features are accessibility and continuity of data in the hub.

In many organizations, a lack of a unified view of their information resources is created by the variety of different systems to support their diverse data-driven goals—e.g., data warehouse for operations reporting, storage systems to keep data available and safe, specialized massively parallel databases for large-scale analytics, archives for cost-effective backup, and systems for finding and exploring information with the ease of a web search engine. Banks, like other businesses, take advantage of new channels and technologies (for example, mobile) and external data sources, especially on customer analytics (marketing research). An open, integrated approach to unifying these systems around a central data hub allows each to share data more easily among the others for analyses that span formats, structures, and lines of business. As well, data hub usage helps to serve more data than previously was possible. Hadoop complements traditional architecture with a high-visibility, integrated into single view of all data, and a Hadoop data hub includes solutions for:

- Security and governance
- Resource isolation and Management
- Chargeback and showback capabilities

A Hadoop big data hub consists not only from Hadoop itself and HDFS file system as storage, security and administration components, such as Yarn, Cloudera manager and Cloudera navigator. In addition, it includes tools not only for data processing or batching, but for data search, tools for analytic, statistics, for machine learning applications and others. Hadoop hub or cluster applies MapReduce programming model, which gives the opportunity to work with data parts in parallel on all cluster machine simultaneously. Table 17.2 presents a list of Enterprise Data Hub technologies, listed by Cloudera in a 2015 white paper (Cloudera 2015).

MapReduce mechanism on Hadoop cluster gives the opportunity to run task on all cluster machines in parallel. In this way, a wide range of tasks—data manipulation, statistics, analytics, search, distributed pattern-based searching, distributed sorting, graph algorithms, inverted index construction, document clustering, machine learning, statistical machine translation and others could be solved. However, all these tasks should be developed using MapReduce specifics. Moreover, data mining methods can be used this way—for example, the well-known K-Means algorithms are successfully ported to MapReduce model.

On top of Hadoop, HDFS and MapReduce framework there is a set of high level applications that simplify data manipulating for non-programmers and non-IT people. They have been designed for analytics, and some of this software extends Hadoop and MapReduce opportunities. Enterprise data hub offers Impala and Apache Spark—the next-generation, open-source processing engine that combines batch, streaming, and interactive analytics on all the data in HDFS using in-memory capabilities—fully integrated with the storage and applications layers of existing data infrastructure to provide fast, complete transformation, calculation, and reporting at a fraction of the cost.

Apache HBase—Hadoop's distributed, scalable, NoSQL database for big data—provides real-time storage of massive data and more descriptive data to enable analysis at much greater scale.

Tools like Apache Mahout provide the opportunity to use a variety of machine learning algorithms: Collaborative Filtering, Classification, Clustering, Dimensionality Reduction and others. It also makes possible to use custom algorithms.

Table 17.2 Technologies included in an enterprise data hub

Name	Description
Apache Hadoop	An open-source software framework for storage and large-scale processing of large data sets on clusters of industry-standard hardware
HDFS	The distributed file system and primary storage layer for Hadoop
MapReduce	The batch processing engine in Hadoop
Cloudera Manager	The first and most sophisticated management application for Hadoop and the enterprise data hub
Cloudera Navigator	The first fully integrated data security and governance application for Hadoop-based systems, providing full discoverability, lineage, and encryption with key management
Cloudera Impala	Hadoop's massively-parallel-processing SQL query engine
Apache Spark	The next-generation, open-source processing engine that combines batch, streaming, and interactive analytics on all the data in HDFS via in-memory capabilities
Cloudera Search	The full-text, interactive search and scalable, flexible indexing component of Hadoop
YARN	A Hadoop-sub-project that allows resource dynamism across multiple processing frameworks
Apache Sentry	The open-source, role-based access control system for Hadoop
Apache HBase	Hadoop's distributed, scalable, NoSQL database for big data
Apache Flume	Hadoop's service for efficiently collecting, aggregating, and moving large amounts of log data
Apache Hive	Open-source software that makes transformation and analysis of complex, multi-structured data scalable in Hadoop

Source: Cloudera (2015)

Apache Spark is a fast and general engine for large-scale data processing. This tool could be used on Hadoop HDFS and other data storage environments like Cassandra, HBase, and S3. A useful feature of Spark is that it also can work on structured data, using Spark SQL module. It's MLib library for machine learning algorithms contains logistic regression and linear support vector machine (SVM), classification and regression tree, random forest and gradient-boosted trees, recommendation via alternating least squares (ALS), clustering via k-means, bisecting k-means, Gaussian mixtures (GMM), and power iteration clustering, topic modeling via latent Dirichlet allocation (LDA), survival analysis via accelerated failure time model, singular value decomposition (SVD) and QR decomposition, principal component analysis (PCA), linear regression with L_1, L_2, and elastic-net regularization, isotonic regression, multinomial/binomial naive Bayes, frequent item set mining via FP-growth and association rules, sequential pattern mining via PrefixSpan, summary statistics and hypothesis testing, feature transformations, model evaluation and hyper-parameter tuning. As well, Spark contains two libraries for stream data and data stored in graphs.

The numerous successful applications of Hadoop hub include fraud detection, risk management, contact center efficiency optimization, customer segmentation for

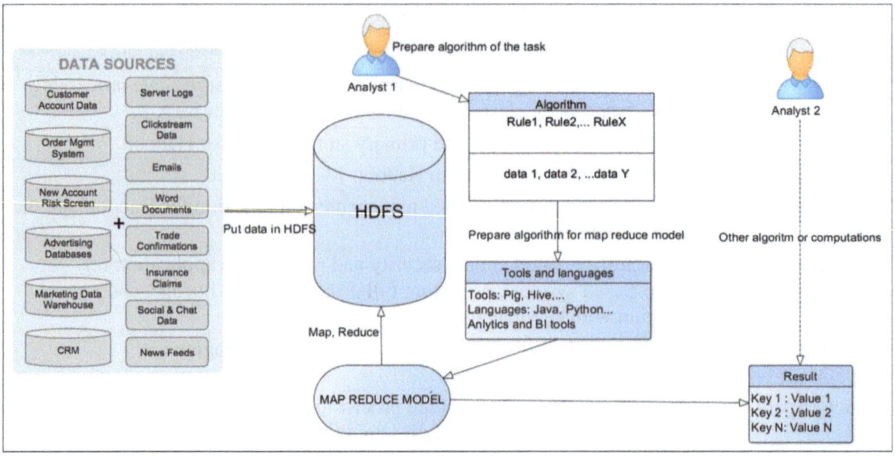

Fig. 17.1 A generic discovery process using Hadoop technology. Source: authors

optimized offers, customer churn analysis, sentiment analysis, customer experience analytics. A generic discovery process using Hadoop technology is presented in Fig. 17.1, where a set of data sources is loaded into Hadoop file system (HDFS) and then submitted to a set of rules in an algorithm presented by an analyst to answer certain business questions. Using software tools (Pig, Hive) and/or languages (Java, Python), the algorithm is "translated" for interpretation and execution in a MapReduce model; the result is a set of values.

Below we present an example set of procedures to be used in financial institutions for fraud detection.

17.4.3 Examples of Fraud Detection

In order to detect financial crime, a range of various techniques is used. However, the core of any system for crime detection is constructed around behavioural profiling. Detection of unusual account activity becomes possible by profiling and tracking individual account behaviour—from initial client onboarding, to the monitoring of transactions and management of customers. The most accurate results are achieved through a combination of behavioural profiling, scenarios of real-time detection, as well as predictive analytics. By using Big Data, financial institutions are now able to provide such services on a scale that simply wasn't possible several years back. In contrast to point solutions that focus only on a single step in the entire process or provide a simple score, financial institutions integrate multiple different capabilities in order to enable to take on fraud throughout the whole claim lifecycle (see Fig. 17.2).

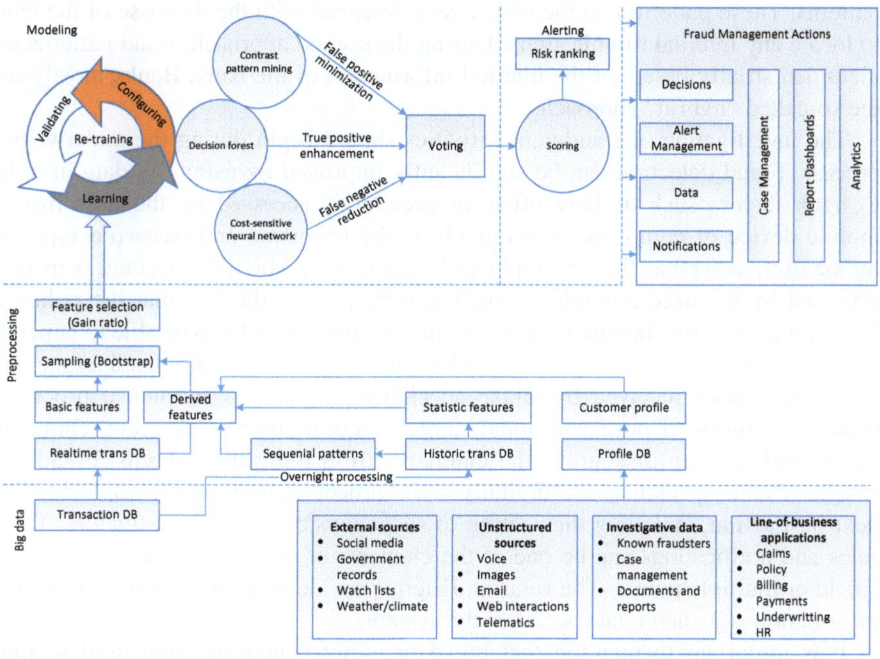

Fig. 17.2 Fraud investigation and evaluation process. Source: authors, based on Wei et al. (2013) and Chen et al. (2012)

It should be noted that all fraud investigation systems used in financial institutions consist of four tiers: database, data pre-processing, modelling, as well as alerting, based on the mining process. Data resources are located by the database tier which then connects them in order to retrieve any related data. Relevant data on the detection of fraud related to online banking is collected from heterogeneous data sources which include historical transaction data, investigative data, line-business applications, as well as external and unstructured sources.

The pre-processing tier controls the accumulation of real-time transactions, maintenance of historical data, data preparation for model training and prediction. In addition to the latter, it is also in charge of the basic feature and deriving feature selection function. The data pre-processing stage includes two major tasks: feature selection and data sampling. The sampling process is necessary prior to the application of any data mining models, due to the fact that data on online banking transactions is highly imbalanced. An increasing concern for many companies is the ability to detect fraudulent actions, and, by using big data analysis, much more fraudulent actions can be detected and reported.

In order to detect fraud patterns, financial institutions developed two different approaches. During the first approach, third party data warehouse (that could contain information on transactions carried out by a number of companies) is tapped by the bank which then uses the analytics programs of big data in order to identify the fraud patterns. These patterns can then be cross-referenced with the database of the bank to locate any internal trouble signs. During the second approach, fraud patterns are identified strictly based on the internal information of the bank. Banks mostly use the so-called "hybrid" approach.

The investigation of fraudulent activities should begin during the underwriting process. Fraud detection can be significantly improved by using big data in order to track factors such as how often an account is accessed by the user from a mobile device or computer, how quickly is the username and password typed in by the user, as well as the geographical location from which the account is usually accessed by the user. Attention should also be paid to the fact that the policy is being purchased for fraudulent purposes in right time in order to be able to generate warnings and notifications to the underwriters, agents or automated processes. Automated alerts received by intake specialists, as well as automated processes make the ability to pose additional, targeted questions to suspicious claimants easier, and this can discourage the claimants from filing their claims. Fraud risk insights enable specialists and digitally based processes to promptly implement the next-best action, including the routing of suspicious claims to investigators. These rules and notifications can be one of the elements of the claim intake process at a single or multiple points. The rules and alerts provide support to responses as new information is gathered and accesses the system.

It is important to mention that big data is not a new measure used to fight fraudulent activities—financial institutions have been using various data sources for quite a while in order to evaluate fraud risk, control correct connections to bank accounts and similarly use not only internal, but also external data. External data sources are particularly essential as they make fraud attempts everywhere visible to the bank. When external sources identify a known fraud threat, the bank must learn about this threat before it comes knocking at the bank's door. When a score, reason code or a flag are used in the course of fraud detection and prevention processes by the bank personnel or bank systems, the latter are probably already leveraging big data coming from a large billion-record database.

Fraud techniques and methods change constantly; therefore, as stated by Smith and Awad (2015), it is necessary for financial institutions to continuously upgrade their tools used for stopping fraud without obstructing the flow of account applications and transactions that are considered to be "good". The stage during which a transaction will be flagged by fraud prevention and detection systems of institutions as a potential fraud depends on the involved individual's or organization's risk profile, and the defences that the financial institutions have in place at any point in time must be flexible enough in order to respond to the evolving financial crime typologies. For instance, it sends the bank a notification when a transaction from a device is carried out for the first time by the credit card holder. However, if several transactions are carried out from different devices in the course of a day, then the generated data is a reasonable justification to be suspicious and raise an alarm. Valid card holders are usually immediately notified by banks and, in some cases, online transactions may even be blocked. Big data makes it easier to detect unusual

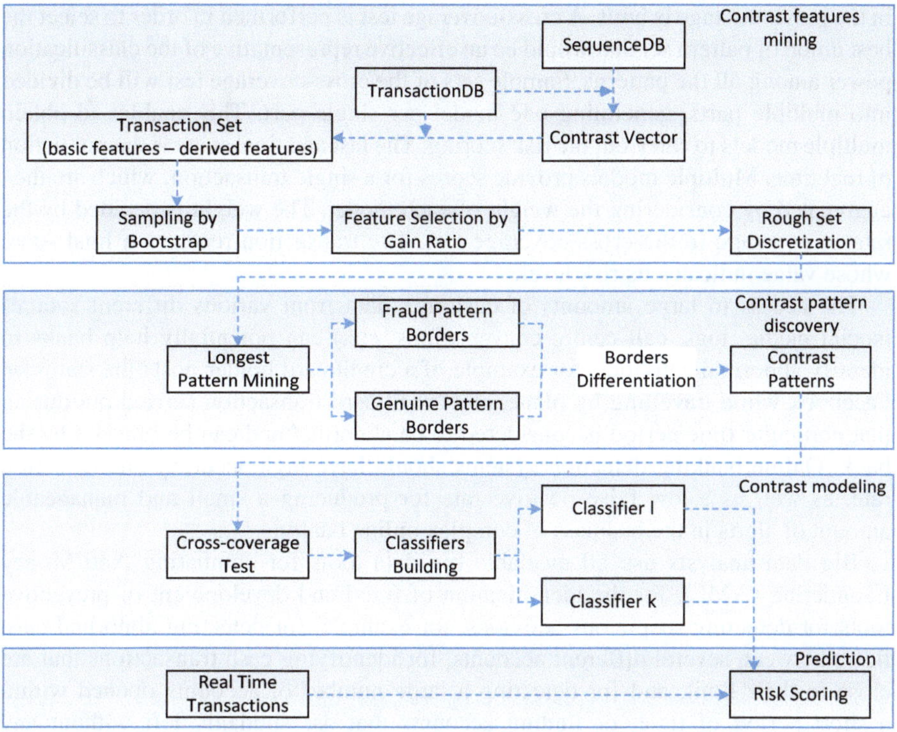

Fig. 17.3 The basic model for identifying contrast patterns in banking. Source: Wei et al. (2013)

transactions. For instance, the bank will receive red flags if the same credit card is used to carry out two transactions at different cities within a short period of time (see Fig. 17.3).

The system described by Wei et al. (2013) is comprised of four stages: contrast feature mining, contrast pattern discovery, multiple method-based contrast modelling and risk scoring by contrast. Stage one of the system focuses mainly on the pre-processing of data. First of all, banking transactions of customers are used to form a sequence for each customer, which is then used for generating contrast vector per each transaction. Afterwards, basic features and contrast vector (derived features) are combined to form the set of raw features. In order to obtain the training set, bootstrap sampling is adopted for carrying out under-sampling. Then an information gain ratio is used for the selection of significant features whose weights are higher than the indicated threshold. And finally, a rough set for the discretization of the selected attributes is introduced. Stage two is the fundamental step to mine contrast patterns. Firstly, pattern borders are created for both fraud and genuine sets by using the border-setting algorithm Max-miner (Bayardo 1988). Processes of border-differentiation and validation are carried out to continuously output the contract patterns. During stage three, a model group powered by the pattern output in the previous stage is built. A cross-coverage test is performed in order to select the best union of pattern set that would be an effective representative of the classification power among all the patterns. Sample sets of the cross-coverage test will be divided into multiple parts, generating one model per single part. This enables to obtain multiple models to carry out the risk scoring. The last and final stage is the prediction of real time. Multiple models provide scores for a single transaction, which are then aggregated by considering the weight of each model. The weight is decided by the rate of coverage in the cross-coverage test. The transaction receives a final score whose value indicates its risk level.

The access to large amounts of customer data from various different sources (social media, logs, call centre conversations, etc.) can potentially help banks to identify abnormal activities, for example, if a credit card holder posts the status on Facebook while travelling by plane, any credit card transaction carried out during that particular time period is considered to be abnormal and can be blocked by the bank. Online banking detection systems should have high accuracy and detection rate, as well as a low false positive rate for producing a small and manageable amount of alerts in the business of complex online banking.

Big data analysts use all available big data tools for combating Anti-Money Laundering (AML), for the identification of fraud and development of preventive tools for detecting suspicious activities, for example, for detecting abnormal cash flows between several different accounts, for identifying cash transactions that are close to their limit, and for detecting a large number of accounts opened within a short period of time, or finding accounts that are suddenly left without any transactions. Thus, most banks use data modelling and neural network interaction in order to detect fraud and to adhere to AML guidelines. AML specialists check documents filled by other bank employees who work directly with customers. The aim of Anti-Money Laundering specialists is to select the necessary information from the entire big data flow. Banks usually assign communication managers for each company, who identify whether the company is active and prepared for analysis. If there is lack of data, inspections are carried out whether the company is active, "frozen", bankrupt or suspended. Also, in addition to the inspection of big data, documents are also checked by reviewing their mutual compatibility. If it is determined that the company halted its operational activities and is no longer active, such information is marked and the company is no longer inspected for the purposes of anti-money laundering prevention (see Fig. 17.4).

Specialists check all the information, whether the information and documents are in order and whether big data provides all the necessary information in order to evaluate if the company's cash flows from and into the bank account reflect the business model and if the company is able to carry out illegal activities. The aim is to check each company, have all the necessary information and make sure that this is reflected in the big data database when representatives of the state supervisory authorities perform inspections on whether the bank is carrying out anti-money laundering prevention. However, in order to reach the desired result, data of a single company is checked by an average of four specialists at once. It takes 5 min to an hour and a half to inspect the big data database of one company, depending on the company's size and its activities.

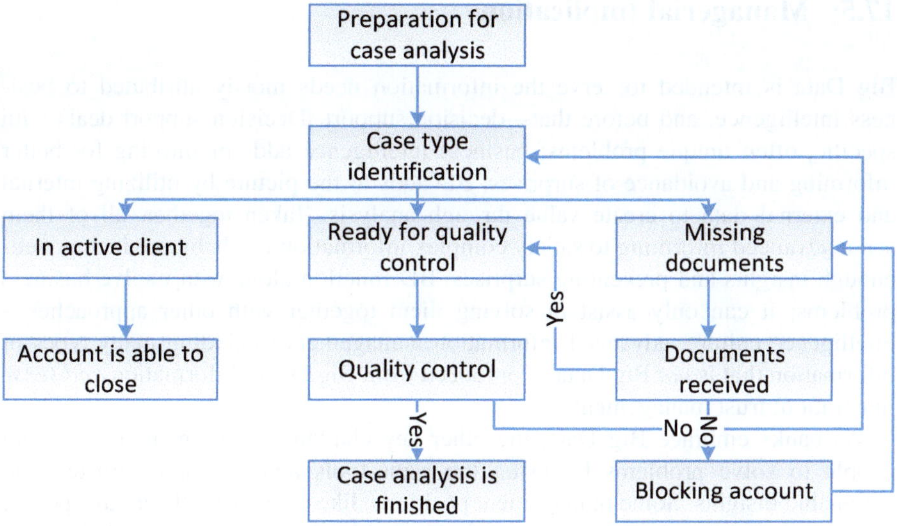

Fig. 17.4 Big data collection for AML prevention. Source: authors, based on Agu et al. (2016)

It should be emphasized that the inspection process is the same or very similar for all the companies. Low-risk companies require much less time to inspect their big data compared to average or high-risk companies, therefore companies are divided according to their risk level. Each company has individual risk signs. For example, a company that sells coffee and carries out transactions with Sierra Leone or often withdraws large amounts of cash from its bank account will look suspicious and will be immediately noticed by a big data analyst. However, such circumstances will not automatically mean that the company is carrying out money laundering, if the company declares that its main supplier is registered in that country. In this case, the company would be classified as high-risk; however, transactions in its bank account would not be frozen. Thus, it can be stated that each business area has its own risk signs; however, some of them are generic, such as money transfers from/to high-risk countries, especially tax havens, large amounts of cash deposited to or withdrawn from bank accounts, excessively high bank account turnover compared to the company's activities, etc. Big data must be sufficiently large, as the main areas that anti-money laundering prevention specialists focus on are supplier allocation, and the nature of the purchased products or raw materials (there is a risk of transactions between companies who have the same owner, between companies that are very different, etc.). As mentioned previously, transactions of large amounts of cash are most often thought to be illegal business, as it is quite hard to trace cash movements. Possible risk signs are encountered continuously and errors are often sent to big data analysts.

17.5 Managerial Implications

Big Data is intended to serve the information needs mostly attributed to business intelligence, and before that—decision support. Decision support deals with specific, often unique problems; business intelligence adds monitoring for better informing and avoidance of surprises. BD adds to the picture by utilizing internal and external data to create value through analysis. Taken together, all of them serve advanced informing to satisfy complex information needs by producing well-enough insights and preventing surprises. BD function alone won't solve business problems; it can only assist in solving them together with other approaches—intelligence culture, advanced information management (including many types of information that is *not* Big Data or produced from Big Data); information *and sense* integration; trust management.

As banks embrace Big Data, the other key challenges are engaging the right people to solve problems by using the right tools and producing trusted and actionable insights. Some management positions, like chief risk officer, are specific to banking and largely defined by analytics; evidently, a permanent collaborative dialogue between the business users and the data scientists is required.

The discussions on the competencies required to effectively extract value from Big Data analytics predominantly accentuate technical and analytical skills, while the ability to ask well-pointed business questions is seldom considered a prime competency. Meanwhile, analytical techniques help in finding patterns and relationships, but the causality has to be explained by smart people. Some sources (Fuller 2015) argue that skills required by data scientists for BD analysis are not exclusively within the formal domain, and hermeneutical skills are going to have an important role.

The concept of BD embraces powerful technology innovations that work in tandem with (or feed) organization's intelligence network that joins together people whose main function is to produce insights. Importance of human issues is stressed by many sources dealing with BD adoption and value creation (CapGemini, p. 5; Fuller 2015; Challenges and opportunities white paper, p. 11; . . .). Analytical or, in a broader sense—intelligence culture is a unifying term for human issues affecting all kinds of analytical innovations. We can suggest a framework for defining intelligence culture from earlier work (Skyrius et al. 2016), which essentially ties together the features proposed by many sources:

- *Cross-functional intelligence activity*: decentralized and horizontal nature without functional borders.
- *Synergy of virtual teamwork*: shared information, insights, mental models; the permeating and participative nature of intelligence community.
- *Lessons* as experience and proof of value; lessons from earlier success stories as well as failures, mistakes; decision-making best practices.
- *Intelligence community* that is motivated, sustainable and growing; self-reinforcing installed base—when users contribute, the user base and the value of system increases; role of change agents and intelligence leadership in an organization.

- *Balance of centralized and decentralized* intelligence conventions and functions; expandable with universal standards, preventing eventual lock-in.
- *Technology management*: simple and useful tools whose benefits (faster data access, easier analytical functions) have been communicated clearly from the beginning; the IT platform stimulates use by being simple, shared and open; easy feed, exchange and use of information and insights. Although BD uses advanced complex technology, a Forrester study (Hopperman and Bennett 2014) states that "financial institutions require solutions that are cost-effective, easy to manage, and highly flexible ...".

An IDC Big Data White paper (IDC 2013) states that in setting the analytics strategy, an enterprise-wide approach to handle Big Data is required, and the key role goes to the business management, while IT is less involved. The creation of inter-silo channels for insight sharing will require some balancing of interests, and companies that have found out how to communicate more transparently without destroying the practical advantages of silos and departments will have a competitive advantage. This point is also supported by Katherine Burger in Banks, Big Data ... (SAS Institute Inc. & Bank Systems & Technology 2012).

A Ziff-Davis white paper (Ziff-Davis 2016) presents an example how insights are gained internally by crossing functional boundaries. A dip in revenues, detected by finance department, is most likely related to a drop of sales, which gets the marketing department to look into their side of business and leads to a discovery of rising customer churn. In its own turn, the key drivers for churn are pricing and customer satisfaction, so a predictive model for churn propensity is developed. The analytic initiative shifts to customer service department, who use the model to identify the most likely candidates for churn. Further operationalization of the developed model into a contact center application allows customized incentive offers at the time of contact, taming the churn rise. The model, together with data on its actual impact, may be further operationalized with new indicators on churn and customer satisfaction and their relation to sales and accounting data. Thus, the analytics chain has completed a full circle, solving a business problem, gaining new insights and upgrading the intelligence activities. It has to be noted here that the ongoing addition of new indicators to the monitoring environment, if unmanaged, might produce an information overload if not contained in manageable boundaries that prevent attention fragmentation.

Espinosa and Armour (2016) present a coordination framework for Big Data analytics, where they describe a "self-fueling" dynamic cycle of coordination effectiveness, and state that "successful organizations were able ignite this cycle and achieve continuous improvement over time". To our opinion, this approach largely overlaps with intelligence culture regarding creation and self-sustainability of intelligence community.

17.6 Conclusions and Implications for Further Research

The potential and use of Big Data in banking has features that are common to most, if not all businesses: efficient internal operations, cost control, product and service management; customer data use by CRM and other related environments, to name a few. Proper Big Data analytics, like all intelligence and analytic activities, should assist managers and decision makers in concentrating their attention on key issues, instead of scattering this attention over many irrelevant issues.

The specific features of banking activities, however, lead to specific areas for using Big Data approaches, and risk management in all forms dominates in this aspect. The permanently morphing risks require flexible analytics leading to emergence of requirements for BDA agility and intelligence culture. On the other hand, the use of predominantly reliable internal sources of data and information in banks, together with reliable time-tested IT platforms is one of the foundation factors for reliable analytics.

Engaging in Big Data should weigh the risks of not doing Big Data analytics versus the risks of doing it. Additional research is required to develop criteria and metrics for supporting decisions regarding the adoption of BDA methods and techniques. Return-on-investment approaches may be of limited use, so the evaluation of possible benefits might be expected to move towards estimating analytical agility and adoption of intelligence culture.

This work is based mostly on literature research and professional experiences in dealing with information activities and challenges in banking. The statements and conclusions presented in this chapter are of inductive nature; however, empirical research is needed to test at least the most important of the statements. The directions of empirical research should include, but be not limited to, the following:

- Research on sources and factors that influence analytical agility to adapt to changing circumstances and maintain competencies for quality insights.
- Research on factors building and strengthening cross-functional intelligence and analytical culture.
- Research on data and information quality, encompassing heterogeneous information integration and sense management.

References

Agu, B. O., Enugu, U. N., & Onwuka, O. (2016). Combating money laundering and terrorist financing—The Nigerian experience. *International Journal of Business and Law Research, 4*(1), 29–38.

Bank for International Settlements. (2012). *Core principles for effective banking supervision.* Basel: Bank for International Settlements.

Bayardo, R. J. (1998). Efficiently mining long patterns from databases. In *Proceedings of the 1998 ACM-SIGMOD International Conference on Management of Data*, (pp. 85–93).

Bholat, D. (2015). Big Data and central banks. *Big Data and Society*, January-June 2015, 1–6.

Cai, L., & Zhu, Y. (2015). *The challenges of data quality and data quality assessment in the big data era.* Interactive: http://datascience.codata.org/articles/10.5334/dsj-2015-002/print/

Cap Gemini Consulting. (2014). *Big Data alchemy: How can banks maximize the value of their customer data?* Cap Gemini Consulting.

Chen, H., Chiang, R. H., & Storey, V. C. (2012). Business intelligence and analytics: From big data to big impact. *MIS Quarterly, 36*(4), 1165–1188.

Cloudera (2015). *Information-Driven Financial Services, Big Data, and the Enterprise Data Hub*. Cloudera White paper. Version: Q115-102.

Daruvala T. (2013) *How advanced analytics are redefining banking*. Interview, April 2013. Interactive: http://www.mckinsey.com/business-functions/business-technology/our-insights/how-advanced-analytics-are-redefining-banking

Davenport T., & Dyche J. (2013). *Big Data in big companies*. International Institute for Analytics.

Deloitte. (2016). *Deloitte analytics: Banking*. Interactive: http://www2.deloitte.com/us/en/pages/deloitte-analytics/articles/deloitte-analytics-banking.html

Deutsche Bank (2014). *Big Data: How it can become a differentiator*. Deutsche Bank white paper. Interactive: http://www.cib.db.com/insights-and-initiatives/flow/35187.htm

Dyche, J. (2004). *The bottom line on bad customer data*. Sherman Oaks, CA: Baseline Consulting Group, Inc.

Espinosa, J. A., & Armour, F. (2016). The big data analytics gold rush: A research framework for coordination and governance. In *Proceedings of the 49th Hawaii International Conference on Systems Sciences* (pp. 1112–1121).

European Banking Authority. (2014) *EU-wide stress test: Frequently asked questions*. European Banking Authority.

Everest Group Report (2014). *Analytics in banking: Conquering the challenges posed by data integration, technology infrastructure, and right talent to operationalize analytics in banking*. Interactive: http://www.genpact.com/docs/default-source/resource-/analytics-in-banking

Evry Innovation Lab. (n.d.). *Big data for banking for marketers*. Evry Innovation Lab white paper.

Fayad U., Wierse A., & Grinstein G. (Eds.). (2002). *Information visualization in data mining and knowledge discovery*. San Francisco, CA: Morgan Kaufmann.

Fuller, M. (2015). Big data: New science, new challenges, new dialogical opportunities. *Zygon, 50*(3), 569–582.

Gandomi, A., & Haider, M. (2015). Beyond the hype: Big data concepts, methods, and analytics. *International Journal of Information Management, 35*, 137–144.

Hasan, S., O'Riain, S., & Curry, E (2012). Approximate semantic matching of heterogeneous events. In *Proceedings of DEBS'12 conference*, July 16–20, 2012, Berlin, Germany.

Hashem, I. A. T., Yaqoob, I., Anuar, N. B., Mokhtar, S., Gani, A., & Khan, S. U. (2015). The rise of "big data" on cloud computing: Reivew and open research issues. *Information Systems, 47*, 98–115.

Heskett J. (2012). How will the "Age of Big Data" affect management? HBS Working Knowledge. Interactive: http://hbswk.hbs.edu/item/how-will-the-age-of-big-data-affect-management

Hoorn, J.F., & van Wijngaarden, T.D. (2010) Web intelligence for the assessment of information quality: Credibility, correctness, and readability. In Zeeshan-Ul-Hassan Usmani (Ed.), *Web intelligence and intelligent agents*. Rijeka: InTech.

Hopperman J., & Bennett M. (2014). *Big data in banking: It's time to act*. Forrester Research.

Huwe, T. (2011). Meaning-based computing. *Online, 35*(5), 14–18.

IBM (n.d.). *Enhance your 360-degree view of the customer*. IBM Institute for Business Value e-book. Interactive: https://www-01.ibm.com/software/data/bigdata/use-cases/enhanced360.html#

IBM Institute for Business Value (2012). *Analytics: The real-world use of big data. How innovative enterprises extract value from uncertain data*. Interactive: https://www.ibm.com/smarterplanet/global/files/se__sv_se__intelligence__Analytics_-_The_real-world_use_of_big_data.pdf

IDC. (2013). *Executive summary: Using Big Data and analytics as the ticket to strategic relevance*. IDC White paper, December 2013.

Jacobs, A. (2009). The pathologies of big data. *Communications of the ACM, 52*(8), 36–44.

Kahn, B., Strong, D., & Wang, R. (2002). Information quality benchmarks: product and service performance. *Communications of the ACM, 45*(4), 184–192.

Kaisler S., Armour F., Espinosa J.A, & Money W. (2013) Big data: Issues and challenges moving forward. In *Proceedings of the 46th Hawaii International Conference on System Sciences* (pp. 995–1004).

Luckham, D. C., & Frasca, B (1998). *Complex event processing in distributed systems* (Computer systems laboratory technical report CSL-TR-98-754). Stanford University.

Marr, B. (2015) *Big Data: Now a top management issue. Interactive*: http://www.forbes.com/sites/bernardmarr/2015/11/30/big-data-now-a-top-management-issue/print/

McCoy, K. (2016). Wells Fargo fined $185 M for fake accounts; 5,300 were fired. *USA Today*, September 9, 2016.

O'Brien, S. (2010). Crisis early warning and decision support: Contemporary approaches and thoughts on future research. *International Studies Review, 10*(1), 87–104.

Patwardhan A. (2016). *The force awakens: Big data in banking*. Interactive: https://www.finextra.com/newsarticle/28541/the-force-awakens-big-data-in-banking

PwC (2013). *Where Have You been all my life? How the financial services industry can unlock the value in Big Data*. PwC FS Viewpoint, October 2013.

Ram, S., & Liu, J. (2008). A semiotics framework for analyzing data provenance research. *Journal of Computing Science and Engineering, 2*(3), 221–248.

SAS Institute Inc. & Bank Systems & Technology. (2012). *Banks, Big Data and high-performance analytics*. SAS Institute Inc. with Bank Systems & Technology.

Skyrius, R. (2015). The key dimensions of business intelligence. In K. Nelson (Ed.), *Business intelligence, strategies and ethics* (pp. 27–72). Hauppauge, NY: Nova Science Publishers.

Skyrius, R., Katin, I., Kazimianec, M., Nemitko, S., Rumšas, G., & Žilinskas, R. (2016). Factors driving business intelligence culture. *Issues in Informing Science and Information Technology, 13*, 171–186.

Smith, A., & Awad, G. (2015). *Top 10 takeaways on leveraging big data for fraud mitigation*. Equifax, March 10, 2015.

SurfWatch Labs. (2014). *Big Data, big mess: Sound cyber risk intelligence through "Complete Context"*. Interactive: http://info.surfwatchlabs.com/big-data-security-analytics

The Banker Editorial (2013). *Will Big Data cause big problems for the banking world?* July 29, 2013. Interactive: http://www.thebanker.com/Comment-Profiles/Will-big-data-cause-big-problems-for-the-banking-world?ct=true

Turner, D., Schroeck, M., & Shockley, R. (2013). *Analytics: The real-world use of big data in financial services*. IBM Global Business Services Executive report.

Varian, H. (2014). Big Data: New tricks for econometrics. *Journal of Economic Perspectives, 28*(2), 3–28.

Walker, R. (2009). The evolution and future of business intelligence. *InfoManagement Direct*, September 24. Interactive: http://www.information-management.com/infodirect/2009_140/business_intelligence_bi-10016145-1.html

Wand, Y., & Wang, R. (1996). Anchoring data quality dimensions in ontological foundations. *Communications of the ACM, 39*(11), 86–95.

Wei, W., Li, J., Cao, L., Ou, Y., & Chen, J. (2013). Effective detection of sophisticated online banking fraud on extremely imbalanced data. *World Wide Web, 16*(4), 449–475.

Wells, D. (2008). *Analytical culture—Does it matter?* Interactive: http://www.b-eye-network.com/view/7572

Wilcock, M. (2013). Building an analytical culture. Sweetspot Intelligence, 2013. Interactive: http://www.sweetspotintelligence.com/en/2013/04/15/building-an-analytical-culture/

Ziff-Davis (2016). *Monetize your data with business intelligence*. Ziff-Davis Custom White paper.

Žilinskas R, & Skyrius, R. (2009) Management decision support by using early warning environments. In *Ekonomika. Mokslo darbai (Research Proceedings)* (Vol. 86, pp. 42–54). Vilnius University.

Chapter 18
Marketing Applications Using Big Data

S. Srinivasan

18.1 Introduction

Marketing their products and services is an important activity for a business. Businesses spend enormous sums of money for marketing expenses. With the advancements in technology, today much of the expenditure is spent on electronic media than print media. According to the CMO Council, a global organization of Chief Marketing Officers, marketers around the world spent $1.6 trillion in 2014 and it is expected to rise to $2.1 trillion by 2019. Spending for digital advertising is growing rapidly at the rate of 16.1% in 2014. Many Chief Marketing Officers expect to spend over 75% of their marketing budget on digital marketing (CMO Council 2016). The use of smartphones has grown very rapidly in all the developed countries. According to data from Statista, smartphone use is expected to increase from 1.5 billion units deployed in 2014 to 2.5 billion units deployed by 2019. Marketers have taken note of this widespread availability and they are spending billions of dollars on mobile marketing. Knowing how people use their mobile phones, much of the mobile marketing is focused on Search Engine Marketing (SEM). The expected share of online spending through SEM is 47% which accounts for several billion dollars of revenue for businesses.

The growth in marketing expenditures is global in nature. In developed markets such as North America, Europe and Japan, there is a significant push towards mobile marketing. In ten of the developing markets—Brazil, Russia, India, China, Indonesia, South Korea, Turkey, South Africa, Argentina and Mexico—the growth of digital marketing is expected to be huge. In both markets print media continues to be significant, with nearly 40% of the market share. One reason for this is that people perceive that it takes significant effort to produce print media and get it to

---

S. Srinivasan (✉)
Jesse H. Jones School of Business, Texas Southern University, Houston, TX 77004, USA
e-mail: srinis@tsu.edu

S. Srinivasan (ed.), *Guide to Big Data Applications*, Studies in Big Data 26,
DOI 10.1007/978-3-319-53817-4_18

the potential customer. Unlike a digital medium advertising, print medium keeps the document in sight and it reinforces the saying "out of sight, out of mind." Moreover, technologies such as QR (Quick Response) codes and NFC (Near Field Communications) have made it possible for the user to get more information on the product using the print media by scanning the QR codes. The QR codes require an app to interpret the image captured using a smartphone. On the other hand, NFC enables the user to tap their smartphone against the print media in order to go directly to the website relevant to the advertising. Thus, print media is playing an important role in marketing in spite of the ease of use with digital marketing. Businesses measure the cost of marketing using Cost Per Impression (CPI) when it comes to non-digital marketing. In measuring CPI, marketing items such as pens, shirts and caps cost $0.002 on average whereas online marketing results in a cost of $0.0025 per impression. Thus, print media continues to be a good alternative to catch consumer attention.

The discussion above points to the diversity of means available to reach individuals. However, in order to know who might benefit most from sending the marketing materials, it is essential to have a wealth of data. Such data come from multiple sources such as email, social media postings and website interactions. All these sources must be scanned regularly and timely information accumulated in order to send the right materials to an individual. This is made possible through tools such as Hadoop available to process very large volumes of data.

In this chapter we will highlight the following marketing approaches that benefit from using Big Data. These are: multi-touch attribution, granular audience targeting, forecasting, predictive analytics and content marketing. We will look into these aspects in greater detail in the following sections.

18.2 Multi-Touch Attribution (MTA)

Attribution refers to identifying a source for bringing in a customer to a business. This technique has been practiced by businesses for hundreds of years where the business gathers information from the customer as to what brought them to their business so that they could thank that source. Usually, the attribution results in no more than an acknowledgement of the source. However, from a marketing perspective when one notices repeatedly that a particular source was responsible for introducing a customer to a business, then that source is co-opted as a partner. This gives greater incentive to the source for finding more new customers for the business. This is greatly facilitated by digital marketing where it is easy to track the places a customer visited prior to coming to a business. The initial entrée in this regard occurs digitally through a visit to the business' website. The business is then able to provide all the information that the potential customer needs and eventually convert the visitor to a customer. As is clear from this scenario, the customer might have visited multiple sites. Many businesses pay attention only to the last-touch, meaning the site that the customer visited prior to settling down on a particular business. It is much easier to use the last-touch model because it is easy to measure. But with today's advances in technology we are able to measure all factors that facilitate a consumer in making a purchase decision after reviewing several types of information. With Big Data, it is possible to gather data that would show multi-touch. Once again, the goal here is to see how Big Data could be analyzed to see which among the many sites had a significant impact on a prospective customer (Shao 2011).

MTA is practiced primarily for market engagement. As we saw above, a "touch" is an interaction with an individual. There are three different ways in which MTA is practiced. The first way is to provide equal weight for all touches since no interaction converted the individual into a customer. The second way is time-dependent. In this approach the largest credit is given to a touch that resulted in having a customer. All other touches along the way are assigned decreasing credits. The third approach is described as the U-shaped model, whereby the first and last touches are each given 40% of the credit and the remaining 20% of the credit is divided among all the other interactions the individual had on the way to becoming a customer for a business. Typically, businesses can gather the necessary data from Google Analytics, their own web analytics, CRM software such as Salesforce and the feedback from their own marketing teams. Google Analytics provides Multi-Channel Funnel reports. All these reports directly pertain to customer conversions from web browsing or other forms of referral. The five different reports from Google Analytics are: Overview Report, Assisted Conversions Report, Top Conversion Path Report, Time Lag Report (very useful in that it shows how long it took for a person to become a customer) and Path Length Report (shows how many interactions the customer had prior to becoming one). The following table illustrates the use of Google Analytics in this regard. Note that Google Analytics can record any number of conversion paths that a customer takes (Table 18.1).

Google Analytics will show the above conversion path as follows:

In the above illustration, Item 1 is the first interaction and Item 6 is the last interaction before the user became a customer. Each of the above exposures to product information is known as an 'interaction' in Multi-Channel Funnel reports. What is important here is to note that the customer might take different paths before becoming a customer. It is up to the business to monitor customer activities and keep their interest in the product live.

Table 18.1 Possible steps in conversion path

1. View Display Ad by business
2. Read a blog post via Twitter or Facebook
3. Click on a Paid Search Ad placed by business
4. View Display Ad by business
5. Click on a generic search result
6. Visit directly website of business
7. Purchase the product

Fig. 18.1 Conversion Path stages

In analyzing channel conversion, Google Analytics uses both Default Channel Labels (some of these are shown in Fig. 18.1) or Custom Channel Labels (the unique ones the customer uses). In order to understand customer search patterns for a product, the Custom Channel Labels are divided further into Branded Keywords or Non-Branded Keywords. This information is useful to the business to know so that they could modify their web pages to reflect the non-branded keywords so that they appear in customer searches directly.

In the discussion above we have highlighted some of the benefits of using Google Analytics in order to benefit from the Big Data available there. Many individuals have developed specialized software that will help tune their Google Analytics report to be more useful to them (Kaushik 2010). Much of this is available for users easily. What is important to note here is that it is becoming increasingly important to use multi-touch attribution model to identify all modes the users are using to gather information before they make a purchase decision. This is illustrated by the results of car rental campaign results shown by R. Berman in (Berman 2015) where over 13 million consumers were exposed to over 40 million ads over a two-month period. Consumers were exposed to eight different channels of ads. Based on the analysis of this campaign data, Berman concludes that there was a 4-fold increase in conversion rate because of the campaign. This shows the power of MTA.

18.3 Granular Audience Targeting

Traditionally in marketing, data is gathered about the number of consumers a business was able to attract based on a marketing campaign. In this section we show the power of granular audience targeting based on the availability of consumer preferences gathered from multiple sources. Thus, Big Data becomes a powerful tool in marketing to target the right people for the marketing campaign. The success of any marketing campaign is measured by the number of new customers acquired. Today, consumers have a variety of choices from TV, radio, online sources, social media, print media and recommendations from friends. With the widespread availability of wireless media at a good rate of speed in US, we find that younger people in the age group of 18–24 rely on the social media and recommendations from friends more (Reichheld 2003). Generally, people under the age of 40 tend to use wireless media more. People under the age of 40 are easy to reach by social media. Since people using social media identify their preferences on various issues quite easily, sites such as Facebook, Twitter, LinkedIn, and Instagram are able to capture plenty of data on personal preferences. When these are analyzed it is easy to

Table 18.2 Generations as percentage of the US population

Cohort	Birth year	Age in 2020	Population (in 1000s)	% of Population in 2020
Matures	Prior to 1946	75+	23,173	6.9
Baby Boomer	1946–1954	66–74	75,560	22.6
Generation X	1955–1980	40–65	60,836	18.2
Millennials/Generation Y	1981–2001	19–39	89,792	26.9
Generation Z/iGen	Post 2001	<19	84,537	25.3

Source: CRMTrends.com and US Census Bureau

identify the segment of the population under 40 that is more likely to pay attention to advertising material sent to them on matters of interest. This is granular audience targeting (Table 18.2).

The US population can be grouped as follows:

This data shows that people under the age of 40 will be nearly half of the population and so it makes more sense for advertisers to concentrate on this population. As mentioned earlier, people have many sources that they can tap into in getting information about the products of services that they are interested in at any point. Marketers will be able to succeed better if they can target their ad spending on the population that is more likely to view it carefully (Berry 1997). This is where granular marketing succeeds because an organization has wasted its resources if it sends advertising material to people who ignore it.

Big Data helps in the ability to analyze data from multiple sources by combining them in different ways and finding the right mix of parameters that would fit with the advertising goal of an organization. Thus, Big Data enables the advertisers to use multiple types of media to bring their informational material to the attention of the prospective customer. Here, in combination with today's technology, Big Data helps to personalize the information delivered to an individual through multiple sources such as email, web pages, banner ads and social media communication. This type of personalization of ad materials is a benefit of granular audience targeting. In order to perform this type of granular targeting, the business must have access to granular interest data that will help with audience segmentation. Then they will be able to engage with such a segmented group with interest in their product or service. Facebook allows businesses to import CRM data of their target audiences. Moreover, Facebook enables businesses to segment their target audience based on behavioral, demographic and socio-economic traits. These are all essential steps a business should take in order to target the right group of people. At the same time, if the wrong type of material is sent to these prospects then they become disengaged quickly with their future ads.

Granular audience targeting requires having the information needed to know the people's interests before they are sent material relevant to their interests. It is possible to do this when data from multiple sources is gathered to identify the interests and pitch the right type of products or services to such people. Big Data helps in this regard in analyzing data from multiple sources quickly and developing actionable intelligence. Hadoop is one such tool available for rapid data processing of all related data, both structured and unstructured. Hadoop is the open source implementation by Apache Software of the Google technology MapReduce. This information would help the business place the right advertisement for the right group. In coming with up such information speed is of essence. Hadoop has the ability to process different types of data and Spark is even faster in processing real time data because it has the capability to handle data in fast memory. Once this information is gathered, then the business has to focus on ad placement so that it attracts a large number of audience. Thus, ad placement in a web page is an optimization problem. This reasoning shows the importance of mathematical processing in order to have the right information for ad placement. This is classically illustrated by the use of Big Data by Red Roof Inn to target the right customers. This hotel chain has many of its hotels near major airports. It leveraged data from flight cancellations which average 1–3% daily. This results in nearly 25,000–90,000 stranded passengers daily needing overnight accommodation. Also, it gathered weather information and combined it with hotel and airport information to know passengers at which airports would need hotel accommodation. It was able to use social media data to know who is at which airport and target them with an ad for overnight stay. Since this is timely from the user perspective, Red Roof Inn was able to benefit. Because of this granular targeting, Red Roof Inn realized a 10% business increase from 2013 to 2014. A look back at the weather data shows that the winter of 2013/2014 was one of the harshest in history causing many flight cancellations. Thus, the algorithmic approach that combines data from severe weather, time of day, flight cancellations, number of stranded passengers, knowing who they are from social media data and targeting them with a room to stay overnight is not only helping the hospitality industry but also helping the travelers.

Another important part of granular audience targeting is the need to keep the customer engaged during the entire process of their pre-customer status. Visitor experience in this period is very critical and it is supported primarily by rapid response with the necessary information that the visitor is searching for. In order to meet the rapid response requirements to the right demographic, in-cache processing of information regarding visitor searches, product and inventory updates and location-based recommendations are essential. This should be supported by high availability of the system. Cloud computing facilitates this aspect of granular audience targeting. An example of such an approach is the decision by Spotify recently to use their vast Big Data sources to identify the right demographic for choice of people's musical interest and send them targeted ads. Spotify collects various types of information on its nearly 70 million users. It is now leveraging that data in a new way. In order to target the ads to the right people, Spotify shares data such as age, gender, location and musical preferences with their advertisers. In real time the advertisers consider this data and choose a particular demographic where they want their ad targeted. By this anonymization process the advertisers are not getting any data that pertains to people. Instead, they are getting general demographic information that enables them to target their ads to a specialized audience. Spotify then uses this information to place the ads to the right audience. People's privacy is protected by this process because the advertisers do not get any details about people in this demographics but select the demographics that they want to target with their musical product and notify Spotify, which in turn uses that information to send the suitable ad to the listener.

Does the need for granular advising override people's desire to keep their information private? The clear answer is no. However, people feel helpless in that they need the services in order to be connected to others and the service providers who provide the service for free, gather the data that they need in order to monetize the data. This ability to gather specific data that Facebook has exploited for many years is exactly what supports granular advertising. The study by Turow et al. show that people are aware of what they are giving up but since they do not have a say in the way their data is used, they are more resigned to it than accept it as a valid trade off (Turow et al. 2015).

18.4 Forecasting

Advertisers want to target the right audience for their marketing campaign. One way to do this is to forecast based on available data from multiple sources about prospective customer interests. In this section we will look at some of the forecasting techniques available for such use. Forecasting is essential in business in order to know future demand and be prepared for it. For example, utility companies generating electricity cannot meet future demand if they did not plan for that and have the necessary infrastructure to generate and carry that electricity to the places where it will be needed. This is usually achieved using statistical techniques such as regression analysis and curve smoothing based on charts drawn based on existing data and future prediction. With Big Data, the technique of forecasting is significantly enhanced because the forecaster now has access to more data and can base the predictions on more reliable data (Myerholtz 2014). For example, in 1974, US electric utilities predicted a 7% annual growth in demand and built new systems for the anticipated demand. However, the annual growth rate was 2% instead of 7% and so the utilities had excess capacity (Barnett 1988). Unfortunately, electricity once generated cannot be stored and must be consumed. This industry was hurt badly because of poor forecasting based on limited data at that time. However, today the businesses have access to a very large amount of data and so they could base their forecast on a better set of input.

The question arises as to how Big Data could be used with forecasting demand. As mentioned earlier, if assumptions are flawed in making the forecast then the forecast would not be of much help. In forecasting demand one should take into account all potential drivers of demand. By gathering data on this aspect from social media communications such as twitter feeds and Facebook posts, the business would be in a better position to see what those drivers would be in the future. This analysis would also help the business know what the barriers to entry would be for any competitor in the future. Sometimes business forecasts fail to take into account how easy it would be for a competitor to enter the market and take away market share after a business had invested significant sums of money to lure new customers. The classic example illustrating this aspect is in the telecommunications sector. This sector was decentralized in 1982 with the breakup of AT&T into seven regional Bell Telephone companies. For years the telephone companies were making their forecasts related to demand for long distance phone calls based on historical trend lines in revenue. This was true when there was a single monopoly but when seven different companies compete, then the trend lines are not a good predictor of future demand. As can be seen from today's explosive growth in mobile services, telephone companies that invested heavily in land lines would suffer the consequences. This example shows how Big Data would be able to help a business use the right type of data in forecasting future demand.

Another technique used in forecasting is to divide the total demand scenario into smaller components that would be homogeneous (Barnett 1988). Based on Barnett's description, the key steps in developing a good forecast are:

1. Define the market.
2. Divide total industry demand into its main components.
3. Forecast demand drivers in each segment and project how they are likely to change.
4. Conduct sensitivity analyses to understand the most critical assumptions.
5. Gauge risks to the baseline forecast.

In defining the market, it is important to know how product substitutions would have adverse consequences on demand. We illustrate the importance of proper forecast using the case of large appliances. Market demand shows that this market is divided into three categories: appliances for new homes under construction, replacement of existing appliances, and appliance penetration in existing homes. Often such data is available from government census data or from industry associations. Gathering such data is essential in order to perform the forecast analysis. With Big Data, this data collection can be even more granular by taking into account the age of existing appliances and the age of consumers. With targeted marketing new types of appliances can enter the marketplace. The following statistical data shows that this a large market segment and benefit greatly from proper forecast (Table 18.3 and Fig. 18.2).

We provide data below on the South Korean company LG Electronics which has been growing steadily in the American Appliance market (Table 18.4).

Table 18.3 2013 global net sales of home appliances

Item	Details
Household appliances	$16.6 billion
Automatic washers	10.03 million units
Electric ovens	677,000 units
Refrigerators	10.96 million units
Total # of units for all appliances	64.61 million units

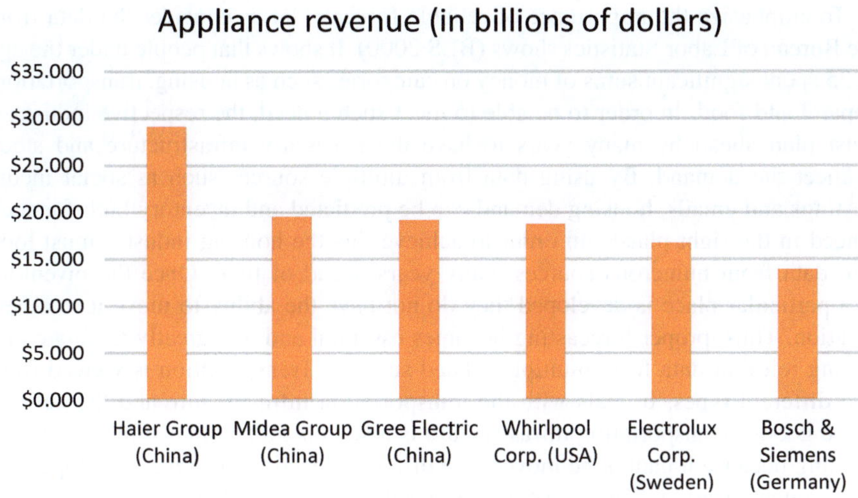

Fig. 18.2 2013 global net sales of home appliances. Source: Statista 2016

Table 18.4 LG electronics appliance sales in US

Year	Revenue from appliance sales (in billions of dollars)
2011	$10.08
2012	$10.21
2013	$10.74
2014	$14.25
2015	$14.05

Source: Statista 2016

Analysis of the statistical data presented above shows that businesses can enhance their forecasts by using all related data that is available from a Big Data application.

In the above paragraphs we have emphasized the importance of forecasting demand. The examples provided show that a wrong forecast unnecessarily diverts the resources of a business to the wrong sector thereby causing disruption in the areas where services are needed. The goal of forecasting is to develop the resources ahead of the demand so that the business can be prepared when demand arises (Chase 2013). The primary inputs for forecasting are the trend and the demographic that would need the product or service. In order to project this properly a business must be able to identify the population that would require their product or service. Thus, targeting the right group becomes a specialized form of a recommendation engine. It is important to note that the data about user preference are structured and simple and often comes from a variety of sources. Hence, the processing tool should be capable of handling such data but also process them rapidly in order to derive benefits for the business. One such tool is Spark, which is similar to Hadoop, except that it is used with in-memory processing and thereby providing up to 10 times the speed of Hadoop in processing.

To emphasize the importance of reliable forecast, let us consider the data from the Bureau of Labor Statistics shows (BLS 2000). It shows that people under the age of 35 spent significant sums of money on categories such as housing, transportation, apparel and food. In order to be able to meet such a need, the respective industries must plan ahead by many years to have the necessary infrastructure and stock to meet the demand. By using data from multiple sources such as social media postings and emails, housing demand can be predicted and inventory built for such a need in the right places. In order to achieve this the housing industry must look into data from numerous sources many years ahead of time. Once the inventory in a particular place is developed they do not have the ability to move it to a new location. Thus, proper forecasting becomes essential and it is greatly facilitated by having relevant data from multiple related sources. Transportation is viewed from two different types. It deals with the transportation infrastructure and the need to move goods. Transportation infrastructure is also constrained in a similar way like housing because it cannot be moved to a different location once it is developed for a particular location. The transportation need to move goods produced to a place where it will be consumed is slightly less restrictive in this matter. Without too much difficulty it could be moved to a different location where there is need but still good forecasting is essential to have all other related infrastructures available. Apparel and food have some flexibility in that some minor changes are possible after a forecast is followed. Once again, these industries also depend on having extensive data about the potential consumers of apparel and food to have the right type of food available in the places where they are needed in the future.

As a practical example of the need to have the right data we gather some information from the McKinsey Global Reports. In one of its reports, McKinsey points out that businesses store 235 terabytes of data, which is quite large. To provide some baseline for the reader to judge this size, the entire US Library of Congress collection is 15 terabytes in size. In 2011, it studied the behavior patterns of Chinese consumers and concluded that people in the age group 55–65 tended to be more conservative in their purchases and were not particular about certain well-known brands. Their 2016 study shows that there has been a significant shift in consumer behavior of people in the age group 55–65 and they prefer well-known brands more now. This change in behavior shows that businesses need to take into account customer attitudes as well in forecasting.

18.5 Predictive Analytics

This approach is gaining momentum in marketing as an important tool. As a technique, many companies have been using predictive analytics for many years using varying levels of success. With the availability of Big Data, predictive analytics usage is significantly enhanced. Before we delve into this topic let us start with a definition of this concept. For this purpose, we use the SAS definition which states that "Predictive analytics is the use of data, statistical algorithms and machine learning techniques to identify the likelihood of future outcomes based on historical data." In using this definition, we are emphasizing the need to have all related data. This approach shows how automation is needed in following through with actions and how machine learning will be used in this approach. We will include examples of how predictive analytics helped some businesses. Some of the areas in which predictive analytics can make a difference are: fraud detection, forecasting and risk reduction.

In order to perform predictive analytics, the business needs extensive data about its customers. Often such data are available within the business in multiple units. So, as a first step, the business must bring together all related data to get a profile of the customer. Today, many third parties gather extensive data on people's web searches and others specialize in gathering social media data. Moreover, with so many users using mobile devices, there is extensive geolocation data. All such data must be gathered and combined effectively. The best use of predictive analytics is in retaining existing customers and thus prevent churn. After all, it is lot cheaper to retain an existing customer than to acquire a new customer. With this in mind we will look at the ways predictive analytics could be used by various businesses.

Predictive analytics approach is good for fraud detection, enhancing operations and reducing risk. In fraud detection, the ability to combine data from multiple sources becomes essential. Also, speed of processing such data is critical to avoiding fraud. For example, when an applicant opens a new account with the same employer phone number from multiple sources, the financial institution has the ability to check that the phone number provided matches something that is already used by someone else. This in itself would not flag fraud since a business might have many employees. However, when the same employee uses the same phone number in multiple branches of the same institution then it would raise a simple flag. When the information relating to the account opener's work or home address is combined and the place where the account is not related to either of them, then it raises a larger scrutiny. This is when the financial institution suspects that the account opener may use the account in fraudulent ways. Predictive analytics comes into play here by combining even more data from the social media and emails that the account opener had indicated in other communications their ulterior motives. Predictive analytics thus helps identify a fraudster. In the case of enhancing operations, predictive analytics is used to forecast demand so that appropriate inventory could be maintained in places where they will be needed. For example, airlines, hotels and car rental companies can leverage data from multiple sources to know where the demand is likely to be occurring and the respective organization could lure a customer with an enticing offer.

Predictive analytics is good at answering the question "what is next?". It does not focus on trying to find out "why things failed?". In exploring this further, it is better to look at the experience of both the retail sector and airlines sector. Both these sectors have a rich collection of customer engagement data. Retail sector uses the data it has about its customers from a variety of sources such as purchase history, coupon usage, reacting to sales promotions, social media communications and emails. These data are combined with their loyalty program to offer more products that the customer would want. This is a direct result of predictive analytics pointing out the customers' preferences on products. For example, Best Buy found out from its analysis of sales data that 7% of the customers are responsible for 43% of the sales. This has enabled Best Buy to focus on their most lucrative customers and redesign their stores to place products in most preferred places by these customers (Cognizant 2011). Thus, retail industry is more customer-centric in its use of predictive analytics information. However, airlines industry seems less customer-centric because it does not offer many privileges to its profitable customers. Globally, airlines industry transports 3 billion passengers annually. Many large airlines have customer data for 30 years. By suitably grouping passengers into cohorts they are able to offer better pricing that attracts appropriate cohorts. Airlines gather data pertaining to travel as business or pleasure. Airlines could use this data to offer more privileges to its business customers because they are the most profitable to the airlines. However, they don't seem to offer any attractive packages such as easy upgrades, priority boarding, seats with more leg room, airport lounges or better food choices. This observation is made in light of the fact that loyal customers form the backbone for a business. Often loyal customers want better service and their loyalty is not tied to price sensitivity. Predictive analytics facilitates gathering more data about their loyal customers and taking care of their preferences.

Businesses that benefit most from predictive analytics are in the following sectors: hotels, airlines, cruise lines, restaurants, car rental companies and online travel agencies. Of these, online travel agencies and cruise lines tend to depend heavily on some middle men and in the other sectors the people are able to make their booking directly using the internet. In this case the predictive analytics should take into account the role of the middle men in promoting their sector. As we discussed earlier, the hotel industry has access to a large volume of data from several sources. That is valid only up to getting the customer to the hotel. In order to achieve repeat visits, the individual properties are responsible to make the customer experience extremely appealing. Major brands are able to take care of attracting the customer but it is up to the individual property to make the customer experience memorable. Data related to this is gathered by the individual properties through surveys. Additional customer satisfaction data is available from third party rating sites. Thus, gathering all such data becomes essential in order to address any customer concerns.

18.6 Content Marketing

Marketers have adapted their toolsets to where consumers go. When internet became popular in the mid-1990s, marketers adapted their techniques from print and TV media to the internet. Google revolutionized people's capability to search all the information stored in the internet. Even though mobile technology attracted people's attention in early 2000, its use in search capability did not become quite popular until 2012 when 4G technology was available. In order to use the mobile phones for search, people needed greater download speeds and that is what 4G enabled. Because of this change in people's search habits, marketers had to adjust their techniques once more to meet the consumer needs. The introduction of Smartphones in 2007 revolutionized the search capabilities significantly and this required the marketing content to adapt itself to be fully available on the mobile devices. Thus started the content marketing revolution. In this regard, it is important to note that the following are some of the important metrics used in content marketing: page views, downloads, visitors, time on page and social chatter. Data for these are available from tools such as Google Analytics (Rancati and Gordini 2014).

Traditionally marketing has focused on a "broadcast" mentality whereby content is made available and the potential customer will seek out what they want. However, with the advancements in technology and the ability to gather relevant consumer preferences, businesses are armed with the knowledge of which customer needs what type of information about the product being marketed. Technology today allows the marketer to cater to this need by exploiting the information gathered about consumers. This is where content marketing differentiates itself by providing the relevant content in an interactive manner, thus enabling the consumer to go as deep as they need in order to understand the product features. Instead of assuming that all consumers would need certain type of information, content marketing helps the business understand that different consumers would need different type of information and it is something that they can pull from the set of available information (An 2016). This is further illustrated by the examples that we provide below about the importance of interactive content being critical to content marketing.

The main thrust of content marketing is to keep the content brief. Otherwise, the users tend to skip the details. People prefer video content and value the recommendations of their friends via social media. Ability to get video content to mobile devices is greatly facilitated today by the availability of higher bandwidth and greater speeds. Content marketing approach is focused on creating very small length videos that can be viewed quickly on smart phones. Also, engaging users on social media is essential and in this context brevity is very important. We now describe some successful applications of content marketing approach by some global brands. Car manufacturer Peugeot, a global brand, had good success with interactive content. Peugeot had 223,000 page views per publication concerning Peugeot cars. Also, 35% of all visits came through mobile devices. Peugeot created seamlessly accessible content across multiple platforms such as mobile, tablet and desktop. Their interactive digital experience known as e-motion provides for an immersive brand experience that engages customers effectively. According to Peugeot, e-motion reaches 92,000 visitors. Another major brand, Target, used for their content marketing its #MadeforU interactive system. This system was heavily used by college students moving into a dorm where they wanted to customize their dorm room. Target created various items that the students could select from and personalize their dorm room. This has been a major success for Target. Target, which has 22 million followers on Facebook, created this microsite to address the college students. Results show that females tweeted more about this than the males.

This information helps Target with the knowledge as to which demographics likes their products more. Thus, content marketing is a good source for finding out the proper audience for new products and services.

Another example of the use of content marketing is from eBay. eBay tried out a content marketing approach to see which of its many postings attract more customers. Using its Facebook page the company was able to find that the following three postings were the top postings for potential customers:

1. Interesting listings (i.e., rare and fascinating listings)
2. Viral products (i.e., products with social traction)
3. Guides (i.e., selection from eBay Guides platform)

Metrics show that some postings had an increase of 225% in click-throughs. Posts with high-engagement showed a 200% increase in interest. These are all attributable to content marketing. Content marketing requires the ability to gather related data from social media and use it for greater impact (Pulizzi 2010). So, Big Data is essential for the success of content marketing (duPleiss 2015).

18.7 Weaving Big Data in Applications

The power of Big Data lies in its ability to process very large volumes of data quickly. It is important to note that in marketing applications the advertiser is expected to target the right content for the right consumer at the right time. Big Data facilitates this by gathering data from multiple sources, especially social media where people disclose plenty of details about their preferences. The technological advancements enable the gathering of social media data quickly. At the same time additional information is needed about the location where the person is at a given time and what they have expressed in other media such as email. Such information is also brought in as part of the Big Data (Brown 2011). Geo location information is obtained from the mobile device of the individual. Now that a business that processed this Big Data quickly is able to know much about the preferences of an individual and send that consumer the targeted ad that addresses their immediate need. This ability to know the need and make the product information available to the consumer makes it very beneficial to the consumer and that person is able to favorably react to the ad containing the information that they are in need of at that time. This is called weaving Big Data knowledge to meet the consumer expectation.

Big Data is known for its ability to process data that comes in large quantities (volume), at a rapid rate (velocity), and in different types (variety). Google's Map/Reduce technology that was implemented as an open source software by Apache Software as Hadoop made it possible to handle such data rapidly. Cloud computing technology enabled the business to have all the necessary computing resources to process such volumes of data. These aspects made the start-ups to focus on their techniques and ideas and let cloud computing make the infrastructure available to them. This confluence of technology and business processes made it possible to provide the useful applications to the users. In this environment when mobile technology grew rapidly, the marketers were able to make available the relevant content about various products to the consumers using the various marketing techniques discussed earlier.

We conclude this chapter with remarks about Big Data. Much of the required data come from organizational data warehouses. This data refers to both the customers and the various products that the business sells. In order to derive actionable information from this extensive data, businesses use sophisticated tools such as Hadoop and Spark. One major concern that arises when combining data from multiple sources is the threat to individual privacy. Businesses will have to develop policies to protect the Personally Identifiable Information (PII) even though it is not directly used when targeted ads are directed to individuals. Often the data gathered includes credit card information of the users. Such data are governed by Payment Card Industry (PCI) requirements. The very fact both PII and PCI information are involved calls for greater privacy protection. In this regard, the Spotify approach seems to be less threatening for privacy since the identities of individuals are never shared with the advertisers. Instead, the ads targeted by the advertisers towards a specific population are then directed by Spotify to those groups. If other advertisers could adapt this model to their advertisement model, then there will be less of a threat to individual privacy.

18.8 Summary

In this chapter we have looked at various marketing approaches based on the use of Big Data to identify the right population. It has been pointed out that sending the right content to the right individual at the right time is critical. Today's advertisement approach varies significantly from the old model of print ad where it was impossible to target a specific group. Technology today provides the ability to target individuals appropriately with the right content. Moreover, the users tend to use mobile devices extensively and so the targeted ad must be delivered to the mobile devices. The advent of cloud computing enables the advertisement approaches to be personalized and the right content sent to the individual. Surveys have shown that people prefer video content with an interactive feature so that they could get details on what they are interested. This is possible using cloud and each user can explore the topic to the level of detail that they desire. In the above paragraphs we have discussed how target marketing, predictive analytics, forecasting, content marketing and Big Data applications are all trying to leverage the knowledge gained from people's preferences and show them the relevant information. These approaches do not specifically have a built-in process for privacy protection.

References

An, M. (2016). *Future of content marketing, hubspot*. https://research.hubspot.com/reports/the-future-of-content-marketing. Accessed 20 October 2016.

Barnett, W. (1988). Four steps to forecast total market demand, *Harvard Business Review, 66*(4), 28–38.

Berman, R. (2015). *Beyond the last touch: Attribution in online advertising, SSRN*. http://ssrn.com/abstract=2384211. Accessed 24 October 2016.

Berry, M., & Linoff, G. (1997). *Data mining techniques: For marketing, sales, and customer support*. New York: Wiley.

Brown, B., Chui, M., & Manyika, J. (2011). Are you ready for the era of Big Data? *McKinsey Quarterly, 4*(1), 24–35.

BLS. (2000). *Spending patterns by age*. http://www.bls.gov/opub/btn/archive/spending-patterns-by-age.pdf Accessed 10 October 2016.

Chase, C. W. (2013). Using Big Data to enhance demand-driven forecasting and planning. *Journal of Business Forecasting, 32*(2), 27–32.

CMO Council. (2016). https://www.cmocouncil.org/facts-stats-categories.php?view=all&category=marketing-spend. Accessed 24 October 2016.

Cognizant. (2011). *How predictive analytics elevate airlines customer-centricity and competitive advantage*. https://www.cognizant.com/InsightsWhitepapers/How-Predictive-Analytics-Elevate-Airlines-Customer-Centricity-and-Competitive-Advantage.pdf. Accessed 10 October 2016.

du Plessis, C. (2015). Academic guidelines for content marketing: research-based recommendations for better practice. In LCBR European Marketing Conference, Lisbon (pp. 1–12).

Kaushik, A. (2010). *Web analytics 2.0*. Hoboken, NJ: Sybex Publishing (a John Wiley Company).

Myerholtz, B., & Caffrey, H. (2014). *Demand forecasting: Key to better supply chain performance, BCG perspectives*. https://www.bcgperspectives.com/content/articles/supply_chain_consumer_retail_demand_forecasting_key_better_supply_chain_performance/. Accessed 10 October 2016.

Pulizzi, J. (2010). *B2B Content Marketing Benchmarks, Budgets and Trends, Content Marketing Institute*. http://contentmarketinginstitute.com/2010/09/b2b-content-marketing. Accessed 26 October 2016.

Rancati, E., & Gordini, N. (2014). Content marketing metrics: Theoretical aspects and empirical evidence, European. *Scientific Journal, 10*(34), 92–104.

Reichheld, F. (2003). The one number you need to grow. *Harvard Business Review, 86*(12), 46–54.

Shao, X., & Li, L. (2011). Data-driven multi-touch attribution models. In *Proceedings of the 17th ACMSIGKDD International Conference* (pp. 258–264).

Turow, J., Hennessy, M., & Draper, N. (2015). The Tradeoff Fallacy: How marketers are misrepresenting American consumers and opening them up to exploitation, University of Pennsylvania Research Report.

Zhang, Y. (2014). Multi-touch attribution in online advertising with survival theory. In *IEEE international conference on data mining* (pp. 687–696).

Chapter 19
Does Yelp Matter? Analyzing (And Guide to Using) Ratings for a Quick Serve Restaurant Chain

Bogdan Gadidov and Jennifer Lewis Priestley

19.1 Introduction

Many businesses, including restaurants, have access to data generated every day from customers on sites like Yelp, but do not take advantage of the data. This has been true for two main reasons. First, the rise of social media data is a recent phenomenon and the tools, skills and technology available to translate this data into meaningful information is evolving, but is still relatively nascent. Second, previous research has indicated that reviews of restaurants in the lowest price points have limited relevance. We challenge this premise.

In this chapter, we seek to answer the question—*Does Yelp Matter in the Quick Serve Restaurants Sector?* Within the context of this study, we also explore operational performance differences between company-owned and franchised outlets, the most frequently used terms associated with 5-star restaurants versus 1-star restaurants and provide future researchers with a guide on how to use the R Programming language to extract reviews for further analysis.

19.2 Literature Review

19.2.1 The Rise of Social Media Data

The Gartner Group defines the concept of "dark data" as "*the information assets organizations collect, process and store during regular business activities, but*

B. Gadidov (✉) • J.L. Priestley
Kennesaw State University, Kennesaw, GA 30144, USA
e-mail: bgadidov@kennesaw.edu; jpriestl@kennesaw.edu

S. Srinivasan (ed.), *Guide to Big Data Applications*, Studies in Big Data 26,
DOI 10.1007/978-3-319-53817-4_19

generally fail to use for other purposes . . ." (Gartner 2016). The term is derived conceptually from dark matter in physics—matter which is known to exist but cannot be experienced directly.

Dark data has historically not been recognized as having value because (a) it was not viewed as data (e.g. security video files), (b) it was recognized as data, but too unstructured and therefore too difficult to translate into meaningful information (e.g., text narratives from surveys) or (c) it was truly "dark" and the target organization was not aware of its existence. Until recently, data derived from social media outlets met all of these conditions.

Because successful utilization of analytics to improve the decision making process is limited to the data available (Halevy et al. 2009), the information embedded in dark data can be massively valuable. Researchers are now able to leverage new and evolving analytical techniques (e.g., machine learning, natural language processing and text mining) and scripting languages (e.g., R, Python) which enable access to and translation of this social media-generated dark data into information. This relatively new phenomena of extracting and leveraging social media analysis for organizational decision making is no longer a marginal "nice to have", but rather a central informational asset.

Economic sectors across the U.S. economy are increasingly extracting previously "dark data" from outlets such as Yelp and Twitter to inform their decision making. For example, rather than waiting for impressions to be driven by traditional media or advertising, CEOs increasingly use Twitter as a medium through which to have direct communication with their customers—sometimes millions of them at the same time. They can then receive immediate feedback reflected in responses, retweets and "likes" (Malhotra and Malhotra 2015).

One sector where social media analysis has become particularly critical is the restaurant sector. This is true because "*Restaurants are a classic example . . . where the consumer has to make a decision based on very little information*" (Luca 2011). The largest restaurant review site is Yelp with 135 million monthly visitors, followed by Open Table with 19 million monthly visitors (Open Table 2016).

Restaurants, particularly chains of restaurants which engage in national and large-scale regional advertising and product launches, can benefit from the immediacy of customer feedback from sites like Twitter and Yelp.

This chapter will examine the specific role of Yelp reviews for a national quick serve restaurant chain.

19.2.2 The Quick Service Restaurant Sector

The quick service restaurant sector plays a significant role in the US economy. In 2015, 200,000 fast food restaurants generated revenue of over $200 billion. Industry estimates indicate that 50 million Americans eat in a quick serve restaurant every single day. This sector also represents an important source of employment across the country—with over four million people employed in quick serve restaurant franchises in 2015, and one in three Americans worked in this sector at one point during their lives (Sena 2016). While the food is often highly processed and prepared in an assembly line fashion, customers of these restaurants have placed value on consistency of service, value for money and speed (National Restaurant Association 2014).

However, quick service restaurant failures are almost epidemic. Although a relatively modest 26 percent of independent restaurants failed during the first year of operation, quick service restaurants fail at substantively higher rate—failure of franchise chains have been reported to be over 57% (Parsa et al. 2014). Failures of quick serve restaurants—like any small businesses—create negative externalities on local economies in the form of unemployment and lost local spending power.

While this failure rate has been attributed to macro factors like economic growth, federal and state legislation of minimum wage rates, new and different forms of competition, as well as to micro factors like access to capital, location, owner incompetence and inexperience (Gregory et al. 2011) only recently has any meaningful attention been paid to the role of online customer reviews (e.g., Hlee et al. 2016; Taylor and Aday 2016; Remmers 2014).

Although some researchers have indicated that reviews do not matter in the quick service restaurant sector because of the low per ticket price point (Vasa et al. 2016), this chapter challenges part of the premise of that perspective.

First, in 2016, there were millions of reviews associated with quick serve restaurants on Yelp. This is an indication that some customers are willing to provide feedback related to fast food experiences. Importantly, this is consistent with the point that other researchers have demonstrated that satisfaction with a restaurant experience is strongly related to perceived, rather than to absolute, value for price paid—across the range of price points (e.g., King 2016; Dwyer 2015). Second, the study highlighted in this chapter, provides some initial evidence for correlation between the number of guests, sales and numeric ratings on Yelp for a quick serve restaurant. Both of these points provide at least directional evidence that analysis of Yelp reviews for the quick serve restaurant sector could improve operational performance and provide meaningful feedback regarding customer experiences, thereby helping to mitigate the high rate of outlet failure.

19.3 Analysis of Numeric and Text Reviews in Yelp

The current study examined both numeric and text reviews for over 2000 locations of a quick serve restaurant chain across the United States. The results associated with the numeric and text results are provided in the following sections.

19.3.1 Description of Numeric Ratings

Numeric ratings taken from Yelp can range from 1–5 stars, where 1 star represents the worst possible rating and 5 stars represents the highest possible rating. Ratings are aggregated at the business (restaurant) level, and can be pulled by calling for the specific rating parameter in the code provided. Unfortunately, only the current numeric ratings can be extracted—meaning that researchers cannot extract ratings at specified past periods to ascertain changes in ratings over time.

Using these ratings, comparisons are made at the state level, and the GMAP procedure in SAS is used to create illustrative maps. Furthermore, data is used to compare franchise owned to non-franchise (corporate) owned restaurants. Transaction data is used for the non-franchise locations, to assess whether the numeric ratings of restaurants correlate to the number of sales, number of customers, or total order size. The transaction data set includes over 4.7 million transactions at over 400 non-franchise locations of this restaurant chain from January through June 2015. A sample of the transaction data can be seen in Table 19.1 below.

In Table 19.1, the variables from the restaurant are provided: the date of the transaction, the store (restaurant location) ID, the order number, the number of guests served, the number of items ordered and to total ticket value are provided. The StoreID column was used to identify the phone number of the location, which in turn was used to differentiate a franchise versus non-franchise location. This transaction data was a separate component to the analysis and obtained directly from the quick serve restaurant company. This information is not found publicly through Yelp, and is not readily available for reproduction of analysis performed in this chapter.

Table 19.1 Sample of transaction dataset

DateOfBusiness	StoreID	Order number	GuestCount	ItemCount	NetSales
1/2/2015	20115	Order #214	1	1	1.99
1/2/2015	20116	Order #215	1	15	26.17
1/2/2015	20117	Order #216	1	5	15.65
1/2/2015	20118	Order #217	1	4	6.29
1/2/2015	20119	Order #218	1	1	1.29
1/2/2015	20120	Order #219	1	1	1
1/2/2015	20121	Order #219	1	1	1.39
1/2/2015	20122	Order #221	1	1	1
1/2/2015	20123	Order #222	1	8	11.68

19.3.2 Comparison Between Non-Franchise and Franchise Locations

Limited formal research has been conducted evaluating the differences in online customer reviews between corporate owned and franchised outlets in the quick serve food sector. Research from the hospitality sector, indicate that franchised outlets typically outperform corporate owned outlets (e.g., Lawrence and Perrigot 2015). This study found evidence that the reverse could be true in the quick serve food sector.

Corporate owned locations for this restaurant chain are located primarily in Southeastern and Midwestern states. A total of 462 locations for which there is also transaction data were found on Yelp. These locations spanned 17 states, and the average rating is calculated at the state level. A map of the U.S. with these 17 states highlighted is shown in Fig. 19.1 below. Note that the numbers within the states represent the sample size of restaurant chains drawn from these states, while the color code represents the average rating within the state; dark blue represents the highest rated states followed by cyan followed by orange, which represents the lowest rated states.

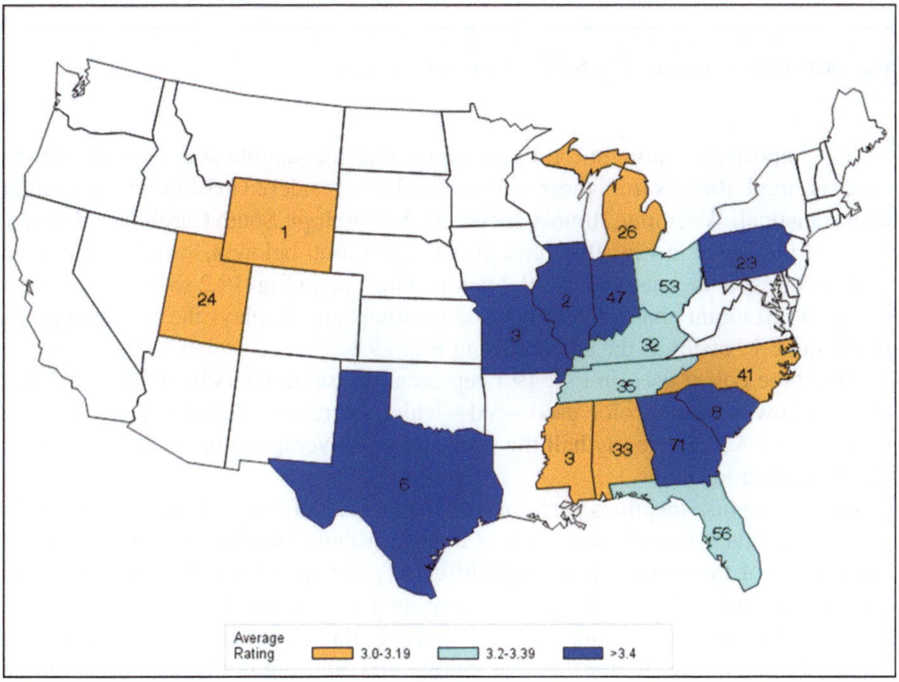

Fig. 19.1 Map of average rating by state for non-franchise locations

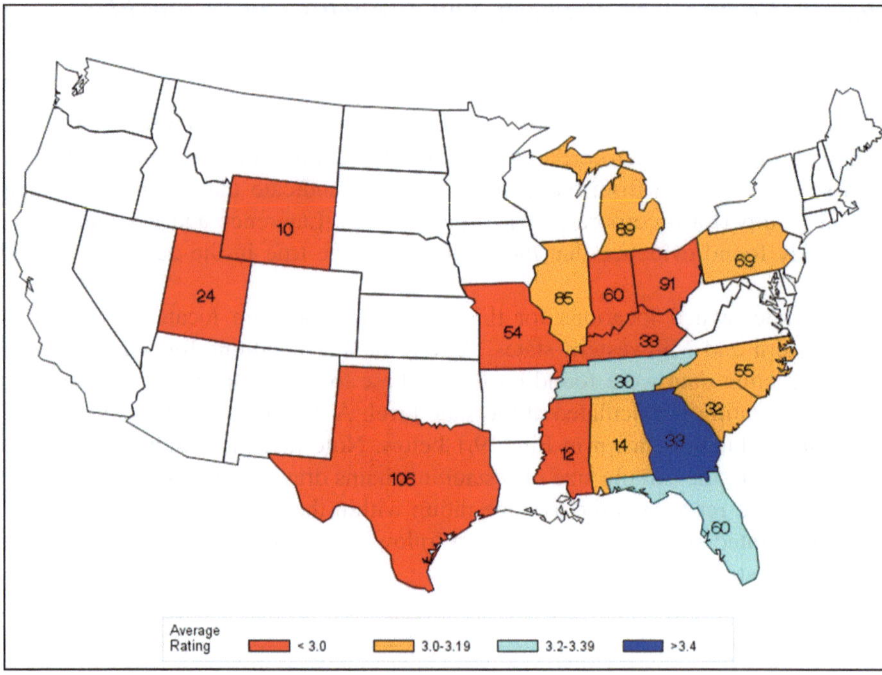

Fig. 19.2 Map of average rating by state for franchise locations

It is important to note in the figure above that the sample size for some of the states is small. Results from these states should be considered to be directional rather than statistical: Wyoming, Illinois, Missouri, Mississippi, South Carolina and Texas. None of the states in Fig. 19.1 have an average rating below 3, which will draw a sharp contrast to the map in Fig. 19.2 below. The map in Fig. 19.2 shows the average rating of restaurant chains for franchised locations and displays the average ratings in the same 17 states as the above map in Fig. 19.1.

The three colors used in Fig. 19.1 represent the same intervals in Fig. 19.2, but there is now a fourth color used—red—which represents states with an average rating below 3 stars. Nearly half the states have an average rating of less than 3 for the franchised locations.

Many franchise locations perform worse than their non-franchise counterparts in each state. This is true of states such as Indiana, Kentucky, Ohio, Pennsylvania, and Utah. Each of these states is colored differently between Figs. 19.1 and 19.2, and the coloring in Fig. 19.2 shows a lower average rating of non-franchise locations as compared to franchise locations in these states. Indiana, for example, has an average rating of over 3.4 for its franchise locations (blue coloring in Fig. 19.1), but has an average rating of under 3 for its non-franchise locations (red coloring in Fig. 19.2). A more detailed state level comparison between the average rating of non-franchise and franchise locations is shown in Fig. 19.3.

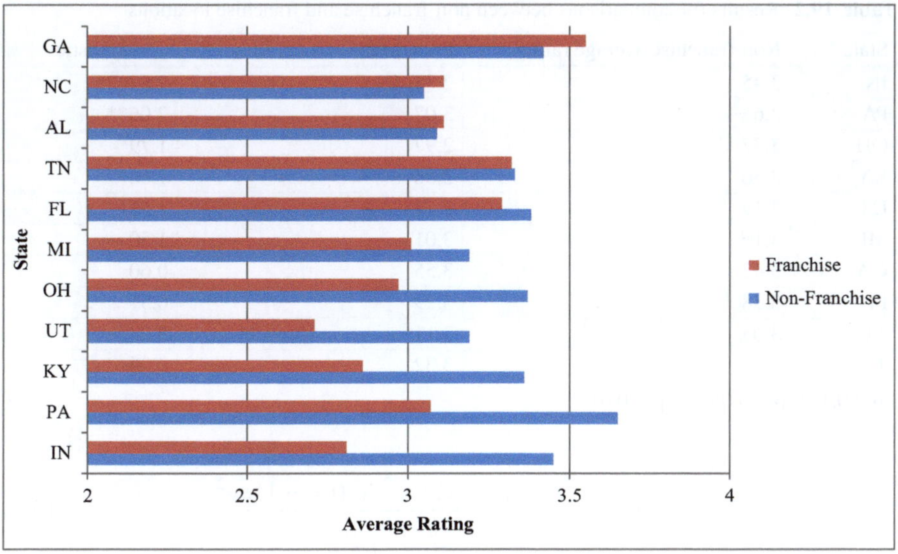

Fig. 19.3 Comparison of average ratings between franchise and non-franchise locations

To better ascertain the state-by-state differences in average ratings between franchise and non-franchise locations, Fig. 19.3 below shows a bar graph illustrating these differences. It should be noted that states which have single digit sample sizes in either Fig. 19.1 or Fig. 19.2 are not included in this graph. For example, a state like Texas which has 106 franchise restaurants and performed very poorly (average rating less than 3) only has six non-franchise locations for comparison. While the six non-franchise locations have an average rating greater than 3.4, it is difficult to draw statistically meaningful conclusions with this imbalance in sample sizes.

It can be seen that only franchise locations in Georgia, North Carolina, and Alabama perform better than their non-franchise counterparts, and the differences are rather small. There are some rather large differences in favor of the non-franchise locations in states such as Indiana, Pennsylvania, Kentucky, Ohio, and Utah, which is confirmed in Fig. 19.3.

The differences in the means between franchise and non-franchise locations can be tested using a two-sample t-test. The sample sizes for the states in Fig. 19.3 are generally large enough (most have at least 30) to make meaningful comparisons. The standard deviations for the testing groups were found to be sufficiently similar to allow for the use of a pooled standard deviation.

The test statistic for a two-sample t-test:

$$\text{test statistic} = \frac{\overline{x_{NF}} - \overline{x_F}}{s\sqrt{\frac{1}{n_{NF}} + \frac{1}{n_F}}} \tag{19.1}$$

Table 19.2 Results for comparisons between non-franchise and franchise locations

State	Non-franchise average rating	Franchise average rating	Test statistic
IN	3.45	2.81	2.87***
PA	3.65	3.07	2.06**
OH	3.37	2.97	1.79*
KY	3.36	2.86	1.66*
UT	3.19	2.71	1.51
MI	3.19	3.01	1.50
GA	3.42	3.55	0.60
FL	3.38	3.29	0.51
NC	3.05	3.11	0.28
TN	3.33	3.32	0.04

*p < 0.1, **p < 0.05, ***p < 0.01

$$\text{where s} = \sqrt{\frac{(n_{NF} - 1)\, s_{NF}{}^{2} + (n_F - 1)\, s_N{}^{2}}{n_{NF} + n_{F-2}}} \tag{19.2}$$

In the equations above, $\overline{x_{NF}}$ and $\overline{x_F}$ represent the mean rating of the non-franchise and franchise locations in a given state, respectively, n_{NF} and n_F represent the number of non-franchise and franchise locations in a given state, respectively and finally, s_{NF} and s_F represent the standard deviation of ratings for non-franchise and franchise locations in a given state, respectively.

The results of the tests are shown in Table 19.2. The test statistics can be used to calculate a corresponding p-value by using a t-distribution with the corresponding degrees of freedom ($n_{NF} + n_F - 2$). The differences in average ratings are statistically significant for Indiana and Pennsylvania. The differences in the average rating between non-franchise and franchise locations in Ohio, Kentucky, and Utah are not statistically significant, but given the relatively large differences (0.4, 0.5, and 0.48, respectively), may be considered to be practically significant.

The remaining states in Table 19.2 have relatively small differences between the non-franchise and franchise locations, and the corresponding p-values suggest that there is no evidence of a statistically significant difference.

These results provided the quick serve company with insight they had not previously understood—specifically that there were differences in customer perceptions and experiences between non-franchise and franchised restaurants.

These findings were quick, inexpensive and easy to extract.

Table 19.3 Correlations between ratings and number of guests

State	Correlation with number of guests	Correlation with average check amount
OH	0.43**	0.03
PA	0.42**	−0.11
MI	0.31*	0.16
AL	0.23	0.05
UT	0.23	0.11
GA	0.22	0.13
NC	0.18	0.03
KY	0.06	0.37*
IN	0.01	−0.09
FL	−0.02	0.08
TN	−0.15	0.29

*$p < 0.1$, **$p < 0.05$

19.3.3 Analysis of Ratings and Transaction Data

The second stage of the analysis was to determine if the ratings of restaurants in a state had any impact on the number of guests or the total amount of dollars spent at a restaurant outlet.

Table 19.3 contains the correlations between numeric ratings (i.e., 1–5) and the number of guests which visited the restaurant, where correlations measures the strength of the linear relationship between the two variables. It can range between −1 and +1, with negative values indicating a negative association between the variables, and positive values indicating a positive association. Correlations closer to −1 or +1 indicate strong correlation while values near 0 indicate weak correlation.

There is moderately positive correlation between the ratings and number of guests in states such as Ohio, Pennsylvania, and Michigan. For all other states, the correlations are weak. This may indicate that, at least in these three states, Yelp does "matter" for customers selecting a quick serve restaurant option.

There was little correlation found with the average check amount. This is likely due to the limited scale of the check values.

19.3.4 Analysis of All U.S. Locations

Analysis of numeric ratings was completed at the state level, regardless of whether the location was a franchise or non-franchise restaurant. The map of results is shown in Fig. 19.4.

Almost every state is represented in this map, except for Rhode Island, Vermont and New Hampshire, where no restaurant outlets were located with reviews on Yelp. Again, the numbers on the states represent the sample size of restaurants drawn from that state and the colors reflect the average numeric ratings.

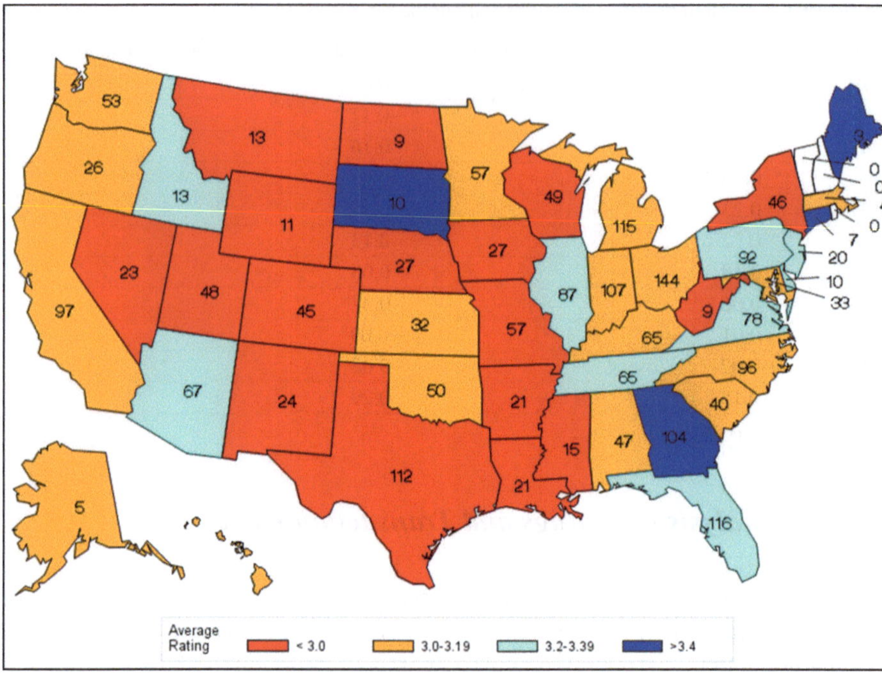

Fig. 19.4 Map of average rating by state across United States

The Midwest and Mountain states have the lowest ratings on average, with many of these states in red, signifying an average Yelp rating of less than 3.0 stars. States in the Southeast and Northeast regions generally have higher numeric Yelp ratings than the Midwest and Mountain states. These two regions exhibit states with similar average ratings, with numerous blue and cyan states in the Southeast and Northeast, signifying average ratings of 3.2 or greater. The Pacific Coast states, including Alaska and Hawaii all have average numeric ratings of slightly above 3.0 stars.

19.3.5 Description of Text Reviews

In addition to the numeric values that are included in a Yelp review, records also include a substantive amount of text. Yelp reviews can include up to 5000 characters. This is in contrast to the 140 character limit for Twitter. Whereas numeric ratings, once extracted, can be analyzed using relatively traditional statistical techniques like t-tests and correlations, text analysis requires a new generation of analytical techniques.

While options for text analytics can include concepts like polarity and sentiment analysis (Wang et al. 2016), frequency analysis of terms and associated word clouds are fairly straight forward, communicative and simplistic options.

A word cloud is simply a spatial image of extracted words where the size of the word indicates its frequency or importance in a text narrative.

Word clouds were generated from the Yelp text reviews for the quick serve franchises in the current study. An example of a word cloud from all reviews is shown in Fig. 19.5 below. The larger the font displaying the word, the more common the word among the reviews.

In the study, word clouds were created to compare franchise locations to non-franchise locations, similar to the comparisons in the actual numeric ratings of the restaurants. Results showed that the word clouds generally had the same pattern, with many customers talking about food items. Word clouds were also created to compare reviews based on the rating of the restaurant (i.e., 1-star rated to 5-star rated restaurant locations).

For restaurants rated 5 stars, it was observed that the most common terms referred to the friendliness of the staff and cleanliness of the restaurant. This is potentially an extension of previous research related to what customers value at quick serve restaurants—service (which may include friendliness of staff), value and speed (Sena 2016). For a customer to give a 5 star review to a quick serve restaurant there is likely something that greatly exceeded their expectations—see Fig. 19.6 below.

Fig. 19.5 Word Cloud created from Yelp reviews

Fig. 19.6 Word cloud for restaurants rated 5 stars

The word clouds shown measure the frequency of single words, but it is also possible to plot combinations of words, referred to as n-grams. In an n-gram, the frequency of "n" consecutive words is measured and then plotted in a word cloud. While not used in this study, a bigram or trigram word cloud is also another simple alternative method for generating word clouds (Garcia-Castro 2016).

Simple word cloud analysis presents quick serve restaurants with an inexpensive and convenient way to ascertain the most frequently occurring words associated with highly rated restaurants versus poorly rated restaurants. This is a potentially powerful method to assess the effectiveness of messaging and advertising for quick serve restaurant chains—to determine which menu items are most frequently mentioned in reviews, and whether those mentions are associated with low ratings or high ratings.

19.3.6 Caution Related to Analysis of Reviews

There is no question that some percentage of reviews on Yelp, like any other social media site, are "fake". This is true because either the owner/manager of the restaurant is providing his own "glowing" reviews or because competitors are unfairly "slamming" the restaurant in an effort to drive traffic to their own restaurant.

Researchers engaged in examining questions related to falsified reviews have found that roughly 16% of restaurant reviews on Yelp are fake. They also found that restaurants are more likely to commit review fraud when its reputation is weak, i.e., when it has few reviews, or it has recently received genuine bad reviews in an effort to "balance out" the ratings. Restaurants are also more likely to be victims of unfavorable fake reviews when they face increased competition. However, these same researchers found that chain and quick serve restaurants are less likely to have fake reviews on Yelp (Luca and Zervas 2015).

While fake reviews do not negate the value of performing analysis of numeric ratings and text comments for restaurants, researchers should be mindful of their presence—particularly when sample sizes are small (Lim and Van Der Heide 2015).

19.4 Guide to Using R to Extract Yelp Data

In addition to providing' research results related to a study in the quick serve restaurant sector, this chapter is also written as an instructional "how to" guide for researchers and practitioners to access and extract Yelp data.

In the current study, the programming language R is used to connect to the Yelp API (Application Program Interface). The Yelp API allows researchers to search and query Yelp for information about rated businesses. In the current study, the R packages used for working with the Yelp API are stringr, httr, jsonlite, and RCurl.

The first step in connecting to the API is creating an account on Yelp. Creating an account provides the researcher with access to the Yelp developers' page (https://www.yelp.com/developers). To initialize access through R, the researcher needs to generate a unique consumer key, consumer secret, token, and token secret. These elements are analogous to a password to connect to Yelp's API. A sample of what these tokens look like is shownin Fig. 19.7 below (Yelp Developers Search API 2015). Some of the characters in the tokens are shaded out in the figure due to the nature of the data, but notice that these tokens can be generated freely by creating an account on Yelp.

Fig. 19.7 Consumer tokens from Yelp API

To create the connection with the API, two lines of code are needed in R. The first of the two lines takes the consumer key and consumer secret, and registers an application. The second of the two lines takes the token and token secret, and creates a signature which can then be used to generate requests from the Yelp API. This process can be thought of as a "handshake" between the R console and Yelp API.

Sample code is shown below:

```
myapp = oauth_app("YELP", key=consumerKey, secret=consumerSecret)
signature = sign_oauth1.0(myapp, token=token, token_secret=token_secret)
```

Once the connection is made, the researcher can now search Yelp for a particular business. This is accomplished through identification of the desired search parameters. The general form of the URL string is http://api.yelp.com/v2/search/. Then, depending on specified parameters such as location, name, or category of the business, the URL string is expanded to contain these search criteria.

In the example of searching for the quick serve restaurant used in this analysis, the search string is created as follows:

```
yelpurl <- paste0("http://api.yelp.com/v2/search/?location=", city, "&term=restaurant_name")
```

The location input into the search string took the variable "city". Notice that the term "city" does not appear in quotes in the search string. This is because it was not static, and varied throughout the process. In order to obtain results for this quick serve restaurant nationally, a list of approximately 2000 U.S. cities is used as arguments in this search string. With each iteration, a different city is passed into this line of code, retrieving results when searching that city. The second part of the search string, coined "term", is the name of the quick serve restaurant. Since the name of the actual restaurant will remain anonymous throughout this analysis, the term "restaurant_name" will signify where the researcher should input the name of the business of interest.

Each time a given search string is passed through the API, the results are restricted to only 20 search results returned for each iteration. While this is a challenge, there is an alternative to using a list of thousands of cities to perform the search. Specifically, there is an optional "offset" feature which can be used; instead of passing a list of cities through the search strings, the offset parameter can be utilized to retrieve the first 20 search results in the first iteration, followed by the 21st to 40th search results in the second iteration, and so on. Either way, a loop is required to either cycle through a list of cities, or through all the search results. In using the former option, the results should be stripped of duplicates. This is true because frequently searching neighboring cities may yield the same business in the returned search results. Up to 25,000 calls can be made through the Yelp API daily.

Once the search string is built for each iteration, the "GET" function can be used in R. The relevant and required arguments in this function include the search string (called "yelpurl") and the signature variable from above. Sample code is shown below:

```
data = GET(yelpurl, signature)
datacontent = content(data)
```

```
yelp.json = jsonlite::fromJSON(toJSON(datacontent))
yelp.df = yelp.json$'businesses'
```

The first line creates an R object which contains the data pulled from Yelp for the given search string. The second through fourth lines of the code above transform the data into a structured file which is more readily analyzable. After the fourth line of code, the object "yelp.df" is a data.frame, which in R is similar to a matrix or table of data. Importantly, this is where the data converts from being "unstructured" to becoming "structured".

The next line of code allows the user to specifically choose the desired attributes about the business or restaurant of interest. For example, if the researcher wants the phone number, rating, name and indicator of whether the business is closed or open, then the following line of code can be used:

```
ScrapeOutput = yelp.df[1:20, c("phone", "rating","name","is_closed")]
```

In this line of code, the researcher needs to select which parameters should be kept from the list of returned parameters from the scrape. Parameters should be listed in quotations just as they appear above (such as "phone" or "rating"). It is useful to get the name of the business, as sometimes search results yield names of different businesses, which can then be removed. Additional parameters which can be selected are shown below in Fig. 19.8. A more comprehensive list of all search parameters, including location data such as the longitude and latitude coordinates of the business, can be found on the Yelp website under the "Developers" section.

Business:

Name	Type	Definition
id	string	Yelp ID for this business
is_claimed	bool	Whether business has been claimed by a business owner
is_closed	bool	Whether business has been (permanently) closed
name	string	Name of this business
image_url	string	URL of photo for this business
url	string	URL for business page on Yelp
mobile_url	string	URL for mobile business page on Yelp
phone	string	Phone number for this business with international dialing code (e.g. +442079460000)
display_phone	string	Phone number for this business formatted for display

Fig. 19.8 Sample parameters available through Yelp API

The first step is to collapse all the separate reviews gathered from the previous section into one string of words (McNeill 2015). The next step is to use built in functions to make all the letters lowercase, remove any punctuation, and erase extra whitespace which appears in the reviews. Some sample code which performs these actions is shown below:

```
r1 <- paste(nonfranchise, collapse=“ ”)
review_source <- VectorSource(r1)
corpus <- Corpus(review_source)
corpus <- tm_map(corpus, content_transformer(tolower))
corpus <- tm_map(corpus, removePunctuation)
corpus <- tm_map(corpus, stripWhitespace)
```

In analyzing the text reviews, it is also useful to suppress common terms which are not of interest. For example, pronouns such as “I”, “my”, or “we” are not typically of interest. There is a built-in function in the tm package which contains “stopwords”. Stopwords is an already built list of 174 terms which include common pronouns and verbs typically ignored in analysis. Some of these terms are shown in Fig. 19.9 below.

It is also useful to add to this list. For example, in searching for a restaurant, it is not of particular interest to see the term “food” appear in a review. To add to this list, one can use the concatenate feature in R to append more words to the existing list of stopwords. A sample line of code which adds the words “food”, “get”, and “will”, followed by code to apply the stopwords, is shown below.

```
mystopwords <- c(stopwords(“english”), “food” ,“get”, “will”)
corpus <- tm_map(corpus, removeWords, mystopwords)
```

The remaining code pertains to finding the frequency of the individual terms after all unwanted terms have been removed. In the last line, the wordcloud function in R takes the top words with their corresponding frequencies to plot them. The number of words which are to be included can be adjusted, as the code shown below will take the top 50 words.

```
> stopwords("english")
 [1] "i"          "me"         "my"          "myself"     "we"
 [6] "our"        "ours"       "ourselves"   "you"        "your"
[11] "yours"      "yourself"   "yourselves"  "he"         "him"
[16] "his"        "himself"    "she"         "her"        "hers"
[21] "herself"    "it"         "its"         "itself"     "they"
[26] "them"       "their"      "theirs"      "themselves" "what"
[31] "which"      "who"        "whom"        "this"       "that"
[36] "these"      "those"      "am"          "is"         "are"
[41] "was"        "were"       "be"          "been"       "being"
[46] "have"       "has"        "had"         "having"     "do"
```

Fig. 19.9 List of stopwords in text mining library of R

```
dtm <- DocumentTermMatrix(corpus)
dtm2 <- as.matrix(dtm)
frequency <- colSums(dtm2)
frequency <- sort(frequency, decreasing=TRUE)
words <- names(frequency)
wordcloud(words[1:50], frequency[1:50],colors=brewer.pal(8, "Dark2"))
```

The second part of this analysis involves using the text reviews given by customers. As described above, the "snippet_text" parameter can be returned to view written comments by customers of the business. However, this only contains the first text review of a business on Yelp. As a result if there are multiple reviews, the remaining reviews are ignored. To work around this, R can be used to perform an HTML scrape. To perform the HTML scrape in R, the httr and XML packages are needed. For the HTML scrape, it is necessary to view the source code behind the web page. This can be done by "right clicking" on a web page and selecting the "View Page Source" option. When doing this, a new page will open with the HTML code. In looking at this page, one needs to find where the comments start. For example, in Fig. 19.10 below, the text "Start your review of" signifies that comments will begin (the name of the quick serve restaurant in this study is blacked out, but would appear in place of the black box). Once this is identified, there are a series of R functions which can be used to parse this information and translate it into a data frame for analysis.

To create the HTML scrape, the "url" parameter (shown in Fig. 19.8) must be returned from the original Yelp API call. This parameter contains the URL of the business of interest. Using this parameter, the HTML functions can be used to go to this specific webpage, and retrieve each of the individual comments left by customers. Some sample code is provided below. In this code, the dataset "test" contains the URL of each individual restaurant searched, along with a corresponding phone number and state to uniquely identify the location. A for loop is run for each URL in the dataset, to iterate through all of them and output the reviews for each individual business which is searched. The functions "htmlParse" and "xpathSApply" are needed to turn the HTML source code into a data frame in R which can then be manipulated by the user. The "grep" function is also required to find the specific location on the webpage where the comments begin. Refer to the example in Fig. 19.10, where the words "Start your review" signify a new review. The "grep" function is used to locate where on the page this begins, and

```
            </div>
            <a class="js-war-text-link" href="/writeareview/biz/DXclCtCDkFmDzS6MI
                Start your review of <strong>[redacted]&#39;s</strong>.
            </a>
        </div>
    </div>
</div>
```

Fig. 19.10 Sample view of HTML source code

the subsequent line begins retrieving results only once this string has been located. The "grep" function is also useful in removing lines which contain something other than a review. For instance, as part of each review, there is a line which asks if the review was helpful. By using the statement "-grep('Was this review helpful')", all lines containing this statement can be removed. Ultimately, the lines selected in this code should each reflect a separate review from a customer. The "cbind" function (column bind) is used to append the individual phone and state identifier to each review. The "rbind" function at the end is used to append results from each iteration to create one master data frame object which contains all the reviews for searched businesses. Notice that in the code below, a "#" symbol reflects a comment in R, which is ignored in processing of code. Sample code is shown below here:

```
finaldata <- NULL # initialize a data frame named finaldata

for (i in 1:nrow(test)){ # data frame test contains the url, phone and state for business
tempurl <- test$url[i] # tempurl is the url in each iteration
doc <- htmlParse(tempurl) # htmlParse function
y <- xpathSApply(doc,'//p', xmlValue, encoding="UTF-8") # turns result into a table
n <- grep('Start your review', y) # grep is used to search for strings
y2 <- y[-c(1:n, (length(y)-2):length(y))] # further subset results
y3 <- y2[-grep('Was this review helpful', y2)] # further subsetting
y4 <- cbind(y4, as.character(test$phone[i]), as.character(test$state[i])) # combine data

finaldata <- rbind(finaldata, y4)} # append results
```

Select SAS Code

SAS v9.4 was also used in this project. Specifically, SAS was used for the statistical analysis as well as the creation of the maps. Sample SAS code is shown below for creating the maps. The format for creating the ranges for the state colors is shown, followed by the PROC GMAP function which creates a map using one of the default maps (maps.us) in the SAS maps library. To put the sample size labels on the state, an annotate option is available within the GMAP procedure, which uses a separate dataset that contains the labels for the states.

```
proc format;
value rating_format low-3='< 3.0'
3.0-3.2 = '3.0-3.19'
3.2-3.4 = '3.2-3.39'
3.4-high = '> 3.4';
run;
proc gmap data=dataset map=maps.us; format average_rating rating_format.;
id state;
choro average_rating/ discrete coutline=black annotate=maplabel;
run;
quit;
```

19.5 Conclusion

This study sought to determine whether ratings on Yelp are relevant to a quick serve restaurant's performance, with particular attention paid to differences between franchised and non-franchised outlets. Both numeric ratings and text reviews were analyzed. The numeric ratings indicated that non-franchise locations of restaurants for this company generally performed better in terms of Yelp ratings relative to the franchise locations. Results were plotted in a series of maps to highlight differences by state.

Given the volume of reviews, combined with the detected variation between franchised and non-franchised outlets and the correlation between numeric ratings with the number of guests, this study also provided evidence for the position that Yelp reviews are relevant to operational performance and evaluation of customer satisfaction in the quick service restaurant sector. This is contrary to previous findings.

Overall, accessing information from review sites like Yelp can provide practitioners with quick, inexpensive (effectively free), valuable information regarding operational productivity, customer perceptions, efficacy of messaging and advertising. The flexibility provided to researchers by Yelp to extract specific data related to location, time period, identified words related to products or menu items, allows for the translation of previously "dark data" into meaningful information to improve decision making.

References

Dwyer, E. A. (2015). Price, Perceived Value and Customer Satisfaction: A Text-Based Econometric Analysis of Yelp! Reviews. Scripps Senior Theses. Paper 715. http://scholarship.claremont.edu/scripps_theses/715. Accessed 2 September 2016.

Garcia-Castro (2016). *R. Example of creating n-gram clouds.*https://rstudio-pubs-static.s3.amazonaws.com/118348_a00ba585d2314b3c937d6acd4f4698b0.html. Accessed 20 January 2016.

Gartner Group (2016). *Dark Data.* http://www.gartner.com/it-glossary/dark-data/. Accessed 3 September 2016.

Gregory, A., Parsa, H. G., & Terry, M. (2011). *Why Do Restaurants Fail? Part III: An Analysis of Macro and Micro Factors*. University of Central Florida, The Dick Pope Sr. Institute for Tourism Studies UCF Rosen College of Hospitality Management.

Halevy, A., Norvig, P., & Pereira, F. (2009). The Unreasonable Effectiveness of Data. *IEEE Intelligent Systems, 24*(2), 8–12.

Hlee, S., Lee, J., Yang, S., & Koo, C. (2016). An empirical examination of online restaurant reviews (Yelp.com): Moderating roles of restaurant type and self-image disclosure. *Information and Communication Technologies in Tourism, 2016*, 339–353.

King, B. (2016). Caught in the middle: franchise businesses and the social media wave. *Journal of Business Strategy, 37*(2), 20–26.

Lawrence, B., & Perrigot, R. (2015). Influence of organizational form and customer type on online customer satisfaction ratings. *Journal of Small Business Management, 53*, 58–74.

Lim, Y., & Van Der Heide, B. (2015). Evaluating the wisdom of strangers: The perceived credibility of online consumer reviews on Yelp. *Journal of Computer Mediated Communication, 20*, 67–82.

Luca, M. (2011). Reviews, Reputation, and Revenue: The Case of Yelp.com. Harvard Business School Working Paper, No. 12–016, September 2011. (Revised March 2016. Revise and resubmit at the American Economic Journal-Applied Economics.).

Luca, M., & Zervas, G. (2015). *Fake it till you make it: reputation, competition, and yelp review fraud*. http://people.hbs.edu/mluca/fakeittillyoumakeit.pdf. Accessed 2 September 2016.

Malhotra, A., & Malhotra, C. (2015). *How CEOs Can Leverage Twitter*. http://sloanreview.mit.edu/article/how-ceos-can-leverage-twitter/. Accessed 3 September 2016.

McNeill, M. (2015). *Text Mining in R: How to Find Term Frequency*. https://deltadna.com/blog/text-mining-in-r-for-term-frequency/. Accessed 3 October 2015.

National Restaurant Association (2014). *Restaurant Industry Forecast*. https://www.restaurant.org/Downloads/PDFs/News-research/research/RestaurantIndustry Forecast2014.pdf. Accessed 3 September 2016.

Open Table (2016). *Our Story*. https://www.opentable.com/about/. Accessed 3 September 2016.

Parsa, H.G., van der Rest, J.P., Smith, S., Parsa, R., & Buisic, M. (2014). Why Restaurants Fail? Part IV The Relationship between Restaurant Failures and Demographic Factors. *Cornell Hospitality Quarterly* October 2014.

Remmers, M. (2014). *5 Ways to Get More Diners with Yelp: Leveraging the Online Review Site Can Maximize Your Restaurant's Exposure*. https://www.qsrmagazine.com/outside-insights/5-5ways-get-more-diners-yelp/. Accessed 3 September 2016.

Sena, M. (2016). *Fast Food Industry Analysis 2016—Cost & Trends*. https://www.franchisehefranchisehelp.com/industry-reports/fast-food-industry-report. Accessed 3 September 2016.

Taylor, D., & Aday, J. (2016). Consumer generated restaurant ratings: A preliminary look at OpenTable.com. *Journal of New Business Ideas & Trends, 14*(1), 14–22.

Vasa, N., Vaidya, A., Kamani, S., Upadhyay, M., & Thomas, M. (2016). *Yelp: Predicting Restaurant Success*. http://www-scf.usc.edu/~adityaav/Yelp-Final.pdf. Accessed 3 September 2016.

Wang, Q., Wu, X., Xu, Y. (2016). *Sentiment Analysis of Yelp's Ratings Based on Text Reviews*. http://cs229.stanford.edu/proj2014/Yun Xu, Xinhui Wu, Qinxia Wang, Sentiment Analysis of Yelp's Ratings Based on Text Reviews.pdf. Accessed 20 October 2016.

Yelp Developers Search API. (2015). https://www.yelp.com/developers/documentation/v2/search_api. Accessed 1 August 2015.

Author Biographies

Dr. Tommy L. Binford Jr. holds a B.S. in physics and mathematics (1998) and an M.S. in physics (2000) from Sam Houston State University (Huntsville, TX) where he performed research on low-temperature superconducting materials. He completed a Ph.D. in computational and applied mathematics at Rice University (Houston, TX) in 2011. In 2000, he joined the oil industry to work in research and development. He is Senior Staff Scientist at RD&E of Weatherford International studying, supporting, and developing logging-while-drilling technology. His current interests include resistivity modeling, sensor development, computational electromagnetics, modeling and inversion, numerical optimization, and high-performance computing.

S. Srinivasan (ed.), *Guide to Big Data Applications*, Studies in Big Data 26,
DOI 10.1007/978-3-319-53817-4

Dr. Ann Cavoukian is recognized as one of the world's leading privacy experts. She is presently the Executive Director of Ryerson University's Privacy and Big Data Institute. Dr. Cavoukian served an unprecedented three terms as the Information & Privacy Commissioner of Ontario, Canada. There she created Privacy by Design, a framework that seeks to proactively embed privacy into design, thereby achieving the strongest protection possible. In 2010, International Privacy Regulators unanimously passed a Resolution recognizing Privacy by Design as an international standard. Since then, PbD has been translated into 39 languages.

Dr. Cavoukian has received numerous awards recognizing her leadership in privacy, most recently as a Founder of Canada's Digital Economy.

In her leadership of the Privacy and Big Data Institute at Ryerson University, Dr. Cavoukian is dedicated to demonstrating that Privacy can and must be included, side by side, with other functionalities such as security and business interests. Her mantra of "banish zero-sum" enables multiple interests to be served simultaneously – not one, to the exclusion of another.

Dr. Jiefu Chen received the B.S. degree in engineering mechanics and the M.S. degree in dynamics and control from Dalian University of Technology, Dalian, China, in 2003 and 2006, respectively, and the Ph.D. degree in electrical engineering from Duke University, Durham, NC, in 2010.

He was with the Department of Electrical and Computer Engineering, Duke University, as a Research Assistant from September 2007 to December 2010. He was a Staff Scientist with Advantage R&D Center, Weatherford International, Houston, TX, from March 2011 to August 2015. Since September 2015, he has been with University of Houston, Houston, TX, where he is currently an Assistant Professor of Electrical and Computer Engineering. His research interests include computational and applied electromagnetics, multiphysics modeling and inversion, electronic packaging, subsurface wireless communication, and well logging.

Dr. Jonathan H. Chen M.D., Ph.D. is an Instructor in the Stanford Department of Medicine. After a Ph.D. in Computer Science, he completed training in Internal Medicine and a VA Research Fellowship in Medical Informatics. He continues to practice medicine with research interests in data-mining electronic medical records for insights into medical decision making. With the support of an NIH Big Data 2 Knowledge Career Development Award, he is systematically extracting the collective wisdom of practicing clinicians from electronic health records. This will translate endpoint clinical data into an executable form of expertise in a closed loop learning health system.

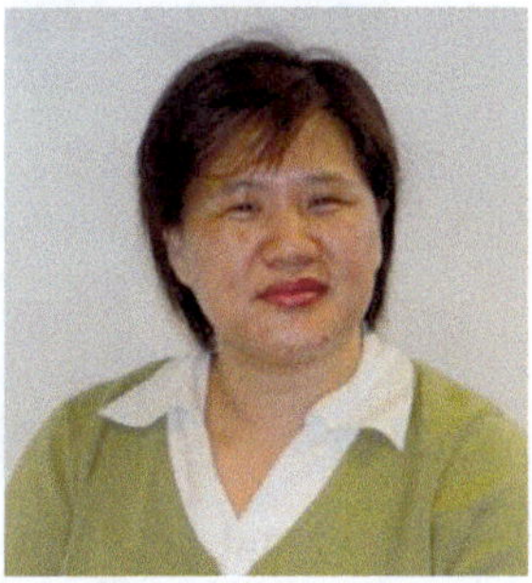

Dr. Hongmei Chi is an Associate Professor of Computer & Information and Sciences at the Florida A&M University. She currently teaches graduate and undergraduate courses in data mining and Cyber Security and researches in areas of big data and applied security. Dr. Chi has published many articles related to data science, parallel computing, and cyber security research and education. Her web page is www.cis.famu.edu/~hchi.

Michelle Chibba is a Strategic Privacy/Policy Advisor at the Privacy and Big Data Institute, Ryerson University. She was Director, Policy Department and Special Projects at the Office of the Information and Privacy Commissioner of Ontario, Canada (IPC). During her 10-year tenure at the IPC, she was responsible for conducting research and analysis as well as liaising with a wide range of stakeholders to support proactively addressing privacy and technology issues affecting the public, otherwise known as Privacy by Design. Michelle received a master's degree from Georgetown University (Washington, D.C.), with a focus on ethics and international business. She is a frequent speaker on Privacy by Design and emerging data privacy/technology issues and has written a number of publications on privacy and technology.

Dr. Wenrui Dai received B.S., M.S., and Ph.D. degree in Electronic Engineering from Shanghai Jiao Tong University (SJTU), Shanghai, China in 2005, 2008, and 2014. He is currently a postdoctoral researcher at the Department of Biomedical Informatics, University of California San Diego. His research interests include learning-based image/video coding, image/signal processing and predictive modeling.

Rishi Divate, Co-founder and Vice President of Engineering, MityLytics Inc. Rishi co-founded MityLytics in 2015 and is responsible for the development of the MityLytics product. Prior to MityLytics, Rishi spent over 16 years of product development and professional services experience at companies such as Sumo Logic, HP, ArcSight, Oracle and Spinway. Rishi has led enterprise software and SaaS deployments in the high performance, security, analytics, middleware and database areas for a variety of customers including Fortune 500 and mid-sized companies, startups, universities and government agencies in North America, Europe and Asia. He has been a session speaker at industry conferences such as Oracle Open World and ArcSight Protect. Rishi received an M.S. degree in computer science from the University of Houston and a B.Engg degree in computer engineering from the University of Pune.

Bogdan Gadidov is a Ph.D. candidate student at Kennesaw State University. He also completed his Master's degree in Applied Statistics at Kennesaw State University prior to enrolling in the Ph.D. program. His undergraduate degree is in Industrial Engineering from Georgia Tech. Prior to enrolling in graduate school, he worked for a year as an implementation engineer at Noble Systems, which specializes in telecommunication solutions for call centers. While at Kennesaw State University, he has enjoyed teaching undergraduate courses in algebra and elementary statistics as a graduate teaching assistant. In 2014, he was awarded as a SAS Analytics Student Poster Winner by the SAS Institute at their yearly Analytics conference for presenting on risk model validation following an internship at SunTrust bank.

Dr. Cuilan Gao is an Assistant Professor of Statistics at the University of Tennessee at Chattanooga (UTC) in USA. She received her M.S. and Ph.D. in statistics from the University of Mississippi in USA in 2010. After graduation, she had been working as a Postdoctoral Research Associate in the Department of Biostatistics at St. Jude Children Research Hospital in Memphis in USA from 2010 to 2012, where she developed statistical methods and conducted the design and analyses of laboratory-based experiments, including genetics and genomic studies for pediatric brain tumors. After joined UTC in 2012, she continues her research interests in statistical methods on computational biology, analysis of high dimensional data and large scale data sets. Dr. Gao is also an experienced collaborator and award–winning teacher. She had wide collaborations with researchers from cancer research, public health and computers science etc. She won the 2016 Alumni Outstanding Teacher award across the University of Tennessee system.

Dr. Gintarė Giriūnienė, Ph.D. is a lecturer of Business Management Systems at Vilnius University in Lithuania. She was educated at Kaunas University of Technology where she took her first degree in Management followed by a Ph.D. in Economics which she completed in 2014. Her research interests lie in accounting, audit and information systems, and she is an author of five handbooks and more than thirty scientific papers.

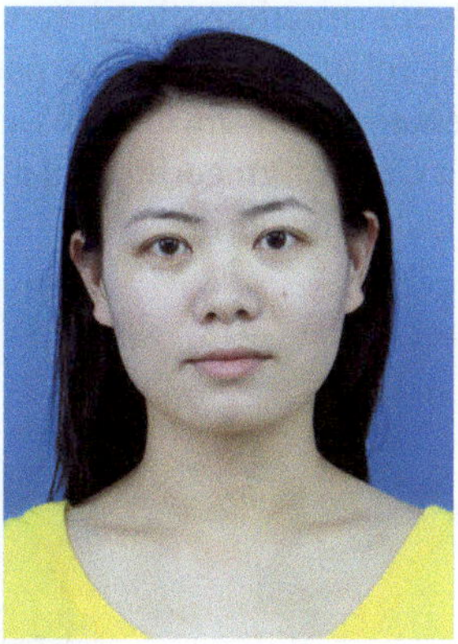

Dr. Yueqin Huang received her B.S. degree in electrical and computer engineering from Jimei University, China in 2005, and her M.S. and Ph.D. degrees in electrical engineering from Xiamen University, China in 2007 and 2011, respectively. During her Ph.D. studies, she spent two years as a visiting scholar in the Department of Electrical and Computer Engineering at Duke University, Durham, NC, USA.

Following her graduation, Dr. Huang worked as an Assistant Professor for one year in Department of Electronic Science at Xiamen University and continued her research in forward and inverse modeling in seismic and electromagnetic wave applications. Since 2015, Dr. Huang has been the owner and a Research Scientist of Cyentech Consulting LLC, Cypress, TX. Her research interests include ground penetrating radar, modeling and inversion of resistivity well logging, and signal processing.

Dr. Dryver Huston has been an engineering faculty member at the University of Vermont since 1987. He has over thirty years of experience in developing and implementing systems for assessing the performance and health of structural systems, along with electromechanical and precision instrument design. Current research projects include developing methods for monitoring and mapping underground utilities, ground penetrating radar methods for detecting buried landmines, developing intrinsic shape sensing networks for inflatable structures, monitoring bridges during accelerated construction, flood scour effect measurements, self-healing wiring systems, soft robotic systems for patient handling and avian lung based extracorporeal oxygenators. Dr. Huston has a PhD (1986) and MA (1982) from Princeton University in Civil Engineering and a BS (1980) from the University of Pennsylvania in Mechanical Engineering.

Dr. Sowmya S. Iyer, M.D., MPH is a Geriatric Medicine physician at the Palo Alto Veterans Affairs Medical Center and a Clinical Assistant Professor (Affiliated) of Medicine at Stanford University. She earned her undergraduate degree in Music and MD at the University of Louisville. She then completed her Internal Medicine residency at Kaiser Permanente Oakland Medical Center. She completed her Masters in Public Health at the University of California, Berkeley and her clinical fellowship in Geriatric Medicine at Stanford University. Her professional interests include improving dementia care throughout health systems, quality improvement, medical education, and medical journalism.

Dr. Shankar Iyer is a Staff Data Scientist at Quora. He works closely with the company's Core Product and Quality Team to conduct data analyses that directly inform product decisions. He also leads the Quora Data Science Team's research efforts. Prior to joining Quora in 2013, Shankar completed a Ph.D. in theoretical condensed matter physics at the California Institute of Technology, where he studied phase transitions in quantum materials.

Pierre Jean has twenty years of oil and gas experience from research scientist to project manager and Asia business manager. Pierre started his career with two M.Sc. degrees, the first one in theoretical physics and the second in microelectronics. He has worked at developing new oil and gas measurement tools (optics, sonic, mass spectrometer, and high pressure), integrating software and real-time measurement at Daniel Industries, Commissariat a l'Energie Atomique, Weatherford and Schlumberger. In the last 5 years before creating Antaeus Technologies, Pierre Jean was software business manager at Schlumberger in Asia - where he grew the business from $500K to 8M over 2.5 years with a very limited team - and in 2014 moved to further develop the software business in North America for Schlumberger. This worldwide technical and business experience gave Pierre Jean the right hindsight as to the market needs in various regions of the world.

Dr. Xiaoqian Jiang is an assistant professor in the Department of Biomedical Informatics, UCSD. He received his PhD in computer science from Carnegie Mellon University. He is an associate editor of BMC Medical Informatics and Decision Making and serves as an editorial board member of Journal of American Medical Informatics Association. He works primarily in health data privacy and predictive models in biomedicine. Dr. Jiang is a recipient of NIH K99/R00 award and he won the distinguished paper award from AMIA Clinical Research Informatics (CRI) Summit in 2012 and 2013.

Pankush Kalgotra is a doctoral candidate majoring in Management Science and Information Systems at Oklahoma State University (OSU). His research interests include healthcare analytics, network science, dark side of IT and neuroimaging in Information Systems. He has more than five years of experience with Data Mining, Texting Mining, Sentiment Analysis and Big Data Analytics. He is proficient in using and teaching Teradata Aster, a Big Data Platform. He is a SAS® certified Predictive Modeler and has been awarded with SAS Student Scholar award in 2013 and SAS Student Ambassador Award in 2014. His team won the SAS Shootout competition in 2014. For his teaching effectiveness, he received Spears School of Business Outstanding Graduate Teaching Associate Award in 2015 and selected as a teaching mentor by the Institute for Teaching and Learning in 2016. He was also awarded the Distinguished Graduate Fellowship in 2015 and 2016.

Dr. Igor Katin, Ph.D. is a lecturer with Department of Economic Informatics, Faculty of Economics at Vilnius University, Lithuania. He gained his Ph.D. from Informatics Engineering Department of Vilnius University, Institute of Mathematics and Informatics. His research and teaching interests include big data analytics, data mining, software systems and modeling, IT technologies, game theory, local and global optimization.

Samsheel Kumar Kathuri is a *Graduate Student* in Management Information Systems majoring in Data Analytics at *Oklahoma State University*. He is enthusiastic and a result-oriented professional, well-versed in analyzing the data and implementing high-impact strategies to target new business opportunities. Over the past 5 years he is been deeply involved in Data Analytics, Business Analysis, Business Intelligence and Reporting. He has worked for 4 years at Tata Consultancy Services and over 1 year as a Graduate Research Assistant at Oklahoma State University. Samsheel was one of the overall winners at *'2016 Teradata Analytics Challenge*' for the work on Health Analytics. He has been offered a position as a *Data Science, Senior Consultant* at CVS Health. He is looking forward for a learning oriented career in the field of Health Analytics.

Dr. Michail Kazimianec, Ph.D. is a lecturer with the Department of Economic Informatics, Faculty of Economics at Vilnius University, Lithuania. He received his Ph.D. degree in Computer Science from Free University of Bozen-Bolzano, Italy. His current research and teaching interests include business intelligence automation technologies and application of predictive analytics as well as of big data analytics in business intelligence.

Mark Kerzner is an experienced/hands-on Big Data architect. He has been developing software for over 20 years in a variety of technologies (enterprise, web, HPC) and for a variety of verticals (healthcare, O&G, legal, financial). He currently focuses on Hadoop, Big Data, NOSQL and Amazon Cloud Services. Mark has been doing Hadoop training for individuals and corporations; his classes are hands-on and draw heavily on his industry experience.

Mark stays active in the Hadoop/Startup communities. He runs Houston Hadoop Meetup. Mark contributes to a number of Hadoop-based projects.

Dr. Elizabeth Le, M.D. is a practicing academic Hospitalist at the Palo Alto Veterans Affairs Medical Center and a Clinical Assistant Professor (Affiliated) of Medicine at Stanford University. She earned her MD at the University of California, Los Angeles and completed an Internal Medicine Residency at the University of California, San Francisco. Current interests include medical education and training at the residency level. Elizabeth currently lives in the Bay Area with her husband and two rambunctious young children.

Dr. Nan Li is an Associate Professor of Key Laboratory of Environment Change and Resources Use in Beibu Gulf at Guangxi Teachers Education University. He currently teaches graduate and undergraduate courses in Marine Microbial Ecology and Genome Data Mining and researches in areas of microbial ecology and bioinformatics. Dr. Li has published many articles related to mining information from huge genomic data, evolutionary analysis and microbial diversity. He researchgate website profile is https://www.researchgate.net/profile/Nan_Li12?ev=hdr_xprf.

Dr. Ron C. Li M.D. is an internal medicine resident at Stanford. He has interests in applied clinical informatics, with a focus on the implementation science of informatics and digital health tools in healthcare systems. He plans on continuing his training as a clinical informatics fellow to better understand how to study and implement innovations that improve the way clinicians make medical decisions and engage with patients.

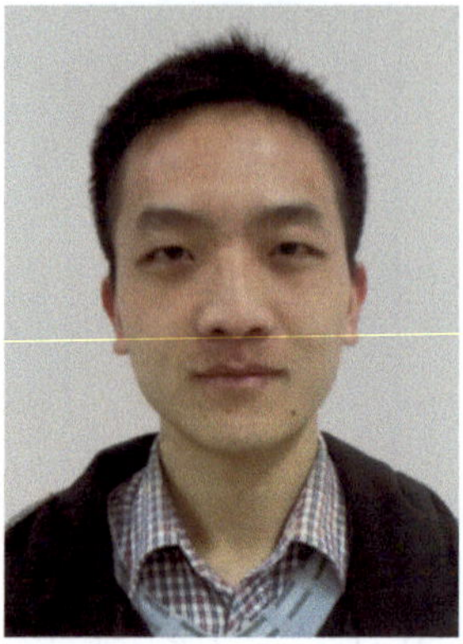

Yang Li received his B.S. degree in Information Security from Northwestern Polytechnical University in 2014. He currently is a Ph.D. student in the same university.

His research interests include Natural Language Processing (word embedding, sentiment analysis, topic model), Deep Learning, etc.

Dr. Yaohang Li is an Associate Professor in the Department of Computer Science at Old Dominion University. He is the recipient of an NSF CAREER Award in 2009. Dr. Li's research interests are in Computational Biology, Monte Carlo Methods, and Scientific Computing. He received the Ph.D. and M.S. degrees in Computer Science from the Florida State University in 2003 and 2000, respectively. After graduation, he worked at Oak Ridge National Laboratory as a research associate for a short period of time. Before joining ODU, he was an associate professor in the Computer Science Department at North Carolina A&T State University.

Dr. Yu Liang is currently working at the Department of Computer Science and Engineering of University of Tennessee at Chattanooga as an Associate Professor. His funded research projects cover the following areas: modeling and simulation, high-performance scientific and engineering computing, numerical linear algebra, the processing and analytics of large-scale sensory data, and computational mechanics. His research work has appeared in various prestigious journals, book and book chapters, and refereed conference, workshop, and symposium proceedings. Dr. Liang is serving in the International Journal of Security Technology for Smart Device (IJSTSD), Journal of Mathematical Research and Applications (JMRA), and Current Advances in Mathematics (CAM) as an editorial board member. . Dr. Liang has a PhD in Computer Science (1998) from Chinese Academy of Sciences, a PhD in Applied Mathematics (2005) from University of Ulster, and a BS (1990) from Tsinghua University.

Dr. Guirong Liu received Ph.D. from Tohoku University, Japan in 1991. He was a PDF at Northwestern University, USA from 1991–1993. He is currently a Professor and Ohio Eminent Scholar (State Endowed Chair) at the University of Cincinnati. He authored a large number of journal papers and books including two bestsellers: "Mesh Free Method: moving beyond the finite element method" and "Smoothed Particle Hydrodynamics: a Meshfree Particle Methods." He is the Editor-in-Chief of the International Journal of Computational Methods, Associate Editor of IPSE and MANO. He is the recipient of numerous awards, including the Singapore Defence Technology Prize, NUS Outstanding University Researcher Award and Best Teacher Award, APACM Computational Mechanics Awards, JSME

Computational Mechanics Awards, ASME Ted Belytschko Applied Mechanics Award, and Zienkiewicz Medal from APACM. He is listed as a world top 1% most influential scientist (Highly Cited Researchers) by Thomson Reuters in 2014, 2015 and 2016.

Dr. Z. John Ma, P.E, F.A.S.C.E., received his Ph.D. degree in civil engineering from University of Nebraska-Lincoln in 1998. He currently serves at the University of Tennessee at Knoxville (UTK) as a professor in the College of Engineering's civil and environmental engineering department. Dr. Ma has conducted research in the area of evaluation of ASR-affected structures; as well as reinforced and prestressed concrete structures including the investigation of shear behavior of thin-web precast bridge I-girders and the development of connection details and durable closure-pour materials for accelerated bridge construction. He has been awarded the NSF CAREER, ASCE Tennessee Section Outstanding Engineering Educator, ASCE Raymond C. Reese Research Prize, and ASCE T.Y. Lin Awards. He is an Associate Editor for ASCE Journal of Structural Engineering and Journal of Bridge Engineering. He has also served as a member on several professional technical committees within ASCE, ACI, PCI, and TRB.

Dr. Ali Miri has been a Full Professor at the School of Computer Science, Ryerson University, Toronto since 2009. He is the Research Director, Privacy and Big Data Institute, Ryerson University, an Affiliated Scientist at Li Ka Shing Knowledge

Institute, St. Michael's Hospital, and a member of Standards Council of Canada, Big Data Working Group. He has also been with the School of Information Technology and Engineering and the Department of Mathematics and Statistics since 2001, and has held visiting positions at the Fields Institute for Research in Mathematical Sciences, Toronto in 2006, and Universite de Cergy-Pontoise, France in 2007, and Alicante and Albecete Universities in Spain in 2008. His research interests include cloud computing and big data, computer networks, digital communication, and security and privacy technologies and their applications. He has authored and co-authored more than 200 referred articles, 6 books, and 6 patents in these fields. Dr. Miri has chaired over a dozen international conference and workshops, and had served on more than 80 technical program committees. He is a senior member of the IEEE, and a member of the Professional Engineers Ontario.

Bhargav Molaka is a Graduate Student in Management Information Systems with a concentration in Business Analytics at Oklahoma State University. He is also a Statistical Analyst at University Assessment and Testing Center, OSU. Bhargav is a SAS Certified Business Analyst, Base and Advanced Programmer with a specialization in data mining and 4 years of professional experience in Business intelligence, ETL, and Data Analysis. Bhargav is one of the Overall Winners of Teradata University Network's Student Analytics Challenge at Teradata Partners Conference, 2016. He was awarded for their Big Data Project on Health Analytics which was done using Teradata Aster, App Center. Recently, he has been offered the position of Senior Credit Analyst from Bluestem Brands, Inc.

Dr. Teja Suhas Patil M.D. is a hospitalist at the VA Palo Alto Health Case System and a clinical instructor of medicine at Stanford University. She has a B.S. in Cell Biology and Biochemistry from the University of California, San Diego and a Masters in Public Health from the University of Michigan, Ann Arbor. Her special interest is in medical education.

Dr. Sharmini Pitter received her Ph.D. in Environmental Science from Stanford University in 2014. During her graduate study, she conducted research in the Department of Environmental Earth System Science in collaboration with the Stanford Archaeology Center. Her dissertation research focused on the link between changes in the paleoenvironment, cultural technology, and agricultural decision-making during the Neolithic period of Turkey. Her research interest focus is on the connections between variables in complex social and environmental systems. Dr. Pitter is currently Project Coordinator for the FAMU Florida IT Career Alliance (FITC), a program that focuses on recruitment, retention, graduation, and career placement of the next generation of Florida's technology workforce.

Hoi Ting Poon is a graduate student in Computer Science at Ryerson University, Toronto, where he is currently a Ph.D. candidate. He also holds a degree in Electrical Engineering and has authored various works in areas related to information security. His research interests include information security, cryptography, authentication systems, Cloud computing, searchable encryption, security in embedded systems and applications of homomorphic encryption in addressing security and privacy issues. He is a student member of the IEEE.

Dr. Jennifer Lewis Priestley, Ph.D. is a Professor of Applied Statistics and Data Science at Kennesaw State University, where she is the Director of the Center for Statistics and Analytical Services. She oversees the Ph.D. Program in Advanced Analytics and Data Science, and teaches courses in Applied Statistics at the undergraduate, Masters and Ph.D. levels. In 2012, the SAS Institute recognized Dr. Priestley as the 2012 Distinguished Statistics Professor of the Year. She served as the 2012 and 2015 Chair of the National Analytics Conference. Prior to receiving her career in academia, Dr. Priestley worked for Accenture, Visa EU, MasterCard and AT&T. She has authored articles on Binary Classification, Risk Modeling, Applications of Statistical Methodologies for Problem Solving, and Data Science Education.

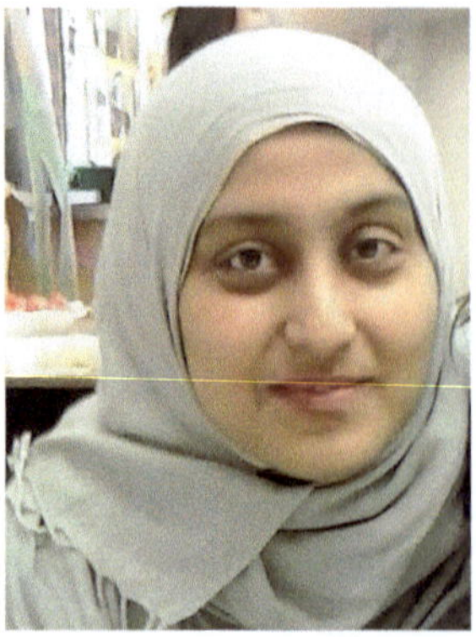

Dr. Fatema Rashid is currently working as a Visiting faculty in Ryerson University School of Continue Education, Computer Science Department, Toronto, Canada. She is also working as an Information Management Analyst in an IT oriented firm in Toronto. She completed her Ph.D. from Ryerson University in 2015 with the thesis chiefly focused on Big Data Security in Cloud computing and different strategies to save storage space in clouds. She also completed her MS from Ryerson University in 2009 with major in Inverse Biometrics for user Authentication.

Sankalp Sah, Founding Engineer, MityLytics Inc. Sankalp was the first engineering hire at MityLytics and is responsible for leading the development of product features for Big Data batch, streaming and query processing technologies. He has over eight years of software development experience in the field of large scale system development and networking. He has worked on products like the Ericsson's Converged Packet Gateway (CPG) which have been deployed in the world's first LTE offerings from Verizon, TeliaSonera and MetroPCS to name a few. Whilst at Ericsson (formerly Redback Networks) he worked on an in house network processing chipset that featured thousands of cores to enable products to scale from 100 Gbps to 1 Tbps. He has also been involved in the mobile handset market with Samsung Electronics, where he helped commercialize phones for the Japan, Brazil and the US market. Sankalp has a Bachelor's degree from the Indian Institute of Technology (IIT) and a Master's degree in computer engineering from Texas A&M.

Dr. Mohamed Sayeed (Ph.D. CE North Carolina State University, 2003) is a computational scientist and a faculty associate at Arizona State University. His area of expertise is in Advanced Computing/Cyber Infrastructure hardware and software for scientific computing including high performance computing, high throughput computing, accelerated, grid and cloud based computing. His research and teaching involves interdisciplinary scientific computing using high performance computing. The current research and tool development efforts are to lower the barriers to use of parallel computing including efforts for automatic parallelization using machine learning techniques. His research interests are parallel computing for big data computation and analytics, dynamical systems modeling, parallel numerical linear algebra, numerical methods, machine learning, inverse modeling and optimization.

Dr. Tavpritesh Sethi (M.B.B.S., Ph.D.) is an Assistant Professor of Computational Biology at Indraprastha Institute of Information Technology (IIIT), Delhi and a Wellcome Trust/DBT India Alliance Early Career Fellow at All India Institute of Medical Sciences (AIIMS), New Delhi, India. He is a clinician and a Data-scientist. With a bridge-expertise in medicine and computer science, he works on developing actionable models for healthcare and his areas of research interests include social networks, machine learning and time series analysis for critical-care and community-health settings. He is a recipient of the MIT-India Young Innovator Award for developing an exquisitely sensitive technology for early detection of small airway disease and Wellcome Trust/DBT India Alliance Early\vadjust{\pagebreak} Career award for supporting his ongoing research on developing machine learning and artificial intelligence models for early detection of sepsis in pediatric abd neonatal Intensive Care Units at AIIMS, New Delhi, India.

Dr. Ashfaque B. Shafique (Ph.D. EE Arizona State University, 2016) graduated from Arizona State University with specialization in control theory. His research involves the detection and control of Epileptic seizures through the use of control theory, chaos theory and signal processing. Additional interests are in control systems and their deployment trough embedded controllers, adaptive control, robust control, system identification and chaos theory. He has worked on various projects involving the application of control theory, namely the design of discrete-time PID controllers through frequency loop-shaping. He has also worked on system identification and control of VTOL model aircrafts and model heating problems.

Dr. Ramesh Sharda is the Vice Dean for Research and Graduate Programs, Watson/ConocoPhillips Chair and a Regents Professor of Management Science and Information Systems in the Spears School of Business at Oklahoma State University. He has coauthored two textbooks (Business Intelligence and Analytics: Systems for Decision Support, 10th edition, Prentice Hall and Business Intelligence:

A Managerial Perspective on Analytics, 3rd Edition, Prentice Hall). His research has been published in major journals in management science and information systems including Management Science, Operations Research, Information Systems Research, Decision Support Systems, Decision Science Journal, EJIS, JMIS, Interfaces, INFORMS Journal on Computing, ACM Data Base and many others. He is a member of the editorial boards of journals such as the Decision\vadjust{\pagebreak} Support Systems, Decision Sciences, and Information Systems Frontiers. He is currently serving as the Executive Director of Teradata University Network and received the 2013 INFORMS HG Computing Society Lifetime Service Award.

Manish Singh, Co-founder, CEO and CTO, MityLytics Inc. Manish co-founded MityLytics in 2015 and is responsible for business and technology strategy and execution at MityLytics. Manish has 17 years product development experience at various Silicon Valley based enterprise software product companies namely GoGrid (acquired by Datapipe), Netscaler (acquired by Citrix), Ascend (acquired by Lucent) and Redback Networks (acquired by Ericsson). He has developed revenue generating features and products used in datacenters at Google, Amazon and several financial companies. He received an M.S. degree in computer science from the University of Houston and a B.S. degree in computer science from Banaras Hindu University.

Dr. Rimvydas Skyrius, Ph.D., is a Professor and head of the Economic Informatics department at the University of Vilnius, Lithuania. He received his Ph.D. in Operations Research and Computer Applications from ASU-Moscow Institute in

1986, and his Master's degree from the University of Vilnius in 1978. His principal research areas are IT-based decision support in business and management, business intelligence and management information needs, and he has published a monograph, a number of articles and conference papers on the subject, as well as co-authored several textbooks in the field.

Dr. Erica Sobel, DO, MPH is a hospital medicine physician at Kaiser Permanente Santa Clara Medical Center. She earned her DO at Touro University California and completed her Internal Medicine Residency at Kaiser Permanente Oakland Medical Center. Erica completed her Masters in Public Health at the University of California, Berkeley. Her professional interests include medical education and health policy.

Dr. S. Srinivasan is the Associate Dean for Academic Affairs and Research as well as the Distinguished Professor of Information Systems at the Jesse H. Jones (JHJ) School of Business at Texas Southern University (TSU) in Houston, Texas, USA. He is the Director of Graduate Programs at the JHJ School of Business. Prior to coming to TSU, he was Chairman of the Division of International Business and Technology Studies at Texas A & M International University in Laredo. He spent 23 years at the University of Louisville (UofL) in Kentucky where he started the Information Security Program as a collaborative effort of multiple colleges. He was Director of the InfoSec program until 2010 when he left for Texas. The program was designated a National Center of Academic Excellence in Information Assurance Education by the US National Security Agency and the Department of Homeland Security. He successfully wrote several grant proposals in support of the InfoSec Program. His two books on Cloud Computing are "Security, Trust, and Regulatory Aspects of Cloud Computing in Business Environments" and "Cloud Computing Basics". His area of research is Information Security. He is the Editor-in-Chief for the Southwestern Business Administration Journal. He has taught Management of Information Systems and Computer Science courses. He spent his sabbatical leaves from UofL at Siemens in their R & D facility in Munich, Germany; UPS Air Group in Louisville, KY; and GE Appliance Park in Louisville, KY. Besides these industry experiences, he has done consulting work for US Army, IBM and a major hospital company in Louisville, KY. He is currently a Cybersecurity Task Force member of the Greater Houston Partnership.

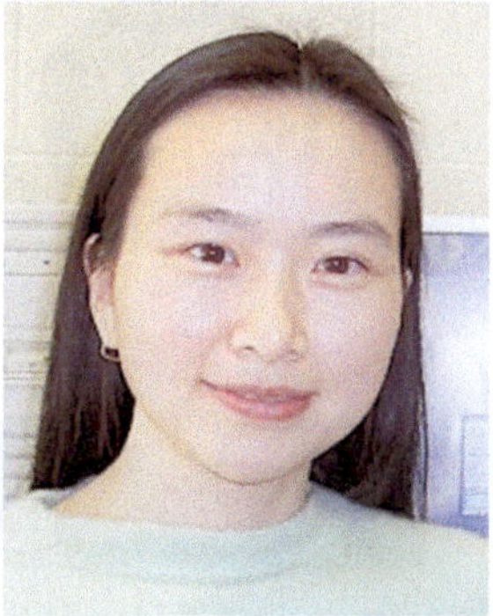

Dr. Haiyan Tian is an Associate Professor of Mathematics at the University of Southern Mississippi. Her research interests include ordinary and partial differential equations, applied analysis, computational mathematics, numerical analysis, and mathematical modeling. She is also actively involved in math education and since eight consecutive years she receives from the US Department of Education, through Mississippi Institutions of Higher Learning, funding for hosting the USM Summer Math Institute for mathematics teachers. Her webpage is https://www.usm.edu/math/faculty/haiyan-tian

Dr. Konstantinos Tsakalis (Ph.D. EE University of Southern California, 1988) is currently a Professor and Undergraduate Program Chair of the Department of

Electrical, Computer and Energy Engineering at Arizona State University. His expertise is in the theory and applications of control systems, adaptive control, system identification and optimization. He has worked on the integrated system identification and controller design and the implementation of high-performance multivariable controllers for semiconductor manufacturing applications. He has also worked on the application of robust control theory, system identification and optimization principles in various industrial problems in collaboration with Honeywell and EPRI. More recently, his activities include power system and biomedical applications, and in particular, prediction and control of epileptic seizures. His educational objectives are to provide students with an operational understanding and hands-on experience with modern system identification and feedback controller design techniques and implementation of embedded control systems.

Dr. Michael Wang M.D. graduated from Harvard College with a BA in Biochemistry. He then obtained his MD from Loyola University of Chicago Stritch School of Medicine before completing his residency in Internal Medicine at Alameda Health System in Oakland, CA. He is currently a clinical informatics fellow at UCSF with an interest in natural language processing, learning health systems, underserved medicine, and genomics.

Dr. Shuang Wang received the B.S. degree in applied physics and the M.S. degree in biomedical engineering from the Dalian University of Technology, China, and the Ph.D. degree in electrical and computer engineering from the University of

Oklahoma, OK, USA, in 2012. He was worked as a postdoc researcher with the Department of Biomedical Informatics (DBMI), University of California, San Diego (UCSD), CA, USA, 2012–2015. Currently, he is an assistant professor at the DBMI, UCSD. His research interests include machine learning, and healthcare data privacy/security. He has published more than 60 journal/conference papers, 1 book and 2 book chapters. He was awarded a NGHRI K99/R00 career grant. Dr. Wang is a senior member of IEEE.

Joe Weinman is the author of the seminal *Cloudonomics: The Business Value of Cloud Computing* (Wiley, 2012) which remains a top-selling book in Cloud Computing over 4 years after publication, and *Digital Disciplines: Attaining Market Leadership via the Cloud, Big Data, Social, Mobile, and the Internet of Things* (Wiley CIO, 2015), which was the Amazon #1 Hot New Release in Computers & Technology. These books have been translated into 3 Chinese editions. He is also the contributing editor for Cloud Economics for *IEEE Cloud Computing* magazine, and has been named a "Top 10 Cloud Computing Leader," among many other accolades.

Dr. Dalei Wu received the B.S. and M.Eng. degrees in Electrical Engineering from Shandong University, Jinan, China, in 2001 and 2004, respectively, and the Ph.D. degree in Computer Engineering from the University of Nebraska-Lincoln, Lincoln, NE, USA, in 2010. From 2011 to 2014, He was a Postdoctoral Research Associate with the Mechatronics Research Laboratory, Department of Mechanical

Engineering, Massachusetts Institute of Technology, Cambridge, MA, USA. Since August 2014, he has been an Assistant Professor with the Department of Computer Science and Engineering at the University of Tennessee at Chattanooga (UTC). His research interests include intelligent systems, networking, and cyber-physical systems.

Dr. Xuqing Wu earned a B.S. degree in Electrical Engineering from University of Science and Technology Beijing in 1995. He also earned a Master of Science in Mechanical Engineering from the University of Alberta in 1999 and Master of Science in Computer Science from the Carleton University in 2001. Dr. Wu received his Ph.D. degree in Computer Science from the University of Houston in 2011. Dr. Wu worked as a software engineer from 2001 to 2007. Dr. Wu finished his 2-years Postdoc Fellowship at the University of Houston. Dr. Wu was a research and data scientist of the Schlumberger-Doll research center before he joined the University of Houston as an Assistant Professor in the Department of Information & Logistics Technology in 2015. Dr. Wu has been actively involved in research areas of Big Data, Predictive Modeling and Forecasting, High Performance Computing, Mobile and Cloud Computing, Probabilistic Multi-Physics Modeling, and Scientific Visualization.

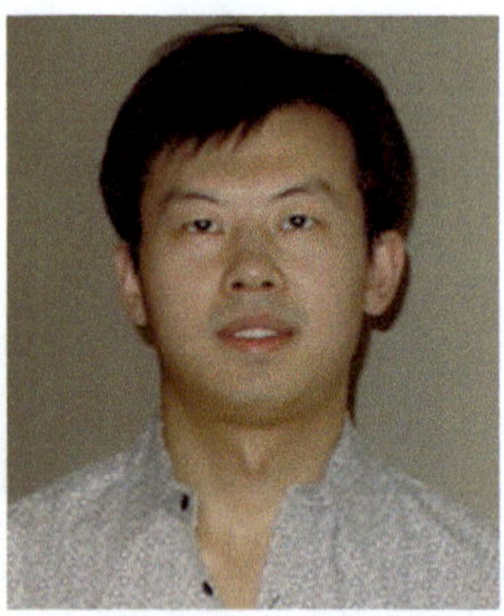

Dr. Tao Yang received the M.S. degree in Computer Science and the Ph.D. degree in Automation Control Engineering from Northwestern Polytechnical University, Xi'an, China, in 2009 and 2012 respectively. From 2009 to 2010, he was a visiting Ph.D. in the Department of Computer Science Engineering, Ohio State University. Before his current position, he worked as a postdoctoral researcher at the Department of Computer Science in Xi'an JiaoTong University.

He is currently an Associate Professor in Northwestern Polytechnical University. His current research interests include data mining methodologies, machine learning algorithms and information security, etc.

His research has been supported by NSFC, National Aerospace Science Foundation of China, and Chinese Postdoctoral Science Foundation, etc.

He has served as an executive at Bell Labs, AT&T, and HP, and was most recently Senior Vice President at Telx (recently acquired by Digital Realty). He currently serves on the advisory boards of several technology companies. He has a BS and MS in Computer Science from Cornell University and UW-Madison, respectively, and has completed executive education at the International Institute for Management Development in Lausanne. He has been awarded 22 patents in a variety of technologies such as cloud computing, distributed storage, homomorphic encryption, TCP/IP multicasting, mobile telephony, and pseudoternary line coding.

Dr. Hongkai Xiong received the Ph.D. degree in communication and information system from Shanghai Jiao Tong University (SJTU), Shanghai, China, in 2003. Since then, he has been with the Department of Electronic Engineering, SJTU, where he is currently a full Professor. His research interests include source coding/network information theory, signal processing, computer vision and machine learning. He has published over 170 refereed journal/conference papers. He is the recipient of the Best Student Paper Award at the 2014 IEEE Visual Communication and Image Processing (IEEE VCIP'14), the Best Paper Award at the 2013 IEEE International Symposium on Broadband Multimedia Systems and Broadcasting (IEEE BMSB'13), and the Top 10% Paper Award at the 2011 IEEE International Workshop on Multimedia Signal Processing (IEEE MMSP'11). In 2014, he was granted National Science Fund for Distinguished Young Scholar and Shanghai Youth Science and Technology Talent as well. He served as TPC members for prestigious conferences such as ACM Multimedia, ICIP, ICME, and ISCAS. He is a senior member of the IEEE (2010).

Dr. Raimundas Žilinskas, Ph.D. is an Associate Professor with the Department of Economic Informatics, Faculty of Economics at Vilnius University, Lithuania. He gained his Ph.D. degree from Vilnius University. His research and teaching interests include Business Intelligence, Information Systems Strategies, and Early Warning Systems.

Index

S. Srinivasan (ed.), *Guide to Big Data Applications*, Studies in Big Data 26,
DOI 10.1007/978-3-319-53817-4

C

D

O

MIX
Papier aus verantwortungsvollen Quellen
Paper from responsible sources
FSC® C105338

If you have any concerns about our products,
you can contact us on
ProductSafety@springernature.com

In case Publisher is established outside the EU,
the EU authorized representative is:
Springer Nature Customer Service Center GmbH
Europaplatz 3, 69115 Heidelberg, Germany

Printed by Libri Plureos GmbH
in Hamburg, Germany